The Social Organization of Work

FOURTH EDITION

RANDY HODSON
Ohio State University

TERESA A. SULLIVAN
University of Michigan

THOMSON
WADSWORTH

Australia • Brazil • Canada • Mexico • Singapore • Spain
United Kingdom • United States

THOMSON
WADSWORTH

The Social Organization of Work,
Fourth Edition
Randy Hodson and Teresa A. Sullivan

Acquisitions Editor: Chris Caldeira
Assistant Editor: Christina Beliso
Senior Editorial Assistant: Tali Beesley
Technology Project Manager: Dave Lionetti
Marketing Manager: Michelle Williams
Marketing Assistant: Emily Elrod
Marketing Communications Manager: Linda Yip
Project Manager, Editorial Production: Cheri Palmer
Creative Director: Rob Hugel
Art Director: John Walker

Print Buyer: Doreen Suruki
Permissions Editor: Mardell Glinski-Schultz
Production Service: Aaron Downey, Matrix Productions Inc.
Illustrator: Maria Sas, NuGraphic Design, Inc.
Cover Designer: Yvo Riezebos
Cover Images: Corbis (globe, school teacher, garment worker);
istockphoto (office worker, construction worker);
Masterfile (welder, nurse and child); Getty (fireman)
Compositor: Integra
Printer: West Group

Library of Congress Control Number: 2007923913

ISBN-13: 978-0-495-00371-7
ISBN-10: 0-495-00371-9

Thomson Higher Education
10 Davis Drive
Belmont, CA 94002-3098
USA

For more information about our products, contact us at:
Thomson Learning Academic Resource Center
1-800-423-0563

For permission to use material from this text or product, submit a request online at
http://www.thomsonrights.com.
Any additional questions about permissions can be submitted by e-mail to **thomsonrights@thomson.com.**

Brief Contents

Contents

v

PART V   Work in the Twenty-First Century   353

15 The World of the Large Corporation   355

Preface for Instructors

Studies of work occupy a core position in sociology and in the social sciences more generally. Workplace studies have an enduring importance because of the centrality of work for both individual and collective well-being defined both materially and in terms of achieving meaning and purpose in life. Workplace studies are made even more exciting by constant and accelerating change in the organization of work. Contemporary areas of change include globalization, gender and racial/ethnic relations on the job, and technology-based transformations of work. Additional excitement (and satisfaction) is provided by the ability to link changes at the national or global level to changes in the lived experiences of work for individuals—one of the core contributions and promises of sociology.

One response to contemporary changes in work has been innovative course offerings. Many departments still offer industrial sociology and occupations and professions, but now departments are also offering courses with titles such as the sociology of work, work and family, women and work, and technology and work. Finding adequate, up-to-date information for these courses often means coordinating a series of monographs and articles and relying heavily on the class lectures and discussion to provide integration.

We have faced these issues in our own courses. Hodson initially taught industrial sociology, and Sullivan taught occupations and professions. As we discussed our classes together, we began to borrow from each other's knowledge and insights. We found it difficult to teach industrial sociology without also providing material on occupations, and difficult to make sense of contemporary changes in occupations without knowledge of changing economic and industrial structures. We discovered vast bodies of scholarship that neither of us had explored but that our students found exciting. We pored over materials on new technologies and changing family-work relations and tried to understand and communicate these changes to our students. We lamented the lack of materials to help our students understand the methods we used to conduct our own research on work. Finally, we began to develop a more unified view of the sociology of work, a view represented in this book. This unified view highlights key themes of technology, class, gender, race and ethnicity, and globalization and allows the book to be adapted to courses focusing on occupations and professions, gender and work, or industrial sociology. Although we collaborated on each of the chapters, Hodson had primary responsibility for Chapters 1, 3, 6–9, and 15–17. Sullivan had primary responsibility for Chapters 2, 4–5, and 10–14.

Students are vitally interested in an analytic approach to work, and with good reason. The social sciences are not merely part of a "liberal arts" education; they are literally "liberating" because they give students a vocabulary and perspective for understanding the world around them. Given the tremendous importance of work in our lives, understanding the work world is both intellectually satisfying and pragmatic.

Although the content of this book deals mainly with substantive issues concerning the world of work, we also hope that the book will be useful in developing students' skills of analysis, reasoning, and argumentation. We have tried to be fair in presenting competing theoretical arguments, but we have also indicated on which side we believe the weight of the evidence lies. You and the students may disagree with us. Our own students often do, and some of our best class sessions are generated from these disagreements. We have tried to identify prejudices and cultural biases that affect perceptions of work and workers. In particular, we have integrated the discussion of women and minorities into every chapter. We have also grappled with the profound changes surrounding the microelectronics revolution and the rapidly changing global economy. We have tried to be frank about the gaps that exist in social scientists' current knowledge and to point out alternative scenarios for future developments.

We also provide support for more general curricular goals by including frequent boxes that highlight cross-cultural issues and by providing tables and graphs to help students develop the skill of interpreting data. Every chapter ends with a list of key concepts and questions for thought. These materials are useful for student review, for written assignments or homework, and for examinations. We also provide a brief annotated list of additional library, Internet, and media sources at the end of every chapter. Students can use these sources for further exploring issues developed in the chapter or for assistance in preparing term papers. Both of us encourage our students to write, and the subject of work lends itself to creative and thoughtful student papers.

Changes to the Fourth Edition

One of the most significant changes we have made in the Fourth Edition is to add more first-hand ethnographic material in which workers speak with their own voices. In addition, we further highlight the conceptual underpinnings of each section through more extensive use of paragraph-level headings. We have systematically updated data, concepts, and sources, and we have rigorously edited the manuscript for length and style so that each chapter can be read in one setting. We have given increased attention to new concerns about overwork, stress, and reduced leisure. We have also written more about social psychological aspects of work, including dignity, bullying, and the pursuit of meaning through work. Toward these goals, we have added a thematic icon to the ones already used in the book identifying "good work"—work that contributes to meaning and fulfillment in life.

Finally, we have developed a new Chapter 4 on "Class, Race, and Gender." This chapter combines the material on discrimination from the previous chapter on "Barriers to Work" with a new section on class inequality at work. This new chapter centralizes the material on inequality and makes it more conceptually integrated and accessible for students. The remaining material on health and disabilities from the previous chapter on "Barriers" has been moved to Chapter 14 on "Marginality," again furthering the goal of conceptual integration for each chapter.

Supporting Materials

An *Instructor's Manual/Test Bank* is also available with the fourth edition; we recommend you contact your local Wadsworth representative or order online at *www.thomsonedu.com/sociology/hodson* to request a copy. *The Instructor's Manual with Test Bank* includes chapter outlines, key concepts, learning objectives, lecture topics, class activities, and a test bank including multiple-choice, short-answer, and essay questions, complete with page references and sorted by question type. The material in the *Instructor's Manual with Test Bank* is also available on the companion website at the URL above.

We hope that instructors will be able to use this book in a variety of educational settings and course titles under both the semester and quarter systems. For a course on *occupations* we recommend Part I, which provides a historical overview and discusses research methods for studying the world of work; Part II, which discusses individual and collective adaptations to work; Part IV, which discusses the major occupations; and Chapter 17, which discusses the future of work. For a course on *industrial sociology* we recommend the same starting sequence, but the substitution of Part III, which discusses organizations, manufacturing, the microelectronics revolution, and service industries, for Part IV. Chapters 15 and 16, on large corporations, mergers, and the world economy, will also fit well into an industrial sociology course, depending on the number of weeks available. For a course on *women and work,* we recommend Chapters 1, 2, 4 and 5: history, methods, class, gender and race, and work and family. Chapter 10 on service work and Part IV on occupations will also be essential for a course on women and work, as well as Chapters 16 and 17 on the global economy and the future of work. For a semester course on the *sociology of work* we recommend the entire book, with about one chapter assigned per week along with whatever supplementary readings the instructor chooses. For a quarter-length course on the sociology of work several chapters can be skipped while retaining the core of the book. Depending on the instructor's preferences, omitted chapters might include Chapter 3, on the experience of work, Chapter 6, on unions, Chapter 7, on organization and technology, Chapter 14, on marginal work, or Chapter 15, on large corporations.

We enjoy teaching, and we enjoy becoming better teachers. If you have questions about our text or if you have ideas for improving the text or for using the material in a particular setting, we would like to hear from you. Our addresses appear at the end of this preface. Our own teaching has been improved by our collaboration, and we are eager to continue the dialogue with others.

We would like to acknowledge our debts to the many colleagues who have unfailingly assisted us. We have not always taken their advice, but we have always appreciated it, and the book has been substantially improved by their contributions. We appreciate the careful editorial work that Meera Dash and Charles M. Bonjean devoted to every chapter. We learned first-hand about corporate acquisition and reorganization when Dorsey Press, our original publisher, was acquired by Wadsworth, which was later acquired by International Thomson Publishing. Paul O'Connell, Serina Beauparlant, Sheryl Fullerton, and Lin Marshall offered us extremely helpful editorial assistance and taught us much about textbook publishing. Many colleagues have shared with us their pedagogical and scholarly expertise by reading and commenting on various chapters. These include Andrew Abbott, Howard Aldrich, Robert Althauser, Ronald Aminzade, James Baron, Vern Baxter, John Bodnar, David Brain, Harley Browning, Phyllis Bubnas, Beverly Burris, Johnny Butler, Catherine Connolly, Daniel Cornfield, Sean Creighton, Tom Daymont, Nancy DiTomaso, Frank Dobbin, Michael Dreiling, Lou Dubose, Sheldon Ekland-Olson, Joe Feagin, Neil Fligstein, Ramona Ford, Eliot Freidson, Omer Galle, Maurice Garnier, Tom Gieryn, Michael Givant, Jennifer Glass, Norval Glenn, Mark Granovetter, Larry Griffin, Ein Haas, Richard Hall, John Hannigan, Heidi Hartmann, Jeff Haydu, Jane Hood, Gregory Hooks, Arne Kalleberg, Jacqueline King, James Kluegel, Judith Langlois, Eric Larson, Kevin Leicht, Laura Lein, Sanford Levinson, Susan Marshall, Garth Massey, Ruth Milkman, Delbert Miller, Joanne Miller, Jeylan Mortimer, Mary Murphree, Jan Mutchler, Janet Near, Annette Nierobisz, Brigid O'Farrell, Toby Parcel, Alan Ponak, Brian Powell, David Rabban, Sabine Rieble, Pamela Robers, Rob Robinson, Nestor Rodriguez, Patricia Roos, Rachel Rosenfeld, Arthur Sakamoto, Paul Schervish, Carmi Schooler, Peter Seybold, James Shockey, Ken Spenner, Suzanne Staggenborg, Robin Stryker, Joyce Tang, Peggy Thoits, Charles Tolbert II, Linda Waite, Michael Wallace, Sandy Welsh, Christine Williams, James Wood, and Gloria Young. We are also grateful to a number of research assistants, whose help has been invaluable: Dick Adams, Bill Brislen, Robert Dixon, George A. Harper, Jr., Laura Hartman, Robert Parker, and Matthew B. Ploeger.

We would especially like to thank the following reviewers who contributed to the development of the fourth edition of *The Social Organization of Work*:

Elaine Draper
California State University at Los Angeles

Marty Laubach
Marshall University

Kenneth J. Mietus
Western Illinois University

Kwaku Twumasi-Ankrah
Fayetteville State University

Milan Zafirovski
University of North Texas

And of course we are grateful to our students, who are both our toughest critics and our greatest supporters. They have provided the essential ingredient that makes teaching, writing, and learning such as a satisfying and rewarding experience.

Finally, we would like to acknowledge the creative insights and support of our spouses, Susan Rogers and Douglas Laycock. They provided detailed comments on every chapter and always supplied whatever we lacked at the moment, whether it was conviction, energy, courage, or just appreciation.

Randy Hodson
Distinguished University Teacher and Editor, *American Sociological Review*
Department of Sociology
Ohio State University
Columbus, Ohio 43210
www.soc.sbs.ohio-state.edu/rdh
hodson.8@osu.edu

Professor Teresa A. Sullivan
Provost and Executive Vice President for Academic Affairs
3074 Fleming Administration Building
503 Thompson Street
University of Michigan
Ann Arbor, MI 48109-1340
tsull@umich.edu

Preface for Students

Most people will work throughout their adult lives. Work will absorb the best part of their days. College students are naturally interested in the world of work, how it is changing, and the implications of those changes for themselves and their families. In our own classes we find that students are very concerned, even worried, about their future roles as workers. We hope that this text will help you explore some of these issues by yourself and with your classmates and instructor.

The intellectual backbone of any course on work concerns the process through which work becomes more and more specialized, the transformation of specialization into stratification and inequality, and resulting struggles over the organization of work. This skeletal framework informs this text, though it will often be part of the only faintly visible background. You will spend most of your time reading about topics such as the impact of the microelectronics revolution, the rapidly changing roles of women at work, and the constantly evolving world economy. Six themes, in particular, are highlighted through the use of boxed materials identified with the following special thematic icons:

- *Technology*
- *Race and ethnicity*

- *Class relations* $
- *Globalization*

- *Gender*
- *Good work* ★

These themes are developed throughout the book from the first to the last chapter.

Part I provides background material for the study of work. Chapter 1 offers an overview of work in past societies and identifies key themes that will be followed throughout the book. Chapter 2 explains how we study work in contemporary society. This chapter will be of value both to those wishing to specialize in the sociology of work and to others interested in understanding research findings based on studies of individuals, groups, and organizations. Part II, made up of Chapters 3–6, deals with our work roles and how these influence our daily lives. The topics covered here include careers, the life cycle, integrating work and family, finding meaning at work, job discrimination, and participation in unions and other collective organizations at work.

Part III, made up of Chapters 7–10, deals with the technology and organization of work. The chapters parallel the major economic sectors: agriculture and manufacturing, high-technology industries, and services. Changes in the technology and organization of work give rise to the transformation of occupations discussed in the next section.

Part IV, composed of Chapters 11–14, deals with the occupational roles that we hold and with the unique sets of skills that are needed to perform these roles. The chapters in this section focus on professionals, managers, clerical workers, sales workers, and marginal workers. (Manufacturing and service workers are discussed in Part III).

Part V, made up of Chapters 15–17, focuses on societal-level consequences of the changing nature of work. The topics covered in this final part include the world economy and the role of huge transregional and transnational corporations in molding the world of tomorrow.

We are pleased that your instructor has adopted our book. Since you have become our student by proxy, we would like to share some of the study tips we give our own students in class. Educational studies show that the more actively you are engaged in reading and reviewing text material, the more likely you are to understand, integrate, and retain ideas. An active reader brings several senses to play in every study session. We recommend that you read and study with a pen or pencil in hand and that you make frequent notes to yourself as you identify and learn new ideas. It also helps to read key passages aloud and to use a tape recorder or note cards to highlight core ideas for review. Try to study regularly. You will enjoy the material more if you set yourself a regular schedule for studying and reading, giving yourself sufficient time to assimilate the material.

When you begin a study session, preview the chapter to learn about its contents. At the end of every section, quiz yourself about the main points of the section, and underline points that you consider important. At the end of every chapter, review the key concepts. If you cannot recall the meaning of a concept, return to the text and reread the relevant paragraph, or look the concept up in the glossary. All the boldface key concepts are defined in the glossary. Every chapter ends with questions for thought. Even if your instructor does not assign them, try to answer them. Some are designed to be easy, and others are hard. Some do not have one correct answer but provide an opportunity to apply the material you have read. We find that students who practice these questions generally write more insightful essay exams and term papers. Read the tables; the information in them is the most current we could find, and data interpretation is an important skill for you to develop, regardless of your occupational destination.

When you review a chapter before an examination, begin with the chapter summary. It is often helpful to develop hypothetical test or essay questions to assist your review and to identify points that you want to bring up in class before the test. Additionally, discussing ideas and concepts with your classmates helps to permanently cement your learning.

The multimedia and Internet resources at the end of each chapter will be helpful if you want to learn more about the topics in the chapter. In addition, at the back of the book there is a list of references detailing the source of every study we have cited; you might want to look at some of these in more detail to deepen your knowledge or to help prepare a paper.

We are college professors by occupation, and we find our work very rewarding. We hope that you, too, will find a place in the world of work that is both satisfying and challenging. And we hope that this book will help you become better prepared for that world. After reading this text, let us know your views, either positive or negative. We are very responsive to suggestions from students. Only with feedback from students will we know if our efforts have been successful. Our addresses are listed at the end of the "Preface for Instructors."

PART I

Foundations

1

The Evolution of Work

There is nothing better for a man, than that he eat and drink,
and tell himself that his labor is good.
ECCLESIASTES 2:24 (NEW AMERICAN STANDARD BIBLE)

Every morning at six I drove myself out of bed, did not shave,
sometimes washed, hurried up to the Place d'Italie and fought for a
place on the Metro. By seven I was in the desolation of the cold, filthy
kitchen, with the potato skins and bones and fishtails littered on the
floor, and a pile of plates, stuck together in their grease, waiting from
overnight. I could not start on the plates yet, because the water was
cold, and I had to fetch milk and make coffee, for the others arrived at
eight and expected to find coffee ready. Also, there were always several
copper saucepans to clean. Those copper saucepans are the bane of a
plongeur's life. They have to be scoured with sand and bunches of
chain, ten minutes to each one, and then polished on the outside with
Brasso. Fortunately, the art of making them has been lost and they are
gradually vanishing from French kitchens.
(GEORGE ORWELL, DOWN AND OUT IN PARIS AND LONDON,
PP. 107–108. COPYRIGHT © 1933 BY GEORGE ORWELL AND RENEWED
1961 BY SONIA PITT-RIVERS. REPRINTED BY PERMISSION OF
HARCOURT, INC.)

These two quotations point out the contradictory nature of work: it is both a
salvation and a curse. Work creates prosperity and meaning in life, but it can
also contribute to poverty and alienation. This chapter will review changes in the
nature of work across time so that you can better understand its possibilities and
limitations. This entire book is an effort to sort out the varied experiences of
workers in order to make sense of work in modern society.

First, what is work? Work is the creation of material goods or services, which may be directly consumed by the worker or sold to someone else. Work thus includes not only paid labor but also self-employed labor and unpaid labor, including production of goods and services done in the home.

KEY TRANSFORMATIONS IN THE NATURE OF WORK

In this chapter we explore some of the ways in which work has changed as well as the consequences of these changes for individuals and for society. The organization of work has varied greatly over time. The *division of labor* and the extent of *social inequality* also vary significantly over time. The changing nature of work thus has important implications for personal satisfaction, for economic security, for relations between men and women, and for our ideas about the meaning of work and its place in social life.

The Social Organization of Work

The social organization of work is the entire set of relations between management, workers, coworkers, and customers. In this section we outline a set of themes that describe the social organization of work and that will be explored throughout the remaining chapters of the book. These themes highlight the organizational and technical aspects of work, the demographics of the workforce, the consequences of work for economic security and social inequality, and the meaning of work for individuals.

The Division of Labor The most fundamental transformation in the nature of work over time has been the increasing **division of labor** (Durkheim, 1966 [1897]). In primitive societies each member engaged in more or less the full range of work activities. The only differences in work activity were those based on age and gender. In later feudal society most workers were engaged in agricultural work, but some specialized in a single type of product so that they became, for example, tailors,

cobblers, or bakers. In modern industrial societies work has become so specialized that each trade is broken down into seemingly innumerable specialties. The meat-packing industry provides a good example of an extremely specialized division of labor: "In the slaughter and meat-packing industry one can specialize as: a large stock scalper, belly shaver, crotch buster, gut snatcher, gut sorter, snout puller, ear cutter, eyelid remover, stomach washer (sometimes called belly bumper), hindleg puller, frontleg toenail puller and oxtail washer" (Wilensky and Lebeaux, 1986:33). Specialization constantly creates new lines of work requiring new skills. Specialization, however, can also reduce the *range* of skills needed to perform jobs. A much narrower range of skills, for example, is needed to be a "gut snatcher" than to butcher and clean a whole animal. Greater specialization can thus have both positive and negative consequences for workers as they struggle for dignity and meaning at work.

The modern division of labor also occurs between different regions and even between different nations—some areas specialize in agriculture, some in different types of manufacturing, and others in service industries, such as health care or banking and investment. For further discussion of the modern division of labor, see Chapters 8, 9 and 10 on manufacturing, high technology employment, and services.

Technology Technological developments have multiplied productivity over time. Starting with only simple hand tools, people advanced through the use of steam and electric-powered tools to the automation of many aspects of production through assembly-line technology and robotics. More recently, computer technologies have again revolutionized work in what many have called a second industrial revolution. The workers and nations that successfully harness the power of computer technologies will be the winners in the competitive global economy of the twenty-first century (Baldoz, Koeber, and Kraft, 2001).

Inequality   Not only do the division of labor and technology change over time, but the ways in which people work together also change. Work always involves social relations between people. Relationships exist between employer and employee, between co-workers, between trading partners, and between suppliers and consumers. These relations are called the **social relations of production**. Social relationships at work can be *cooperative and egalitarian,* as in primitive societies, or *competitive, hierarchical, and unequal,* as in industrial societies. In primitive societies people decided jointly how to proceed with a given task and shared equally in its results. Cooperative arrangements were grounded in the reality that most skills were held in common. In societies with a more advanced division of labor, such egalitarian arrangements are replaced by more hierarchical ways of organizing work, in which some skills are held as more important than others and in which some societal members have vastly greater power and wealth than others. The relationship between peasants who till the land and landowners is among the earliest hierarchical organizations of work. The most important contemporary form of hierarchy at work is the relation between owners and employees (Perrucci and Wysong, 2003).

Bureaucracy   The nature of business enterprises is also crucial to the meaning and experience of work and to the success of enterprises. In the contemporary workplace hierarchical relations often take the form of **bureaucracy**. A bureaucracy is a hierarchical system with clearly designated offices and responsibilities and a clearly defined chain of responsibility leading to the top position. The behavior of all parties in a bureaucracy, no matter how high up, takes place within the dictates of clearly stated rules. Bureaucratic procedures characterize modern corporations and governments and are the major way in which work is organized in contemporary society (Scott, 1998). Post-bureaucratic forms of work organization based on greater worker involvement and initiative are,

however, emerging today and are expected to have a defining influence on the experience and meaning of work in the twenty-first century.

Women, Minorities, and Immigrants   Who works, and in what capacity, is also key to understanding the nature of work and its consequences. During much of industrial society, men have been more likely than women to leave home to work in factories and offices. This difference is rapidly eroding today with women making up over 46 percent of the labor force in the United States and Canada. Minority ethnic populations in many societies have traditionally been segregated into lower-paying occupations and trades. The spread of various forms for protective legislation for minority populations addresses these inequalities although the attainment of full equality has often been elusive (Wilson, 1997). New immigrants to a country also typically occupy the lowest rungs of the occupational ladder with succeeding generations climbing to greater heights. The accelerated movement of peoples around the world in the twenty-first century has increased the significance of immigrant populations and workers in many nations and the challenges of assimilating these workers and their families (Jasso et al., 2000). These dramatic changes have spurred the development of new concepts in sociology, such as occupational segregation, tokenism, gender-typing, and comparable worth. Chapter 4 on class, gender, and race focuses on these and related issues.

The Professions   The hierarchy of authority in the workplace is further complicated by the growing significance of highly educated professional workers. These workers claim special rights and privileges based on their possession of specialized knowledge gained through long study. At the beginning of the twentieth century, only 4 percent of the labor force in the United States was made up of professional workers. At the beginning of the twenty-first century, 20 percent of the labor force is made up of professional workers, making them one of the largest occupational categories. Many students in college are following courses of study that prepare them to become members of a particular profession. The professions are discussed in greater detail in Chapter 11.

Meaning and Dignity in Work People's ways of thinking about the role of work in their lives also change over time. In primitive societies people did not experience work as an activity separate from the broader round of daily events. In agricultural societies work was seen as an inevitable burden, made even heavier by the abuses of greedy landlords, bad weather, and variable market prices. Capitalism saw the emergence of a work ethic that identified work with piety and grace. Many fear that the work ethic grounded on frugality and unquestioning effort has been lost in contemporary society. Perhaps it has. But if so, it has been replaced by a vision of work as a route of upward social mobility (Ospina, 1996). These different visions of the meaning of work become parts of social **ideologies**—systems of ideas that justify the economic and political arrangements of a society as appropriate and desirable. In all settings, however, workers desire autonomy and respect in order to experience dignity in their daily lives at work (Hodson, 2001). For further discussion of the importance of meaning at work see Chapter 3.

Globalization An increasing share of economic exchange occurs between nations. In addition, large corporations located in industrially advanced nations typically have many branch plants and joint ventures outside their home nation. The world economy is thus characterized by dense networks of economic links between nations and between transnational corporations. These new realities increase world competition, pushing down prices for many commodities. But they also allow corporations to transfer production to areas with lower-priced labor, thus placing the workers of each nation in ever sharper competition with each other and creating downward pressures on wages, health and safety protections, and environmental protections (Chase-Dunn, Kawano, and Brewer, 2000). The realization of a truly integrated global economy beginning in the late twentieth century is possibly the most significant transformation in the world of work since the Industrial Revolution and is an important background reality for most, if not all, of the issues discussed in this book.

Consequences of Work for Individuals

Individual workers often seem reduced in importance in the large-scale, bureaucratic world of modern organizations. Their individual contributions seem interchangeable or even expendable. This reality can lead to a sense of *alienation* from work, in which people feel detached from their activity, from one another, and, eventually, from their own selves.

The modern organization of work also has positive consequences for individuals. Many individuals receive a share of the expanded productivity of modern industry. Industrial societies produce a much wider range of material goods and services than pre-industrial societies; these goods include more and better food, as well as many items that would have been considered luxuries in previous societies or were completely unavailable, such as central heating and television. Improved services include better medical care and higher education. For many people, work experiences also continue to be a primary source of fulfillment and self-realization (Hochschild, 1997).

Consequences of Work for Society

Self-Interest The very nature of society has been fundamentally altered by the changing organization of work. The most significant change is the transformation from rural society, based on deeply felt *bonds of commonality,* to urban society, based on more fragmented, fluid, and changing relationships, often grounded in *self-interest.* Traditional rural societies placed a high value on conformity and on maintaining solidarity in the face of external threats. These values were necessary because of the harshness and vulnerability of peasant life. In industrial societies the relationships between people are based on distinct yet interdependent contributions rather than on commonly shared abilities and positions (Durkheim, 1966 [1897]). Modern societies thus make greater allowances for, and may even encourage, diversity and competition among their members.

Traditional world's emphasis on:

Community and kinship
Rural and village life
Work on land or in small manufacture
Landed interests
Monarchy
Tradition
Religion

Reformation, Enlightenment, French and Industrial Revolutions

Modern world's emphasis on:

Individual and immediate family
Urban life
Factory and large-scale bureaucratic work organization
Business and industrial interests
Democracy
Reason
Science

FIGURE 1.1 Traditional Societies versus Modern Societies

SOURCE: Tony J. Watson, 1980. *Sociology, Work and Industry*. London: Routledge and Kegan Paul, p.8. Reprinted by permission of International Thomson Publishing Services. Ltd.,

Church-mediated social thought

Science-based social thought, including sociology

1500   1600   1700   1800   1900   2000

Organizational Size   The greater *size* of economic organizations has also produced changes in how both work and society are organized. Products once manufactured in millions of households and, later, in thousands of small factories are manufactured today for the entire world by a few giant corporations operating in a handful of locations. Giant companies that produce automobiles, tobacco products, petroleum, and computers provide some of the clearest examples. Such corporations have tremendous power over workers, customers, suppliers, and even nations. The increasing size of organizations further contributes to the loss of individuality and intimacy at the workplace. Direct personal relationships are often replaced in these corporations by bureaucratically administered rules. In this context, the way in which individuals experience work depends to a significant extent on whether their smaller work groups provide opportunities for them to realize positive self-images. We explore the controversial role of the large corporation in modern society further in Chapter 15.

Markets   The role of markets in the organization of work has increased over time. In primitive societies several people got together for the hunt, or several people joined together to dig roots or gather edible plants. All members of the group, perhaps fifteen to twenty in number, then shared equally in the proceeds of these labors. Today, thousands of people labor together at one site to manufacture, for example, saline solution bags for hospitals, while tens of thousands of other employees of the same corporation work in different parts of the world to manufacture other pharmaceutical products. These workers are integrated in a network of markets in which they exchange their earnings from manufacturing saline solution bags for a diversity of goods and services such as housing, food, and entertainment. These workers may never have occasion to make use of the products they themselves manufacture. Modern industrial systems have brought about the organization of the world into a single interconnected economic unit. Today we live in a truly global economy. As Figure 1.1 points out, these transformations in social arrangements are paralleled by changes in how we think about and understand not only work but other aspects of our lives as well.

Greater organizational size and bureaucracy allow increased productivity because of efficiencies associated with producing a great number of similar or identical things. These sorts of efficiencies are called **economies of scale**. Rationalized planning further increases productivity. Giant economic organizations produce standardized products in quantities unthinkable under previous industrial systems. The resulting power of large corporations over people, the environment, and society often necessitates at least some government regulation of many economic activities and industries.

Social Stratification

Modern forms of work produce abundant goods and services, but they distribute these goods and services unequally. This unequal distribution of rewards and power in society is known as **social stratification**. Members of society receive shares based on their positions, and their families' positions, in the division of labor. In this way, the organization of work in modern societies influences not only our work lives but also our entire lives from birth to death. Some members of society labor long days and weeks, perhaps even holding down two or more jobs, but receive relatively little for their efforts. Some cannot find work and suffer poverty throughout all or parts of their lives. Many work regularly and receive a reasonably good living for their efforts depending on their exact location in the division of labor. Some do not have to work at all but have inherited riches unimaginable to the majority of people.

THEORIZING WORK

Understanding the nature and consequences of work is a core conceptual project in the social sciences. The three founding figures in sociology, Karl Marx, Emile Durkheim, and Max Weber, provided path-breaking ideas on the nature of work that set the stage for modern theories of society. Each theorist conceptualized industrialization as entailing profound challenges to human dignity and to a fair and just society. For Marx, the central challenge was the control of labor by capitalists and their exploitation of workers, resulting in alienation from meaningful work. For Durkheim, the focal point was the breakdown of social norms governing workplace relations due to the drive toward

endless expansion generated by modern industry. For Weber, the key issue was the imposition of bureaucratic rationality in the world of work and the resulting stifling of human creativity.

Karl Marx on Alienation and Exploitation

According to Marx, the exploitation of workers by capitalists and the resulting alienation from work result in the denial of workers' humanity (Marx, 1959). Our species' humanity is realized through meaningful work, and the alienating nature of work under capitalism robs workers of their inherent potential for human growth and development. The exploitation of workers arises because capitalists own the means of production (the technology, capital investments, and raw materials) and treat labor as if it were just another inanimate factor of production. For a capitalist, labor is to be hired as cheaply as possible, used up, and discarded.

Marx argued that the exploitation and misery of workers results directly from the laws of capitalism in which the market system demands that every capitalist buy labor as cheaply as possible in order to produce and sell goods and still turn a profit. If capitalists do not exploit their employees, they will be undercut by other capitalists who do (Marx, 1967). In Marx's vision, capitalism robs workers of the creative, purposive activity that defines their nature as human beings by subordinating their productive activity to profit making and thus reducing it to a means to acquire material sustenance. Marx saw the solution to the problems of industrial capitalism as the overthrow of capitalists by workers and the imposition of control by a new class—but one resting on a broader base than the capitalist class.

Emile Durkheim on Social Disorganization and Its Resolution

Writing several decades after Marx, Emile Durkheim likewise saw industrial society as being exploitative and abusive. He believed that economic life in modern society was in a state of normlessness *(anomie)* because of its unrelenting drive toward economic expansion. In primitive society, in contrast, people had been held together by shared values based on their common position in a relatively undifferentiated division of labor—since most people did basically the same tasks, they shared common values about how to organize society and how to divide the available resources. Durkheim referred to societal cohesion based on shared values arising from common positions as *mechanical solidarity* (Durkheim, 1984).

In Durkheim's vision, the division of labor into finer and finer parts with workers having less and less power as their tasks were made ever simpler had simply outstripped the development of the appropriate moral regulations for economic life. He saw this state as a painful transitional stage. The solution to the problems of industrial society is not, in this vision, the violent overthrow of capitalism. A proletarian revolution would only replace control by one group with control by another group. Rather, the solution is for capitalists and workers to develop a new shared moral order that orients the behavior of both groups toward common goals. These new shared values would arise from different but *interdependent* positions in the division of labor. Durkheim named this new basis for social cohesion *organic solidarity,* reflecting the integration of different parts with different functions within a unified system. For Durkheim new values defending meaning in work and a just society would emerge through participation in voluntary workplace associations that collectively organize work and that provide safeguards against abuse, exploitation, and overwork. Such associations would include industry associations of employers and trade and professional associations among workers. The new values of organic solidarity would be hammered out in negotiations within and between these associations.

Max Weber on Bureaucratic Rationality

Max Weber, writing at the dawn of the twentieth century, envisioned the evils of modern society in terms of excessive rationality and bureaucracy, rather than in terms of exploitation, like Marx, or the breakdown of norms and values, like Durkheim.

According to Weber, the essence of modern capitalism is the rational calculation of profit and loss. This formal rational calculation replaces earlier less rational motives, such as those based on allegiance to traditional values or traditional authority (Weber, 1946). To Weber, capitalism is the science of applying formal rationality to economic life.

Weber believed that the coercive organization of economic life typical of slave and feudal societies, and of the early stages of capitalism, was not a viable basis for organizing modern society. Formal rationality, as manifest in bureaucratic rules and procedures, however, did provide a workable, though flawed, solution for how to organize the modern industrial economy. Only bureaucratic forms of organization can engage in the long-term planning and integration necessary for a modern economy and society. As a result of its technical efficiency, the extension of bureaucracy to more and more spheres of economic and political life seemed inevitable to Weber.

Bureaucratic principles thus displace coercion, favoritism, and nepotism as less efficient foundations for economic life (Perrow, 1986). But, according to Weber, these same bureaucratic principles also lead to a depersonalization of social life. Along with Durkheim, Weber believed that the solution to the problems of modern industrial capitalism lie in the reintroduction of moral values. Weber and Durkheim differ, however, in the proposed source of new values. Durkheim saw these new values as emerging from occupational groups. Weber saw new values as emerging from charismatic leaders.

Socio-Technical and Interactionist Theories

Two contemporary theoretical traditions also provide important foundations for understanding the modern world of work: the *socio-technical* school and the *interactionist* school. The socio-technical school rests on the observation that the specific, detailed social and technical arrangements at work set the stage for meaning, satisfaction, and productivity. Such issues as being treated with civility and having positive relations with coworkers can create the preconditions for an effective and humane working experience. While these issues are of somewhat smaller scope than the issues of exploitation, normlessness, and bureaucracy that motivated early theorists, they can still be extremely important in the day-to-day experience of work. The insights of the socio-technical school are widely used today in designing work and form the basis for much of organizational psychology. We discuss the socio-technical and human relations approaches in greater detail in Chapter 7 on technology and organization.

The interactionist school developed from studies of social life in natural settings in the early decades of the twentieth century. Various social activities were studied, with the workplace playing a pivotal role. These studies observed that people creatively made meaning out of their interactions with others through the use of symbols, such as styles of speech and manner of dress. These meanings then guided their behaviors and interactions with each other. Applied to the workplace, these insights gave rise to two principal lines of study. One line investigates various *occupational subcultures* and studies how people draw meaning out of their work, even work that might be considered tedious, mundane, or even meaningless to outsiders (Trice, 1993). The second line of inquiry investigates how the rules and norms that guide work arise as a *negotiated order*. This line of inquiry rests on the observation that rules are not simply imposed from the top, but are constantly negotiated and renegotiated in application by those responsible for carrying them out. The interactionist approach has provided the theoretical underpinning to a wealth of ethnographic studies of work life as it is actually lived and experienced (Fine, 1999). These studies, and other approaches to understanding work and the workplace, are described in greater detail in Chapter 2.

Classic and contemporary theories of work provide a solid foundation for understanding the nature, problems, and joys of work in modern society. However, the nature of work is constantly changing—seemingly at an ever-accelerating pace. Classic and even contemporary theoretical models of work are thus constantly being challenged by changes in the nature of work. This constantly changing terrain is part of what makes the study of work so exciting.

A HISTORY OF WORK

The theoretical model underlying the following history of work argues that the *social organization of work* and the *technology* at a given stage of history determine the nature of society, including its degree of social inequality. These factors also create contradictions and limitations that set the stage for the next level of development (Lenski, 1984; Marx, 1967 [1887]). An understanding of these broad changes in the nature of work will help you understand the evolution of work and its possible futures.

Hunting and Gathering Societies

By about 300,000 BC the human species, *Homo sapiens,* had evolved to its present form. Humans lived exclusively as nomadic hunters and gatherers until about 8,000 BC. Thus, the hunting and gathering stage includes about 97 percent of the collective life of our species and continues in isolated areas even today. Hunters and gatherers did not experience "work" as a separate sphere of life. Activities necessary to secure sustenance took place throughout the day and were not clearly distinguished from leisure activities—as people gathered berries, dug edible roots, and hunted game, they did so in a relatively leisurely manner, depending on the circumstances of the moment. People did not work hard because there was no point in creating a surplus. Surpluses of food or possessions could not be stored or transported for future usage. Work, leisure, and socializing formed an integrated flow of activities.

The Band   A hunting and gathering band consisted of fifteen to twenty members, depending on how many people the vegetation and animal life could support. The group's hunting and gathering activities eventually depleted the resources in the area immediately around the encampment, and the group was forced to move on. These nomadic movements were cyclical; the group would move through the same areas year after year, following the seasons and the food supply.

Technology was very simple. The most important elements of technology were the various skills possessed by each member of the band. Without the skills to make and bank fires and to locate and harvest edible plants and animals, the band would have quickly starved. The technology also included hides, lodge poles, and other equipment necessary for survival, such as bone needles, stone cutters and scrapers, and wooden spears.

Most skills were commonly shared so that any single member could do all or most of the tasks required of the group as a whole. However, there were rudimentary forms of the division of labor based on gender and age. Young people tagged along with their elders and performed helping functions, such as gathering wood or picking berries. In this way they received the equivalent of modern on-the-job training. Older people who lacked the stamina or mobility for hunting and gathering tended the fire and prepared food or made tools.

The Gender Division of Labor A division of labor based on gender initially resulted from biological differences between men and women. Few people lived beyond childbearing age, so women spent much of their adult life either pregnant or nursing. These activities restricted their physical mobility, so they were less often involved in hunting large game. Women specialized in gathering roots, berries, and other edible plants and in hunting small animals. Considerable evidence shows that women's gathering and small-game hunting were often more productive economic activities than men's hunting of large game—in most hunting and gathering societies, roots, berries, and small game made up a larger part of the diet than did large animals (Lee, 1981).

The social position of women throughout most of human history has been subordinate, to some degree, to that of men. Gender inequality was less extreme in hunting and gathering societies than in many later societies. In hunting and gathering societies, the greater power of men rested on their monopoly over large-game hunting, which provided rare periodic surpluses of meat, and on their central role in the important arenas of contact, trade, and conflict with other groups (Suggs and Miracle, 1999).

Sharing The early forms of the division of labor based on age and gender did not typically result in any great inequality of material rewards among the members of a society. All members shared equally in the food secured from the environment. This arrangement produced optimal benefits for everyone because of the unpredictability of hunting and gathering activities. If one person or family was successful on a certain day, they could not store or transport the surplus. Through equal sharing, however, all members were assured a share of the bounty of others when their own efforts were unsuccessful. Because food and other possessions could not be accumulated in quantity, hunting and gathering societies did not develop distinct social classes based on possession of different amounts of wealth. Instead, equal sharing of resources prevailed and provided a fundamental precondition for the continuation of these societies—in hunting and gathering societies there was more to be gained by sharing than by hoarding.

Few true specialists existed in such societies. Some members may have taken on a leadership role more often than others, but even this role was often rotating. Incumbency in this role was based on personality traits or leadership ability and carried with it few, if any, privileges of position. Similarly, some individuals may have received recognition for their ability to minister to the sick, forecast the weather, or predict the movements of the animals. These individuals sometimes took on the role of a *shaman* or *medicine man*.

Motivation to Work In hunting and gathering societies the motivation to work was straightforward. The band lived a day-to-day existence. If one did not engage in purposive activity on a regular basis, then one either went hungry or relied on others to share a portion of their food. Hunger and social pressure to participate in the group's activity provided a ready motivation to work. In an account of the Andaman Islanders of the Indian Ocean, the anthropologist A. Radcliffe-Brown describes these norms in the following way: "Should a man shirk this obligation, nothing would be said to him, unless he were a young unmarried

BOX 1.1 Acorn Gathering in Sacramento Valley

The Maidu were one of the principal tribes of the Sacramento Valley and adjacent sierras. Their country followed the eastern banks of the Sacramento River and encompassed the modern city of Sacramento, California. The collection and preparation of acorns for food were among the most important industries of the Maidu, in common with most of the Central California tribes. At the time in the autumn when the acorns are ripe, everyone is busy. The men and the larger boys climb the trees and, by the aid of long poles, beat the branches, knocking off the acorns. The women and the smaller children gather these in burden baskets, and carry them to the village, storing them in granaries or in the large storage baskets in the houses.... In addition, eels were speared, split and dried. In preparing them for food, they were usually cut into small pieces, and stewed. Salmon were split, and dried by hanging them over a pole. When thoroughly dry, the fish was usually pounded till it was reduced to a coarse flour, and kept in baskets. Deer and other meat was cut into strips and dried. Usually this was done in the sun; but occasionally a fire was lighted under the drying meat to hasten the process, and to smoke the product slightly. Except on their hunting trips, the Maidu seem not to have been travelers. They rarely went far from home, even on hunts. The Northeastern Maidu traded with the Achoma'wi Indians, getting chiefly beads, and giving in exchange bows and deer hides. Those in the higher sierra traded for beads, pine nuts, salt, and salmon, giving in exchange arrows, bows, deer hides and several sorts of food.

SOURCE: Excerpted from Roland B. Dixon, 1977, "The Northern Maidu." In *A Reader in Cultural Anthropology*, edited by Carleton S. Coon, pp. 262–291. Huntington, New York: Krieger. Reprinted by permission of the publisher.

---

man, and he would still be given food by others, but he would find himself occupying a position of inferiority in the camp, and would entirely lose the esteem of his fellows" (Radcliffe-Brown, 1922:187).

Hunting and gathering people thought of work in ways that would seem quite foreign in the modern world. They did not view work as a distinct activity; life as a whole was seen in a sacred context in which the various forces impinging on the group were held in awe and reverence. Groups that were seasonally dependent on a particular plant or animal, for example, might consider this item sacred and worship its spirit to assure its continued presence. Such entities, called *totems,* were seen as representing and protecting the group. The worship of such totems was one of the earliest forms of religious expression. Other daily activities of hunting and gathering were also undertaken with appropriate piety and observation of ritual. In this sense primitive people experienced work in a spiritual context.

The differentiation of society into classes did not take place until the group was able to settle in one place and accumulate a surplus of goods. In rare situations hunting and gathering societies settled and flourished in environments with an abundance of naturally occurring food sources. Such settings were almost totally restricted to fishing communities, such as those in the Pacific Northwest of North America. In general, settled societies awaited the development of agriculture. The nature of economic life in hunting and gathering society is illustrated in Box 1.1.

Early Agricultural Societies

The major difference between settled and nomadic societies is that settled societies can accumulate a surplus of food and other goods. Accumulation becomes possible because agricultural and pastoral activities are more productive than hunting and gathering and because the group is able to accumulate and store surpluses over time.

Agriculture developed independently in several places around the world from 9,000 to 3,000 BC. These areas include Southeast Asia, the Persian Gulf, and Mesoamerica. The development of agriculture started with the harvesting of wild grains, such as wheat, barley, and corn, and wild tubers and the eventual development of techniques to encourage the growth and yield of these plants. The technologies included the use of the digging stick and, later, the hoe.

The First Surplus With the development of agriculture and the domestication of animals came tremendous changes in the organization of society. A surplus of food was produced, though at first it was quite small. The work of perhaps eighty to one hundred farmers was required to support one nonfarmer. On this scanty basis a new social order came into being. Instead of everyone in society occupying the same role, specialized positions came into being with differentiated activities. These positions included warrior, priest, and, eventually, other official positions such as scribe and tax collector. The production of everyday goods was still carried out mainly by the agricultural worker. Later, craft positions specializing in the production of religious, civic, and military goods also developed.

The life of the agriculturalist differed little from that of the hunter and gatherer, though it perhaps lasted a little longer because of the better protection from famine afforded by greater ability to store food. Children still helped with basic work activities until they were able to take on a fuller role, and the elderly returned to a helping role as dictated by declining strength and stamina. The relative positions of men and women also changed little. "Since men had been hunting, men were the inventors of systematic herding. Since women had been gathering plants, women were the inventors of systematic agriculture" (Deckard, 1983:199). Based on their continuing contributions to the household economy, men and women enjoyed roughly equal access to the goods and services produced by society. With the development of the plow drawn by a team of draft animals—typically handled by men—women's relative contribution and position may have declined somewhat.

The orientation of the agriculturalist to work differed from that of the hunter and gatherer. For the farmer, the land took on a sacred status similar to that of animals and plants for hunters and gatherers.

A new element in the peasant's orientation to work was a focus on the importance of bountiful harvests. Agriculturalists were tied to one spot all year in order to protect their investment of time and resources in planted fields and to wait for and protect the harvest. They thus depended on the land they farmed to produce a surplus that would sustain them throughout the year.

Plunder and Warfare With the accumulation of surplus also came the possibility of plunder by outside groups. This possibility spurred the creation of a warrior class. Some sociologists believe that this increased importance of warfare accounts for women's more subordinate position in agricultural societies. Because men assumed the principal responsibility for warfare, their role in society grew in power and importance (Sanday, 2002). The warlord and priest became powerful roles in the organization of agricultural society. The focus of religion in agricultural society shifted away from personalized spirits toward more abstract, powerful, and distant deities that could be approached only through privileged religious intermediaries.

The Birth of Specialization Improvements in agricultural technology gradually allowed more and more people to leave agricultural work. These improvements included terracing and irrigation, the use of animal and human fertilizers, and advances in metallurgy that led to the proliferation of metal tools (Lenski, 1984). Box 1.2 on the Gheg of Albania describes the life of a typical agriculturalist. Agriculture continued to be the dominant form of economic activity in Western Europe until well into the Middle Ages. Throughout this long period few changes occurred in the life of the average person. This period, however, included the births and deaths of what are known as the classical civilizations and the emergence of feudal society.

Imperial Societies

Imperial societies were based on the subjugation of smaller and weaker agricultural societies by larger and more militaristic societies and the extraction of

---

BOX 1.2 Agriculturalists: The Ghegs of Northern Albania

In the mountainous regions on the eastern shore of the Adriatic Sea the Ghegs lived as agriculturalists until recent times. Over 90% of the Ghegs are farmers.... Cutting tools are of iron. They are made by blacksmiths in the market towns in which the Ghegs do their trading. With hammer and tongs, a small anvil, and a pair of foot-pump bellows they forge nearly all of the metal objects needed by the mountaineers for farming, herding, transport, and household carpentry. They make small anvils, hammers, nails, axes, adzes, knives, ploughshares, shovels, hoes, toothed sickles, door hinges, horse shoes, and horse hobbles.

The Gheg at home in his mountains is a jack-of-all-trades. He may not be able to manufacture the special objects listed above, but he can adze out beams for his house, put together a new plough during the winter months when there is little work out of doors, or build one of his massive chairs. Professional carpenters are rare, although some men show more skill in this than others.

The interiors of most houses are quite bare, with no carpets on the hewn plank floors, and little decoration on the walls. In the older houses a fireplace covers about 10 square feet of floor area, and the smoke finds its way out through the thatch or roof tiles. In so doing, it cures meats hung in the rack above the fire.

The basic cloth is woven at home. The farmer gives his wife the necessary wool from his sheep, both white and black. She cards it with a homemade device, either a flick bow or a nail studded card, and spins it.

Her spinning kit consists of a distaff, a spindle, and a basket. Holding the distaff in one hand, she twists the wool with the other, and winds it onto the spindle in the basket. Whenever she has nothing else to do, or when she is walking along the trail, the Albanian housewife dutifully spins, and her spinning has soon become a reflex action, like knitting or bead-telling.

Farming is a family affair. The man buys the iron implements which he needs in town, and makes the rest of wood. He breaks the soil with a spade, and ploughs it with oxen. As he ploughs each furrow, his wife walks behind him with a basket of seed, sowing it. Later the women will do the weeding, and the whole family comes out to reap.... Women have charge of milking the cows and making butter, curds, and cheese. Small boys are usually employed as shepherds, and lead their flocks high on the mountains. In the summertime many Gheg families drive their cattle up to the Alpine meadows on the mountain passes, and keep them there weeks at a time while they make butter and cheese. While on these heights they live in small temporary houses.

Animal husbandry furnishes the Ghegs not only with much of their food, but also with a supply of energy, for they use oxen in ploughing, and horses for travel and the transport of goods. In wintertime they keep their cows and horses indoors.

SOURCE: Excerpted from Carleton S. Coon, 1977, "The Highland Ghegs." In *A Reader in Cultural Anthropology*. Huntington, New York: Krieger, pp. 347–356. Reprinted by permission of the publisher.

---

food, goods, and slaves as tribute. Based on the subjugation of these smaller societies and on modest improvements in agricultural technology, the classical empires grew to immense size. For instance, historians estimate that the Inca Empire included 4,000,000 people at the time of the Spanish conquest (Lenski, 1984). Other examples of such civilizations include the Mayans and the Aztecs of Central America, the Azande of East Africa, the Phoenicians and the Egyptians of the Mediterranean, and the Imperial Chinese (Lenski, 1984:149). Imperial societies gave rise to the first large cities. In these cities, several thousand people lived off the agricultural surplus of the surrounding areas. New

craft trades emerged in the cities to produce more refined products for the rising tastes of the empires' rulers, officials, and attendants.

Slave and Free Labor Perhaps as much as two-thirds of craft work in the classical empires was done by slave labor (Cartledge, Cohen, and Foxhall, 2002). Urban workshops were typically quite small, employing fewer than a dozen artisans. Most were smaller, being composed of an artisan who hired free labor or bought slave labor to increase the output of pottery, cloth, woodenware, or metal tools. Because of the ready availability of slave labor, few technological advances occurred in craft

---

B O X 1.3 Slave Labor in the Roman Mines

The miner of ancient times was nearly always either a slave or a criminal. This explains why the means used remained almost unchanged for thousands of years.... It was considered unnecessary to make the work easier for the slave, whose hard lot inspired no sympathy, although it kept him to the end of his days buried in the gloomy depths of the earth, suffering all sorts of torments and privations. There was mostly a super-abundance of slaves. After [military] campaigns there were usually so many that great numbers of them were massacred. So there was no dearth of labour. And so it happened that in almost all the mines of the ancients only the simplest means were adopted.

The tunnels constructed in the rock by these sim-ple means are often of astonishing length. It has been computed and confirmed by observing the marks of wedges that in even relatively soft stone the progress made amounted to about half an inch in twenty-four hours. This low efficiency was compensated to some extent by making the tunnels very low, by working only along the seams of the ore and by avoiding as far as possible the removal of unnecessary stone. Conse-quently, the galleries and tunnels were so narrow that a slave could squeeze himself through only with great difficulty. In many mines, in particular those of the Egyptians, Greeks, and Romans, children were employed, so that as little stone as possible would have

to be removed. Although the slaves must have become weakened by their sojourn in the mines and by the unhealthy posture during work, as well as through sickness—in lead mines particularly through lead poisoning—they must often have used very heavy tools. Hammers have been found that weighed between 20 and 26 pounds.

At the same time there were no precautions against accidents. The galleries were not propped up and therefore often collapsed, burying workmen beneath them. In ancient mines many skeletons have been found of slaves who had lost their lives in this way while at work. Nor were attempts made to replenish the supply of air or to take other steps for preserving health. When the air in the mines became so hot and foul that breathing was rendered impossible the place was abandoned and an attack was made at some other point. These conditions must have become still more trying wherever, in addition to the mallet and chisel, the only other means of detaching the stone was applied, namely fire. The mineral-bearing stone was heated and water was then poured over it. There was no outlet for the resulting smoke and vapors.

SOURCE: Excerpted from Albert Neuberger, 1977, "The Technical Skills of the Romans." In *A Reader in Cultural Anthropology*, edited by Carleton S. Coon, pp. 517–518. Huntington, New York: Krieger. Reprinted by permission of the publisher.

---

production during the period of classical civiliza-tion. Slaves possessed neither the opportunity nor the motivation to be innovative at work. Such advances would await the development of free trades and free labor in the Middle Ages. Box 1.3 describes the harsh conditions of slaves working in Roman mines.

In Rome, free craft workers eventually formed guilds to regulate the standards of their trade and provide religious and social services for their mem-bers. During the decline of the Roman Empire and throughout much of the Middle Ages, in an effort to stabilize production, the authorities forbade leav-ing a guild or refusing to follow in one's father's trade (Lilley, 2002). "The Venetian government, for instance, strictly prohibited the emigration of

caulkers, and from a document of 1460 we learn that a caulker who left Venice risked six years in prison and a two-hundred-lire fine if apprehended" (Cipolla, 1980:189). Guilds thrived during this per-iod and would take on an important role in orga-nizing production throughout the Middle Ages.

The End of Classical Civilizations   In the centu-ries following the birth of Christ, Roman civilization declined because of the difficulties of maintaining a worldwide empire and because of direct challenges from Germanic tribes from Northern Europe and Mongolian tribes from Western Asia. The end of the period of classical civilizations is typically dated substantially later, sometimes as late as the fall of Charlemagne's empire in Western Europe in the early 800s. People left the large cities and returned to rural areas during these centuries. However, agricultural work was no longer undertaken by independent cultivators who were members of agricultural societies or by slaves laboring in large holdings. Instead, agriculture was organized around large estates in which local landlords ruled from fortified manors and the cultivators were legally tied to the land.

Feudal Society

In many ways **feudal society** was simply an extension of agricultural society. The majority of people still tilled the land in the same traditional ways. However, the way in which agricultural surplus was extracted changed. In simple agricultural society and in imperial societies, peasants had given up a portion of their crops as tax to feed the rulers, priests, and warriors, or they were forced to work as slaves. In feudal society landlords extracted surplus both as a share of the peasants' crops and in the form of forced labor on the landlords' land. The latter imposition was called *corvee labor,* and peasants working under the feudal system were called *serfs.* Forced labor averaged three days per week. The burden was lessened to the extent that the labor could be performed by any member of the peasant's family, but it was increased to the extent that requirements were greatest at planting and harvest when the peasants most needed to tend their own

fields. The movement of peasants off the manor was either forbidden or might result in a fine. Fines might be levied, for example, when a daughter married away to a different manor or a son moved to the city.

Extreme Inequality   Incremental improvements in technology expanded agricultural productivity during the Middle Ages. These advances included the horseshoe, the padded horse collar, the wheeled plow, and the three-field system of crop rotation. These developments resulted in a gradual growth of population, but they did little to improve the position of the peasant who on average subsisted on a scant 1,600 calories per day (Lilley, 2002).

The ruling class absorbed the additional surplus in what was perhaps the most extreme period of inequality in human history. Historians have estimated that between 30 percent and 70 percent of serfs' crops were expropriated in the form of taxes or duties by feudal lords or by the Catholic Church. In much of the feudal system as practiced in Western Europe, "when a man died, the lord of the manor could claim his best beast or most valuable movable possession, and the priest could often claim the second best" (Lenski, 1984:269).

Artisans and Guilds   What is most significant about the Middle Ages from the standpoint of changes in work is the growth of a new class of producers, the free artisans. **Artisans** were typically the sons and daughters of serfs who had escaped the rural servitude of their parents and had moved to a town or "bourg." The citizens of medieval towns had a large degree of self-governance and were free from the obligations that characterized the rural manors. The trades of artisans included baking, weaving, and leather working. Instead of performing these diverse activities as part of their daily round of duties, the artisans specialized in a particular trade, producing superior goods for other town dwellers who included church officials, soldiers, and merchants engaged in intercity trade.

The practitioners of the various crafts formed **guilds** for their mutual benefit. These guilds regulated the quality of goods, acceptable hours of work, and even prices. For example:

> Chandlers had to use four pounds of tallow for each quarter pound of wick. Makers of bone handles were forbidden by their guilds to trim their products with silver lest they pass them off for ivory. . . . In the interest of preserving fair competition, work on Sundays and saints' days was banned, thus preventing the impious from gaining an unfair advantage over the pious. Night work was forbidden both in the interest of fair competition and because poor lighting compromised meticulous workmanship. (Kranzberg and Gies, 1986:2–3)

Merchants, too, organized themselves into guilds to regulate and standardize their activities. Prices, hours, and first rights to bid on cargo were among the many regulated practices (Lilley, 2002). Medieval guilds were similar to those of the classical civilizations, but they played a considerably broader role in organizing the economic and political life of the medieval city. In the classic civilizations the king or the emperor and powerful landowners held political power. In the Medieval cities, by contrast, the guilds were major actors in both economic and political life.

The training of new artisans was also strictly controlled by the guilds, again with the goal of regulating quality and thus protecting the reputation and status of the guild. An additional goal was ensuring work for existing guild members by limiting the number of people who could practice a trade.

Apprentices were recruited from the extended families of artisans within the trade, from other artisans' families, and from rural areas. After years of on-the-job training, the apprentice would produce what was judged to be a masterpiece and would then be admitted formally to full guild membership as a master craftsman. Over the course of the Middle Ages, however, opening one's own shop became increasingly difficult as more and more craftsmen competed for limited markets. As a result, guilds became more restrictive in their membership criteria, and a new position emerged called the *journeyman*. A journeyman artisan had successfully completed his apprenticeship training but did not own his own shop. Instead, he traveled from artisan to artisan, and sometimes from city to city, looking for work as an artisanal helper.

An additional major change in work that occurred during the Middle Ages was the disappearance of slave labor. Massive public works were constructed during this period, just as during the period of classical civilization. These monuments, cathedrals, and monasteries were financed by surplus extracted primarily from feudal serfs and secondarily from taxes on intercity trade. The labor, however, was not provided by slaves or by serfs. Rather, these structures were constructed by skilled artisanal workers, such as masons, carpenters, blacksmiths, and glaziers. The use of free skilled labor for these construction projects provided an important impetus to the growth of the artisanal classes during the Middle Ages.

A New Vision of Work   The advent of free artisanal work brought with it important new perspectives on the meaning of work and on relations between members of society. Artisans strongly supported traditional, highly skilled ways of work. Producing large quantities of goods quickly was not the overarching goal because there were no mass markets; the artisans served a very restricted clientele. The guilds therefore encouraged group solidarity to lessen the danger of being undercut in limited markets by price-cutting or by the sale of shoddy merchandise. As members of a guild, artisans identified their interests with preservation of high quality standards rather than with getting ahead

FIGURE 1.2 Time Line of the Organization of Work

of other artisans in their guild. These ideals of equality and solidarity would later help inspire the revolutionary demands of the artisans and the peasants as they sought to overthrow feudal society and install a society based on freely producing craft labor.

Economic Expansion and the End of Feudal Society   Many changes led to the passing of feudal society and the transition to modern industrial society. This transition provides a fulcrum for much of the social history written since that time. It was also a central concern for the forerunners of modern sociology—Karl Marx, Max Weber, and Emile Durkheim. The transition from feudal society to industrial society was brought about by an expansion of population, trade, and markets. Between the years 1000 and 1500, more than a thousand new towns sprang up in Europe. Connected by usable roads, these towns provided the basis for *regional specialization,* that is, for a division of labor based on the unique resources of different regions (Lilley, 2002). The Crusades also spurred the growth of important new markets for European goods, especially woolen goods. The exploration of the New World and the opening of trade routes to Asia were also boons to the European economy. Finally, the huge store of precious metals seized from Indian civilizations in Central and South America allowed a dramatic growth in the money supply and facilitated the expansion of credit and trade.

The period between feudal society and industrial society was one in which increased trade provided the impetus for changes in the organization of work. This intermediate period, called the Age

of Merchant Capitalism, lasted from the fourteenth century to the advent of the first modern factories in England in the mid-eighteenth century. Figure 1.2 provides a timeline that summarizes major historical transformations in the organization of work.

Merchant Capitalism

The earliest form of capitalism grew not as a way to organize production but as a way to organize trade. Previously, craft workers had bought their own raw materials and retailed the finished goods from their own shops. Under **merchant capitalism** the merchant capitalist increasingly took on these networking roles. This change frequently occurred because of the merchant's monopoly over lucrative intercity markets for finished goods or for agricultural products. For example, a leather merchant might have had a corner on the purchasing of hides from a cattle-raising region or on the market for leather shoes sold in neighboring areas. Local craftsmen thus had to work through the merchant to participate in these markets.

The Merchant as Labor Contractor   The system of production under merchant capitalism was called the **putting-out system** because the merchant would "put out" the raw materials to be worked up and would later collect the finished products to be sold.

> In England the typical form of cottage or domestic industry was wool and, later, cotton weaving. . . . Merchants brought raw materials to rural cottages and then picked up the woven cloth which they had finished in towns or large

villages. By having cloth woven in the countryside, the merchants managed to escape the control of the guilds. (Tilly and Scott, 1978:14)

In essence, craft workers became subcontractors for merchants and were paid a piece rate for their work. The system also brought many non-guild workers into production because the merchant capitalist would put out simple tasks, such as preparing and softening leather or cleaning and carding wool, to less-skilled workers, whose labor was cheaper. These workers included seasonally underemployed peasants, recent immigrants from rural areas, widows, young women waiting to marry, and wives of underemployed husbands (Gullickson, 2002). Apprentices and journeymen who could not find employment as artisans because of the encroachment on craft markets by the merchant capitalists were also recruited into the putting-out system. In rural areas this system was called *cottage industry*. In urban areas it was called *sweatshop* production because the work typically took place in the often hot, cramped, and dirty attics of people's homes.

Putting-out arrangements were the earliest form of the system of wage labor typical of industrial production throughout the world today. The system was successful because it undercut the pricing structure of guild regulations. The artisan made a full range of goods in his trade and wanted to be paid accordingly. A tailor, for example, would want to make a fair day's wage for his labor, regardless of whether he worked that day on petticoats, shirts, or jackets. However, some tasks, such as making petticoats, did not require the full range of the artisan's skills and could be done reasonably well by less skilled workers. The merchant would put out such work to a seasonally underemployed peasant or to a journeyman tailor unable to set up his own shop, and would pay this worker less than an artisan expected to receive as a living wage. On this basis, the merchant capitalists undercut the artisans' prices and encroached further and further into their markets.

Guild Resistance to Merchant Capitalism The guilds resisted the putting-out system by implementing civic laws regulating the number of journeymen or apprentices that one person could employ. However, the merchant capitalists' control of the intercity markets and their flexibility in putting out work to rural areas afforded them options unavailable to the craft workers of any given city. As a result, the putting-out system eventually replaced the guild system in the manufacture of many basic commodities, most importantly in textiles. The craft system of work typical of the guilds lingers on to this day in such areas as specialty tools and the manufacture of some luxury items. The legacy of the guilds is also felt in modern unions and professional associations. However, artisanal production would never again be the way in which the majority of goods were manufactured in society.

The social relations of work were profoundly transformed by the putting-out system. In place of free artisans, two classes emerged with distinct and even antagonistic relations: the merchant capitalists and those whom they employed in the putting-out system. The merchant capitalists sought to pay as little as possible for each type of work they put out. Those who worked under this system sought to secure a living wage for their labor, a goal often hampered by the availability of cheaper labor in another city or region. These relations set the stage for the historic conflict between *capital* and *labor*.

The daily lives of artisans were dramatically affected by the advent of merchant capitalism. Wages fell for craft workers as they were forced to cut their prices to retain a share of markets (Gullickson, 2002). They were forced to work longer hours, and in general their positions as members of the middle class were gravely threatened. Even the average age of marriage among journeymen increased substantially at this time because of the difficulty of securing a position as a master craftsman who could support a family (Aminzade, 2001). Box 1.4 describes the working and living conditions of a family employed as subcontracted labor to make cloaks in their urban apartment.

Under the guild system, women had worked as helpers in their husbands' crafts and sometimes as members of their own guilds, though generally at lower earnings. Under the putting-out system women were often employed directly by merchant

---

BOX 1.4 Sweatshops—Home Work in Early Capitalism

The three main features of the sweatshop have been described as unsanitary conditions, excessively long hours, and extremely low wages. The shops were generally located in tenement houses. As a rule, one of the rooms of the flat in which the contractor lived was used as a working place. Sometimes work would be carried on all over the place, in the bedroom as well as in the kitchen. Even under the best of conditions, this would have made for living and working in grime and dirt.... A cloak maker used one room for his shop, while the other three rooms were supposed to be used for domestic purposes only, his family consisting of his wife and seven children. In the room adjoining the shop, used as the kitchen, there was a red-hot stove, two tables, a clothes rack, and several piles of goods. A woman was making bread on a table upon which there was a baby's stocking, scraps of cloth, several old tin cans, and a small pile of unfinished garments. In the next room was an old woman with a diseased face walking the floor with a crying child in her arms.

SOURCE: Excerpted from *America's Working Women*, edited by Rosalyn Baxandall, Linda Gordon, and Susan Reverby, pp. 101–102. Copyright 1976 by Rosalyn Baxandall, Linda Gordon, and Susan Reverby. Reprinted by permission of Random House, Inc.

---

capitalists, who used their low wages as leverage to undercut artisanal wages. By working at home, women were able to combine various forms of productive activity including paid work, domestic activity, and care for children. Because paid work was only one part of their productive activity, they were often willing to undertake this work for lower wages than were skilled artisans who needed to secure their entire livelihood in this way (Crowston, 2001).

Spiritual Grace and Worldly Success Merchant capitalism also witnessed the emergence of new theologies based on the thought of Martin Luther and John Calvin, whose writings gave voice and content to the Protestant Reformation. These theologies suggested a new vision of work sometimes called the **Protestant work ethic** (Weber, 1958 [1904]). This vision identifies successful pursuit of one's occupational calling with spiritual grace. If one prospers through diligent work, this prosperity is seen as evidence of being among those chosen to go to heaven. The Protestant work ethic was well matched with the emerging worldview of the merchant capitalist, who was engaged in a competitive struggle for success on earth. This ethic provides a justification and a motivation for hard work in the pursuit of worldly success. The Protestant work ethic also encourages savings and thus lays a foundation for the accumulation of capital to be reinvested in business and commerce. Savings are encouraged by the ethic's call for the realization of worldly asceticism through frugality, austerity, and plain living. The ethic also provides a basis for despising less successful individuals because their lack of success implies that they are among those whom God has forsaken. Although few would subscribe completely to this view today, elements of the early Protestant work ethic are retained in many aspects of modern Western culture (Chalcraft and Harrington, 2001).

The Industrial Revolution

The transition from merchant capitalism and putting-out industry to industrial capitalism was a violent one. Known as the **Industrial Revolution**, it involved the forcible movement of large numbers of peasants off the land and into factories. In the words of Karl Marx, it is a history "written in letters of fire and blood." The Industrial Revolution took place first in England, significantly because the English were deeply involved in the expanding woolen trade with Flanders (today's Belgium). England's role in the woolen trade set the stage for the forcible removal of peasants from the land and their replacement by grazing sheep.

This displaced peasantry, in turn, provided a ready pool of labor for the early factories. The woolen trade also generated capital for investment in new factories and machinery.

Replacing Agriculture with Industry Peasants were forced off the land through the **enclosure movement**, in which land previously held in common by the peasants and the landlord and used for grazing livestock, was enclosed with fences (Hobsbawm, 1975). The land was then used for raising sheep. This change caused a dramatic deterioration in the situation of the peasants, who were no longer able to use this land to support their few farm animals. Marx notes the actions undertaken by the Duchess of Sutherland between 1814 and 1820 as an example of this process:

> [The] 15,000 inhabitants, about 3,000 families, were systematically hunted and rooted out. All their villages were destroyed and burnt, all their fields turned to pasturage. British soldiers enforced this eviction, and came to blows with the inhabitants. One old woman was burnt to death in the flames of the hut, which she refused to leave. Thus this fine lady appropriated 794,000 acres of land that had from time immemorial belonged to the clan. (Marx, 1967 [1887]: 729–730)

After the peasants had been forced off the land to make room for sheep, they were further hounded as vagabonds until they entered the early factories, often as forced labor.

The government of England enforced the movement of displaced peasants into the early factories through what Marx called "bloody legislation against vagabondage":

> Beggars old and unable to work receive a beggar's license. On the other hand, whipping and imprisonment for sturdy vagabonds. They are to be tied to the cart-tail and whipped until the blood streams from their bodies.... For the second arrest for vagabondage the whipping is to be repeated and half the ear sliced off; but for the third relapse the offender is to be exe-

cuted as a hardened criminal and enemy of the common weal.... Thus were agricultural people, first forcibly expropriated from the soil, driven from their homes, turned into vagabonds, and then whipped, branded, tortured by laws grotesquely terrible into the discipline necessary for the wage system. (Marx, 1967 [1887]: 734–737)

Vagabondage laws, poor laws, and head taxes were used to force displaced agricultural workers into the early factories. The earliest factory workers were thus frequently the victims of penal sanctions rather than freely hired wage labor.

The emergence of these factories was the final blow against the feudal guilds and their apprenticeship systems for recruiting and training skilled workers. Indeed, many laws prohibiting the existence of guilds, unions, and other combinations of workers—and even defining union membership as a crime punishable by death—were passed in England in the sixteenth, seventeenth, and eighteenth centuries (Hobsbawm, 1975). Factories also signaled an end to merchant capitalism, which was based on expanding production by putting out work to more and more home-based workers, and ushered in the next stage of society and economic history—industrial capitalism.

The Factory System The defining characteristic of industrial capitalism is the development of the factory system. What made the early factories so terrible that people refused to enter them except under the force of law? To understand the position of the early factory workers, it is helpful to know something about how these factories operated. Workers were centralized under one roof in the factory system. Such centralization avoided the costs of transporting partly finished goods from one location to another as required by the putting-out system. Importantly, it also forced workers to labor according to the dictates of the owners rather than according to their own pace and rhythm. The centralization of work thus meant that in order to have access to any work at all, a worker had to be willing to work the hours and days demanded by the employer. The result was an expansion in the length of the working day, an increase in the intensity of work, and a decrease in the number of religious and personal holidays allowed (Pollard, 1963).

The massing of many workers in one location also allowed the development of machinery to do repetitious tasks. A seemingly endless variety of tasks could be broken down into their simplest components and mechanized. The widespread introduction of machinery had not been feasible under the putting-out system because too few repetitions of a task were performed at any one location to warrant the expense of machinery. The centralization of work in the factory thus facilitated the introduction of machinery.

The push toward longer working days was further intensified by the factory owner's desire to run the new machinery as many hours per day as possible in order to justify its purchase. The machinery heightened productivity but ironically resulted in lower wages for workers because fewer skills were needed. Increased productivity and decreased costs allowed factory owners to sell their products at prices that drastically undercut artisanal producers and even merchant capitalists. These dynamics resulted in the rise of factory production over older social organizations of work (Thompson, 1967).

The Centrality of Textiles   The Industrial Revolution in England had begun in the woolen industry. The focus shifted to cotton textiles as the import of cotton from the New World grew and markets for British textiles expanded in Asia and North America. It is noteworthy that even the beginnings of the Industrial Revolution depended on the availability of global markets. Several inventions also facilitated the rapid growth of the textile industry, including the flying shuttle (1733), the spinning jenny (1767), the water frame (1769), the spinning mule (1779), the power loom (1787), and the cotton gin (1792) (Faunce, 1981:14). These developments generated tremendous growth in British industry in a very brief period. "Output of printed cotton rose from 21 million yards in 1796 to 347 million yards in 1830. Pounds of raw cotton con-

sumed increased from 10.9 million in 1781 to 592 million in 1845" (Faunce, 1981:13). This boom in cotton cloth production also sparked the rise of cotton plantations in the American South and fueled the *slave trade* as a source of cheap labor for the cotton fields.

The consequences of growth in the textile industry spread rapidly to other British industries. Coal was needed to provide steam for power looms. Steel was needed to build railways to carry the coal. More coal was needed to make coke for smelting iron and steel. Machine tools were needed for building and maintaining textile, mining, and railway machinery. Ship manufacture advanced perhaps more rapidly than any other field because of the dependence of the Industrial Revolution on foreign trade. By the end of the nineteenth century, huge steel-bodied, steam-powered ships would replace the slower, smaller sailing vessels allowing a further expansion in the tonnage of goods shipped around the world.

A Detailed Division of Labor   The division of labor into finer and finer activities also advanced more rapidly during this period than at any other in history. The English economist Adam Smith described this process for the simple trade of pin manufacture:

> One man draws out the wire; another straightens it; a third cuts it; a fourth points it; a fifth grinds it at the top for receiving the head; to make the head requires two or three distinct operations; to put it on is a peculiar business; to whiten the pin is another; it is even a trade by itself to put them into the paper; and the important business of making a pin is in this manner divided into about 18 distinct operations, which in some manufactories are all performed by distinct hands, though in others the same man will sometimes perform two or three of them. . . . Ten persons, therefore, could make among them upwards of 48,000 pins in a day. (Smith, 1776)

Under the artisanal system a pin maker might have repeated each of the 18 distinct operations several times before proceeding with the next step, thus increasing efficiency by not having constantly to take up and put down each task. However, these tasks would never have been assigned to different workers. Social theorists have offered strongly divergent appraisals of the consequences of this detailed division of labor. Smith applauded the system because it lowered the price of pins. Marx condemned it because it reduced the skills and lowered the wages of the workers who made the pins.

In the early factories, work was organized under the supervision of foremen. These foremen were more similar to subcontractors than to modern day supervisors. They hired their assistants, trained them in their tasks, set them to work, supervised them on the job, and paid them out of the piece-rate earnings they received for the goods produced. This system was typical of the textile industry, iron making, gun making, saddle making, coach building, and most other trades until the beginning of the twentieth century (Jacoby, 2004).

Forced Labor   The position of workers deteriorated rapidly during the early stages of the Industrial Revolution. The earnings of weavers fell to as little as a tenth of what they had been before the introduction of machinery (Hobsbawm, 1975). This resulted partly from the reduced need for skilled workers in the early factories. The early factory workers were more likely to be women, children, indentured laborers, vagabonds forcibly placed in poor houses, and, in the New World, slaves and indentured servants. **Indentured laborers** were workers under contract to work for a certain amount of time—typically eight to ten years—for a set price or as part of their penalty for being found guilty for a crime such as petty theft or vagabondage.

The rights of those who labored in the early factories were minimal. Slavery, forced labor, and indenturement reappeared and were relatively common, especially in the New World. The role of indentured laborers in settling the new lands started as early as 1607 with their use by the Virginia Company. Conditions for these workers were terrible. They suffered high levels of mortality and were treated with great cruelty. In response some indentured workers ran away to live with the Indians. The Virginia colony's governor dealt firmly with recaptured laborers: "Some he apointed to be hanged Some burned Some to be broken upon wheles, others to be staked and some to be shott to death" (quoted in Galenson, 1984:4). A *quarter of the labor force* in the United States in the early 1800s was made up of slaves. From the standpoint of the slave owners, this provided a workable solution to the problem of labor shortages in the New World and, simultaneously, a solution to the problem of retaining and controlling their workers. Wage laborers thought less well of this solution as it set a harsh standard against which their own labor was measured. Slaves were not consulted on the system.

Government involvement in the economy during the period of early industrialization occurred in the form of military support for the establishment of British colonies and the expansion of trade with these colonies. Indeed, this period of British political and economic history is sometimes called *mercantilism* to indicate the pivotal importance of establishing international markets. Government involvement did not, however, include any but token efforts to establish minimum working or safety standards in British factories.

Inside the factories the round of daily life was extremely monotonous. The strict routine was quite unlike preindustrial rhythms of work, which had been based on seasonal variations and a degree of personal discretion in organizing one's daily tasks (Hobsbawm, 1975). The hours were extremely long. Neither before nor after have people worked longer or harder than during early industrial capitalism: "In some huge factories from one fourth to one fifth of the children were cripples or otherwise deformed, or permanently injured by excessive toil, sometimes by brutal abuse. The younger children seldom lasted out more than three or four years without some illness, often ending in death" (quoted in Faunce, 1981:16). Beginning in the early 1800s, the working day was further extended from twelve hours to as many as sixteen hours by the use

BOX 1.5   Complaint by Lowell Factory Women, 1845

The first petitioner who testified was Eliza R. Hemingway. Her employment is weaving—works by the piece.... She complained of the hours for labor being too many, and the time for meals too limited. In the summer season, the work is commenced at 5 o'clock, a.m., and continued til 7 o'clock, p.m., with half an hour for breakfast and three quarters of an hour for dinner. During eight months of the year but half an hour is allowed for dinner. The air in the room she considered not to be wholesome. There were 293 small lamps and 61 large lamps lighted in the room in which she worked, when evening work is required.... About 130 females, 11 men, and 12 children (between the ages of 11 and 14) work in the room with her. The children

work but 9 months out of 12. The other 3 months they must attend school. Thinks that there is no day when there are less than six of the females out of the mill from sickness. Has known as many as thirty. She herself is out quite often on account of sickness.... She thought there was a general desire among the females to work but ten hours, regardless of pay.... She knew of one girl who last winter went into the mill at half past four o'clock, a.m., and worked till half past 7 o'clock, p.m. She did so to make more money.

SOURCE: Excerpted from *America's Working Women*, edited by Rosalyn Baxandall, Linda Gordon, and Susan Reverby, pp. 49–50. Copyright 1976 by Rosalyn Baxandall, Linda Gordon, and Susan Reverby. Reprinted by permission of Random House, Inc.

---

of gaslight. As a result, workers labored from before dawn well into the night (Hobsbawm, 1975).

Women and Children   In 1838 only 33 percent of the workers in British textile factories were adult males (Hobsbawm, 1975). Women and children thus played a central role in the early Industrial Revolution. They were employed because it was acceptable to pay them much less than men and because they were easier to bully into the harsh discipline of mechanized production. Young women were also considered more expendable to agricultural work and were thus more likely to be available for factory work (Crowston, 2001). Given the conditions of early factory work, it would be inaccurate to consider the participation of women at this time as a sign of their emancipation. The harsh working conditions of women in the early factories are illustrated in a petition to the state of Massachusetts that the women of Lowell brought against their employers (see Box 1.5).

Industrial Cities   The cities in which this emerging working class lived were no more healthy or pleasant than the factories in which they labored:

The industrial town of the Midlands and the North West was a cultural wasteland....

Dumped into this bleak slough of misery, the immigrant peasant, or even the yeoman or copyholder, was soon transformed into a nondescript animal of the mire. It was not that he was paid too little, or even that he labored too long—though either happened often to excess—but that he was now existing under physical conditions which denied the human shape of life. (Polanyi, 1957:98–99)

Not until the 1890s did sewage systems begin to catch up to the need for them, even in the largest industrial cities (Lilley, 2002). Epidemics of cholera and typhoid took an appalling toll on workers in the early industrial cities. Domestic cooking and heating as well as factory production depended on coal, and a thick layer of soot settled over everything in the industrial city. As a result, thousands died from tuberculosis and other respiratory ailments—or had their lives shortened and crippled by these diseases (Hobsbawm, 1975).

The spread of factories separated work from home. If people wanted to find gainful work, they had to leave their families and venture out alone. The removal of work to the factory undermined the family's function as the primary unit of economic production. To the extent that women were unable to leave the house in the search for work because of the demands of domestic activity and child rearing, their relative contribution to the household's economy was undermined (Kessler-Harris, 2001). Women often supplemented their contribution, however, by taking in boarders, doing laundry or sewing at home, or performing other income-generating activities compatible with their domestic responsibilities (Crowston, 2001).

The Skilled Trades   Centralized work in factories under close supervision, with machines dictating the pace, robbed workers of the skills and autonomy necessary to take pride in their work. It is no wonder that early factory workers were alienated and resentful. It would be inaccurate, however, to characterize all workers as having experienced a loss of skill and pride in their work at this time. Older, skilled ways of work prevailed among blacksmiths, shoemakers, stonemasons, and the new machinist trades whose growth was spurred by the Industrial Revolution. These workers continued to experience strong feelings of *pride in work* and *solidarity* with their fellow craft workers.

Industrial Capitalists   The new class of factory owners, called *industrial capitalists,* as well as those who aspired to become members of this class, developed new ideologies to represent their view of work. One popular ideology among the capitalist class in North America in the latter part of the nineteenth century was Social Darwinism, patterned after Charles Darwin's evolutionary theory of the survival of the fittest (Perrow, 1986). The Social Darwinists viewed extreme competition and inequality not only as just but also as in the best interests of society because they would ensure that the most able individuals would rise to the top. Similar lines of reasoning were embodied in the philosophy of *laissez-faire,* which argued that economic systems work best when undisturbed by government regulation. The growth of such individualistic ideologies and the waning of craft ideologies stressing pride in work and solidarity left little middle ground for the development of a shared workplace ethic suitable for reconciling the

two opposing classes of industrial capitalism. Attempts at reconciliation would await the development of postindustrial society.

In the long run, industrial capitalism allowed for some improvement in the position of the working class. In the modern world "nearly every indicator of physical well-being corresponds to the extent of industrialization" (Tausky, 1984:30). Increased well-being is based on greater agricultural and manufacturing productivity and on workers' abilities to demand a share of these rewards. The latter, in turn, is based on the continuing need for new skills as technology advances and on the organization of workers into unions and their successful petition for legal rights and safeguards in the workplace.

Early Trade Unions   In 1824 the illegality of trade unions in England ceased, though every effort to destroy them was still made by employers (Hobsbawm, 1975). By the 1870s other important legal changes were also occurring in England. These included the passage of child labor laws and a law limiting the working day to ten hours. By the 1880s all children under the age of ten were required to be full-time students. Factory inspections were also increased to provide at least some enforcement for these laws. These changes occurred about fifty years later in North America, which experienced a slower process of industrialization and a later emergence of unions. In North America a variety of union forms would emerge, including local craft unions, the Knights of Labor, and national craft unions organized under the banner of the American Federation of Labor. We discuss these early attempts at unionization further in Chapter 6. In aggregate, these changes produced an increase in life expectancy for workers and improved standards of living.

Monopoly Capitalism

By the early years of the twentieth century, *industrial capitalism* was displaced by *monopoly capitalism* (Jacoby, 2004). The greatly expanded size of companies at this time provided them with immense power over competitors, suppliers, and consumers. Large companies could utilize more efficient types of mechanization and more powerful marketing techniques than smaller companies and gradually came to replace them in more and more industries. Economic power became centralized in an *oligopoly* of a few large companies in many industries. Other industries became dominated, at least regionally, by a single company—a *monopoly*. Today, railroads and power companies are good examples of complete or almost complete regional monopolies.

Along with greater oligopoly and monopoly power came an increased involvement of government in the productive process. Government involvement grew for a number of reasons. For one, there was a heightened need to establish secure markets for the greatly expanded productivity of the large companies. Industrial nations struggled for a share of world markets, and governments took on an important role in securing such markets through diplomatic and military action (Hobsbawm, 1975). The First and Second World Wars were, to a significant extent, a result of such global economic competition. In addition, competing demands on the state by monopoly capitalists, small capitalists, and the growing working class encouraged greater government involvement in the economy to stabilize class relations and to mediate between these competing interests. This intervention was frequently less than evenhanded, with powerful groups successfully using the government to support their interests.

Huge Centralized Factories   The end of the nineteenth century and the beginning of the twentieth was a time of great change in the organization of enterprises. Increased *size and centralization* were the order of the day. Factories increased in size, sometimes employing many thousands of workers, and single companies acquired control of multiple factories. "In 1851 there were fifty telegraph companies; in 1866 Western Union operated fifty thousand miles of line and had bought up all but a few remaining telegraph companies. . . . In the 1890s Carnegie's steel factories controlled two-thirds of steel production in the United States" (Tausky, 1984:46–47). American Tobacco Company controlled over 90 percent of cigarette sales by 1890. The Westinghouse

Electric Company and General Electric Company controlled electrical equipment. The F.W. Woolworth Company held a huge share of sales in general merchandise stores. Sears, Roebuck and Company, and Montgomery Ward and Company dwarfed their competitors in the mail-order business. Several dozen early car manufacturers were overshadowed and forced out of business by the wildly successful Ford Motor Company. E.I. DuPont de Nemours and Company dominated chemical production, and the Standard Oil companies, owned by the Rockefellers, dominated petrochemicals. These successes were partly based on efficiencies associated with economies of scale and on the ability to develop and deploy new technologies. However, increasing centralization was also based on the use of economic and political power to undercut competitors. Smaller companies had a difficult time competing against these corporate giants. They were forced to operate in regions or product lines as yet unconquered by the large companies, or to become suppliers of parts and services to the giant corporations (Adams, 2004).

Along with greater size and concentration of companies came increased bureaucracy. Bureaucracy entails the systematic use of rules and the creation of formalized job positions with clearly delineated duties. Standardized procedures had to be developed as the internal workings of companies became more complex and there was less and less reliance on various forms of subcontracting to foremen. Foremen no longer hired their work crews, for example. Instead, a central personnel office did the hiring. For workers, such bureaucratic procedures can have positive as well as negative consequences. Standardized practices protect workers from some of the worst abuses and favoritism of foremen, but they also tend to further depersonalize the work environment.

Mass Production   The immense size of economic organizations at the beginning of the twentieth century brought about the possibility of new forms of technology and new means of organizing work and controlling workers. The most notable of these was the **assembly line**. The full-fledged use of the assembly line occurred first in the automobile industry, but it built directly on existing forms of mechanized production being used in other industries at the time:

> An unskilled operator snapped engine blocks, for example, onto specially designed tables and watched a machine mill them automatically and accurately. Made this way, parts such as cylinder heads and engine blocks could be fitted together without the need for hand scraping of surfaces during assembly.
>
> Once hand fitting had been eliminated and the specialized machines had been arranged in the sequence of manufacturing operations, the next and revolutionary steps were to reduce the transit time of work pieces from machine to machine and to systematize their assembly.... The assembly line may have been inspired by the disassembly lines in Chicago slaughter houses, which circulated carcasses from butcher to butcher, or by Ford's own gravity slides and conveyors.... The first crude moving line cut the time needed for final assembly from just under 12.5 to about 5.8 man-hours. (Sabel, 1982:33)

With assembly-line production, job skills become highly specific to the technology and procedures used in a given plant. Such jobs are considered *semi-skilled* because they require a specific skill but one that can be learned in a relatively short time, often one to two weeks. A smaller portion of workers are required to have broader based skills, such as those of the machinist, electrician, or tool and die maker. The organization of production around an assembly line and the resulting deskilling of jobs are sometimes referred to as the "Fordist" organization of work in recognition of its origins in Ford's automobile factories.

The assembly line sets a rapid pace for workers and keeps them at required tasks much more closely than even the harshest foreman (Jacoby, 2004). Assembly lines and other forms of advanced mechanization are organized under the principles of *scientific management.* Scientific management is identified with the work of Frederick Taylor, an American industrial engineer, who believed that all

thinking should be removed from the realm of the worker. Instead, the worker was to execute diligently a set of motions engineered to ensure the most efficient performance of a given task. This so-called scientific plan came from first observing how workers did the task, then progressively redesigning the task to increase efficiency. The theory of scientific management was highly compatible with the new assembly-line system of production. In combination, however, these new forces sparked fierce resistance from workers, who felt that the new production systems treated them like automatons rather than human beings.

The daily life of production workers under monopoly capitalism was somewhat better than it had been in early industrial capitalism. The working day was progressively shortened because of heightened productivity and because of pressure from working-class trade union and political activity. However, the monotony of work, if anything, was increased. In North America the emergence of monopoly capitalism coincided with a period of heavy immigration of workers from Western Europe, in the 1870s, and from Eastern Europe, in the 1910s. These immigrants, both male and female, provided labor for many of the expanding mass-production industries. Ethnic groups often moved through the factories in successive waves—as one group moved upward toward craft and white-collar jobs, another group replaced them on the factory floor.

The shared situation of large numbers of semi-skilled workers in the mass-production industries facilitated solidarity and the identification of common grievances among these workers. Workers in such industries as automobiles and steel organized into industrial unions to promote their collective welfare. In North America there was a dramatic growth in union organizing during the Great Depression of the 1930s. Demands for higher wages, however, were not the primary motivation for these organizing drives. Rather, the workers resented the dehumanizing practices of scientific management, the speed-ups imposed on the already rapidly paced assembly lines, and the associated lack of job security. We discuss these union organizing drives in greater detail in Chapter 6.

A More Complex Class Structure The **class structure** of monopoly capitalism was significantly different from that of early industrial capitalism. Society remained strongly divided between owners and workers, who were themselves divided between skilled craft workers and less skilled workers. However, there was a rebirth of the middle class as a result of a heightened need for clerical and mid-level managerial workers in the new giant companies. The demand for professional services, such as education and health care, also expanded as a result of a rising standard of living—occupations which are the focus of Chapter 11. Managers and administrative support occupations are discussed further in Chapters 12 and 13.

Ideas about the meaning of work were also changing among capitalists, and especially among the new class of hired managers. The old ideology of "survival of the fittest" provided little guidance on how to placate angry workers or encourage their productivity. As a result, new ideologies focusing on persuasion and cooperation emerged. One ideology, associated with Elton Mayo and Chester Barnard, early industrial sociologists, came to be known as the *human relations* approach to industrial management (Perrow, 1986). Workers were seen as tractable and as desiring recognition and personal treatment. Such an approach reflected the reality that workers had been robbed of such individual recognition and personal treatment by the advent of industrial capitalism and the development of mass-production systems.

Postindustrial Society

The transition from mass production to the present stage of industrial society, sometimes called **postindustrial society**, resulted directly from the immense productivity of mass-production systems. This transition began with two decades of stable economic growth in the world following the Second World War. By the 1960s a smaller and smaller percentage of the labor force was needed to manufacture goods.

Similarly, farm productivity continued to rise. As a result, new employment growth has taken place mainly in clerical, service, and professional work. Some of these jobs are rewarding and exciting; others are as monotonous as factory work and frequently pay worse. In combination, their growth signals the advent of postindustrial society.

Service Industries A growing proportion of workers in the industrially advanced nations are employed in service industries, such as transportation, wholesale and retail trade, finance, insurance, real estate, professional and business services, public administration, entertainment, and health. In 1940 less than half of the United States labor force was employed in service industries. By the early years of the 2000s the proportion exceeded 80 percent.

Jobs in **service industries** vary greatly. Some are low-skill jobs in food service and health care. These jobs are predominantly filled by women and minorities. Other jobs entail highly skilled professional work. Examples include computer programming and systems analysis. The overall effect of this variability in service jobs is an increase in the *diversity* of employment situations rather than a clear improvement or deterioration in the quality of available work. In addition, the pace at which new skills are needed has accelerated dramatically as a consequence of rapid technological changes based on the widespread use of microprocessors and computers in the workplace. This results in an additional set of problems of adjustment and retraining for workers. We explore some of the challenges of providing training for rapidly changing careers in Chapter 5 on work and family.

Professional Jobs The number of highly skilled professional workers has also grown in postindustrial society. Professional workers include accountants, lawyers, and doctors and other health professionals in professional service industries, as well as engineers, chemists, biologists, and related scientists in increasingly sophisticated manufacturing industries. These workers hold a privileged position in the division of labor based on their possession of knowledge and expertise not widely available without

---

BOX 1.6   Engineering the Future

Developing new product lines is the glamorous work. This is seen as the essence of creative engineering, what engineering is all about. It is high-pressure work: crunches, slips, and other forms of organized hysteria accompany the pressure to be creative, to produce, to be smart. In development, engineers typically work on projects.... Within development, engineers sort themselves out by the type of work they do and their perceived skill. Engineering is a highly competitive arena in which formal statuses are supplemented by informal ratings. Informally, engineers are categorized by their skill. There are the "brilliant" and the "geniuses," their status sometimes debated ("the only way he made the list of 100 brightest scientists is if he mailed coupons from the back of cereal boxes") and sometimes acknowledged ("Peter is brilliant. There is no question about that; he is a crackerjack engineer"); and there are journeymen (and the occasional journey women), who might be "solid citizens—no rah rah."

SOURCE: Excerpted from Gideon Kunda, *Engineering Culture*, Philadelphia, Temple University Press, pp. 39–40.

---

rigorous, extended study and preparation. As a result of their expertise, they command relatively high wages as well as a certain degree of autonomy in decision making (Sullivan, 2005). Box 1.6 describes the competitive work environment of engineers in a high-technology company.

Class Diversity The proportion of skilled and semi-skilled manual workers in the labor force has remained relatively stable in postindustrial society. However, a new class of low-paid service workers has come into being. Their wages and conditions are considerably worse than those of skilled craft workers and those of most semi-skilled workers. Their conditions more closely resemble those of the declining segment of unskilled labor. In many ways their actual jobs are also similar to those of unskilled laborers. Service workers move hamburgers around the fast-food outlet or bedpans around the hospital, just as laborers move materials around the factory floor or construction site.

The class structure of postindustrial society is thus the most diverse of any society to date. It includes a capitalist class, a managerial class, a large professional class, a large manual class, and a large low-wage service class. The disparate situations of these classes include both opulence and continuing poverty. It was once believed that postindustrial society would bring about an end to poverty. Con-

temporary social scientists now believe that such an expectation was grossly exaggerated (Bluestone and Harrison, 2000). We discuss continuing problems of poverty and marginality at greater length in Chapter 4 on class, gender, and race and Chapter 14 on marginal jobs.

Work Motivation What motivates people to work in postindustrial society? Some observers assert that the work ethic has died in recent years and that apathy rather than motivation typifies the work force. But this vision only represents a partial truth. For people whose jobs yield only minimal rewards, there is little reason to work enthusiastically. And even for these workers, dignity is often achieved by taking pride in their work. Alternatively, for those with advanced education, more desirable job opportunities are available. These workers are motivated by the high level of rewards they can expect, by the pride associated with advanced training in a professional specialty, and by the expectation of their profession and their employer that they will be committed to their work (Leicht and Fennell, 2001).

In postindustrial society work can take on an overriding importance in people's lives, overshadowing family and community attachments that prevailed in previous periods. Many factors support this tendency. Greater demands for geographic mobility preempt family attachments and curtail friendships of long duration. Fear of unemployment intensifies competition for available positions. Retirement, access to health care, and social status are all attached to one's job. In postindustrial society, work has become a "master status" determining a person's overall position in society and his or her enduring sense of dignity and identity.

Women's Liberation?   The position of women has significantly improved in postindustrial society. Women have entered jobs that were once exclusively male preserves, including those of airplane pilot, fire fighter, and heavy-equipment operator. The improved position of women is partly the result of a reduction in the demands on women to perform homemaking duties. These roles have declined in significance because of delayed marriage, reduced birth rates, labor-saving technologies, and the substitution of paid services for work that women previously performed at home—such as child care. Women have also secured better jobs because of their high rate of college graduation in combination with the increased importance of higher education in postindustrial society. In spite of these gains, however, women's earnings for full-time work are still only about three-quarters of men's.

Racial and Ethnic Diversity   The movement of people around the world in the search for greater economic opportunities is accelerating in the twenty-first century. As a result, the economies of both developed and developing countries are characterized by increasing racial and ethnic diversity. This diversity represents both a challenge and an opportunity. The *challenge* is to train and integrate the new work force while minimizing discrimination and resentment from other workers. The skills and motivations of these new workers, however, also represent an important *opportunity* for heightened economic growth and prosperity for all. Chapter 4 describes some of the special challenges faced by female, minority, and immigrant workers.

Globalization   The world today is characterized by an increasingly international economy, which has significantly displaced local markets. The economic fates of nations and individuals are highly dependent on their position in this global economy. Thus, the nature and rewards of work are not determined solely by social and economic relations within one's work group or employing organization. Rather, the nature of one's work is importantly determined by its location in the global division of labor.

The heightened international division of labor has intensified competition between nations. Some nations appear trapped in the role of agricultural production or mineral extraction; others are becoming centers of low-wage assembly work reminiscent of early industrial society; still others are characterized by the growth of service and professional employment. The postindustrial world includes all of these different types of work and societies. The international division of labor and the relations between the industrially developed and the industrially developing world are key to understanding the contemporary world of work. The growth of manufacturing in developing nations is directly linked to the decline of manufacturing in industrially developed nations and the growth of the service sector in these societies, with all its positive and negative consequences. A report on Malaysian electronics workers in Box 1.7 illustrates how the conditions of the early factories are being reproduced today in manufacturing establishments in developing nations, even those producing high technology products for global markets. The nature of the modern global economy and its influence on work is the focus of Chapter 16.

Work and Leisure

Do people work harder today than in the past? The answer is complex and perhaps surprising. Workers in postindustrial society typically work many more hours than did hunters and gatherers. How can this be, given the vastly greater technological efficiencies of modern industry? For hunters and gatherers it made sense to kill only one or two antelopes, not ten or twenty. They had only limited technologies to preserve, store, or transport any surplus, and so much of it would have gone to waste. Modern workers *can* profitably

BOX 1.7   Electronics Assembly in Malaysia

The large-scale entry of labor-intensive industries such as garment manufacturing, food processing, but especially electronics assembly greatly increased the absorption of rural Malay women into the industrial labor force.... They represented a fairly well-educated labor pool, which was often over-qualified for the mass semiskilled factory occupations. About 50 percent of these workers had at least lower secondary education, and many had aspired to become typists, secretaries, trainee nurses, or teachers....

Maximum product output was extracted from these rural women by the factories. Quickly exhausted operators were replaced by the next crop of school leavers. By keeping the wages low, the factories motivated the operators to work overtime on a regular basis, to take on more unpleasant tasks (which exposed them to fumes and acids), or to work at an increased pace in order to earn special cash allowances. Freshly recruited workers were routinely assigned to the production processes that required continual use of microscopes. Thus most workers, by the end of a couple of years, suffered from eye strain and deterioration of their eyesight....

The industrial firms not only exploited the workers in this manner but they also attempted to limit their employment to the early stage of adult life, a strategy that ensured fresh labor capable of sustained intensive work at low wages. In one factory new workers were employed on six-month contracts so that they could be released or rehired at the same low wage rates. Government legislation for the protection of pregnant female workers has had the unintended effect of reinforcing factory policy to discourage married women from applying, although employed workers who got married would stay on. Married workers were given advice on family planning and provided with free contraceptives by the factory clinic.

The rapid exhaustion of the operators also resulted in most of them leaving on their own accord after three to four years of factory employment, although an increasing number remained working, even after marriage. Operators leaving the factories have not acquired any skills which would equip them for any but the same dead-end jobs.

SOURCE: Aihwa Ong, "Global Industries and Malay Peasants in Peninsular Malaysia," from *Women, Men, and the International Division of Labor*, pp. 429–431, edited by June Nash and Maria Patricia Fernandez-Kelly. Copyright 1983 State University of New York Press. Reprinted by permission of the publisher.

work longer hours in order to earn more money to acquire more possessions or buy an increasing array of services, including luxury services—such as travel and entertainment. The result is an increase in the duration and intensity of labor in modern society. Advertising, which accentuates the desire for goods, and ideologies, which value hard work, further contribute to a culture of overwork (Schor, 2004).

The lure of more goods and luxuries is only part of the story though. Owners and managers also want employees to work longer and harder in order to get maximum output and to minimize costs associated with adding more employees. And, because owners control who works and who does not, they can demand hard work and fire those who do not comply. Thus, workers in modern society often face the prospects of highly demanding work *versus* no work at all—with all the attendant insecurities and privations of unemployment.

The history of leisure, however, is not a linear one. The greatest intensification of work occurred during the height of the Industrial Revolution when the newfound power of factory owners was unmediated by any competing forces, such as trade unions or limits and protections later developed by democratic governments. Thus, the end of the sixteen-hour day in the late 1800s, the growth of trade unions, and the emergence of legislation specifying hours, overtime, and safety rules brought about a gradual decline in working time and intensity. This period of declining hours appears to have concluded, at least in North America, with the end of the post-Second World War economic boom in the mid-1970s. Since that time, average annual hours in the United States have begun to climb again, increasing from around 1,700 to nearly 1,900—Americans have simply had to work more hours to maintain their lifestyle in an increasingly competitive global economy (Jacobs and Gerson, 2004). These additional 200 hours translate into an additional five weeks of work per year.

Much of the growth in annual hours worked is due to women, particularly married women and women with children, increasing their paid hours of work. The result is a growth not just in individual time at work but also in total family time given to paid work and a decline in time available for home and leisure. Thus, many families experience greater time pressure and a sense of continuous multitasking in order to get needed chores done. People are *not* spending less time with their children. Rather, the losers appear to be personal leisure, time for interaction with one's spouse, and sleep (Bianchi, Robinson, and Milkie, 2006; Sayer, 2005). People *do* desire greater leisure and it competes with paid work for their time and attention, but today the desire for leisure wins out for fewer people than the desire (and need) for money. We will explore other aspects of time, work, and leisure in Chapter 5 on work and family and in Chapter 16 where we compare work and leisure internationally.

The Future

Work is here to stay. The future, however, will in all likelihood bring vast changes in its nature. The international division of labor among nations specializing in agricultural and extractive products, manufacturing products, and services will generate continuing changes, tensions, and conflicts in the international organization of production (Chase-Dunn et al., 2000). Automation will increase productivity but reduce the number of workers needed in many industries. Automation will also lead to the creation of new products and new jobs in other industries. The future seems destined to bring greater participation for workers in determining the direction of their enterprises. But the nature and degree of this participation remains very uncertain. The future may bring greater or lesser safety, security, satisfaction, and dignity for workers. It may bring greater equality for women. As productive members of society, all of us will determine by our actions which of these possibilities will be realized. The choice between these different futures is the focus of the concluding chapter of the book.

In Chapter 2, we explore the nature of the information and data that can be used to address the questions and themes developed in this introductory chapter. A brief grounding in the methods of scientific sociology and in the data available for the study of work, workers, and workplaces is an essential starting point for understandings grounded on available facts.

SUMMARY

The nature of work has changed dramatically over time. The division of labor has consistently marched ahead. The level of social inequality has increased and then declined. These changes have given rise to, and have sometimes been a result of, the development of new technologies. The position of women deteriorated in agricultural and industrial society relative to hunting and gathering society but has improved again in postindustrial society.

Drastic changes have also occurred in the family, which has lost many of the functions it once had as a center of economic production. Market forces have entered almost all areas of social life and exert a profound influence on the way we live in modern society. Most recently, this means that our lives as producers and citizens are being profoundly influenced by our position in the international division of labor. Finally, the increasing division of labor and the growth of large, complex organizations have encouraged the growth of complex governmental structures charged with the function of ensuring the stability and success of a highly complex economic system.

Changes in the nature of work have brought affluence for many, but they have also produced alienation and continuing poverty for many others. Changes in the organization of work have been a pivotal force moving the world from a past dominated by localism and tradition to a modern society dominated by rapid change and an interconnected world economy.

KEY CONCEPTS

division of labor

social relations of production

bureaucracy

ideology

economies of scale

social stratification

feudal society

artisan

guilds

merchant capitalism

putting-out system

Protestant work ethic

Industrial Revolution

enclosure movement

indentured labor

assembly line

class structure

postindustrial society

service industries

QUESTIONS FOR THOUGHT

1. What are the key characteristics of a global economy, and why has it come into being? What do you think will be the most important developments in work between now and 2060 when today's college students will begin reaching retirement age?

2. What are some of the ways your work roles have influenced your life? How have your parents' work roles influenced your life?

3. Compare the social organization and lifestyle of a hunting and gathering band to the social organization and lifestyle of a modern work group in a large business enterprise along as many dimensions as possible.

4. What changes in people's lives resulted from the development of agriculture? What are the most important consequences of these changes?

5. How did the wealthy extract surplus from those who produced food in feudal society? How is this different from how the sources of wealth emerge today?

6. Describe two ideologies that have been used to define the meaning of work at different periods in history. What different aspects of work do they highlight?

MULTIMEDIA RESOURCES

Print

Carleton S. Coon (editor). 1977. *A Reader in Cultural Anthropology.* Huntington, New York: Krieger. A

wealth of fascinating examples of work and life in primitive societies.

De Graff, John (editor). 2003. *Take Back Your Time: Fighting Overwork and Time Poverty in America.* San

Francisco, CA: Berrett-Koehler. Essays on why workers in advanced economies continue to work such long hours and strategies for how to combat this.

Hobsbawm, Eric J. 1975. *The Age of Capital.* New York: Scribner. The most accessible and well-written history of the industrial revolution in England.

Gerhard Lenski. 1984. *Power and Privilege.* New York: McGraw-Hill. Considered by many to be the best sociological account of the progressive development of more and more complex societies.

Alice Kessler-Harris. 2001. *In Pursuit of Equality: Women, Men, and the Quest for Economic Citizenship in 20th Century America.* New York: Oxford University. A detailed history of the experiences and struggles of working women in North America.

Karl Marx. 1967 [1887]. *Capital, Volume 1.* New York: International Publishers. The original critical interpretation of the industrial revolution and the rise of capitalism.

Adam Smith. 1937 [1776]. *The Wealth of Nations.* New York: Random House. The original positive interpretation of the expanding division of labor and its benefits for society.

Internet

American Sociological Association (ASA). *www.asanet.org* Official web site of the professional association for sociologists in the United States. The latest news, research, and events in American sociology.

Canadian Sociology and Anthropology Association. *www.csaa.ca* The latest news, research, and events in Canadian sociology.

Organizations, Occupation and Work section of the ASA. *www.northpark.edu/sociology/oow* The ASA section focusing specifically on issues concerning work.

Jobs and Employment Practices. *www.workindex.com* Human resources information on hiring, compensation, benefits, employment law, training, and retirement. Valuable for employees, employers, and human resources professionals.

Economic History. *www.eh.net* Information and services for students and researchers interested in economic history.

RECOMMENDED FILM

The Good Earth (1937). A profound and moving depiction of the challenges of peasant life in pre-Second World War China. Based on a novel of the same title by Pearl Buck.

2

Studying the World of Work

Tinker, tailor, soldier, sailor,
rich man, poor man, beggar man, thief,
doctor, lawyer, merchant, chief.

This counting rhyme reminds us that children are aware of only a few occupations, usually through exposure to their parents' work and to workers whom they have met in education, medicine, and other social services. But as you saw in Chapter 1, modern industrial economies are specialized. Even adults may know relatively little about the world of work beyond their own jobs. Because of evolving technologies, new jobs are rapidly being created while old jobs become obsolete. Studying the rapid changes in the workplace poses a challenge for social scientists just as it does for all workers.

This chapter examines how sociologists study the world of work, especially when that world is expanding and changing. The chapter will explain some concepts and techniques used to gather information about work and workers. Finally, it will look at problems that researchers encounter when studying work. Anthropologists, economists, psychologists, and other social scientists use some of the same methods to study work. The techniques and tools discussed may also help students or workers changing jobs to find useful information about the labor market.

TECHNIQUES OF ANALYSIS

Work is so varied and important an aspect of human life that sociologists need many methods to study it. Any social science method should be both **valid** and **reliable**. A valid method yields accurate information about the phenomenon being studied. A reliable method produces the same results if it is used repeatedly or if different investigators use it. The strong emphasis on valid and reliable methods is one important distinction between social scientific studies of work and journalistic accounts in newspapers or magazines.

This section presents three major techniques that sociologists use to study work and workers: ethnographies, case studies, and sample surveys. These techniques do not exhaust the available methods. For example, some sociologists may study work or workers using experiments (Valian, 1999), historical approaches (Bernstein, 1997), or comparative studies of societies (DiPrete, 2002). But the three methods described here are among the most important.

Ethnographies

One way to learn what workers actually do on the job and how they interact with their fellow workers is through an **ethnography**, a careful analysis of a work situation written by a knowledgeable observer after many months or years of observation. The observer seeks not only to explain the work from the worker's perspective but also to describe and explain larger patterns that may be invisible to individual workers. This narrative account of work seems familiar because it superficially resembles friends' or relatives' accounts of life on the job. It is different, however, because the trained observer is sensitive to subtle features of the job and interactions among the workers. Their observations are also more detached. Each ethnography helps social scientists understand a work role or work settings. Evidence cumulated from many ethnographies provides an even stronger basis for conclusions (Hodson, 2001).

There are several types of ethnography. In **participant observation** the observer actually becomes a worker for a period of time. This technique gives the observer an intimate, everyday familiarity with the job content and the actual interactions among the workers. Sociologist Everett C. Hughes and his students at the University of Chicago in the 1950s and 1960s popularized participant observation with provocative studies of medical students, janitors, and dancers, among other occupations. Box 2.1 provides a participant

observer's account of work at a fast-food restaurant. These studies are valuable for their rich detail about working and about interactions among workers and between workers and supervisors.

The validity and reliability of participant observation have limitations. Participant observers can typically study only a limited range of jobs. It is unlikely that the sociologist observer would have the skills or access necessary to participate, even for a short time, in highly technical jobs or in top-level management positions. In addition, the participant observer may inadvertently choose an atypical work site or join a work group that is atypical. Different observers of the same work situation might also interpret aspects of the job quite differently because of their different backgrounds, predispositions, or experiences.

In **nonparticipant observation** the trained observer does not actually become a part of the work group. One famous example of nonparticipant observation is the study of the Bank Wiring Room in the Western Electric Company plant in Hawthorne, Illinois (Roethlisberger and Dickson, 1939:379–408). Fourteen men worked in this room, wiring, soldering, and inspecting electrical boards. An overt nonparticipant observer sat with them for a number of days, watching their work and interactions. Initially, the observer noticed how the workers joked with and teased one another, or occasionally helped one another with their work.

The observer also noticed that the group's productivity was basically constant, despite company efforts to increase it. The observer eventually learned that the small work group had developed an informal norm defining an appropriate level of productivity. A norm is a rule that a group develops for thinking, feeling, or behaving. Laws are examples of formal norms, but most norms are informal. A worker who could not reach the normatively defined productivity level would be helped by others, but a worker who produced too much would be teased, called a "speed king," or eventually be subjected to "binging" (a thump on the upper arm). The observer reasoned that the workers, concerned about job security as many workers were during the Great Depression, feared that

---

BOX 2.1   An Ethnography of Hamburgers

Ester Reiter conducted participant observation at a Burger King in Toronto. Fast food is popular in Canada, but the industry is dominated by multinational companies headquartered in the United States. The Toronto Burger King outlet where she worked opened in 1979, and by 1980 it was the highest-volume Burger King in Canada.

Fast-food establishments standardize their products in part by standardizing the way in which workers do their jobs. One objective of Reiter's participant observation was to learn the impact of this standardization on the workers. Making hamburgers, she found, was one of the most enjoyable jobs.

The store had two conveyors that could broil up to 835 patties per hour. Near the meat conveyor were two bun chains that toasted the buns and dropped them into a chute near the cooked patties, a process that took about thirty seconds.

A worker keeps the freezer near the broiler filled by hauling boxes stored in the walk-in freezer located on the other side of the kitchen. During busy times, one worker keeps the chains of the conveyor belt broiler filled with meat and buns, while another worker stands at the other end. The worker at the "steamer" end of the belt uses tongs to pick up the cooked patties as they fall off the belt, and places them on the "heel" of a bun (the bottom half). The bun is then "crowned" with the top half and the ungarnished hamburger is placed in a steamer, where, according to Burger King policy, it can

remain for up to ten minutes. Jobs at the broiler-steamer are often assigned to new workers as they can be quickly learned.

The burger board, where the large and small hamburgers are assembled, is made of stainless steel and can be worked from both sides. When the store is busy, the larger hamburgers, called "[W]hoppers," are produced on one side, and the smaller hamburgers on the other....

First the [W]hopper cartons are placed printed side down on the table, and the patty removed from the steamer. The bottom half, or the bun heel, is placed in the carton, and the pickle slices spread evenly over the meat or cheese. Overlapping the pickles is forbidden. Then the ketchup is applied by spreading it evenly in a spiral circular motion over the pickles, starting near the outside edge. The onions (½ oz.) are distributed over the ketchup. Mayonnaise is applied to the top of the bun in one single stroke and $\frac{3}{4}$ oz. of shredded lettuce placed on the mayonnaise, holding the bun top over the lettuce pan. Then the two slices of tomato are placed side by side on top of the lettuce. If the tomatoes are unusually small, the manager will decide whether or not three tomato slices should be used that day.

SOURCE: Excerpt from Ester Reiter, 1991, *Making Fast Food: From the Frying Pan and into the Fryer*. Montreal: McGill-Queen's University Press, pp. 99–100.

---

increased productivity would become an excuse for laying off workers.

Nonparticipant observation is useful to sociologists who cannot study a job as participant observers. It would be difficult, for example, to be admitted to medical school to do a participant observation of medical students. Box 2.2 describes a study of young physicians who were learning to become surgeons. The sociologist who studied them was allowed into operating rooms and conferences but was not himself a physician. Nonparticipant observation has disadvantages, too. The nonparticipant observer may

have a more difficult time than the participant observer in winning the confidence of the workers being observed, and workers who appear to be acting naturally may nevertheless be quite conscious of the observer.

Workers may also change their behavior to please the observer, a phenomenon similar to **experimental bias**. An example of experimental bias may have occurred in early research at the Hawthorne plant. The researchers had designed experiments to test the relationship between levels of lighting and worker efficiency. Workers maintained their productivity even under conditions of very low light

---

B O X 2.2 **Putting on the Hair Shirt: A Nonparticipant Observation of Surgery Residents**

Sociologist Charles Bosk is now a professor of sociology and medical ethics at the University of Pennsylvania. As a graduate student, he conducted an eighteen-month nonparticipant observation in a hospital he identifies only as "Pacific Hospital." He was studying how recently graduated physicians learn to become surgeons in the course of their surgical residency. Although not medically trained himself, he was allowed to be present at surgical interviews and rounds with patients, and he observed the interactions between the "attendings" (the fully accredited surgeons who ran the surgical residency) and the house staff (residents).

An important feature of the surgeons' work is the weekly Mortality and Morbidity conference at which any mistakes or complications from the previous week's operations are discussed. Particularly high-risk surgeries that had bad results were "expected failures," but the conferences focused on the "unexpected failures," when a patient was expected to recover without complication and something went wrong. Surgeons were expected to admit their mistakes and accept the criticism of their peers, a ritual called "putting on the hair shirt." In the conference reported below, a resident tried to suggest

that an unexpected failure was in fact an expected failure.

Andrew, a resident, had just explained that the death of an elderly lady following a gallbladder removal was caused by her old age and general physical weakness. He was immediately challenged by an attending. "That's not what I would call thinking real hard. I mean, you didn't exactly scratch your head until it bled on that one, did you, Andrew? You can't stand there and tell us this lady died from old age. If she was going to die from old age, why operate on her to begin with?" Dr. White, the attending on the case, rose to his resident's defense: "You're not exactly being fair. You know well enough that things like this can happen any time. It was just one of those unpredictable catastrophes." The other attending answered, "That's bullshit. It doesn't sound like the treatment of this lady was very well thought out." White replied, "C'mon now, you have your share of cases like this. She was a very strange old woman."

SOURCE: Charles L. Bosk, 2003, *Forgive and Remember: Managing Medical Failure*, 2nd edition. Chicago: University of Chicago Press, pp. 134–135.

(Roethlisberger and Dickson, 1939:14–18; Schwartzman, 1993). One interpretation is that the factory workers were trying to please the experimenters, regardless of the lighting conditions. The phrase **"Hawthorne Effect"** has come to refer to an experiment in which participants try to do what the experimenters want. Experimental bias is difficult to detect, and there is dispute over whether it really occurred at the Hawthorne plant (Jones, 1992; Gillespie, 1991). In principle, however, if workers change their behavior because of the observer, both the validity and the reliability of the observations are endangered.

Regardless of whether the researcher is a participant or a nonparticipant observer, the observation may be either overt or covert. If the observation is overt, the other workers may know the observer's true "cover story," which is often that the observer is writing a book. The workers may initially feel

uneasy or suspicious at the presence of an overt observer, but many researchers find that after a few days their presence is no longer noted.

In covert studies the workers do not know that their fellow worker is an observer. Alternatively, the covert observer may be disguised as a customer, an inspector, or some other stranger with a right to be in the workplace. Covert observers may avoid experimental bias, but this enhancement of the validity of the study is counterbalanced by the ethical issues raised by subterfuge and pretense. Covert observers may not be able to ask clarifying questions, and so their interpretations may be superficial or incorrect.

Case Studies

Ethnographic studies are usually limited to fairly small work groups during a specific period. The reader of an ethnography typically learns the point of view of one group of actors in a workplace. A **case study** attempts to bring several perspectives to understanding a workplace issue, such as the views of supervisors, customers, suppliers, and union leaders, in addition to the workers. Thus a case study is usually larger in scope and uses more types of data. Case studies may also use various methods. A case study typically examines a work site using combinations of personal interviews, analyses of written documents, and observations. Both official documents and personal records of workers may be consulted. Case studies frequently analyze entire companies or large divisions within companies. The findings and conclusions emerge from all the materials and people that the researcher consults.

Ethnographies typically present a work group at a particular point of time—the time frame in which the observer was there. By using written documents, the case study can provide information about the history of a work site and how existing arrangements came about (Feagin, Orum, and Sjoberg, 1991). Case studies may illustrate how an organization solves a problem, or they may identify new problems faced by workers. Case studies are often used to examine the effects of recent job changes. For example, a case study might examine a work site before and after the introduction of computerized workstations. Because different management teams implement innovations in different ways, a researcher might develop case studies to compare the effects of the innovation in different work settings.

Sociologist Rosabeth Moss Kanter presents a case study of gender roles in a large company in her book *Men and Women of the Corporation* (1977). Box 2.3 is a brief selection from her work illustrating the types of conclusions that can be drawn from a case study. Because case studies use several kinds of information, the researcher can search for agreement and disagreement among the various sources. This cross-checking tends to improve both the validity and the reliability of the evidence. Many case studies are based upon interviews of informants along with studies of documents. The semi-structured interview consists not only of questions covering certain specified content but also allows the interview

subject to address the issues in multiple ways or to introduce new content. Such interviews may last an hour or longer.

A good case study nearly always requires the cooperation of the employer. The researcher is unlikely to receive access to written records in any other way. Some companies are so eager to have the research conducted that they will commission and pay for it. Even these companies, however, often insist that published research refer to the company using a disguised name. For example, Kanter refers to the company she studied as "Indsco."

Some workplaces, however, do not welcome research. They may place certain documents off-limits to the researcher or allow access only if their documents are not quoted. Case studies are especially threatening to companies that are in fiercely competitive economic situations, are closely regulated by the government, have a record of hostile labor relations, or are suspected of wrongdoing by citizen's watch groups, environmentalists, or others (Cornfield and Sullivan, 1983). These, of course, may be the very companies that are of greatest interest to the sociologist.

Multiple Methods

Researchers may seek to enhance the validity and reliability of their studies by using multiple methods. Historical methods look at specific data or at case studies from several different time periods. Comparative methods look at specific data or case studies that have been completed in different countries or different workplaces. Box 2.4 discusses "rape work," which refers to a set of jobs whose members interact with rape victims. This study drew its conclusions after comparing information from a large number of interviews at a large number of sites.

Sample Surveys

The **sample survey** is widely used to study many social phenomena, including work. A survey is conducted by asking a uniform set of questions of

BOX 2.3 Gender: Praise-Addiction Among Secretaries at Indsco: An Example of a Case Study

Rosabeth Moss Kanter served as an outside consultant for several years to the company she calls the Industrial Supply Corporation (Indsco). During this period she collected materials and developed a network among employees. Convinced that a case study of a large corporation was needed, she began to analyze many sources of information, including group discussions, conversations, and documents. She also used participant observations of meetings, and she analyzed data from employee surveys (Kanter, 1977:293–298). She could then check each source of information against the information available from other sources. Specific incidents were reported to illustrate the more general principles that she developed.

She uses the term praise-addiction *to describe a condition she observed and heard others use to describe secretaries. Kanter's identification of praise-addiction is one sort of finding that can result from a case study.*

The emotional-symbolic nature of rewards in the secretarial job; the concern of some bosses to keep secretaries content through "love" and flattery; and the continual flow of praise and thanks exchanged for compliance with a continual flow of orders—all of these elements of the position tended to make some secretaries addicted to praise. Praise-addiction was reinforced by the insulation of most secretaries from responsibility or criticism; their power was only reflected, the skills they most exercised were minimal, and authority and discretion were retained by bosses. Thus, many years in a secretarial job, especially as private secretary to an executive, tended to make secretaries incapable of functioning without their dose of

praise. And it tended to make some wish to avoid situations where they would have to take steps that would result in criticism rather than appreciation. Their principal work orientation involved trying to please and being praised in return.

One older executive secretary with long tenure at Indsco was a victim of praise-addiction. Though happy as a secretary, and well-respected for competence, she accepted a promotion to an exempt staff job because she thought she should try it. After a year and a half, it was clear to her and to those around her that she could not take the pressures of the new job. Her nervousness resulted in an ulcer, and she asked to return to the secretarial function. In the exempt job she had supervisory responsibilities and had to make decisions for people—sometimes unpleasant ones, such as terminations. Her manager thought she spent much too long making such decisions, "moaning" afterwards even if she knew she had made the right decision. But she felt herself to be in an intolerable position. She had a feeling she was not appreciated. No one said "thank you" for her work in the new job. As the manager put it, "She was used to lots of goodies from her boss—'Hey, that's a good job.' Here we have to be of service to managers as well as subordinates. The managers feel we're one of them, so they don't go out of their way to thank us. And subordinates don't thank managers. So she was missing something she had been used to."

SOURCE: Excerpt from Rosabeth Moss Kanter, *Men and Women of the Corporation*. Copyright ©1977 by Rosabeth Moss Kanter. Reprinted by permission of Basic Books, Inc., Publishers.

a systematic sample of people. The people who answer the survey are called *respondents*. They are selected according to the principles of a branch of mathematical statistics known as sampling theory, so that they will be representative of the population, the larger group from which they were selected. The **sample** may be selected to represent all workers in the United States, all employers, the workers in a particular workplace, or any other population of interest.

The set of questions, or questionnaire, may include questions of fact ("How long have you worked at your current job?") and questions of opinion ("How satisfied would you say you are with your current job—very satisfied, somewhat satisfied, or not at all satisfied?"). Unlike the semi-structured interview, however, the sample survey places great emphasis on asking the respondents the same questions in the same ways, with relatively little opportunity for the respondents to introduce

---

B O X 2.4   Learning about Rape Work: An Example of Multiple Methods

Patricia Yancey Martin, a sociologist at Florida State University, set out with several colleagues to learn about the service needs of rape victims in the state of Florida. As part of this work, she visited twenty-eight Florida communities and interviewed and observed 124 different settings, including police offices and sheriff's offices, hospitals, prosecutor's offices, and rape crisis centers. Her research required interviewing 145 people, as well as observing rape trials, collecting financial reports and statistics, and reading technical reports and policies from other researchers and state agencies. Rape victims encounter a number of workers whose jobs range from providing medical assistance and consolation, to collecting evidence and conducting criminal trials. Her findings about this type of work are summarized in a book-length study (Martin, 2005:4–5).

Rape is a violent act that arouses strong emotions, including in the people who come into contact with the victim. One of Martin's findings was that different workers must manage their emotional reactions in different ways. Some of the workers, such as the rape crisis center workers and victim witness advocates, are expected to develop emotional closeness with victims. Other workers, such as the prosecutor, are expected to maintain emotional distance. One of her informants explained the concept:

[(Interviewer)Do prosecutors become emotionally involved with the victims?] No ... they don't. They can't afford to. They are supposed to stay objective and concentrate on the victim so we can help [her]. They don't become personally that involved. It's not really their job. And they wouldn't be very good at their job if they did. [(Interviewer): so you people, the victim advocates, help them?] In a way. They tell us to tell the victim something or find out something. So we're the ones always in touch. Their job is to look at the facts and see what they have and prosecute. (prosecution victim advocate, white woman, age forty-six) (Martin, 2005:187)

Perhaps not surprisingly, the rape crisis center workers are more likely to feel distress, powerlessness, and sadness, whereas the prosecutors are more likely to feel skepticism and anger. To understand the services that rape victims receive, Martin could not study just one occupation, nor just one workplace. Instead, she interviewed a number of people in different workplace settings and also studied other information about the organizations to reach her conclusions.

SOURCE: Patricia Yancey Martin, *Rape Work: Victims, Gender, and Emotions in Organization and Community Context* (New York: Routledge, 2005).

---

new topics. In addition, only a brief time is typically required to complete the questionnaire. Researchers administer questionnaires in three basic ways: (1) personal interviews are conducted face-to-face by a trained interviewer; (2) telephone interviews take place over the respondent's home telephone, again with a trained interviewer asking the questions; (3) self-administered questionnaires are handed, mailed, or e-mailed to respondents who answer the questions at a convenient time.

A cross-sectional survey is administered once to a sample of respondents. The same questionnaire might be administered again to a different sample of respondents. The repeated use of cross sections is useful for detecting trends in job satisfaction, work

commitment, and so on. In a longitudinal or panel study, the researchers return several times to survey the *same* sample of respondents. Longitudinal studies are useful for such things as studying job changes that occur during the work life of the sample of workers.

Sample surveys are extensively used in many countries to study work. Every month, the U.S. Bureau of the Census and Bureau of Labor Statistics conduct the Current Population Survey (CPS) (http://www.bls.census.gov/cps/). This survey of about fifty thousand households asks various questions about whether its members who are over the age of sixteen are looking for work or have jobs. Those with jobs are asked additional questions about hours of work, type of work, and earnings. This is a cross-sectional survey, but some of the respondents are reinterviewed eight months after their first interview, so it also has a longitudinal component.

The Bureau of the Census also conducts the American Community Survey (www.census.gov/acs/www/index.html). This annual cross-sectional survey of three million households collects information about individuals and households, including information about age, race, education, commuting time to work, and some information about individuals' jobs. It is designed to provide information for local labor market areas and for states as well as for the nation.

The National Longitudinal Surveys (NLS), supported by the Department of Labor, contain interviews of the same samples of workers several times over a period of years to examine changes in employment, earnings, and work-related attitudes (www.bls.gov/nls/).

Various **establishment surveys** sample employers to ask questions concerning the characteristics of their companies and employees. An **establishment** is the location of an employer; it is also the work place to which an individual worker reports. The Bureau of Labor Statistics conducts the Current Employment Statistics Survey, which is used to develop estimates of earnings. The National Organizations Survey (http://webapp. icpsr.umich.edu/cocoon/ICPSR-STUDY/04074.xml) is conducted by asking half the respondents to a large annual survey (called the General Social Survey) for the names and addresses of their employers. The employers are then asked for information on topics such as company structure, employment policies, the effects of workplace stress on worker productivity, and health insurance.

Sample surveys may be designed to represent certain groups of workers or certain regions or states. Professional associations or unions survey their members on workplace issues. Trade associations survey employers or owners who are members. Many businesses survey their customers to evaluate customer satisfaction. Innovations based on the use of laptop computers, e-mail, and web-based surveys are extending the range of contemporary survey practices.

Compared with ethnographies or case studies, surveys have the advantage of being more easily generalized to the population they were designed to represent. Sampling theory allows the researcher to estimate by how much the survey results are likely to vary from the "true" answer, which is the answer that would have resulted from interviewing the entire population. By directly questioning workers, a survey can measure subjective indicators, such as job satisfaction. Changes in facts and attitudes can be traced and studied if the same question is asked in repeated surveys.

Potential disadvantages also arise from the use of survey methods. One problem is selection bias in which only certain types of people respond to a survey. Respondents may mistrust the interviewer or fear what use might be made of their replies. Some respondents refuse to cooperate at all with surveys. If such refusals cluster within an important subgroup, the resulting sample is no longer representative of the population. For example, if rich people refuse to answer questions about their income, estimates of overall income will be too low.

Another common problem is **response error**, resulting from a respondent's misunderstanding a question or intentionally giving a false answer. Response error may happen if the questionnaire contains ambiguous or double-barreled questions. A double-barreled question includes more than one issue so that the answer cannot be clearly interpreted: "Have your hours of work or your working conditions recently changed?" A "yes" answer might mean that either hours of work or working conditions or both had changed. Selection bias and response error may also result if the questions pry into areas that respondents consider sensitive or confidential. Response error is difficult to detect, and it threatens the validity and reliability of the information gathered.

The collection and analysis of survey information forms a specialized area within the social sciences. Sociologists, political scientists, and economists all make use of survey data.

UNITS OF ANALYSIS

Ethnographies, case studies, and sample surveys are examples of how sociologists study work; specific units of analysis are what and whom they study. The unit of analysis may be the individual worker or groups of workers. Or the unit of analysis may not be individuals at all—it may be groups or organizations. For example, the sociologist may study unions, businesses, factories, or corporate networks. The world of work may look quite different from one perspective—say, that of an individual worker—than it does from the perspective of the managers of a large corporation.

Individual workers are quite concrete units for the analyst. Other units are more abstract and highly aggregated. One important unit of analysis is the **labor force**, a collective term for all the workers within a country. Outside the United States an equivalent term is the *economically active population* (Sullivan, 2005).

The Worker and the Labor Force

The most straightforward unit of analysis is the individual worker. Workers can be analyzed in terms of their background, or demographic characteristics, which consist of **ascribed characteristics** and **achieved characteristics.** The worker does not control ascribed characteristics, such as sex, race, or age, although employers and coworkers may react strongly to them. The worker does have some control over achieved characteristics, such as educational background, work experience, and skills.

The U.S. Bureau of Labor Statistics reports the size and composition of the labor force every month by using information from the CPS. Anyone is eligible to be counted in the labor force who is aged sixteen or older, who is not on active duty in the Armed Forces, and who is not institutionalized (for example, in a prison or a residential hospital). Members of the labor force can be either employed or unemployed.

According to the government's definition of employment, employed people in the labor force are those who in the week preceding the survey (1) worked at least one hour for pay or profit, or (2) worked at least fifteen hours without pay in a family business, or (3) were temporarily not working because of illness, vacation, or similar reasons. The unemployed are not merely those without jobs; rather, they are people who are not employed but who actively sought work during the four weeks preceding the survey and were currently available to take a suitable job. In addition, people are counted as unemployed if they are temporarily laid off or are waiting to report to a new job in the near future (U.S. Bureau of Labor Statistics [BLS], 2001b). An eligible person who does not fall into either of these categories is termed **NILF** (not in the labor force). Most NILF people in the United States are students without jobs, retirees, people who are chronically ill or have disabilities, or people who are keeping house.

Using these concepts, the Bureau of Labor Statistics publishes every month two rates to describe the status of the labor force. The first rate, the civilian **labor force participation rate**, is the number of persons in the labor force divided by the number of persons eligible to be in it, multiplied by 100 to convert to a percentage. This can be expressed as:

$$\text{LFPR} = \left(\frac{\text{labor force}}{\substack{\text{all noninstitutionalized} \\ \text{persons aged } 16\ +}} \right) \times 100$$

The labor force participation rate indicates what proportion of the eligible population is economically active. In August 2006 the U.S. participation rate was 66.2 percent (BLS, 2006b). Trends in the rates for certain groups, such as women, teenagers, and elderly people, indicate their levels of incorporation into the economy. Nearly every industrialized country has experienced a phenomenal increase in women's labor force participation rates since World War II. In August 2006 the rate for U.S. women aged twenty years or more was 60.4 percent, compared with 76.2 percent for men (BLS, 2006b).

The **unemployment rate** is the number of unemployed people divided by the number of

---

BOX 2.5 How to Read a Table

Sociologists frequently present their data in tables, which condense a great deal of information within a small space. Because of their concentrated information, however, tables can be difficult to read and understand. We will be presenting many tables in this text, and this box is designed to provide a method of reading tables to glean the maximum information.

Table A presents some information about unemployment rates in the United States and other countries from North and South America. The table has also been marked to indicate the principal parts of a table. Reading the table in the order of the numbered parts will convey efficiently and accurately the information in the table.

Parts of a Table:
1. Headline
2. Headnote
3. Stub
4. Column headings
5. Footnote
6. Source note
7. Entries

① **TABLE A** **Unemployment Rates for Selected North and South American Countries, by sex, 2003**

② (annual data, except as noted)

③

Country	Unemployment Rates		
	Total	Male	Female ④
Argentina	16.4%[a]	NA	NA
Brazil	12.3	10.1	15.2
Canada	7.6	7.9	7.2
Colombia	14.2	11.0	18.5 ⑦
Chile	8.5	7.9	9.7
Mexico	2.4	2.3	2.6
USA.	6.0	6.3	5.7
Venezuela	18.0	16.0	21.1

⑤ [a]Monthly data for April, 2003

NA Not available

⑥ SOURCE: International Labour Office, http://laborsta.ilo.org

---

people in the labor force, multiplied by one hundred. This may be expressed as:

$$UR = \left(\frac{\text{unemployed}}{\text{labor force}} \right) \times 100$$

Box 2.5 presents recent data on labor market indicators, and it includes important information on how to read statistical tables. These data indicate overall economic activity as well as the differing labor market experiences of workers from various demographic groups.

A rise in the unemployment rate often indicates that the business cycle is about to enter a downturn; conversely, a decline often indicates economic improvement. The unemployment rate is high in economically depressed areas and lower in prosperous ones, so unemployment rates indicate local labor market conditions. Historically, the unemployment rate for black workers has been at least twice the rate for whites, and the Hispanic unemployment rate has been intermediate between the rates for white workers and the rate for black workers.

1. The *headline* tells the reader which data are presented in the table, for which groups. The headline in Table A indicates that the table contains unemployment rates, and data will be provided separately for men and women. Sometimes a headline specifies the time and place in which the data were collected. This headline notes that the data are for 2003 and for selected countries.

2. The *headnote* is a parenthetical expression that contains information important in interpreting the table. The headnote in Table A indicates that the data in the table are annual averages, unless otherwise noted.

3. The *stub* is the left-hand column of a table. The categories indicate which data appear in the horizontal rows of material in the table. In this case, the stub contains the names of the countries for which data will be presented.

4. The *column headings* indicate which data are given in the vertical columns in the table. In Table A, the first column presents the unemployment rate for the total national labor force, and the next columns present separate rates for males and females, respectively.

5. A *footnote* contains information that is important for interpreting some, but not all, of the entries of the table. Not every table has a footnote. In Table A, the footnote indicates that for Argentina, the data represent a single month and not an annual average. The notation NA, not available, also applies to the row of data for Argentina; in 2003, the source did not have unemployment rates by sex available.

6. The *source note* is important because it tells the reader where the data were obtained. The reader can refer to the original source for additional information or to check the accuracy of the data.

7. The *entries* of the table should be read last. The reader draws conclusions by carefully reading the entries and comparing them across rows and down columns. Notice, for example, that the unemployment rates differ a great deal among these countries. Also notice the difference by sex. In many countries, such as Brazil and Colombia, the unemployment rate for women is much higher than the rate for men. But in Canada and the United States, men have higher unemployment rates than women.

What other conclusions can you draw from the data in Table A?

As useful as labor force statistics are, the definitions used by the government trouble some observers (National Commission on Employment and Unemployment Statistics, 1979). The definition of the labor force parallels in some ways the measurement of the gross national product (GNP). The gross national product is the value of all the goods and services produced for the market during a year. People who produce goods or services for sale in the market are included in the labor force. The labor force definition excludes many people who perform useful services outside the market economy. For example, homemakers and volunteers perform needed services but not for pay or profit. If there were no homemakers or volunteer workers, then families, churches, and hospitals would have to hire workers to perform those duties or leave them undone. Although the newly hired workers would be in the labor force, homemakers and volunteer workers doing the same work are not in the labor force.

The measurement of unemployment is also controversial. The Bureau of Labor Statistics considers about 1.6 million persons to be marginally attached to the labor force; these people wanted and were available for work and had looked for a job sometime during the year prior to the survey. In the four weeks prior to the survey, however, they had not actively searched for work and so they were not counted as unemployed. Among the marginally attached workers, there are about a half million **discouraged workers**, who were not currently looking for work specifically because they believed no jobs were available. By excluding the marginally attached workers, the measured unemployment rate is arguably too low.

Others argue that many unemployed people conduct only halfhearted searches for work, perhaps because they are required to look for work to continue unemployment compensation benefits. Others may reject available jobs and prolong their search looking for more ideal work. In the view of such critics the measured unemployment rate is too high, for it includes people who could hold jobs if they changed their behavior.

One might wonder why data on unemployment insurance benefits are not used to estimate unemployment (Bureau of Labor Statistics, 2001b). The principal reason is that not every worker is covered by unemployment compensation. In 2006 the number of people receiving unemployment benefits amounted to only about 34 percent of the number of unemployed people. Some types of work, such as agricultural labor, are not covered. Unemployment insurance laws require that a worker have been employed in a covered occupation for a specified length of time, and some workers have not worked long enough to qualify. For example, young people seeking their first jobs are unemployed, but they are not eligible for unemployment compensation. Other workers continue to seek work after exhausting their unemployment compensation benefits. Consequently, unemployment benefit records seriously underestimate total unemployment.

Industry

Industry provides another unit of analysis. **Industry** refers to a branch of economic activity devoted to the production of a particular good or service. The good or service may be quite specific; thus, we might speak of the fast-food industry, but it, in turn, can be considered part of the restaurant industry or part of the even larger personal service industry.

Knowing a worker's industry is important for several reasons. First, conditions of economic competition tend to be quite specific to industries. Some industries experience heavy pressure from foreign competition, for example, while others do not. Some industries are closely regulated by the government, and others are unregulated. Second, the nature of production varies by industry. An industry with an electronically automated production process differs substantially from an industry that still requires large inputs of hand labor. The industrial process determines which specific jobs will be available and what the working conditions will be—specifically, what hazards workers may face on the job, what skills are needed for employment, how much training is needed, and so on.

Finally, workers experience economic consequences from their industries (Sullivan, 1990). Declining industries are often less productive and provide sporadic, lower-paid work. Growing industries are more likely to be productive and to provide better wages, benefits, promotion opportunities, and job security. In 2006 the earnings in industries varied from $27.54 an hour in utilities to $9.50 in leisure and hospitality (BLS, 2006a).

Industrial codes enable researchers to contrast the characteristics of workers in different industries. The North American Industry Classification System (NAICS), which was jointly developed by the governments of Canada, Mexico, and the United States, classifies 1,170 industries based on the activity in which they are primarily engaged. NAICS attempts to ensure that the same code will be used for establishments using similar raw material inputs, similar capital equipment, and similar labor. The code consists of six digits that classify all economic activity into twenty industry sectors. The first two digits indicate the sector. Five sectors are mainly goods-producing sectors and fifteen are entirely services-producing sectors. Each sector is then divided into subsectors by the use of the third, fourth, and fifth digits. For example, the "information sector" consists of communications, publishing, motion picture and

T A B L E 2.1 Industries with the fastest growth in the number of employees, 2004–14

Industry description	2002 NAICS[a]	Thousands of jobs		Change, 2004–14	Average annual rate of change, 2004–14
		2004	2014		
Home health care services	6216	773.2	1,310.3	537.1	5.4
Software publishers	5112	238.7	400.0	161.3	5.3
Management, scientific, and technical consulting services	5416	779.0	1,250.2	471.2	4.8
Residential care facilities	6232,6233,6239	1,239.6	1,840.3	600.7	4.0
Facilities support services	5612	115.6	170.0	54.4	3.9
Employment services	5613	3,470.3	5,050.2	1,579.9	3.8
Independent artists, writers, and performers	7115	41.9	60.8	18.9	3.8
Office administrative services	5611	319.4	449.9	130.5	3.5
Computer systems design and related services	5415	1,147.4	1,600.3	452.9	3.4
Outpatient, laboratory, and other ambulatory care services	6214,6215,6219	836.1	1,160.4	324.3	3.3

SOURCE: U.S. Bureau of Labor Statistics, available at *http://www.bls.gov/emp/empfastestind.htm*
[a]North American Industry Classification System

sound recording, and online services. Each sector can be subdivided into subsectors, such as the computers and electronics subsection of manufacturing (BLS, 2004a). Respondents in a sample survey can be assigned an industrial code based upon their responses to questions about what good or service they produce at their worksite; these codes can then be used to compare information about workers in different industries.

A company's sales, profits, or production can be expressed as a proportion of all the sales, profits, or production within its industrial code. The larger a proportion attributable to a single company, the more dominant or concentrated that company is within its industry. The *four-firm concentration ratio* is the proportion of all production, sales, or receipts accounted for by the largest four firms within an industry, and it can be calculated for large industry groupings or much finer groupings, depending upon the industry code. For example, in 2002 the food manufacturing industry (NAICS code 311) was a $458 billion industry, with a four-firm concentration ratio of 16.8 for the value of shipments. That is, the top four food man-

ufacturing firms accounted for nearly 17 percent of the value of all shipments. But within this large industry, the four-firm ratios for specific industries varied a great deal: 64.2 for dog and cat food manufacturing (code 311111), 53.6 for flour milling (code 311211), 78.4 for breakfast cereal (code 311230), and 69 for chocolate and confectionery manufacturing (code 311320) (U.S. Bureau of the Census [Census], 2006, Table 2, pp. 3–4).

Table 2.1 presents information on the employment changes projected to 2014 for rapidly growing industries, with the NAICS codes provided for reference. Chapters 7 through 10 will analyze important industries in greater detail.

Occupation

Industry differs importantly from occupation. Industry identifies what a worker helps to produce, but occupation identifies the specific kind of work a worker does. More formally, an **occupation** is a cluster of job-related activities constituting a single economic role that is usually directed toward making a living. Because the distinction between occupation and industry is sometimes difficult to understand, it is helpful to see how they are related to each other. Some occupations are found in every industry. Nearly every industry, for example, requires administrative service workers, maintenance workers, and managers. Other occupations are heavily concentrated within a single industry—for example, nurses within the health-care industry or lawyers within the legal services industry. Even these examples have their exceptions, however; some nurses work in factories, camps, or schools; and many lawyers work as house counsel for firms in manufacturing or service industries. A few occupations work only in a single industry; an example would be taxi drivers in the transportation industry. As a rule, however, we must consider both occupation and industry for a full understanding of working life.

Some workers have several occupations because they have more than one job or are able to do more than one kind of work. Occupations are also an important unit of analysis. Chapters 11 to 14 will examine specific occupational groups.

White-Collar/Blue-Collar   Perhaps the simplest occupational classification is the white-collar/blue-collar division. This classification is simple but it is also increasingly outdated and misleading. Blue-collar workers—mostly factory and craft workers—once did only manual labor. White-collar workers—office workers and most professionals—had clean working conditions that made it possible for them to wear white shirts. Traditionally they earned more than blue-collar workers, but today a factory or craft worker may earn more than a clerical or sales worker.

The white-collar/blue-collar classification is less useful today for several other reasons. First, there are now many service workers, some of whose work resembles blue-collar jobs and some of whose work is more like white-collar jobs. For example, the cook in a fast-food restaurant and the elite chief of police in a large city are both service industry workers. The fast-food cook may experience factory-like

conditions reminiscent of blue-collar work. The police chief has job training and autonomy on a par with other white-collar management jobs.

Second, some jobs presently classified as either blue-collar or white-collar may seem misclassified when the actual work conditions are considered. Technicians, for example, are considered white-collar workers. Many of them are highly educated, like other white-collar workers, but they spend most of their day working with machinery, as do blue-collar workers. Some factory operatives, on the other hand, work in industrial laboratories that are not just clean, but sterile. Their day-to-day job responsibilities may look very much like those of the technicians, but they are classified as blue-collar workers.

Finally, the white-collar/blue-collar distinction ignores the so-called pink-collar workers. These workers labor in occupations traditionally filled by women, such as nurse, secretary, or child-care worker. Pink-collar jobs are usually characterized by relatively low pay given their specialized skills. Elementary school teachers, for example, are classified with professional workers, but their pay may not be as high as that of other upper-level white-collar workers. Such jobs are difficult to classify as either white collar or blue collar.

Standard Occupational Classification   A more precise and detailed classification for analyzing occupations was released in 2000 by the U.S. Government. The Standard Occupational Classification (SOC) System provides comparable information for many users of occupational data, such as government program managers, students considering career training, job seekers, vocational training schools, and employers wishing to set salary scales or locate a new plant. The SOC is designed to cover all occupations in which work is performed for pay or profit, reflecting the current occupational structure in the United States.

The 2000 SOC classifies workers at four levels: (1) 23 major groups, which are subdivided into (2) 96 minor groups, which in turn are subdivided into (3) 449 broad occupations, which in turn are further classified into (4) 821 detailed occupations.

Each detailed occupation is designated by a six-digit code. The first two digits of the SOC code represent the major group; the third digit represents the minor group; the fourth and fifth digits represent the broad occupation; and the detailed occupation is represented by the sixth digit (BLS, 2004b).

For example, your college professors would be classified within major group 25-0000, "education, training, and library occupations," and within minor group 25–1000, "postsecondary teachers." Your social science professors would be classified within the broad occupation 25–1060, "social sciences teachers, postsecondary," and the authors of this book, who teach sociology, would be in the detailed occupation 25–1067, "sociology teachers, postsecondary." Because of the dynamic shifts in the American workplace, the SOC is revised every ten years.

Detailed occupational information is valuable for analyses of workplaces and worker skills. Occupation, however, has many social ramifications beyond its instrumental and economic consequences. Sociologists often consider occupation as a proxy for one's position in the social class structure. People of similar occupation, besides having similar incomes and work experiences, often pursue similar patterns of leisure and consumption, share distinctive lifestyles, and are perceived in similar fashion by other members of the society (Trice, 1993). It is for this latter characteristic—how occupations are perceived by others—that sociologists have developed measures of **occupational prestige**.

One of the earliest prestige scales was developed using survey techniques, and it is often called the NORC scale because the National Opinion Research Center carried out the research (North and Hatt, 1947; Hodge, Siegel, and Rossi, 1964). Numerous survey respondents were asked to rate occupations in terms of how much standing members of that occupation have in the community. The ratings were combined and transformed into a ranking of the occupations on a one hundred-point scale. Supreme Court justice received the highest ranking (89), and shoe shiner the lowest

(27). Similar prestige scales have been developed in many countries, and the findings in one country tend to approximate closely those in other countries (Treiman, 1977). Studies in both the United States (Nakao and Treas, 1994) and Canada (Goyder, 2005) indicate that over time an individual occupation may shift up or down, but the overall standing of occupations tends to be stable.

Occupational prestige, education, and income tend to be closely related, but there are exceptions. For example, members of the clergy may receive relatively low incomes despite their extensive education and considerable prestige within the community. Nevertheless, occupation, prestige, and income are so closely related that we can often predict the general status of an occupation by knowing the average education and earnings of its members. Such predictions are called *socioeconomic status (SES) scores.* Data on occupation, education, and income collected from censuses or in periodic surveys are combined statistically to develop the SES scores (Duncan, 1961; Stevens and Cho, 1985).

Occupation is an important concept for sociologists, for studying both work life and life off the job. For this reason there are several ways to study occupations, ranging from simple dichotomies (blue-collar/white-collar) to the complex SOC scores, prestige scales, and SES indicators. Industry and occupation intersect in a specific job, and jobs are found in specific workplaces.

Workplaces

Many workers go to work each day for enormous corporations. Some of them may not even know from day to day just which corporate entity is their employer because of reorganizations, acquisitions, or mergers (see Chapter 15), but workers have a good idea of approximately how many people work at their particular work site. Sociologists are interested in both the local work site and its position within the larger organizational context, often using the workplace as a unit of analysis.

Some employing organizations have only one **establishment**; others may have many. For most workers the establishment is important, because it is where they perform their daily tasks and interact with other workers. Even for workers whose jobs require travel, the establishment serves as a base of operations. The establishment can be distinguished from the **firm**, which is the employing organization. Sometimes the firm may have only one establishment, but other firms may have many establishments.

Firms may be organized as corporations, partnerships, professional practices, or sole proprietorships. A number of firms may be bought or controlled by a **parent company**, which is a firm that owns other firms. The firms that are owned are called **subsidiary** companies. Burger King, the company described in Box 2.1, was at the time of the study a subsidiary of Pillsbury. Pillsbury's traditional products—flour, baking mixes, and other convenience foods—were in industries somewhat related to the fast-food industry. If the subsidiary firms are in unrelated industries, the parent company can be called a **conglomerate**. If the establishments or firms are located in different countries, as Burger King is, the parent company is called a **multinational company** or MNC. Multinational companies are discussed in more detail in Chapter 16.

Firms are often linked to one another through complex networks of suppliers and customers, subcontracts, credit lines and other financing agreements, and "tie-ins" of one product line with another. A new movie, for example, may have "tie-ins" with a toy company, various magazines, television network, recording studio, Internet provider, and other organizations in addition to the customary movie theaters. An additional source of links is **interlocking directorates**, which occur when a director of one corporation is also a director or officer of another corporation.

Sociologists collect information on workplaces directly from workers or employers, and information on many firms is available from annual reports, government regulatory agencies, and other sources. Social scientists analyze data on firms to provide information on conditions that affect workers. In later chapters, we will be discussing the effects of work organizations on workers and their jobs.

Other Units of Analysis

Sociologists also analyze other social units that affect work. Examples include unions and professional associations representing groups of workers. **Trade associations**, which are organizations of firms within the same industry, are significant in understanding the economic conditions and technological considerations affecting an industry. Government agencies, especially those with local, state, or federal regulatory power, are also important units for sociologists to study. The web sites for some of these government agencies appear in the resources at the end of the chapter.

PROBLEMS IN STUDYING WORK

Work is a complex human phenomenon with far-reaching effects. Although researchers continue to refine their techniques and expand their studies, there remain many aspects of work life about which we have little information. This section examines some of the problems researchers encounter in studying work.

Lack of Information

Even with the many sources of information we have already discussed, sociologists have many gaps in their knowledge and understanding of the world of work. Partly because of the definition of the labor force used by most of the world's governments, we have only begun to examine nonmarket work such as homemakers (Bose, Bereano, and Malloy, 1984; Davies, 2005), volunteer workers (Kendall, 2002), neighbors who exchange labor, and so forth. In addition, we have little useful information on the production of illegal goods and services—such as prostitution, gambling, or illegal drugs, to mention just a few.

The employment generated by other legal goods and services is sometimes hidden to avoid taxation. For example, some workers are paid cash to avoid income tax withholding and the payment of employer's and employee's social security tax. Illegal aliens are sometimes employed in this fashion. Smuggling goods or bartering goods and services also constitute unmeasured economic activity and employment. These uninspected aspects of employment, which are not captured in official labor force statistics, will be discussed in greater detail in Chapter 14.

Hard-to-Measure Characteristics

Some characteristics of work are important but difficult to measure. Social scientists are very interested in issues such as job commitment and underemployment, but there is little agreement about how they should be measured (Sullivan, 1978; Hodson, 1991). One reason such characteristics are hard to measure is that they have both objective and subjective elements. For example, workers who are subjectively bored with their jobs may consider themselves uncommitted to their jobs and underemployed. A researcher might reach the same or a different conclusion for these workers by looking at indicators such as rates of absenteeism (for job commitment) or hours and wages (for underemployment).

Even productivity, a relatively easy concept to measure in manufacturing industries, is difficult to measure in service industries. Is a service worker more productive because more customers have been served, or because fewer customers have been served but have greater feelings of satisfaction about the service they received? Developing methods to measure and study such characteristics is an important frontier for research on the sociology of work.

SUMMARY

Sociology and the other social sciences seek to develop valid and reliable information on the world of work. In advanced industrial societies, work is complex and heterogeneous, and the study of work involves subjective elements. For these reasons, sociologists have devised different ways to examine the world of work. Although each method has its limits, each also illuminates certain aspects of work situations.

Three important ways to examine the world of work are ethnographies, case studies, and sample surveys. Using different methods, sociologists study individual workers or collective groups of workers such as the labor force. Labor force studies include the study of the demographic characteristics of workers and their rates of labor force participation and unemployment. Social scientists also study occupations, industries, firms and other workplace units, unions, and government regulatory agencies. Although some aspects of work are not yet being adequately studied, the existing methods have yielded important and substantial findings about the complex modern world of work. In subsequent chapters we will present some of these results.

KEY CONCEPTS

validity	sample survey	NILF	firm
reliability	sample	labor force participation rate	parent company
ethnography	establishment surveys	unemployment rate	subsidiary
participant and nonparticipant observation	establishment	discouraged workers	conglomerate
experimental bias	response error	industry	multinational company
Hawthorne Effect	labor force	occupation	interlocking directorates
case study	ascribed and achieved characteristics	occupational prestige	trade associations
		establishment	

QUESTIONS FOR THOUGHT

1. According to the definitions used by the U.S. Bureau of Labor Statistics, what was your labor force status last month? Try to locate the most recent unemployment rate for your state or locality. (Note: check the federal statistical website below for regional data.)

2. Give an example of how workers in the same occupation, but in different industries, might have different working conditions.

3. As the children's counting rhyme at the beginning of the chapter illustrates, most people know about relatively few types of work. How can a student or worker learn about more jobs?

4. A researcher wants to discover what happens to workers who are laid off when their factory closes. Compare and contrast the advantages of using an ethnography, a case study, or a sample survey to study this issue.

5. "To understand work, you need only understand the workers." Do you agree with this statement? Why or why not?

6. Ponder the economic future of the United States, Canada, or Mexico. What sources of data might be useful for planning our collective future and achieving as positive a future as possible?

MULTIMEDIA RESOURCES

The U.S. Bureau of Labor Statistics offers a wide variety of information at *www.bls.gov*. Information for students considering careers is available at *www.bls.gov/k12/index.htm*. You can search the Standard Occupation Classification at *www.bls.gov/search/soc.asp* and search the industry classification NAICS at *www.census.gov/epcd/naics02/*. The Occupational Outlook Handbook is provided online at *www.bls.gov/oco/home.htm* and provides a great deal of information for career preparation, including projections of the supply and demand for workers in various occupational groups.

The Bureau of Labor Statistics is an agency within the U.S. Department of Labor *(www.dol.gov/)*. The Department of Labor also has regulatory agencies such as the Mine Safety and Health Administration and the Occupational Safety and Health Administration (OSHA).

Statistics Canada/Statistique Canada *(www.statcan.ca/)* provides detailed information on concepts and measurements used in Canada's census and labor reports.

International Labour Office, Geneva *(www.ilo.org/)* collects and analyzes information on work issues throughout the world. It has a useful data base of international labor statistics available at *http://laborsta.ilo.org/*. The online "World of Work" is available at *www.ilo.org/public/english/bureau/inf/magazine/index.htm*

RECOMMENDED FILM

Erin Brockovich (2000, Universal Studios). Directed by Steven Soderberg and starring Julia Roberts, this film is loosely based on a real story about a divorced woman who uses her research skills to bring a large company to account. Rated R for language.

PART II

The Personal Context of Work

For small children, work is nearly indistinguishable from play. For adults, however, there is a sharper distinction between work and leisure, with work often separated by time and place from family and leisure activities. As you saw in Part I, the characteristic types and places of work have changed throughout history, and social scientists have devised many ways to identify and analyze work.

Many people view work as their most important activity. Their earnings provide food, clothing, and other necessities, but work fulfills other needs as well. In this section we will examine how work affects the worker's life and family. Work can be an avenue for expressing creativity, perfecting knowledge and skills, and interacting with others. But every job also has its negative aspects, and many jobs are unremitting drudgery. Unsatisfying jobs and unemployment are major sources of tension and stress. Although there are many types of work within the economies of advanced industrial societies, all workers face certain common problems.

Table A shows how median income varies with gender, race, and family status. Married-couple families have the highest median incomes, while families supported by women have the lowest median incomes. Moreover, there are striking differences among racial/ethnic groups, even when controlling for marital status. Black families supported by women have a median annual income that is more than $10,000 a year less than the income of white families supported by women.

Such data raise questions about the context of work. Because marital status shows such a consistent association with labor force activity, analysts must explore the issue of *how* families affect work. The data also indicate gender and race differences in work. Why do these patterns persist despite laws to eliminate discrimination? Analysts cannot tell from Table A alone whether employers are

TABLE A

	Total Families	White Families	Hispanic Families[a]	Black Families	Asian Families
Median Annual Income[b]					
All families	$54,061	$57,340	$36,837	$36,499	$57,518
Married-couple families	63,630	64,091	42,082	56,834	67,699
Families supported by women	26,964	30,535	21,400	19,718	28,688
Number of Families[c]					
All families	83,869	62,724	9,333	8,785	3,027
Married-couple families	63,367	50,260	6,367	4,180	2,560
Families supported by women	15,814 *	9,236	2,240	3,991	347

[a]Hispanics may be of any race.
[b]Reported in dollars.
[c]Reported in thousands.
SOURCE: U.S. Census, 2006, Statistical Abstract of the United States.

discriminating or whether women supporting households might have lower skill levels, less job experience, or greater geographic immobility than men. Moreover, many women may prefer shorter workweeks because of family responsibilities. An analyst must explore such factors in greater detail before drawing a conclusion.

Social scientists begin to answer such questions by studying workers: The four chapters of this section introduce important problems that affect workers. In each chapter we will show how recent work trends, while alleviating some problems, are also creating new problems for workers. Chapter 3 examines the personal significance of work, including meaning, job satisfaction, and stress at work. Chapter 4 discusses inequalities of class, gender, and race in the workplace. Chapter 5 outlines problems that accompany fitting work into the life of the worker and the worker's family. Chapter 6 explores the efforts of workers to organize themselves collectively to solve work-related problems.

3

Meaningful Work

> There used to be a lot of malicious teasing. I cried a lot....It was a
> supervisor—a woman—who was the mean one....She yelled at us for
> brushing against the clothes that were hanging up in the changing
> room....And if you were sewing labels on garments and you asked for
> some more, she would take a bunch and throw them at you, so they'd
> fall all over the place, and then you'd have to pick them up. It takes time
> to pick them up, and then you'd have to rush like crazy to catch up to
> your quota. I cried a lot.
>
> (ROBERTS, 1994:61)

A good job provides the material necessities of existence. But it also contributes to workers' self-esteem, identity, and sense of order, which are important for meaning and satisfaction in life. Work often falls short of providing workers with self-esteem and a positive identity. As the above quote from a worker in a Japanese apparel plant clearly suggests, on-the-job experiences can cause workers to be deeply demoralized.

In this chapter we discuss the major sociological theories of self-actualization *versus* alienation at work. We also discuss the causes of job satisfaction and dissatisfaction, describe attitudinal and behavioral responses to work, and explore the future of satisfaction and meaning at work.

WHAT IS JOB SATISFACTION?

Job satisfaction is the summary evaluation that people make of their work, whether positive or negative. People can be either satisfied or dissatisfied with their work—they can find their work either highly engaging or utterly meaningless. People's levels of job satisfaction are the result of their job tasks, the characteristics of the organization in which they work, and individual differences in needs and values. These different experiences give rise to different theories that focus either on alienation or self-actualization (Gamst, 1995). This section attempts to make sense of these theories and their principal conclusions.

Alienation occurs when work provides inadequately for human needs for identity and meaning. Work is alienating to the extent that one does it only from economic necessity, not for its intrinsic pleasures. **Self-actualization** occurs when work contributes to the fulfillment of these broader human needs. A self-actualizing job provides for material needs, but if one's material needs were met in some other way, one would want to continue the work anyway for its own rewards.

Theories of Alienation

Modern understandings of alienation owe a large debt to the early writings of Karl Marx. He observed that as the world of material goods increases in value, the value placed on individuals seems to diminish (Marx, 1959 [1844]). Early industrial capitalism brought into being not only unprecedented productivity but also some of the most wretched living and working conditions in human history (see Chapter 1). Marx believed that these conditions resulted from denying workers the right to control their work activity and the products they produce. Thus, he described work under capitalism as "wage slavery."

Four Aspects of Alienation Workers are alienated in four ways under industrial capitalism, Marx argued. First, they are alienated from the *products* of their labor. They no longer determine what is to be made, nor how to dispose of it. In primitive societies workers had a direct relationship to the products of their labor. These products became an important part of their lives. In industrial societies workers no longer have such a direct relationship to their products. Work on these products becomes a *means to an end*; that is, rather than being an end in itself, work is a means to acquire money to buy the material necessities of life. Because workers are robbed of a meaningful relationship to the products of their labor, the products they produce seem disconnected from their lives, rather than manifesting their ideas, skills, and efforts.

Second, workers are alienated from the *process* of work. Someone else controls the pacing, patterns, timing, tools, and techniques of their work. Because workers no longer control their moment-to-moment activity, work becomes less meaningful to them. When workers are emotionally separated from their activity on the job, they become alienated from key aspects of themselves and of their identity as human beings. Their identity is found only outside the workplace through leisure or family pursuits. Non-alienating work, in contrast, is virtually indistinguishable from leisure. The worker experiences work and hobbies with the same enthusiastic absorption.

Marx's third aspect of alienation highlights the removal of creativity from work. He believed that the unique, defining characteristic of humans is the ability to be creative. The capacity for self-directed *creative activity* to meet changing needs is what distinguishes human beings from animals. If workers cannot express their *species being* (their creativity), they are reduced to the status of animals or machines. Extreme specialization, such as working on a highly repetitive task on an assembly line, makes each worker's contribution to the final product obscure. The opposite of non-creative work would be the artist who conceptualizes and creates a work in its entirety.

Fourth, alienated labor is an isolated endeavor, not part of a collectively organized effort to meet a group need. As a result workers are alienated from *others* as well as from themselves (Marx, 1959 [1844]). Human beings are social animals by nature, and their work always involves others, either directly or indirectly. When labor is alienated, control of these social relations is removed from the worker, who interacts with others on the job only as directed by someone else. Alienated workers are not part of an integrated team engaged in collectively determining the nature and goals of their activity. As a result, even workers who labor side by side may nevertheless be isolated from one another.

Alienation not only isolates the worker while at work but also affects the nature of his or her involvement in the broader society. Those who control the process and products of labor are able to control the direction in which society develops. The alienation of workers, which starts at the workplace, thus extends to society as a whole, and workers find no place, no thing, and no experience that they can truly call their own. Thus, according to Marx, industrial capitalism denies meaningful participation for the vast majority of employees.

When Marx wrote about work in 1844, the physical conditions of labor were much harsher than they are today. However, his writings remain the touchstone for studies of alienated labor because harsh working conditions have not disappeared and because alienation persists even in more physically pleasant settings. For example, a telephone receptionist's work area might be comfortable and well lit, but she may still be alienated because of the unceasing flow of superficial interactions. In some ways Marx's writings are even more relevant today because of increased mechanization, standardization, and bureaucratization.

The Erosion of Meaning   Contemporary researchers have extended Marx's theory by analyzing the *subjective* experience of alienation. The American sociologist Melvin Seeman (1993) hypothesized that workers experience alienation in terms of powerlessness, self-estrangement, meaninglessness, isolation, and normlessness.

Seeman's components of the subjective experience of alienation correspond roughly to Marx's components of alienating conditions. *Powerlessness* is the expectation that one cannot control the events in one's life. This idea corresponds roughly to Marx's concept of alienation from one's products. *Self-estrangement* refers to the individual's lack of rewarding and engaging activities. It roughly parallels Marx's concept of alienation from the process of work. *Meaninglessness* arises when workers feel that they cannot adequately predict the future and when their efforts seem to have few worthwhile results. Meaninglessness corresponds roughly to Marx's concept of alienation from human creativity. *Isolation* arises from a disparity between the goals, values, and expectations of an individual and those of the rest of society. It closely parallels

Marx's notion of alienation from others. Seeman added the concept of *normlessness* (French, *anomie*) from the work of the French sociologist Emile Durkheim. Normlessness indicates a state in which either the appropriate standards of behavior are unknown or there is insufficient reason to abide by them. The concept extends the analysis of the subjective experience of alienation outside the workplace and into broader society. In this way it parallels Marx's own effort to link workplace alienation to alienation from broader society.

Contemporary writers continue to be concerned about the potential negative consequences of meaningless, insecure, and unfulfilling work. Social commentators fear that such work does not provide a firm foundation for personal identity and for a sense of values sufficient to provide guidance throughout life. This vacuum leaves people adrift somewhere between shallow consumerism and anxiety about their future security (Sennett, 1998).

Theories of Self-Actualization

Maslow's Hierarchy of Needs   Contemporary theorists have balanced the analysis of work by including a consideration of self-actualizing aspects of work as well as alienating ones. Much of this work was motivated by American psychologist Abraham Maslow's theory of *hierarchical need satisfaction* (1999). This theory argues that human beings have

- *physiological* needs for such things as food and sex;
- *safety* needs for a secure physical and emotional environment;
- *belongingness* needs for acceptance and friendship;
- *esteem* needs for recognition, attention, and appreciation; and
- *self-actualization* needs for developing to one's fullest potential.

These needs are arranged hierarchically; that is, individuals are concerned with their physiological and safety needs first, and only after these lower-order needs are met do they turn their attention to higher-order needs.

In Maslow's theory the most alienating jobs are those that do not provide for minimum levels of material sustenance and security. Less alienating jobs may satisfy these needs but still fail to provide a feeling of self-worth. The best jobs are those that not only provide material necessities—security, belongingness, and esteem—but also help workers develop to their highest potential.

Maslow's Followers   Maslow's theory has been applied to the world of work by several contemporary social theorists including Frederick Herzberg and Chris Argyris. Herzberg (2003) observes that workers consider different issues when determining whether they are satisfied or dissatisfied. Satisfied workers stress professional growth, achievement, recognition, responsibility, and advancement. Dissatisfied workers stress negative factors, such as managerial incompetence, close supervision, low wages, and poor working conditions. Such differences in emphasis led Herzberg to develop a *two-factor theory of job satisfaction*. Workers are dissatisfied if they have to work in unpleasant physical or social settings. Herzberg referred to these characteristics as "hygiene factors." Such negative conditions have to be removed or neutralized to avoid alienation. However, the removal of these conditions is not sufficient for satisfaction. To be satisfied, workers must have their needs for personal development met. Herzberg refers to these factors as "*motivators*." He includes such things as autonomy, challenge, recognition, and opportunity for developing new skills.

Argyris (2003) developed a similar theory of job satisfaction based on the principles of human development throughout the life course. Argyris believes hierarchical and bureaucratic work environments frustrate the natural development from passivity to activity and from dependence to independence and that this frustration is a major source of job dissatisfaction.

Models of job satisfaction based on theories of hierarchical needs have some important limitations. These theories rest on unproven assumptions about human nature. It has not been empirically demonstrated that so-called lower-level needs have to be met before higher-level needs can be addressed. Research attempting to evaluate these assumptions has provided only limited support for their relevance or accuracy. It may well be that human beings attempt to meet whatever of their needs can be met in any given situation, with little inherent ranking into higher-level and lower-level needs. Nevertheless, it is clearly useful to consider a range of needs and motivations if we are to understand the complex underpinnings of meaningful work.

Marx's theory about objective conditions that produce alienation, Seeman's theory about the subjective experience of alienation, and Maslow's and his followers' theories about self-actualization have provided the basis for a vast body of research on job satisfaction. Contemporary researchers retain these early theorists' insights into the all-embracing nature of the experience of work, but they have extended their analysis by looking at the detailed characteristics of jobs and personal expectations. It is to these issues that we now turn.

GOOD AND BAD JOBS

In this section we discuss some of the characteristics of work and of workers that lead to alienation or self-actualization. Factors that determine the degree of satisfaction include the nature of job tasks, technology, organizational characteristics, workers' participation in decision making, individual differences, and prior expectations. A general model of job satisfaction and dissatisfaction is presented in Figure 3.1.

Self-Direction

The nature of one's daily tasks on the job is the most important determinant of self-actualization at work. Work that has autonomy, complexity, and diversity can be self-actualizing. **Job autonomy** is the extent to which the worker controls his or her own work and relations with others at work,

Causes	Alienation or Self-actualization	Consequences

Nature of job tasks
- Autonomy
- Repetition
- Complexity

Technology

Organizational structure

Participation

Individual differences

Expectations

Job attitudes
- Dissatisfaction or satisfaction
- Commitment

Behavioral responses
- Enthusiasm
- Withdrawal
- Focus on informal interactions
- Absenteeism
- Quitting
- Passive resistance
- Sabotage
- Theft

FIGURE 3.1 Causes and Consequences of Alienation or Self-Actualization

including both co-workers and supervisors. The *complexity* of the task is equally important. Men and women whose job responsibilities are more complex and allow for greater self-direction are less psychologically distressed and more intellectually flexible, and they come to value self-direction more for both themselves and others (Kohn, 1990). Diversity of tasks is also a significant factor—*repetitive* work, lacking in variety, done at a forced pace, is among the most brutally alienating forms of work.

The assembly line is a classic example of work organized in an alienating way. An assembly line in a medical products company reveals what work is like under these conditions:

At Biomed, the assembly process was highly segmented. Each worker performed a limited number of movements, over and over. Much of the work at Biomed, and all of the unskilled work, required little, if any, use of one's mental faculties. The only obligation at Biomed was that one "learn" the proper technique of one's assigned job and then "learn" to perform it quickly. At Biomed, mental fatigue occurred much sooner than physical fatigue, due to the repetitive nature of the infinitesimal actions performed (Devinatz, 1999:58).

Because such jobs lack complexity and diversity, they provide few social or psychological benefits.

This work is truly alienated, allowing for no human creativity and for little sense of unique personal identity or social connectedness.

Clerical work, as well as blue-collar assembly-line work, can also be repetitive and alienating. Automation has made some clerical work just as routinized as the most repetitious manual work. Because numeric data can be standardized and regimented more easily than many physical products, some clerical work may be in the process of becoming even *more* alienating than many manual jobs. In Chapter 13 we discuss contemporary changes in the nature of clerical work.

By contrast, pride in work based on significant self-direction is also a common workplace experience. The work of engineers and scientists who have great latitude in self-directing their work provides an example. An ethnography describing the development of the polymerase chain reaction (PCR), which profoundly transformed the human potential to identify and reproduce segments of genetic code, reports on the extreme pride that the chemists and biologists engaged in the project felt in their work. The development of PCR has made possible not only cloning but a vast array of genetic interventions in the areas of medicine, agriculture, biology, and related fields. A young biochemist recalls:

Probably I worked harder in that time in my life than I ever had in terms of hours. I was

---

BOX 3.1   **Friendship and Solidarity in Female and Male Work Groups**

While working as a participant observer in an electronics assembly plant with a predominantly female labor force, an ethnographer reports being overwhelmed by the generosity of the poorly paid women with whom she works:

[After returning from a two-week sick leave without pay,] I was talking to Anna when she stuffed a £10 note in my trouser pocket so quickly I wasn't even really sure what it was. She was giving it to me because I would be short, having lost two weeks' wages....I was quite overwhelmed by her generosity; the gift was completely genuine, and she really didn't want the money back. The whole attitude toward money and seeing that others had got enough was so different from my previous job where although we earned much more, people remembered who owed whom a cup of coffee. All the women were very generous, sharing out sweets and crisps and whatever they bought for themselves. (Cavendish, 1982:62)

Group solidarity is similarly evidenced in an ethnography of an underground mine with a predominantly male *labor force. The ethnographer reports the following episode in which a lead worker and his men gather at the head of a mineshaft to search for coworkers trapped by a fire:*

Suddenly Jimmie Isom picks up a mask from the jeep. "Put one on me, Dan," Jimmie says. Dan stares at his friend, with the deep-etched lines from his heart attack. Dan usually works Jimmie on the outside crew these days, afraid of working him inside. Now Jimmie is volunteering to go into the smoke. Dan doesn't know how to turn him down. (Vecsey, 1974:190)

SOURCE: (1) Excerpted from Ruth Cavendish, *Women on the Line*, p. 62. London: Routledge and Kegan Paul. Copyright 1982. Reprinted by permission of the publisher. (2) Excerpted from George Vecsey, *One Sunset a Week: The Story of a Coal Miner*, p. 190. New York: E.P. Dutton. Copyright 1974. Reprinted by permission of the publisher.

---

very interested in the job. It was really fun to learn how to synthesize DNA.... It was the heyday of biotechnology. There were all kind of bold ideas floating around all the time...and there was absolutely no constraint on the imagination.... The company was really fun.... [We] were right in the middle of something that was a red-hot area. (Rabinow, 1996:90)

Belongingness

Self-actualization is also influenced by the extent to which meaningful interaction is possible on a job. Most people prefer to work as a member of a group rather than in isolation. Peer support and solidarity are essential building blocks for a meaningful experience of work in many settings. Workers report that positive relations with coworkers are an essential characteristic of good jobs (Hodson, 2001). Box 3.1 describes the important role of positive co-worker relations in both female and male work groups.

Technology

Because the nature of job tasks is such an important cause of job satisfaction, a great deal of research has focused on technology as both a cause and a potential solution to the problem of alienation. The technology one uses depends primarily on one's occupation. Accordingly, occupational differences in job satisfaction have frequently been used as indicators of job quality. When workers are asked how satisfied they are with their job as a whole, farmers indicate the greatest satisfaction, with 67 percent reporting that they are very satisfied. By contrast, only about 40 percent of clerical workers and machine operatives report that they are very satisfied with their jobs (General Social Survey, 2002).

Blauner's Theory of Technology and Alienation
Robert Blauner, in his famous book about industrial society, *Alienation and Freedom* (1964), explored the possibility that technology would bring an end to alienation. Blauner based his analysis on case

studies of four manufacturing industries at various stages of technological development: printing, textiles, automobiles, and chemicals.

The printing industry was organized along traditional craft lines. Job tasks were complex and rewarding. Printers were very involved in their work and experienced a good deal of self-actualization. Workers in the textile industry were more alienated than printers. The work was organized around machine tending, with workers being responsible for highly repetitive tasks such as keeping threads attached to moving spindles. The repetitive assembly-line technology of the automobile industry produced even greater alienation among workers because of the increasing size of organizations, the heightened intensity of assembly-line work, and the location of the new factories in regions often far removed from workers' hometowns.

By contrast, workers in the chemical industry were less alienated. Blauner argued that there was less alienation because the automated **continuous-process technology** of the chemical industry allowed a less repetitive arrangement of work, greater worker autonomy, and greater worker responsibility for the care and maintenance of expensive automated equipment. Based on these observations, Blauner proposed an **inverted-U curve of technology and alienation** (see Figure 3.2). He believed that alienation increases as machine pacing and assembly-line technologies replace the craft organization of work, but that it decreases again with more advanced continuous-process technologies. In this view self-actualization at work can be achieved either in craft settings or in settings that use advanced technology, but not in mass-production manufacturing settings.

Modern Technological Developments Developments since Blauner's analysis have neither confirmed nor convincingly refuted his conclusions that advanced technology reduces alienation. Subsequent research has suggested "that workers accept technological change as normal and that they like their machines" (Form and McMillen, 1983:175). Other researchers have found that although advanced technologies may reduce alienation for some workers,

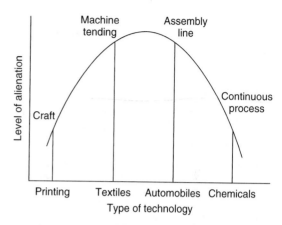

FIGURE 3.2 Blauner's Inverted U-Curve of Technology and Alienation

SOURCE: Adapted from Robert Blauner, 1964, *Alienation and Freedom*. Copyright © 1964 The University of Chicago. Reprinted by permission of The University of Chicago Press.

technology undermines the skills of other workers and increases their sense of alienation (Shaiken, Lopez, and Mankita, 1997). In recent decades the printing industry, which Blauner selected to represent the craft organization of work, has experienced a technological revolution due to widespread use of electronic typesetting. This revolution eliminated the jobs of many skilled craft workers and replaced them with somewhat less skilled data entry and programming jobs. We discuss controversies about the role of advanced technology in determining the nature of work further in Chapter 9.

The most reasonable conclusion that we can draw from these findings is that technology has no single, unidirectional effect on alienation. Rather, alienation can increase or decrease, depending on the nature of the technology and the nature of the jobs the technology creates, eliminates, or changes. For example, clerical work can be *routinized* by computerized systems that narrow a worker's range of activities, or it can be *humanized* by selecting technologies that expand workers' capabilities. Too often, the emphasis on control and monitoring of workers results in the design and selection of technologies that restrict self-actualization at work (Noble, 1997). It is up to all of us as members of society to design and promote technologies that improve working conditions. The possibilities are there, but the outcome is by no means assured.

Organizational Structure and Policies

Tasks and technology are important determinants of alienation, but workers are also strongly influenced by the characteristics and policies of the organizations in which they work. The size of an organization and its pay and promotion policies are especially important in this regard.

Pay   Being paid a living wage for one's work is a necessary condition for self-actualization. Workers consistently rank pay as a crucial characteristic of a good job (Freeman and Rogers, 1999). High wages may not be sufficient to compensate for an alienating job, but the provision of wages adequate to meet basic needs is a fundamental requirement before a job can be experienced as rewarding and meaningful.

Security   Job security is equally important as pay. As people advance in their careers and take on family responsibilities, the sense of security in one's job becomes more and more important. For many employees, however, the threat of layoffs, corporate bankruptcy, or the loss of jobs to offshore locations is a constant reminder of their vulnerability to forces beyond their control. In societies where health care, retirement, and virtually all necessities of life are tied to employment, such insecurities can gnaw on one's sense of well-being and become a chronic source of dissatisfaction (Sennett, 1998).

Size   Workers prefer small companies to large corporations. Large size produces feelings of powerlessness and isolation, because workers have difficulty in identifying the overall purpose and direction of their organization and in feeling that they are a significant part of that purpose. Workers thus report greater satisfaction across a variety of dimensions in small, locally owned companies than in regionally based companies or in the nation's largest corporations (Hodson and Sullivan, 1985).

Promotion Policies   Workers can also feel alienated by a lack of opportunity for promotion. Because almost all complex organizations are organized as a hierarchical pyramid, the ratio of high-level jobs to beginning jobs is generally low, and regular promotions can occur only for a limited number of workers. Workers who have poor prospects for promotion may come to feel disenchanted with their jobs. For a job to be fulfilling, it is necessary to have not only rewarding tasks but also a meaningful career trajectory. This problem is even more acute when people define their self-worth in terms of their advancement at work, as is so common today (Ospina, 1996).

Dignity and Respect   One of the most essential foundations for good work is being treated with dignity and respect. Abusive bosses who yell at workers, fire them without just cause, and mismanage the workplace demoralize their workforces. Conversely, bosses who respect workers' rights and treat them with dignity find that workers are capable of great enthusiasm and loyalty (Hodson, 2001; Ferraro, Pfeffer, and Sutton, 2005). One of the most important contributors to organizational productivity and to a meaningful experience of work is having supervisors who are competent, dependable, and trustworthy (Cook, 2001).

Bullying by supervisors can have devastating consequences for the experience of work and for one's sense of personal security, as illustrated in the following quote from a civil servant forced into early retirement by bullying:

> I retired from work on grounds of ill health with a diagnosis of posttraumatic stress disorder. The total economic cost of my situation must have cost a fortune. There were the costs due to sick-leave, an irritable bowel syndrome investigation, visits to my local [general practitioner], counseling by a clinical psychologist and eventually, psychiatric outpatient care. (Rayner, Hoel, and Cooper, 2002: 44)

Bullying is commonly defined as covering at least three major types of behavior: obstructionism, expressions of hostility, and overt aggression. Obstructionism includes causing intentional delays in someone else's work, interfering with their work, or refusing to share resources or equipment. Hostility includes such behaviors as staring, dirty looks, belittling, obscene gestures, spreading gossip or rumors, and ridicule. Overt aggression includes threats, assaults, destroying needed resources, or destroying the person's personal property. The underlying goal in bullying is inflicting "dignitary harm" on the victim "by humiliating, intimidating, tormenting, pressuring, or mocking" (Ehrenreich, 1999:6–7).

Organizations *can* make work more fulfilling and less emotionally damaging. Such efforts would entail a reduction in the division of labor so that jobs become more diverse and interesting, rather than less so. Technologies can be selected that are compatible with a reduced division of labor. Organizations can also provide job rotation and other forms of continuous learning on the job. Managerial competence can be trained and a respectful work-life experience can be ensured. All of these changes would encourage greater involvement on the part of employees and would contribute both to employees' self-actualization and to organizational productivity.

Participation   Organizations can also facilitate self-actualization (and increase productivity) through allowing workers to participate in making decisions. Worker participation is a multifaceted phenomenon that is difficult to nail down and define in a single formula. It can include an informal openness on the part of supervisors to the complaints and ideas of workers. At the other end of the continuum, it also includes formal participation in determining the overall direction of an organization, as among the members of worker-owned cooperatives (Soupata, 2005). The positive effects of participation in an automobile factory are presented in Box 3.2.

Involvement in an organization can include any combination of formal, informal, and representative participation. The more forms of participation

included, the more likely alienation is reduced (Rothschild and Ollilainen, 1999). Direct participation reaches a peak in collectively owned cooperatives, such as those in the plywood industry in the Pacific Northwest (Greenberg, 1980). In these cooperatives the workers are responsible for the governance of the enterprise. They hold regular shareholder meetings and informal group meetings on the mill floor. As a result, only a fourth as many supervisors are needed in cooperative mills as in privately owned mills. In addition, the supervisors primarily coordinate the flow of materials rather than scrutinizing the work of their fellow workers. Spontaneous cooperation and informal job rotation are the rule in the producer cooperatives, providing workers with active participation not only in the overall direction of their enterprise but also in the day-to-day content of their jobs. The possible forms of participation open to employees today is a primary focus in Chapter 17.

Stress and Overwork

Job stress is a widespread social problem involving millions of workers worldwide. Its manifestations include stress-related health concerns such as high blood pressure and ulcers as well as self-destructive behaviors such as smoking, increased alcohol consumption, and conflicts at home. Seven in ten workers report that job stress causes frequent health problems (Anderson-Connolly et al. 2002). Job stress is also a problem for organizations. Excessive job stress results in absenteeism, tardiness, turnover, low morale and motivation, communication breakdowns and distortions, impaired judgments, and deteriorating workplace relations with increased distrust and animosity, as well as many opportunities lost because of initiatives not taken.

Job stress appears to be on the rise. Increased international competition has created heightened demands on workers (see Chapter 16). As a result, workers are working longer and harder than in past decades. Hours worked are up and vacations are down (Mishel, Bernstein, and Schmitt, 2005). These pressures are further heightened by the increasing prevalence of dual-earner families and problems of

BOX 3.2 Participation and Job Satisfaction at a Volvo Factory in Sweden

Throughout the company we have a hierarchy of works councils, with representatives from both management and the employees. Some of these councils have been required by collective agreement. Others, like the Corporate Works Council, have been created on a voluntary basis to meet our own needs for consultation.

Bedeviled by an inherently noisy and dirty process, body-shop workers and managers got together and chose a working group to assess the various problems, suggest some solutions, and figure out the costs of the alternatives....

The results were promising. The architects suggested ways to cut the noise from jigs and grinding machines. They also proposed a fundamental color scheme with red, orange, blue, and green. These suggestions were put together in a special exhibition in the shop, and employees came to see and decide for themselves. Many had further suggestions and related problems to offer. The response was quite positive. It took several years to implement all the proposals, but today the body shop is one of the brightest spots in the corporation....

In the upholstery shop job rotation was the opening wedge. It started for very practical reasons. Employees complained of sore muscles from doing the same operation over and over. They discovered that if they paired off and traded jobs every day or so, they were able to use different sets of muscles. From that ergonomic, down-to-earth beginning have grown most of the other changes that focus on the quality of work life more generally....

As these changes took place, rather gradually, there was a concomitant increase in team spirit and individual commitment. Employee turnover and absenteeism changed dramatically, and the upholstery quality improved because team members understood the entire process and felt much more responsible for the product of their work.

SOURCE: Excerpted from Pehr G. Gyllenhammar, *People at Work*, pp. 81–91. Reading, MA: Addison-Wesley. Copyright 1977. Reprinted by permission of the publisher.

child care, which can contribute to schedules that are seemingly impossible to meet. In combination these factors result in a heightened sense of *time poverty* for many workers. Possible solutions include greater access to part-time work, more affordable child care, and wider use of flexible schedules.

Job stress can be further aggravated by *conflicting* role demands on employees. For instance, a worker in a fast-food restaurant may be told by management to always ask customers if they want a drink or dessert to go with their order. However, in their relationship to the customer, this may make them seem pushy and artificial rather than truly helpful. Thus the daily tasks of the fast food worker can entail **role strain** between the expectation to sell as much food as possible and the expectation to have a pleasant interaction with the customer. Role strain resulting from the experience of such contradictory expectations within a single role can increase pressure on the job, regardless of the volume of work or other considerations.

Similarly, many jobs entail expectations that may conflict with those arising from other roles or values of the worker. For instance, a nurse may experience a contradiction resulting from the size of the patient load she is expected to service. With too many patients (and too much paperwork) she may not be able to give individual patients the care that they expect and deserve, and that she wants to give them. To the extent that she went into nursing with the goal of serving patients' needs, such a situation may generate a **role conflict** between expectations on the job and expectations arising from other roles or values she holds. We will revisit the issue of role conflict in Chapter 5 where we discuss important tensions between the demands of work and family.

Individual Differences in the Experience of Work

Workers' feelings about their jobs depend not only on the nature of the work but also on the background, values, and needs that they bring to the job. Whether workers are satisfied with a job thus depends to some extent on what they hope to get out of it. Gender, race, and age are important differences between individuals that influence the needs and values they bring to the workplace.

Gender Female workers tend to have lower-paying jobs, more repetitive work, fewer chances for advancement, and fewer opportunities to exercise high levels of skill than male workers. Despite these differences, research consistently indicates that women are about as satisfied as men with their jobs (Tolbert and Moen, 1998). What accounts for this paradox? The explanations that have been offered include the idea that women are socialized to be more acquiescent, and, therefore, simply do not verbalize their complaints. The explanation receiving the most support, however, is that women evaluate their jobs on a different basis than men—that is, relative to those of other women, not relative to those of men. Thus, they do not feel particularly deprived (Ross and Wright, 1998). The use of other women as a **comparison group** is learned in childhood and applied later in school and on the job. The use of different comparison groups by men and women for evaluating their jobs is further reinforced by occupational segregation into different lines of work. Many women work in strongly female-dominated occupations (see Chapter 4). This occupational segregation increases women's contact and feelings of a shared work situation with other women while decreasing their opportunities to make comparisons with men. In the future, given the increased employment of women in jobs that were once the exclusive domain of men, women may increasingly come to evaluate their jobs relative to those of men, ironically increasing their sense of dissatisfaction.

Race Members of racial and ethnic minorities are also more likely to be stuck in poorer paying, less skilled, and less rewarding jobs (Wilson, 1997). Unlike women, however, they often voice strong discontent with this situation. For instance, the negative job satisfaction of black workers relative to that of white workers is among the strongest contrasts in job attitudes found between different social groups (Mueller et al., 1999). Two differences between blacks and women account for their different reactions to inferior jobs. First, blacks, unlike women, do compare themselves with dominant groups when they evaluate their jobs. Second, most blacks believe strongly that they should be treated equally with whites. This ideology of racial equality gives voice and legitimacy to their anger at being stuck in inferior jobs.

Age Young workers are often more unhappy with their jobs than older workers (Weaver, 2002). Part of the reason for this difference is that younger workers are more educated and thus expect more from their jobs. Another part of the explanation is that older workers have either been able to find work adequate to their needs or have made downward adjustments in their aspirations.

Tenure Length of time on the job also makes a difference in how workers feel about their work. New workers are more interested in the importance of their job than are workers with greater tenure, but they are not as concerned about having a high degree of autonomy. Once workers have been on the job for several years, their attitudes and desires begin to change, and they become more interested in autonomy and variety and less interested in the centrality of their task within the organization (Hall and Tolbert, 2004).

Great Expectations

The expectations that workers bring to their jobs arise from their prior experiences and expectations. Socialization for work begins in childhood and continues throughout life. Workers are more or less satisfied with what their job offers depending on the weight that their prior socialization leads them to give to specific aspects of work, such as autonomy, congenial co-workers, fringe benefits, or the chance to develop new skills. Early occupational experiences can also create values and dispositions that influence later occupational choices and satisfaction with those choices (Kohn, 1990). Satisfaction is thus determined not only by the characteristics of the job, but also by the job-worker fit—by the match between a *worker's values* and the *job characteristics*. Prior values have been shown to be one of the most important determinants of overall satisfaction with one's job (Kalleberg et al., 1996).

Education influences job satisfaction through creating expectations about the rewards that the workplace should offer. Thus, holding job quality constant, more educated workers are actually less satisfied! Why does this occur? In college, students have a great deal of autonomy to decide their area of study and to pattern their study habits. Later, when they enter the full-time world of work, many college graduates find themselves in highly routinized white-collar jobs, which have few if any of these characteristics. Instead, their jobs may be boring, lack challenge, and use few of their skills (Daday and Burris, 2001). After graduating from college, students may find a mismatch between what they had come to expect in college and the jobs that are actually available in the world of work (Kalleberg, 2006).

RESPONSES TO WORK

In this section we analyze the nature of workers' attitudes toward work and their behaviors at the workplace. We start again with the general concepts of alienation and self-actualization and then move to more specific attitudes and behaviors.

Attitudes toward Work

Sociologists have developed a variety of measures for appraising workers' attitudes toward their work. These include measures of *intrinsic and extrinsic job satisfaction* and *commitment* to the job.

Social scientists typically use two related approaches to measure job satisfaction. In one approach they ask respondents how satisfied they are with their jobs or with specific aspects of their jobs. Common phrasings of such questions are "How satisfied are you with your job as a whole?" or "How satisfied are you with your chance to use your abilities on your job?" An alternative approach is to ask whether the respondents would stay in their current job if other opportunities were available. Common phrasings include "If you had it to do over again, would you go into the same line of work?" or "Would you continue to work if you won the lottery?" These two approaches produce markedly different proportions of reportedly satisfied workers. About 75 percent of workers report being "fairly" or "very" satisfied with their jobs when asked directly about their level of satisfaction. In addition, as many as 20 percent of workers consider themselves workaholics. Only about 40 percent of workers, however, would continue to work if other options were available (Hamilton and Wright, 1986).

What accounts for these discrepant responses? For one thing, phrasing a question in terms of satisfaction with a current job limits respondents' options. That is, it forces them to consider the current constraints on their options. In this situation, most respondents give cautiously positive evaluations of their jobs. When questions offer workers broader possibilities by asking if they would enter the same job again or would prefer not to work at all, they voice greater reservations about their current jobs. In reality, such options do not exist for many people because of limits imposed by such factors as training, family responsibilities, and regional ties. But given the opportunity, most people would forsake their current job for something better. "For most workers it is a choice between no work [and work] burdened with negative qualities. In the circumstances, the individual has no difficulty with the choice; he chooses work, pronounces himself moderately satisfied, and tells us more only if the questions become more searching" (Kahn, 1972:179). What this means is that people can be satisfied and dissatisfied at the same time. Most people try to make the best of their situation at work and report being moderately satisfied with it. Verbalizing a mildly positive attitude about work is also socially more acceptable and psychologically safer than admitting that one is dissatisfied.

People also have a tremendous capacity to gain at least some satisfaction and meaning from even

B O X 3.3 San Francisco Scavengers

With all the problems and discontents in the work of the garbage collector, there are plenty of scavengers who will answer positively when you ask them, "What's good about this job?" . . .

"I guess you have to go into the psychology of it," Lenny said. The "psychology of it" had a great many different aspects. One of these was what might be called the variety of the work, as unlikely as that may seem to the uninitiated. Ron told me, "Thing about this job is that you do different things." If he had taken the somewhat comparable job of pick-up man on a city street-cleaning crew, Ron said, he would "go crazy. That would be monotonous." The tasks on his route were varied enough to break the monotony of the work's routine—operating the blade, solving problems with customers, blanket work and can work, driving the truck, and so forth. But there was also a different kind of variety, the unexpected in the human events of the workday—like

the burglar on that day with Freddie and his crew. When I met Freddie's young son working with his father, he told me that he had refused a nine-to-five job that paid more: "Staying in one place all day? Naw!" . . .

Another part of "the psychology of it" was being able to set one's own pace, Freddie had said. "You are your own boss." A crew could work as fast or as leisurely as they wished. As I observed, they usually worked as swiftly as humanly possible—if only to go home early or, as Lenny told me from his own experience, to be in a bar waiting for another crew to come in and be able to say, "What's been keeping *you*?" . . .

Being outdoors is a big advantage that even workers in less clement climates emphasize. To be indoors all the time is a drag.

SOURCE: Excerpted from Stewart E. Perry, *San Francisco Scavengers,* pp. 110–117. Berkeley, CA: University of California. Copyright 1978. Reprinted by permission of the publisher.

the most tedious jobs. It is possible to build lives, careers, and even communities around work that is, in fact, quite alienating:

> Workers have a number of ways of dealing with monotony. . . . One of the older women on the floor had a routine she followed religiously. Every day at morning coffee break she went to the corner store and bought a newspaper. She brought it to her table and then went to the bathroom for a paper towel that she spread on her table. She then proceeded to eat half of her sandwich, no more, no less, every working day. There were numerous other examples of women "setting up" their meager possessions— radios, cigarettes, and coffee cup—in similar fashion. (Juravich, 1985:56)

Even garbage collectors can take pride in their work and realize a significant degree of self-actualization from it (Perry, 1978). Are workers satisfied or not? It depends on what we mean by satisfaction and how we ask the question. Box 3.3 describes the work of garbage collectors in a San

Francisco cooperative and how they find meaning in work that many would avoid.

Intrinsic and Extrinsic Satisfaction   Researchers sometimes divide the sources of job satisfaction into intrinsic and extrinsic factors. **Intrinsic rewards** are realized on the job and include such things as the freedom to plan one's own work, the chance to use one's abilities, the absence of close supervision, and positive relations with coworkers. **Extrinsic rewards** are realized off the job and include such things as pay, fringe benefits, and job security.

Satisfaction with both intrinsic and extrinsic factors shows a moderate decline in North America since the 1950s, when social scientists first began to chart job satisfaction (Hamilton and Wright, 1986). What accounts for declining job satisfaction? It does not appear that a declining quality of jobs is responsible. Changes *have* occurred in the content of many jobs, but these changes have been gradual and offsetting. A more likely interpretation relates to greater dissatisfaction among young workers. Over time, with increasing prosperity and educational attainment, expectations have risen. As a result, young people tend to voice greater dissatisfaction with their jobs today than in the past.

Commitment   Above and beyond being satisfied with their jobs, workers can also be more or less committed to their jobs. **Commitment** develops when workers perceive that their own needs will be met through continued employment in the job and when they perceive that the goals and values of the occupation or the employer are compatible with their own goals and values. Commitment implies a willingness both to retain long-term membership in the group and to give one's full energy and abilities to the group's ongoing tasks (Kalleberg et al., 1996).

Other reasons for commitment exist besides a compatibility of goals and values. Lengthy association with a profession or an organization increases the number and importance of benefits that are dependent on continued group membership. For example, pension plans, medical benefits, and friendship networks may all depend on retaining one's current job. As a result, people are often reluctant to leave a job that they have held for a long time, in spite of dissatisfactions with it, because their lives have come to depend on it in many different ways. Japanese corporations have intentionally heightened these "side-bets" by directly tying such peripheral aspects of workers' lives as housing and access to vacation retreats to continued employment with the company. This strategy has resulted in increased commitment among Japanese workers, but it has not resulted in greater satisfaction among Japanese workers than among American workers (Kalleberg et al., 1996).

Evidence suggests that, as with job satisfaction, commitment to employing organizations may be declining. A college-educated worker doing routine clerical work in a large corporation reports the following sentiments: "People don't care about the company. They could care less if [it] burned down tomorrow" (Burris, 1983:144). This reduced commitment occurs at a time when increased commitment and extra effort are being expected of employees in order to increase productivity. We

explore the consequences of this dilemma further in the concluding chapter of this book.

Behavioral Responses to Work

Workers' attitudes toward their jobs, in combination with the options available to them, determine how they will respond to the workplace. Workers, including both those who are quite enthusiastic about their jobs as well as those who are chronically disgruntled, make behavioral adjustments at work to maximize their benefits and minimize their costs. These adjustments range from leaving the job to various forms of accommodation and resistance. A common response to alienating work is passive resistance through making work into a game (Burawoy, 2000), restricting one's output (Roy, 1952), or focusing on aspects of work life tangential to the main productive activity (Collinson, 2003). For instance, workers often adjust to alienating situations by focusing on interactions with their peers. Managers label such behavioral responses "poor performance." However, such behaviors do not necessarily result from incompetence or laziness; rather, they may be straightforward responses to having a job that is tedious, repetitive, or alienating. These responses are difficult to predict from workers' levels of job satisfaction or commitment. Workers who are very committed to their work may be the ones most likely to resist alienating conditions. Those who are less committed may simply exit or grudgingly suffer in silence.

Pride and Enthusiasm   Positive responses to work are likely to occur when the job is adequately rewarding in terms of extrinsic and intrinsic satisfaction and when the worker is committed to the profession or the organization because of shared goals and values. In such situations, employees may work with exceptional enthusiasm, often going beyond the requirements of the job (Walsh and Tseng, 1998). Work above and beyond requirements has been termed "organizational citizenship behavior" and is one of the most widely studied topics in organizational behavioral because of its positive implications for organizational productivity (Turnley et al., 2003). People are vastly more creative when they are working enthusiastically, and their productivity is likely to be immeasurably higher than that of workers who withhold their best efforts because of alienation.

Pride in work is exemplified in an ethnography of ironworkers. The worker responsible for maintaining the crane used to lift the heavy steel girders evidences great pride in his daily chores:

> Most oilers are nearly invisible, fueling and lubricating their rigs before the day begins for the rest of us, vanishing to God knows where during the bulk of the day, reappearing at 4:00 to preside over putting the rig to bed. Beane, however, was not of that stripe. He fussed over the crane like a stage mother, constantly wiping away puddles of oil or grease, touching up scratches with fresh paint, agonizing loudly whenever a load banged into the stick. (Cherry, 1974:166)

Jobs that provide self-actualization and encourage enthusiasm can also have positive consequences for workers outside the workplace. For example, self-actualizing work can result in greater community involvement and in improved relations with family members. Conversely, alienation arising at the workplace can color a worker's whole life. Alienated workers are more likely to abuse alcohol and drugs and to have troubled interpersonal relations outside the workplace, including divorce, spouse and child abuse, and social isolation (Mirowsky and Ross, 2003).

Absenteeism   Absenteeism is a highly visible response to alienating work. On any given day, 3 to 4 percent of the workforce does not show up for work (Ose, 2005). Each worker is absent between seven and twelve days a year, and only a relatively small part of this absenteeism is due to illness. In some industries the daily absenteeism rate runs as high as 10 percent to 20 percent.

Quitting   Quitting is also a common response to an unrewarding job. Turnover rates average about 15 percent per year for the labor force as a whole,

and for some industries they are much higher. For example, nearly half the nation's secretaries leave their jobs within two years, representing a 25 percent annual turnover rate (Vagg, Spielberger, and Wassala, 2002). In the hotel industry, and in other service industries that employ large numbers of part-time workers, turnover is estimated to be even higher. Quitting occurs most frequently where job rewards are lowest. If wages are high and workers have greater resources, such as scarce skills, they are more likely to form a union. In this way they seek to improve their jobs through collective action rather than leaving through quitting (Freeman and Rogers, 1999).

Resistance and Sabotage   Alienated workers who do show up for work are often on the lookout for ways to vent their frustrations. Where workers have no opportunity for redress of their grievances in consensual ways, the behavioral responses of sabotage and theft become increasingly attractive options.

The word *sabotage* originated in the 1400s in the Netherlands, where workers would throw their sabots (wooden shoes) into the wooden gears of the textile looms to break the cogs. The purpose of this activity was to force concessions from their employers. The Luddite movement of the early 1800s in England practiced organized sabotage as a bargaining tactic with employers to forestall the introduction of larger looms that were causing widespread unemployment (Randall and Charlesworth, 2000). Thorstein Veblen, one of the founding figures in American sociology, coined the phrase "conscious withdrawal of efficiency" to describe the phenomenon of sabotage (1921).

In the contemporary economy, sabotage can take a variety of forms. Lower-level administrative workers may intentionally slow up a bureaucracy by scrupulously sticking to regulations, by intentionally destroying or misplacing files, or by refusing to fill out required paperwork (Roscigno and Hodson, 2004). Factory workers may intentionally break machinery or use it in ways that hasten its breakdown. For example, a welder in an automobile factory reports the following events:

BOX 3.4   Refusing to Work on One's Knees

"I didn't do everything those folks told me to do. Some I did and some I didn't. They would tell me to get on my knees and scrub the floor, and I didn't do it. I didn't mess up my knees. I told one lady, 'My knees aren't made for scrubbing. My knees are made to bend and walk on.'" . . .

Ms. Ryder has no hint of servility or obsequiousness about her; on the contrary, she has a self-possessed quality that could be, one would imagine, threatening to some employers. Ms. Ryder enjoyed telling me stories of her resistance to exploitation. Her self-respect was sometimes expressed in overt feistiness: "I remember one Sunday morning, this woman told me to scrub her kitchen floor on my hands and knees. I got mad at her and said to her, 'You sit right down there and wait until I scrub it.' So I got a whole lot of ammonia and Clorox and the stuff with the twin kids on the box. And I just poured it over the floor. And then half wiped it up. And you know what it looked like when I got finished! She just looked at that floor. That floor looked bad for two or three days. I wouldn't wash it 'cause I told her I'd already scrubbed it. That floor was so bad I didn't even like to walk on it 'cause it was muddy and sticky. . . . She had wanted me to get on my knees and scrub it. And I wasn't *thinking* about getting on my knees and scrubbing it. And, after that, I could just mop it up and it would look nice. No, my knees weren't made for walking all over the floor!"

But if they make us work without any time off, then we wreck a gun and take a few minutes while it is being repaired. This happens all through the shop. Many times the guns could be easily repaired. A worker sees his gun going bad. He has no interest in saving it so he'll let it go completely wrong and burn clear up before calling the repairman. Many times we know what is wrong and if we feel good we repair it ourselves. The workers put things in their guns or break them on purpose. A worker was fired for this not long ago. Every time he got mad he would take his knife and cut the rubber hose. He would put something on it to make it look as if it had burst. (Denby, 1978:139–140)

Workers in some alienating situations thus routinely practice sabotage. One of the authors worked at a sawmill where workers frequently tore metal casings off machinery and threw them down the exit chute that took scrap wood out of the mill. The metal casings broke the teeth on the chipper that shredded the scrap wood, resulting in brief unscheduled breaks. Box 3.4 describes how a domestic worker resists what she perceives to be excessive demands by her employers.

Theft   Theft is also a common response to unaddressed grievances at the workplace. The number of employees engaged in some sort of workplace theft has been estimated at 28 percent in manufacturing, 33 percent in the hospital industry, and 35 percent in retail (Shover and Wright, 2001). Common forms of theft include taking merchandise and tools, getting paid for more hours than were worked, purposely undercharging a friend or co-conspirator, and being reimbursed for more money than was spent on business expenses. Theft is especially common among young workers and is more common among men than women.

In some service jobs, theft is an expected part of the employment arrangement between worker and employer. The sociologist Elliot Liebow describes this situation for a dishwashing job held by one of the subjects in his study of street corner society:

Tonk's employer explained why he was paying Tonk $35 for a 55–60 hour workweek. These men will all steal, he said. Although he keeps close watch on Tonk, he estimates that Tonk steals from $35 to $40 a week. What he steals, when added to his regular earnings, brings his

---

BOX 3.5 The Pastry Scam

"The newcomer or trainee has a choice: he can accept the fiddle [scam] and by so doing implicitly recognize the whole landscape of fiddling; or he can reject it, and by so doing draw attention to an individual act of dishonesty.

There's one woman ... who runs the canteen at the tech. college. She orders trays of pies. Every day she orders three, every day she sells three, and for years she's always thought there were sixty pies to a tray when there should be seventy-two. This fiddle was passed on to me when I took the job on, and the bloke who showed me had it passed on to him when he started. It must have been going on for years because that woman had been there thirteen years when I took her on, and that was two years ago. The fiddle is possible

because she'll always sign for three trays with no mention on the voucher of the number of pies.

A salesman who could not accept this handed-down fiddle might well alert the customer to the fact that she had consistently been given thirty-six pies less than she should have been given every day for thirteen years. In doing so he might clear his own conscience though he would also, at the very least, lose his company a valuable customer. But it says much for the careful acclimatisation of newcomers, at least to this company, that the subterfuge should have lasted for so long without disruption."

SOURCE: Excerpted from Gerald Mars, *Cheats at Work*. London: Aldershot, pp. 121–122. Copyright 1994 by Gerald Mars. Reprinted by permission of the publisher.

---

take-home pay to $70 to $75 per week. The employer said he did not mind this because Tonk is worth that much to the business. But if he were to pay Tonk outright the full value of his labor, Tonk would still be stealing $35–$40 per week and this, he said, the business simply would not support. (Liebow, 1967:37)

Workplace theft has been estimated at about 1 percent of the gross national product (Shover and Wright, 2001). In spite of the magnitude of this figure, most employers prefer to look the other way, discounting theft as inventory or operating losses. For example, only a small minority of supervisors indicate that they would report a suspected pilferage to their boss if it were the first incident for the employee involved.

The English urban anthropologist Gerald Mars (1994) has developed a typology of workplace theft. The typology is based on closeness of supervision and on the presence or absence of a strong work group. Workers in closely supervised jobs with a weak work group ("donkey jobs"), such as cashiers, are likely to attempt to retake control of their jobs through stealing time and through pilfering small amounts of money. Workers in closely

supervised jobs with strong work groups ("wolfpack jobs") are more likely to engage in elaborate and systematic pilferage. In longshoring, such practices are called "lightening the cargo." Workers in weakly supervised jobs with strong group structures ("vulture jobs"), such as cab drivers and delivery workers, are often successful in organizing stable systems of pilferage, misaccounting of funds, or both. Box 3.5 gives Mars's account of pilferage in a "vulture job" involving pastry delivery. Finally, workers in weakly supervised jobs with a weak work group ("hawk jobs"), such as professionals and managers, are likely to abuse expense accounts and to spend company time and equipment on private endeavors. An example would be a stockbroker who uses privileged "insider" information to build his or her own investment portfolio. Similarly, a professor might use money from a research grant to buy a computer for a private consulting business. The more privileged of these workers may be motivated less by alienation than by simple greed.

Employees can be quite inventive in devising systems of pilferage. One of the authors worked for several years at a truck stop ("donkey job") where he observed an accounting scam involving several employees. The scam involved misreporting the license plate numbers of out-of-state trucks to the cashier in order to secure an "in-state" discount. The money from the discount was pocketed, and neither the cashier nor the out-of-state truck driver was the wiser. The scam lasted for about six months, at which point it was discovered. The originator of the scam, who had been doubling his take-home pay, was fired but the other practitioners were neither identified nor disciplined.

THE FUTURE OF JOB SATISFACTION

Marx argued that alienation reaches its zenith under capitalism and would be reduced under communism because worker's basic needs would be met, and the class division between capitalists (who control the work process) and the working class (who are controlled) would be eliminated. However, alienation did not appear to have been eliminated in socialist societies such as those in Eastern Europe (Burawoy, 2001). Marx correctly perceived that alienation is a fundamental characteristic of capitalist society; however, the replacement of the capitalist class by state directors did not eliminate alienation. Extreme division of labor, hierarchical forms of organization, centralized control, bureaucracy, and pressures toward economic growth typified socialist societies as well as capitalist ones and produced alienation from work regardless of the underlying economic system.

The elimination of alienation is a multifaceted problem entailing more than changes in control of society at the highest levels. It must also involve concrete opportunities to participate in day-to-day decisions about how work is organized. Without providing opportunities for meaningful participation at the workplace, industrial economies will not be able to eliminate alienation.

In North America the recent past has witnessed decreased satisfaction with work. Work could possibly become even more alienating in the future. Indeed, as people become more and more educated, alienation will tend to increase unless the quality of jobs also improves. In addition, low-level jobs in the service sector are expected to increase, and these are some of the least rewarding jobs. On the positive side, professional jobs are also expected to increase, and these jobs are typically more rewarding and satisfying.

The future of job satisfaction is difficult to predict under these circumstances. However, there are hopeful signs that we are learning how to create jobs that produce greater self-actualization while also increasing productivity. Industrial societies are evidencing increased interest in redesigning jobs to increase worker participation, involvement, and commitment. We discuss such strategies for organizational redesign in Chapter 17.

SUMMARY

Alienation at work arises from repetitive working conditions, lack of autonomy, large bureaucratic organizations, and blocked opportunity. Job attitudes depend not only on alienating conditions but also on individual differences among workers. Prior expectations based on educational attainment and comparison groups constitute important criteria that individuals use to evaluate the quality of their jobs.

Workers' attitudes toward their jobs are based on complex sets of evaluations, involving both intrinsic and extrinsic rewards of the job. Substantial evidence exists showing these attitudes can vary independently so that workers can be simultaneously satisfied and dissatisfied with their jobs. Workers respond to their job conditions in various ways. Most typically these responses involve greater or lesser enthusiasm for work, but they can also involve absenteeism, turnover, and even sabotage and theft in order to redress felt grievances.

Contrary to Marx's prediction, alienating working conditions are not a temporary situation that will automatically pass as economies grow and mature. Levels of job satisfaction in the future are difficult to predict. Expectations will continue to rise because of increased education. Whether the quality of jobs will improve to a commensurate degree is an unanswered question.

KEY CONCEPTS

job satisfaction

alienation

self-actualization

job autonomy

continuous-process technology

inverted-U curve of technology and alienation

bullying

job stress

role strain

role conflict

comparison group

job-worker fit

intrinsic rewards

extrinsic rewards

commitment

sabotage

QUESTIONS FOR THOUGHT

1. What job factors are needed for self-actualization? What jobs have you held that have been most rewarding?

2. Have you ever been bullied, or seen others bullied at work? What were the consequences? What can be done to prevent bullying at work?

3. What creative behavioral strategies have people devised to cope with their jobs at places where you have worked?

4. How is Marx's theory of alienation useful for understanding the rewards and frustrations of work in modern society? What are the theory's limitations?

5. Outline a future scenario resulting in greater self-actualization for workers. Outline a future scenario resulting in greater alienation.

MULTIMEDIA RESOURCES

Print

Melissa Everett. 1999. *Making a Living While Making a Difference: A Guide to Creating Careers with a Conscience*. Gabriola Island, British Columbia: New Society. A great guide to meaningful alternative careers.

Mary Lizabeth Gatta. 2001. *Juggling Food and Feelings: Emotional Balance in the Workplace*. Lanham, MD:

Lexington. A lively account of how fast-food workers reconcile low-wage work with the expectation that they always act pleasant and cheerful with customers.

Ben Hamper. 1991. *Rivethead: Tales from the Assembly Line*. New York: Warner. A sometimes bawdy account of one worker's struggle to maintain his sense of individuality and self-respect in an automobile assembly plant.

Tom Juravich. 1985. *Chaos on the Shop Floor*. Philadelphia: Temple University. A theoretically informed and well-written account of the author's experiences working in a wiring harness factory.

Melvin L. Kohn. 1990. *Social Structure and Self-Direction*. Cambridge, MA: Blackwell. The leading scholarly account of the effects of working conditions on intelligence, creativity, open-mindedness, and other aspects of cognitive functioning.

Gerald Mars. 1994. *Cheats at Work*. London: Aldershot. A fascinating account of the variety of scams practiced at work.

Internet

Take Back Your Time. *www.simpleliving.net/timeday* Strategies for taking back control of our lives from employers and mindless consumerism.

Social Science Information Gateway. *sosig.ac.uk* An information clearinghouse for practitioners and researchers in the social sciences, business, and law in the United Kingdom.

Institute of Labor and Industrial Relations, University of Michigan. *www.ilir.umich.edu* High-quality research on a wide range of workplace issues.

RECOMMENDED FILM

Office Space (1999). Ron Livingston as a disaffected computer programmer whose bad attitude is read as organizational genius by his supervisors. Promotions aside, he is busy plotting revenge.

4

Class, Race, and Gender

...[P]eople set up systems of social closure, exclusion, and control.
Multiple parties—not all of them powerful, some of them even victims
of exploitation—then acquire stakes in those solutions.
(TILLY, 1998:8)

One of sociology's major accomplishments in the last quarter of the
twentieth century was establishing that race and gender matter at
work. We have been far less successful in explaining why workers' sex
and race affect their employment outcomes, however, especially why
jobs are segregated by sex and race, and why whites outearn people of
color and men outearn women.
(RESKIN, 2000:707)

Sociologists use the term **master social status** to refer to a person's social position
that affects many of the opportunities available to that person. In this chapter
we investigate how three examples of master status—class, race, and gender—affect
the opportunities available in the labor market. We will also examine specific
difficulties of discrimination and harassment experienced by some workers.

SOCIAL CLASS

Social classes are groups within society defined by possession of different amounts of wealth and power. Social class is a major indicator of inequality. In his classical discussion of capitalism, Karl Marx argued that there would ultimately be two classes: a small upper class that controlled the means of production (that is, wealth-producing assets) and a large class of workers, or proletariat, that worked for wages. Max Weber argued that in addition to the means of production—or economic power— social class was also defined by political power and inherited social status. Weber's work implies that an advanced industrial society could also support a larger or smaller group of people in the middle class who would control some power. Weber's analysis also implies that the class structure of a society is somewhat dynamic and is a function of both political and economic structures. We will discuss the upper, middle, and lower classes in advanced industrial societies to simplify the conversation, while recognizing that class structures vary from country to country.

The upper classes are characterized by a great deal of economic power in terms of owning land, businesses, and other wealth-producing assets. The possession of *wealth* (everything one owns) usually results in a large *income* (the amount of money one receives in a given time period). The owner of a large firm, for example, may receive a large salary from the firm as well as profits. If the firm is publicly held and issues stock, then the shareholders may also receive dividends in addition to owning shares of stock that have value and can be resold. The possession of wealth also provides access to political power and social influence. Members of the upper class are typically in an excellent position to pass on many advantages to their children, including education, powerful social networks, and the inheritance of wealth. Their children, in turn, will typically have a number of possible job opportunities available and substantial economic security even if they do not work.

The lower classes, by contrast, have very little economic security because they have little or no cash savings and own very few assets. They also typically possess few specialized job skills. The term **underclass** refers to those in the lower class who have access only to the most insecure and poorly paid jobs. Lower-class income is dependent upon finding employment, which might not include fringe benefits such as health care, and the quality of health care and education they have available for their children may depend upon the quality of available government services. If lower-class workers are laid off or injured, the continued well-being of their family may be in jeopardy. The lower classes may have few advantages to offer children, and their children in turn may face the same insecure job opportunities their parents faced. The political power of the lower classes is limited because politicians are aware that lower-class people are less likely to register to vote and to actually vote if registered.

The *middle class* refers to people who are neither wealthy nor poor and who have some measure of economic security either through ownership of some tangible assets, or through the possession of a valuable and marketable skill, or through access to important social networks. A labor union, for example, can be a social network that enables a worker to command high enough income to assure middle-class status. In democratic societies the middle class also possesses some measure of political power. An important challenge for the middle class is the ability to pass on to their children sufficient education or training to maintain a middle-class position in the next generation (Ehrenreich, 1989; Leicht and Fitzgerald, 2007).

Why Does Social Inequality Matter So Much?

Why are economic outcomes so different for people in different occupational and social positions? Two broad answers are possible to this question: a *functionalist* answer and a *conflict* answer. The functionalist answer stresses the useful outcomes deriving from inequality. Inequality motivates people to strive and to perform at their best (Turner, 2001). China, during the Cultural Revolution, experimented with completely ending pay differentials between skilled and unskilled workers, and managers made employment decisions based on the assumption that skills were irrelevant, and anyone could be assigned to any task with a minimum of training. The results were economically disastrous with sustained declines in productivity and manufacturing output (Lynch, 2004). The functionalist answer thus appears to have much to recommend it—differential rewards appear necessary for the effective functioning of a modern economy with widely differentiated tasks.

Proponents of the conflict theory would counter that yes, some inequality may be necessary, but this does little to explain why in industrially advanced societies those in top positions receive *so much more* than those in lesser positions—sometimes many thousands of times more (see Table 4.1 on executive salaries). Conflict theorists argue that the answer lies in differential power (Bartos and Wehr, 2002). The salaries of chief executives (and managers more generally) are set by other chief executives sitting on boards of directors who can set salaries at astronomical levels. The salaries of those at the top are thus protected from market forces that drive the salaries of ordinary employees to the lowest possible level. In a sense, both the functionalists and the conflict theorists are right: some inequality is required but not nearly as much as exists. In the following sections we explore some useful concepts for understanding this complex reality: *skills, technology, markets,* and *power.* We also explore in greater detail the actual degree of inequality evidenced in society and some of the key trends in inequality.

Skills and Technology   The demand for different skills and occupational specialties is determined by the mix of technologies used to produce goods and services (see Chapter 7 on Technology and Organization). Thus, making steel in the 1900s required a large number of workers strong enough to shovel mounds of coke into open blast furnaces. Making steel in the 2000s demands fewer workers, but workers trained in reading dials, gauges, and computer readouts and who punch buttons or type commands to add fuel or other components to new oxygen-fueled furnaces are needed. The skills are obviously different. And the latter workers *seem* more skilled in the sense that they need greater training to do the tasks. But would they really be paid more? The answer based on skills and technologies is important but is incomplete until we also consider other factors. Chief among these is the availability of workers with the relevant skills and the demand for the product—factors commonly called *labor markets* and *product markets.*

Labor and Product Markets   Do the steel workers shoveling coke or the ones pushing buttons make more money? Both are absolutely necessary for making steel, each in their own time. An important part of the answer lies in the availability of other people to do equivalent work at the same or a lower price. If many people have the requisite skills, the work will not demand a high wage. Thus, if there are many workers strong enough to

shovel coke, the work will pay only as much as is necessary to get the most desperate workers to take on the task. If there are few workers, perhaps because there are other or better paying jobs available, then the job will pay more. Thus, wages are set not just by skills but also by the availability of people who have these skills. If many people have the requisite training to read dials and type simple computer commands and few alternative jobs are available, then the modern steel worker may make little more, relatively speaking, than the steel laborer of one hundred years ago.

Similarly, if the demand for steel is high, then companies will seek to increase output and will try to hire more workers. To do so, they may well have to pay a premium, such as higher wages or a hiring bonus, to lure workers away from other employment, and wages for steel workers will be high. If the demand for steel is low, companies may lay off workers, who then become potential competitors for the remaining jobs, putting a downward pressure on wages. Thus, skills are important, but only inasmuch as they operate within situations defined by labor market supply and product market demand to set the actual wages for a given job.

In the contemporary global economy (see Chapter 16) labor and product markets have become global in their reach. This means that workers now compete with workers on the other side of the globe for employment. If a factory uses technology requiring skills widely available in another country at a lower wage rate, then the owning company may well move the factory. Only if the technologies demand skills that are rare or available only locally, is the factory likely to remain in place.

The rapid pace of change in contemporary society suggests that conditions of stability are likely to be short-lived and that workers from different locations around the globe are competing with each other. Further, rising education levels around the world mean that global competition for jobs increasingly involves highly skilled workers, such as engineers and scientists, and not just semiskilled assembly workers. The consequence is a world labor market characterized by constant pressure to increase skills and output and restrain or even lower costs.

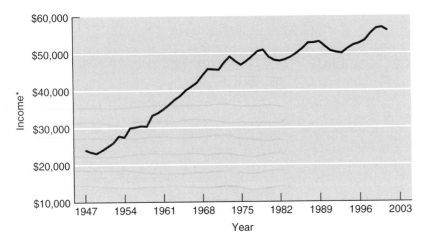

FIGURE 4.1   Median Family Income, 1947–2003

SOURCE: U.S. Census.
*2005 constant dollars.

The consequences of increasingly competitive labor markets for the incomes of American families are displayed in Figure 4.1. This figure reports the median income of American families across the last sixty years. The incomes are in constant dollars to control for the effects of inflation. The figure tells a dramatic story well worth reviewing. In the quarter century between 1947 and 1974, the real incomes of American families doubled. This epoch-defining outcome was a result of American dominance in world product markets following the Second World War. Most of the industrial capacity of the other industrialized nations of the world (Japan, Germany, Great Britain, and the other European nations) had been destroyed by war. The industrial capacity of the United States, in contrast, had rapidly expanded during the war as its plants churned out the war materials to achieve allied victory. Following the war, this industrial capacity was redirected from tanks and trucks to automobiles and refrigerators and from uniforms to consumer clothes, and the United States became the core supplier of manufactured goods to the world for over two decades.

By the mid-1970s, however, France, Germany, Japan, and Great Britain had rebuilt their industrial capacity, and many other more recently industrializing nations were also producing manufacturing goods. Often these goods, including those from recently industrializing nations, competed well with American-made goods because of cheaper prices based on cheaper labor. As the quality of these goods improved, they competed even better and took over a larger and larger share of the world market and even the American market. With shrinking demand for American-made goods, wages stagnated in the United States. This condition has prevailed for the last three decades. Future scenarios debate whether the stagnation will continue or become a sustained downturn. Few realistic scenarios suggest a return to a sustained growth of real wages in the United States typical of the third quarter of the twentieth century. There are simply too many nations and workers competing with American workers for this to happen, and these workers are increasingly skilled and educated.

Power Versus Markets   The image of rising and then stagnant wages for the average American worker is only part of the story. Another important part of the story is presented in Figure 4.2. This

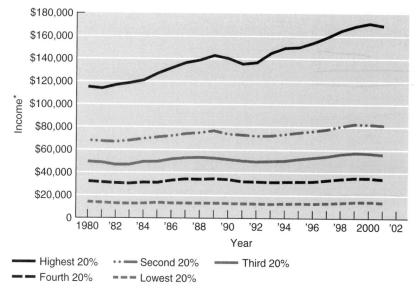

FIGURE 4.2 Family Income Trends by Quintile, 1980–2005
SOURCE: U.S. Census.
*2005 constant dollars.

figure displays the trends in median family income separately for quintiles of the income distribution for the period from 1980 to 2005—the period of general earnings stagnation. In many ways the figure is quite shocking. What it shows is that the earnings of the bottom four quintiles have been broadly stagnant, but the income of top earners has actually increased handsomely during this period of supposed stagnation—by something in the order of 50 percent. In general, the better-off groups have improved while those earning less have suffered. How did this happen?

The answer lies in the ability of privileged groups to shelter themselves from market forces and the inability of other groups to do so. As a result, the largest share of any income gains for American workers for the last third of a century has been bottled up by the highest earners. These workers, chiefly managers and professionals, have been successful in protecting themselves from the sort of market forces that have battered the wages of other Americans. They have protected themselves from labor market competition by restrict-

ing access to the labor market for their occupations. An example is provided by medical doctors who have achieved a chronic labor shortage for doctors by restricting admissions to medical school, thus driving up wages for medical services. In addition, the nature of the product market for many professional services is not easily globalized. For example, medical services are not as easily globalized as apparel manufacturing because those in need of medical attention do not necessarily want to travel extensively to seek medical services. Thus, less flexible product markets for professional services and high barriers to entry into professional labor markets have combined to keep many professional salaries high and rising. Similarly, managers have taken advantage of their positions of power in wage setting to protect themselves from market forces and have simply administratively given themselves high wages. Thus, professionals and managers have often been *price makers* who set the prices for their labor, while most other workers are *price takers* who must accept whatever wages are offered.

"Employment at Will"   Other sources of differential power also amplify inequalities between privileged and less privileged workers. In the United States, workers without a written employment contract can generally be fired for good cause, bad cause, or no cause at all. This principle is called *"employment at will"* and represents one of the most significant and enduring bases for differential power between employees and employers in the United States. Over 60 percent of private sector workers are "at will employees" (Werhane et al, 2004).

There are two major exceptions to the legal principle of employment at will. First, professional and managerial employees are often covered by explicit employment contracts spelling out the conditions that would entail appropriate grounds for dismissal. Employees with such contracts cannot be fired unless they violate these standards of behavior or performance. Employees without such contracts can be fired for any reason, even their supervisor's whim, and have no legal recourse. The second major exception to employment at will is the set of protections provided by Equal Employment Opportunity laws, which are described in greater detail later in this chapter. These laws prevent employers from firing workers or otherwise discriminating against them (such as in pay or promotions) based on their race, ethnicity, or gender. These laws have provided an important exception to the "employment at will" doctrine, especially as the coverage of protected groups has expanded and as even majority workers increasingly make claims of discrimination based on their social status.

Outside of these two important exceptions, however, the doctrine of employment at will has provided a legal basis for the arbitrary use of employer power to discipline, punish, and intimidate employees and creates a situation of chronic employment insecurity for workers across a wide range of occupations and employment settings.

Executive Wages   An additional vantage point on inequality is provided by executive wages. Table 4.1 presents the compensation of the twenty highest-paid executives in the United States. These executives earn between $132 million and $1.6 billion per year. A large part of these compensation packages is in the form of stock options and other perks that are often tax free or are taxed at a reduced rate. The highest of these compensation packages is equivalent to the annual earnings of over 58,000 workers.

Some wealthy individuals have attained their status by hard work and individual achievement. For many others, however, wealth is largely inherited. Individuals tend to be born into the same income quintile they occupy throughout life. Average persistence across generations within wealth quintiles is estimated at about 40 percent with persistence being greatest at the ends of the continuum: 45 percent of the poorest quintile were born into their positions, and 55 percent of the richest quintile were born there. Although these numbers also show significant mobility, the largest portion of those who move do so only a short distance. For all quintiles, at least 75 percent of people were born within one quintile of their adult position (Keister, 2006: Table 2.10).

Consumption patterns are even more persistent across generations than wealth, with estimates ranging as high as 90 percent persistence within quintiles (Aughinbaugh, 2000). These latter estimates suggest that not only wealth, but also houses, cars, vacations, and other consumer goods are passed on directly between generations and that these material inheritances constitute a significant component of inherited wealth. The lives of those who are born to wealth typically include significant contact with the world of work. The meaning of that contact, however, is likely to be significantly different without the pressure of economic survival.

Occupational Differences   Although inheritance is important for the wealthy, for most people in contemporary society, the greatest part of their livelihood comes from their own employment. Occupational income differentials are thus a crucial determinant of inequality. Table 4.2 reports average incomes for men and women within the major occupational groups. Managers and professionals have the greatest earnings, with all other occupations

TABLE 4.1 The 20 Highest Paid Executives in U.S. Publicly Held Corporations $

Company	Industry	Chief Executive	Compensation	Equivalent worker salaries
UnitedHealth Group	Health Care	William W. McGuire	$1,600,000,000	58,862
Oracle	Internet	Lawrence J. Ellison	474,900,000	17,471
Capital One Financial	Finance	Richard D. Fairbank	266,600,000	9,808
Toll Brothers	Residential Construction	Robert I. Toll	256,700,000	9,444
Caremark R$_x$	Health Care	Edwin M. Crawford	248,700,000	9,149
Vornado Realty Trust	Real Estate	Steven Roth	219,600,000	8,079
Yahoo!	Internet	Terry S. Semel	214,600,000	7,895
Countrywide Financial	Finance	Angelo R. Mozilo	204,000,000	7,505
Occidental Petroleum	Energy	Ray R. Irani	182,100,000	6,699
Aetna	Insurance	Ronald A. Williams	174,400,000	6,416
FedEx	Transportation	Frederick W. Smith	169,900,000	6,250
IAC/Interactive	Internet	Barry Diller	167,000,000	6,144
ConocoPhillips	Energy	James J. Mulva	165,200,000	6,078
eBay	Internet	Margaret C. Whitman	165,100,000	6,074
TD Ameritrade	Brokerage	Joseph H. Moglia	158,700,000	5,838
Gilead Sciences	Pharmaceuticals	John C. Martin	147,800,000	5,437
Marriott	Hotels and Restaurants	J. Williard Marriott	143,500,000	5,279
W.R. Berkley Corp.	Insurance	W.R. Berkley	141,500,000	5,206
United Technologies	Manufacturing	George A.L. David	138,100,000	5,081
Target	Retail Sales	Robert J. Ulrich	133,100,000	4,897
Median earnings of full-time workers in the United States (2006):			**$27,182**	

SOURCES: www.forbes.com 2006. "The $100 million CEO club." Also Bureau of the Census, 2006, *Statistical Abstract of the United States*. Washington, DC: U.S. Government Printing Office.

earning considerably less. Although this table is limited to workers who worked full-time and year-round, there is still a substantial gap between men and women, a topic to which we return later in this chapter.

The work situations of people across employment settings, in economic, technological, market, and human terms are the principal focus of the core sections III and IV of this book on industries, technologies, occupations, and professions.

TABLE 4.2 Average Occupational Incomes, by sex, 2004 (for full-time, year-round workers)

Occupational Group	Women	Men
Total, all occupations	$31,223	40,798
Management and business occupations	42,342	61,982
Professional and related occupations	41,367	57,833
Service occupations	30,113	26,226
Sales and office	28,115	40,301
Natural resources, construction maintenance	28,942	34,672
Production, transportation, materials moving	23,203	32,677
Armed forces	(a)	40,815

SOURCES: U.S. Census Bureau, Statistical Abstract, 2007, Table 631.

Notes: (a) data not shown.

Social Mobility

Social class plays a major role in matching workers to the best jobs, and a number of mechanisms reinforce the power of the elite. Nevertheless, there is still substantial *occupational mobility*, or movement among jobs. There are two kinds of occupational mobility. The first kind, **intragenerational social mobility**, also called *career mobility*, occurs when a worker moves from one employer to another, from one position to another, or even from one occupation to another. One longitudinal study showed that younger baby boomers (those born between 1957 and 1964) had held on average ten jobs by the time they reached age forty (BLS, 2006b). This job movement is often an effort to secure a better job.

Social class is relevant to this process of hiring and sorting. Workers acquire signals of social class, such as the way they speak and behave, as a result of their interactions in their homes and their education as children. These signals may subtly affect how suitable job applicants appear to the employers. Eventually, the worker's own job history becomes a sort of class-based signal to the employer and is used to slot job-changers into jobs. Thus, an able and ambitious worker may have opportunities to advance in each job move.

Intergenerational social mobility refers to workers' pursuit of an occupation different from that of their parents. A child of lower-class or middle-class parents who eventually becomes a corporate executive would be an example of upward intergenerational mobility. A child of well-educated parents who drops out of high school and pursues a series of low-paid jobs would typically be considered an example of downward intergenerational mobility. Wealthier parents are able to slow downward mobility of their children, but they cannot always prevent it. Large-scale shifts in intergenerational mobility, however, are typically not a function of the personalities of the workers, but of larger social changes in the types of jobs available. The **opportunity structure**—that is, the kinds of jobs available—sets the limits for occupational mobility.

Opportunity Structure   Much occupational mobility results from large-scale changes in the economy, in technology, and in the mix of goods and services produced (Blau and Duncan, 1967; Hauser and Featherman, 1977). A good example of intergenerational mobility emerges from the increased productivity of farms beginning in the late 1800s. All the children of farmers were no longer needed to produce crops and livestock. If only one or two adult children remained on the farm and used the new agricultural equipment, they could produce as much as ten workers did in their parents' generation. Simultaneously, in the cities the demand for factory labor grew. Farmers' children flocked to the cities to work in factories and foundries. By the time their children and grandchildren were ready to work, opportunities were opening in entirely new fields of communication, information technology, and so on. When many prestigious jobs are being created, more young workers can attain prestigious jobs. Most of the intergenerational occupational mobility in the United States in recent years may be attributed to shifts in the opportunity structure (Hauser and Featherman, 1977).

Occupational mobility (both intergenerational and intragenerational) is often conceptualized in terms of vertical distance. Upward mobility refers to movement into more prestigious, better paid, or more responsible positions. Upward intergenerational mobility occurs when an adult daughter or son enters an occupation of higher standing than that of the parents. Upward intragenerational mobility occurs when a worker is promoted to a more important or responsible job. The shifts in the opportunity structure within the advanced industrial countries have generally supported upward mobility by creating desirable jobs. As we shall see below, there are also forces creating less desirable jobs with the possibility for more downward mobility.

How Much Upward Mobility Is There? The United States has more upward intergenerational mobility than some other industrial countries and a little less than others. A study released in 1985 indicated that 41 percent of a sample of U.S. men had experienced upward mobility, compared with 44 percent of French and Austrian men (Haller et al., 1985:589). American men's mobility is greater than that of men in Italy (Pisati, 1997). Men and women in the United States have similar intergenerational mobility (Li and Singelmann, 1998). Canada also has relatively high mobility (Wanner and Hayes, 1996). One reason for this upward mobility is the continuing shift in the opportunity structure, so that there are more jobs calling for high levels of skill.

Workers may find that they are not competitive for the best jobs because of a lack of technological skills or education. Upward mobility may require additional preparation for advancement or promotion.

> [Sandra] worked for five years in a variety of low-level jobs in semiprofessional engineering work. At the end of that time she was working side-by-side with a man who had an engineering degree. "Of course he was getting two to three times more than I was. I realized we were doing the same thing. And I was thinking, 'If this yo-yo can do it, I am sure I can.' So I said, "I am going back to school." But she was worried that at thirty-five she was too old to be successful. . . . She told her advisor at the university, "Well, I'm thirty-five." He reassured her, "Oh, that is nothing. I went back when I was forty-two." (Dinnerstein, 1992:190–191)

North Americans value upward occupational mobility for their children. Many parents hope that their children will "do better than I have done and have it easier than I have had it." Nevertheless, the desire for upward mobility for their children may cause a dilemma for the parents. They can prepare their children only for the types of jobs they have known. In preindustrial societies a son or daughter could learn virtually every needed skill from a parent or relative, and occupational inheritance was commonplace. In industrial societies parents must work together with schools to provide formal training for their children. With rapidly changing technologies, many parents cannot even imagine the skills and occupations that might be available for their children. Many parents now find that they do not really understand the jobs held by their adult sons and daughters.

Downward mobility refers to movement into less prestigious, less well paid, or less responsible positions. Downward intergenerational mobility occurs when an adult son or daughter enters an occupation of lower standing than that of the parents. Downward intragenerational mobility may occur as a result of a disabling condition, formal demotion, layoff, or the "bumping" of high-level workers to lower-level jobs instead of laying them off. Firm downsizing beginning in the 1980s led to substantial amounts of downward mobility.

> When David Patterson's boss left frantic messages with the secretary, asking him to stay late one Friday afternoon, his stomach began to flutter. Only the previous week David had pored over the company's financial statements. Things weren't looking too good, but it never occurred to him that the crisis would reach his level. He was, after all, the director of an entire division, a position he had been promoted to only two years before. But when David saw the pained look on the boss's face, he knew his head had found its way to the chopping block. He was given four weeks of severance pay. (Newman, 1988:1–2)

The social class and social mobility differences we have described so far in this chapter are further compounded by differences based on ascribed characteristics such as race, ethnicity, and gender. Segregation, discrimination, and differential treatment present special employment problems for minorities and women that are discussed in the following sections.

RACE

In North America and in many other parts of the world, race or skin color is a visible and permanent master social status. The **cultural division of labor** refers to the historical or cultural pattern of assigning certain jobs to certain groups. Race or skin color, and some related statuses such as nationality and language, have been part of the cultural division of labor. Throughout most of the history of the United States, African Americans, Hispanics, and other minorities have been restricted to low-paying agricultural, factory, and service jobs (Wilson, 1997). Both enslavement and then legally required segregation provided legal limits on the types of jobs that blacks could hold. Under the system of Jim Crow in the post-Reconstruction South, for example, black school

teachers could teach only in the schools limited to black children, and this legal segregation in turn affected the salaries and promotions they could expect. Black construction workers were excluded from apprenticeship programs in the skilled trades. There were some predominantly black unions, such as the Brotherhood of Sleeping Car Porters, and in some cases black and white workers constituted separate locals of the same union. Box 4.1 discusses such an example.

During the Civil Rights Movement of the 1960s and 1970s, lawful discrimination was legally overturned, and a number of efforts were undertaken to ensure equal opportunity in hiring and promotions, as well as in other areas of life. Intense competition for the best jobs and the remaining legacy of the cultural division of labor, however, have limited the progress made by black workers. Appendix Table 1 indicates the proportion of various occupations held by black workers.

Indians in the United States and First Nations people in Canada, besides being visibly members of minority groups, have also experienced restrictions in the labor market, some of them as a result of legal restrictions and some of them as a result of limited educational and job opportunities.

Immigrants have traditionally been recruited into unskilled physical labor in both agriculture and manufacturing. The children of these immigrants have usually experienced upward mobility, often assisted by their greater ability to speak English. This process seemed to work best, however, for those groups that were visibly most similar to the native-born majority—white Americans. Visibly identifiable groups—especially those with darker skins—may continue to experience the cultural division of labor even beyond the immigrant generation. Asian workers who entered the United States in the 1800s were initially limited to laboring jobs and officially excluded from some types of work. Asian immigration was legally restricted between 1880 and 1965. Puerto Ricans on the East Coast, Mexican Americans in the Southwest, and Latino immigrants often experienced difficulty in finding employment outside traditional areas of agriculture and poorly paid service work. Appendix

BOX 4.1 Racially Divided Union Locals on the Gulf Coast

There were two longshoremen's locals in Corpus Christi—both International Longshoremen's Association, to be sure—one composed of white men and the other of black men. The white local was ILA Local No. 1224 and the black was 1225. "White and colored" was the expression used to describe them. The work was done by solid gangs of white and black. Each local had its own hiring place and conducted its own business. There was an arrangement where the one local would work the forward hatches of one ship, then the aft (or "after") hatches of the next, and so on. The usual freighter of the day was a five-hatch vessel with three

hatches forward of the smokestack. If a ship having four or six hatches was to be worked, the hatches were split two and two or three and three.

The two locals would hold a joint meeting now and then, but there was never any arrangement for scheduled meetings of both memberships. This segregation lasted until the year 1983, when a federal court decreed that black and white locals having the same work jurisdiction in the same port should merge into one local union.

SOURCE: Gilbert Mers, 1988, *Working the Waterfront: The Ups and Downs of a Rebel Longshoreman.* Austin: University of Texas Press, p. 11.

Table 1 also shows the proportions of occupations held by Asian-Americans and by Hispanic Americans.

GENDER

Women have historically been restricted to clerical and service work and selected professional areas such as nursing and teaching. The cultural division of labor channeled men and women toward different occupational roles. For women, the cultural division of labor is further reinforced by the assumption that women will carry the main responsibility for child care and housework (Parcel, 1999)—an assumption deeply ingrained in social, family, and workplace relations. One important manifestation of the cultural division of labor for men and women is the gender-typing of occupations into jobs thought most appropriate for men, such as construction and engineering, or most appropriate for women, such as secretarial work, nursing, and primary school teaching. Gender-typing of employment has thus channeled women into a restricted set of relatively poorly paid occupations, at least in part because they have been predominately filled by women (Reskin and Padavic, 1994; Charles and Grusky, 2004). Appendix Table 1 shows the representation of women in a large variety of occupations.

Employers may assume that women, as a group, do not possess the interests and abilities to do jobs characteristically filled by men, simply because these jobs have historically been filled by men. Box 4.2 shows that gender differences in work are widespread throughout the world.

HOW CLASS, RACE, AND GENDER INTERACT

Social class, race, and gender interact in ways that affect employment opportunities. Because of the historic association of race or ethnicity with poorly paid and insecure jobs, members of minority groups are proportionately more likely to be in the lower classes and to have lower levels of economic and political power. Race and gender may interact so that minority women are limited not only to jobs considered appropriate for women, but also to jobs that are considered appropriate for members of their minority group (Smith, 2002; Browne and Misra, 2003).

Occupational steering is the channeling of women or minorities into jobs considered "appropriate" for them under the prevailing ideas of the

BOX 4.2 Working Women around the World

The level of female labor force participation is highly dependent upon cultural attitudes. In the traditional Muslim culture predominating in Western Asia and North Africa, women are discouraged or prohibited from leaving the safety and sanctity of their homes to work for others. In Egypt less than 10 percent of women are engaged in labor force activities according to official statistics. In sub-Saharan Africa and in some Asian countries, women traditionally are involved in agriculture and market activities, and labor force participation is much higher. In Latin America female activity has risen sharply since 1950. About 20 percent of Latin American women of working age were in the labor force at the beginning of the 1990s, compared with 33 percent in South Asia and sub-Saharan Africa [Figure A].

Politics also plays a role. In many socialist countries, such as China and Vietnam, women are expected to contribute equally with men. Therefore, female labor force rates are high.

Some of the regional differences in female labor force participation reflect variations in labor statistics. Where culture frowns upon women being involved in economic activity, people underreport it in labor surveys and censuses. This exclusion occurs most often in agricultural societies in which the lines between household chores and economic activity may be blurred. Women may consider raising chickens or vegetables part of their housework, but if they sell or trade the proceeds of their work, their labor is "economic activity" under most definitions. Agricultural work is often seasonal, which can affect statistical reporting. A 1976 labor survey taken during Indonesia's busy harvesting season estimated the female agricultural labor force at 14 million, while one conducted the same year, but during the slack period, estimated it at 10 million. Cultural attitudes, along with misunderstandings or disagreements about what constitutes economic activity, cause an undercount of female labor force participation in developing countries. A careful labor survey in Egypt, for example, estimated that women made up 42 percent of agricultural workers in 1960, not the 4 percent reported in the census.

The shift toward more productive industrial and service sectors is decreasing, not increasing, the female share of the labor force in some developing countries. Despite the gains in female education and lower fertility, which were associated with increased female employment in industrial countries, women comprise

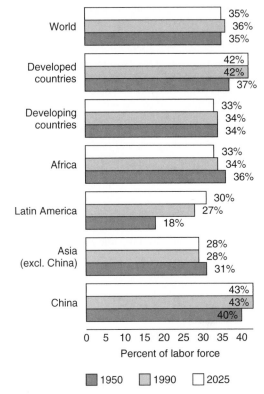

FIGURE A Female Share of the Labor Force in Major World Regions, 1950, 1990, and 2025

SOURCE: International Labour Office (ILO), 1986, Economically Active Populations. Geneva: ILO.

only about 24 percent of the non-agricultural workforce in developing countries. Between 1950 and 1985, the female share of the labor force has remained at 34 to 35 percent.

As the agricultural share of an economy shrinks, the share of women in the labor force may also decline. Further education of women increases their chances of employment in the non-agricultural sectors of the economy only if there are sufficient employment opportunities for both sexes.

SOURCE: David E. Bloom and Adi Brender. 1993. "Labor and the Emerging World Economy." *Population Bulletin* 48, 2 (October): 8–9.

BOX 4.3 Occupational Steering

Malcolm X (born Malcolm Little) was an American black leader during the early 1960s. In his influential Autobiography of Malcolm X (Little, 1965:36) he describes an example of such "steering" that occurred while he was a high school student in Lansing, Michigan:

[Mr. Ostrowski, my English teacher] probably meant well in what he happened to advise me that day. I doubt that he meant any harm. It was just in his nature as an American white man. I was one of his top students, one of the school's top students—but all he could see for me was the kind of future "in your place" that almost all white people see for black people.

He told me, "Malcolm, you ought to be thinking about a career. Have you been giving it thought?"

The truth is, I hadn't. I never have figured out why I told him, "Well, yes, sir, I've been thinking I'd like to be a lawyer." Lansing certainly had no Negro lawyers—or doctors either—in those days, to hold up an image I might have aspired to. All I really

knew for certain was that a lawyer didn't wash dishes, as I was doing.

Mr. Ostrowski looked surprised, I remember, and leaned back in his chair and clasped his hands behind his head. He kind of half-smiled and said, "Malcolm, one of life's first needs is for us to be realistic. Don't misunderstand me, now. We all here like you, you know that. But you've got to be realistic about being a nigger. A lawyer—that's no realistic goal for a nigger. You need to think about something you *can* be. You're good with your hands—making things. Everybody admires your carpentry shop work. Why don't you plan on carpentry? People like you as a person—you'd get all kinds of work."

Because this event made Malcolm X aware of how his race seemed to be more important than his achievements, he referred to it as the "first major turning point" in his life.

SOURCE: Excerpt and quotations from Malcolm X, with Alex Haley, *The Autobiography of Malcolm X*, p. 36. Copyright © 1965 by Alex Haley and Betty Shabazz. Reprinted by permission of Random House, Inc.

cultural division of labor. Occupational steering may be based on gender, class, race, or ethnicity. Little girls, for example, are sometimes encouraged to play with "appropriate" toys or to take music or art courses instead of science and mathematics. Eventually, girls may be steered into a fairly narrow range of occupations considered suitable for women. Occupational steering may be directed toward working-class students because middle-class teachers and counselors may hold stereotyped ideas of their abilities and interests. For example, students from working- or lower-class backgrounds may be encouraged to take vocational-education programs instead of college-preparatory programs, even if their academic abilities would predict success in college. Box 4.3 shows an example of the occupational steering of Malcolm Little—later known as the civil rights leader Malcolm X—that was based upon both race and class.

There are several ways to counter occupational steering. One important method is to provide *role models,* people whose occupational roles become an example for others. As Box 4.3 indicates, Malcolm X did not know any black lawyers or doctors "to hold up an image I might have aspired to." As more minority group members and women enter nontraditional occupations, they become a group of potential role models for younger workers. Other ways to overcome occupational steering include better training of counselors and teachers, providing better information about careers, and making students and parents more aware of the possibly restrictive consequences of occupational steering.

Occupational steering may be kindly intentioned as a way to spare young people the painful experience of discrimination. A better approach, however, is to eliminate the discrimination itself. The section below discusses hiring discrimination and the efforts to eliminate it. In a subsequent section we discuss progress and shortfalls in the struggle for equal pay and promotions once a female or minority worker has been hired.

DISCRIMINATION IN HIRING

In recent years women and members of racial and ethnic minorities have made important gains in employment in the United States and Canada. Legal actions brought by female and minority employees against companies that have discriminated in hiring and promotions have been one important tool in promoting these changes. Multimillion-dollar settlements were secured against American Telephone and Telegraph Company in the 1970s, General Electric in the 1980s, and Ford Motor Company in the 1990s to resolve histories of past discrimination. These settlements typically include restitution to the individual parties damaged by the discrimination as well as a commitment that a certain proportion of new hires will be women or minorities as dictated by the specifics of each situation (Powell, 1999).

These highly visible settlements between large companies and female and minority workers have helped reduce employment discrimination. Unfortunately, employers can still discriminate in many ways that are hard to eradicate. This section explores the gains made by women and minorities as well as the continuing problems they face in gaining equal access to good jobs.

Equal Rights Legislation

The legal basis for the elimination of employment discrimination is contained in Title VII of the U.S. Civil Rights Act of 1964:

> It shall be an unlawful employment practice for an employer (1) to fail or refuse to hire or to discharge any individual or otherwise to discriminate against any individual with respect to his compensation, terms, conditions, or privileges of employment, because of such an individual's race, color, religion, sex, or national origin; or (2) to limit, segregate, or classify his employees or applicants for employment in any way which would deprive or tend to deprive any individual of employment opportunities or otherwise adversely affect his status as an employee, because of such individual's race, color, religion, sex, or national origin.

Title VII made it illegal for employers to discriminate against women and minority workers in hiring and promotions and thus outlawed the deliberate **occupational segregation** of female and minority workers into restricted sets of job positions. The act also established the Equal Employment Opportunity Commission to enforce the law. Although written primarily with minorities in mind, the law has had far-reaching implications for female workers as well. It has also fueled a general expansion of due process in the workplace (Edelman, 1990).

Height and Weight Requirements From the standpoint of women, one of the most noteworthy legal changes resulting from this legislation has been the removal of formal height and weight requirements as employment criteria if the requirements are irrelevant to performing the job. A minimum height requirement of five feet two inches would exclude about one-third of women in the United States while excluding only about 1 percent of men. A minimum weight requirement of 120 pounds would exclude about 22 percent of women but only about 2 percent of men (Wallace, 1982:6). Employment standards including such provisions have been successfully challenged as women increasingly enter such nontraditional jobs as prison guard, firefighter, and police officer.

Affirmative Action United States government policy is to seek to remedy the consequences of past discriminatory employment practices. Various executive orders calling for **affirmative action** attempt to compensate for past discrimination through hiring goals, preferential consideration among otherwise equal candidates, or active recruitment of women or minority workers. The most important government statement supporting affirmative action is Executive Order 11246, issued by President Lyndon B. Johnson in 1967. It mandates the Federal Office of Contract Compliance to issue government purchasing contracts only to organizations that are making efforts to remedy the effects of past discrimination. This mandate has had far-reaching implications because many companies sell at least some of their products to the federal government. Thus, defense contractors, utility companies, computer and electronics manufacturers, and many other businesses have been required to develop affirmative action hiring plans. Such plans can give preference to qualified minorities or women if the purpose of the plan is to erase "a manifest imbalance in traditionally segregated job categories."

Affirmative action plans can be instituted voluntarily by an employer or jointly by an employer and a union. Alternatively, a plan can come about as a result of a discrimination lawsuit. The majority of affirmative action plans have been adopted in large firms with predominately white-collar labor forces. Such plans have the potential to redistribute some desirable jobs to previously excluded female and minority workers. Many of the plans, however, have been extremely modest in conception and have called for only the most minimal adjustments necessitated by law (Reskin, 1998). In addition, the lack of such plans in smaller firms, where the most blatant forms of discrimination exist, means that minority workers continue to face significant limits on employment in these sectors. It is thus a mistake to assume that the enactment of equal rights legislation automatically eliminated discrimination.

Continuing segregation is evidenced in the different occupational profiles of men and women and whites and minorities. Appendix Table 1 shows the representation of women, African Americans, and Hispanics across detailed occupations. African Americans are underrepresented in managerial, professional, and craft occupations. They are overrepresented in service, machine operative, and laboring jobs. Hispanics are also underrepresented in managerial and professional occupations. They are overrepresented in laboring and agricultural jobs. Women are underre-

presented in skilled craft jobs, though not in professional jobs because of the large number of nurses, teachers, and other *female-typed* professions that have long been identified as appropriate for women. Women are overrepresented in clerical and service work.

Affirmative action has had positive consequences for some female and minority workers and has helped break down sexual and racial hiring barriers (Tomaskovic-Devey, 1993). These gains have become apparent as women and minorities have secured jobs in a variety of settings. In addition, equal rights legislation has encouraged the creation of decentralized state, county, municipal, and organizational affirmative action programs. The integration of women into many professional and managerial occupations has thus significantly increased, although the representation of women in the ranks of chief executive officers (CEOs) of major corporations is still very low (Powell, 1999).

Affirmative action programs have also had a positive impact on the economy as a whole through allowing talented individuals to serve in positions from which they would previously have been excluded. Such individuals bring new skills to their positions and valuable new insights based on their unique social backgrounds. Increased workforce diversity thus has the potential to encourage the economic growth that benefits all workers (DiTomaso and Hooijberg, 1996).

An additional, and largely unintended, consequence of affirmative action has been that hiring and promotion decisions are made with increasing openness and rigor. Job openings are more widely advertised than in the past, and these openings are also posted for current employees to consider. Personnel officers use more explicit hiring criteria than in the past, and they frequently employ detailed rating systems for ranking qualified applicants. Such hiring systems displace previous systems that included a greater role for cronyism, favoritism, and nepotism. Rigorous hiring and promotion systems favoring merit over other characteristics are thus advantageous for all individuals, whether they are minority members, majority members, men, or women (Dobbin et al., 1993).

Affirmative action has not, however, been a cure-all for the problems faced by minorities and women in the United States. For instance, companies have put much effort into hiring African Americans for highly visible positions. As a result, the wages of black college graduates have risen faster than those of white college graduates (Mishel et al., 2005). Meanwhile, the wages of black high school graduates have fallen even further behind those of their white counterparts. Affirmative action has thus helped create a black middle class, but it has done little to help the large black underclass (Wilson, 1997). These problems have been aggravated by the concentration of African Americans and Hispanics in regions and urban areas with high unemployment rates (DeFreitas, 1993).

A White Backlash? Affirmative action has sparked resistance based on the belief that increased opportunities for minorities and women must come at the expense of opportunities for whites and men. Thirty years of affirmative action have clearly not eradicated the inequalities resulting from three hundred years of legal restrictions on blacks in America or thousands of years of gender-based differentiation. From the vantage point of a white male job applicant facing a racially or gender-based hiring goal, however, such considerations may seem secondary to their own needs. The increasing employment challenges faced by all Americans in recent decades have also tended to make whites less sympathetic to the problems of minority workers. Consequently, support for affirmative action has fallen among whites, and even among some blacks (Sears, Sidanius, and Bobo, 2000). Some whites assume that every employed black they encounter owes his or her job to federal pressure rather than to personal qualifications and hard work (Kluegel and Bobo, 1993). Such assumptions can be very unfair to minority workers who have struggled and sacrificed to secure their education or training. Affirmative action policies are also increasingly being challenged by referendum and in court. As a result of these legal challenges, such plans increasingly focus on efforts to *actively recruit* qualified minority or female applicants, rather than on quotas or hiring goals.

The likelihood of further progress toward greater equality of employment opportunity remains uncertain. Such progress depends to a significant extent on the political climate of the next decades. Box 4.4 depicts a factory setting in the United Kingdom in which cooperation between the races has become the norm.

Continuing Forms of Hiring Discrimination

Many aspects of discrimination in hiring are subtle and difficult to eliminate through legislation. One such form of discrimination arises as a result of informal, word-of-mouth methods of recruitment. For example, access to apprenticeship programs in the skilled trades often occurs through personal contacts among union members and their immediate friends and families (Fernandez et al., 2000). Such continuing forms of segregation help maintain dual labor markets in which women and minorities are sorted into lower-paying jobs and white males have greater access to higher-paying and more secure jobs (Nelson and Bridges, 1999).

Subjective Screening Criteria Subjective screening criteria also play an important role in eliminating job applicants with personal characteristics that differ from those of employers or current employees. Such criteria are rarely spelled out, leaving ample room for evaluators' subjective impressions of "intelligence," "appearance," "self-confidence," and "vigor." These impressions are easily biased by the evaluators' prior stereotyped notions about women, minorities, or social class. Indeed, standards may even develop that are shaped to match the characteristics of majority group candidates and to exclude women and minorities. Such standards may include hair and clothing styles or aspects of personal demeanor.

Professional jobs in previously male-dominated fields pose special problems of access to women. The role performance expected in these jobs often includes behavioral styles characteristically associated with men. Women in these jobs face a dilemma. They can either conform to the expected role behavior of the job and appear unfeminine (which may make their male colleagues, as well as themselves,

BOX 4.4   Cooperation in a Racially Diverse Workforce

There were eight Irish women on my line, six West Indians, and me. Most of them had worked at Universal Mechanical and Electrical Components for years and knew each other well. It was known as a friendly line, despite some major conflicts—the West Indians said it was friendly because there were so many of them. They addressed each other as "ladies" and referred to themselves collectively as "girls." ... The Irish women were all under thirty, except for the two who were deaf and dumb—they were in their early forties. The black women were older and most had grown-up children.

The cooperation between the women on our line made it more efficient. If someone in front had forgotten to put a small clip or peg on an electrical product, we would shout to them to send one along the line and attach it ourselves, or take it up to them to put right. What we were supposed to do, however, was to put out the part as a reject, marking down the fault both on it and on a sheet of paper. Then the reject operator would collect it, mend it, and return it to where it had left the line, also marking down the fault twice. Of course, all that took much longer....

In the wages and personnel offices, practically all the clerical workers were English, including some young black women. It didn't seem to matter what color the clerical workers were so long as they'd gone to school here and reached whatever standard of spoken and written "Queen's English" the personnel office required.

SOURCE: Ruth Cavendish. 1982. *Women on the Line*. London: Routledge & Kegan Paul, p. 18.

---

uncomfortable), or they can follow the expected role behavior of their gender (in which case they may be acting out of character for their occupational role). This dilemma creates a catch-22 situation in which women have limited options for achieving acceptance and recognition. For example, the American social scientist Barbara Deckard conducted a survey of male lawyers and found about half saying that women lawyers were "tough and masculine." The other half characterized women lawyers as too "weak and feminine" and said that many go to law school only to "catch a man" (Deckard, 1983:128).

Statistical Discrimination   Women and minority workers may also suffer from **statistical discrimination**. This sort of discrimination arises when an employer bases decisions on the *average qualifications* of a group or entire category of people, rather than on an individual's qualifications. Thus, an employer may choose a white high school graduate over a Hispanic graduate because of a belief that white high school graduates *on average* are better qualified. A highly qualified Hispanic job candidate may thus suffer discrimination based on stereotypes about Hispanic job candidates as a group. Statistical discrimination need entail no prejudice on the part of the employer, only a willingness to attribute a group's average characteristics to every member of that group. When past discrimination has been pervasive, as in the case of African Americans and Hispanics, many group characteristics (such as education, training, and work histories) will on average favor the majority group. Real disparities between groups in average characteristics make statistical discrimination particularly hard to eradicate. Statistical discrimination is especially likely in situations where employers do not take the trouble to carefully assess applicants' qualifications on a case-by-case basis.

Gender-Typing of Jobs   Female workers face additional hurdles, based on the gender-typing of jobs, that make occupational segregation into lower-paying and less secure jobs particularly acute for them (Browne, 1999). Occupations involving helping and serving others, such as nursing, have been identified with the female role because women have historically been responsible for these activities

T A B L E 4.3   Percentage of Workers Paid at or Below the Minimum Wage, by Sex and Age Group, 2005

	Paid at $5.15/hour	Paid below $5.15/hour
Men, aged 16 and over	0.5%	1.2%
Aged 16–24	1.6	2.7
Aged 25 and older	0.2	0.8
Women, aged 16 and older	0.8	2.5
Aged 16–24	1.9	6.1
Aged 25 and older	0.5	1.5

SOURCES: U.S. Census, 2007. *Statistical Abstract of the United States, 2007.* Washington, DC: Government Printing Office. Table 634, p. 417. The federal minimum wage in 2005 was $5.15 an hour.

within the family. Some occupations are almost exclusively female, such as secretaries (97%), receptionists (92%), dental hygienists (97%), and childcare workers (95%) (BLS, 2006a). Unfortunately, such occupations are often relatively poorly paid, and their incumbents are treated as subordinate— a subordination that parallels the traditionally subordinate role of wives in relation to husbands. In other female-typed occupations, such as waitress or receptionist, physical appearance may be more important than acquired skills and competencies. Not only is physical beauty fleeting, it does not accumulate over time as a package of skills, experiences, and abilities necessary for promotion to more responsible and more highly paying positions (Williams, 1995).

By contrast, women compose only a very small percentage of the skilled trades: carpenters (1.9%), brick masons (0.9%), electricians (2.6%), and drywall installers (0.8%). On the other hand, they fill 25 percent of lower-paid assembly positions. Table 4.3 shows a comparison of the proportions of men and women employed at or below the federal minimum wage. Among women aged sixteen to twenty-four, over 6 percent were employed below the federal minimum wage, compared with 2.7 percent of men the same age.

Female professionals face special problems of gender-typing. One problem results from the channeling of women into a relatively narrow range of professions: nursing, teaching, library work, social work, and other "helping" professions. As a result there is substantial overcrowding in these professions, which leads to lower wages (Blau et al., 1998). The fact that many female professionals marry male professionals and managers also creates the possibility that geographic moves will be made that facilitate the husband's career rather than that of the wife. Such decisions may be rational to the extent that the husband's job pays more than the wife's. The result, however, may be a cumulative erosion of the wife's career trajectory, which makes future decisions even more likely to be weighted toward the husband's career.

Gender differences in occupational placement are not well explained by differing educational qualifications. Male and female workers both have an average of 12.7 years of education. Women exceed men in two-year and four-year degrees, but men are more likely to complete advanced degrees, a difference that is rapidly being eroded (Buchmann and DiPrete, 2006). Women have obviously prepared themselves educationally for the world of work, both in the past and increasingly in the present, but they still experience restricted employment opportunities. Why does gender-based occupational segregation persist in the face of continuing pressures for change?

One reason that gender-based occupational segregation persists is due to discrimination. Women face additional obstacles, however, that arise outside the workplace. The first obstacle is childhood socialization that encourages women to aspire only to certain types of careers. The second is marriage, which results in additional constraints on women's career options, constraints imposed both by husbands and by wives themselves. Even harder to overcome are the constraints arising from bearing and raising children and the assumption that women will take primary responsibility for these activities.

Recent years have also seen increased attention given to the rights of sexual minorities in the workplace. Employment discrimination against people on the basis of sexual orientation was not prohibited by the original Civil Rights Act of 1964. Instead, homosexuals are protected by a hodgepodge of federal, state, and local laws that leave many loopholes. Gays and lesbians have also had difficulty getting their partners covered under health benefits and retirement packages available to married heterosexual couples. Laws protecting homosexuals and recognizing (or prohibiting) homosexual marriages vary among states, creating many ambiguities and difficulties for gays and for gay couples in the workplace. In recent years, however, many large corporations have started to extend health coverage to "significant others," regardless of sexual orientation or marital status (Raeburn, 2002).

DISCRIMINATION IN PAY AND PROMOTIONS

Equal access to promotions and pay remains an issue for female and minority workers. Although some progress has been made in opening up access to previously closed occupations, the ratios of black-to-white earnings and female-to-male earnings remain very unequal, even in recent decades. The median family income for blacks remains only about 61 percent of that for whites. Full-time female workers make about 75 percent of the earnings of full-time male workers. In 2005 college-educated women who were between thirty-six and forty-five years old earned only 74.7 cents in hourly pay for every dollar that men in the same group did. This was a decline of one cent per hour from 1995 (Leonhardt, 2006). In this section we explore the reasons why such differences in pay and promotions persist.

Racial Discrimination

Racial discrimination in pay and promotions occurs in a number of ways. Decisions about promotions are sometimes made informally, behind closed doors by an executive group composed predominantly or exclusively of majority whites. Subjective recommendations by immediate superiors play an important role in such decisions, and there are rarely explicit, written standards for such evaluations (Fernandez and Davis, 1999). Because executives feel more comfortable trusting people with whom they have much in common, they are more likely to decide in favor of white candidates over minority candidates. Decisions made in this way rely on stereotypes of minorities and fail to give proper weight to more objective criteria. Such selection becomes more likely the higher one goes in the authority hierarchy. As a result, minority gains in the professions, management, and the skilled trades have been particularly slow (Baldi and McBrier, 1997).

Overt expressions of racism have declined in the United States (Kluegel and Bobo, 1993). More subtle everyday expressions of racism, however, are still common. These include efforts to marginalize minorities, to identify minorities as the cause of social problems, and to reject complaints by minorities about prejudice and discrimination as invalid (Essed, 1991). Equal treatment for minorities often occurs only when those within an organization actively demand such treatment (Kalev, Dobbin, and Kelly, 2006).

Prior discriminatory practices may also mean that minority workers have less seniority than majority group workers. As a result, even when employers use a relatively objective criterion such as seniority for allocating promotions and layoffs, minority workers will still often be at a disadvantage (Barnett et al., 2000). Thus, the courts have sometimes mandated the use of separate seniority queues for female or minority workers to allocate promotions or layoffs.

As a result of these continuing problems, the position of African Americans in the United States has improved only moderately, if at all, in recent decades. Black family income grew 1.3 percent per year in the closing decades of the twentieth century, while white family income grew 1.1 percent per year, thus reducing the relative income gap. However, because black income was much lower initially, the absolute gap in family income actually widened during this period (Massey, 1996).

Gender Discrimination

Households headed by women are about 62 percent more likely to live in poverty than households headed by men, and nearly four times more likely to be in poverty than married-couple families (Dalaker, 2005). What factors account for such a large difference, and how are these similar to or different from the factors involved in racial inequality? Some phenomena, such as statistical discrimination, subjective hiring criteria, and restrictions on training and apprenticeship programs, affect women as well as minorities.

Tokenism A problem for both female and minority workers is that they are highly visible representatives of their group when they enter new occupational fields (Kanter, 1977). In such situations of **tokenism** a worker is under a spotlight and may experience exaggerated pressure to overachieve. There may also be open hostility from majority workers who feel their position threatened by the incursion of "lower-status" workers (Williams, 1987).

In traditional female occupations, by contrast, the consequences of discrimination against women are sometimes referred to as a **glass ceiling**—an image that depicts clerical workers being shut out of opportunities to move into male-dominated managerial and professional positions in large modern corporations (Maume, 1999). Several unique barriers also confront female workers. These barriers result from conflicts between women's roles in the family and in the workplace.

Home Duties Many women are expected, and expect themselves, to take principal responsibility for home duties, such as cooking, cleaning, shopping, arranging social engagements, and most importantly, child care. These duties take a tremendous amount of time and energy, leaving less time available for career commitments (Parcel, 1999). Even if a woman does not plan to shoulder these burdens, those in charge of pay and promotions may fear that she will eventually "get married and quit" or "have children and be less willing to give her full commitment to the company" (Benokraitis and Feagin, 1986). Similar stereotypes do not hinder the promotion of men on the basis of possible marriage or family plans. Indeed, marriage plans by a man may be viewed as exerting a favorable influence on his "stability and commitment" (Kanter, 1977:676).

Pregnancy Leave Many women experience difficulties in arranging their work duties around pregnancy. The Pregnancy Leave Act, enacted in 1993, required employers to allow female workers to take a six-week unpaid pregnancy leave. Sometimes women can continue to receive pay during this period under sick leave policies. Most sick leave policies, however, do not cover the six-week minimum recommended convalescence for a normal delivery. As a result many women risk substantial loss of pay when they give birth—a time when extra bills are sure to accumulate (Waldfogel, 1997).

Comparable Worth The Equal Pay Act of 1964 made it legally possible to challenge situations in which a minority worker is paid less for doing the *same job*. It has been much more difficult to resolve the issue of *artificially depressed* wages in an occupation because the work has typically been done by female or minority workers. This issue is particularly important for women because of the extreme occupational segregation they have experienced.

The idea that jobs of equal value or contribution should be paid equally is called **comparable worth**.

Comparable worth discrimination occurs when companies base their pay on existing market rates for jobs, and such market rates build in discrepancies in pay between jobs that have typically been filled by men and those that have typically been filled by women. These discrepancies may have little to do with the skill required in the job. The concept of comparable worth is not really new or revolutionary. Job evaluations establishing the comparable worth of different positions are widely used by large enterprises in both the public and private sectors. "Job evaluation consists of a formal set of procedures for hierarchically ordering jobs on the basis of their relative skill, effort, responsibility, and working conditions for the purpose of establishing relative pay rates" (Hartmann, Roos, and Treiman, 1985:5). The criteria used may include the education or intelligence required, on-the-job training, verbal aptitude, numerical aptitude, complexity of the task, perceptual skills, dexterity, motor coordination, physical strength, and such social skills as speaking persuasively, supervising, instructing, negotiating, and mentoring. About 30 percent of the difference in pay between men and women would be eliminated if such skills were consistently used in determining pay rates for men and women (England, 1992).

The legal basis for challenging comparable worth discrimination has been debated in a series of court cases dating back to the 1970s and continuing through the 2000s. In 1983 the proponents of comparable worth achieved a major victory when a U.S. Federal District Judge found "overwhelming" evidence that "intentional and pervasive" discrimination had caused jobs held mostly by women to be paid an average of 20 percent less than jobs held by men in Washington State public sector employment (Walsh, 1985). The total settlement for the fifteen thousand state employees involved approached $1 billion, including back pay and benefits. This and subsequent cases have been important in forcing employers to reexamine their pay scales and eliminate blatant differences between predominantly male and female jobs. Some of the results from the job evaluation study that established the existence of comparable worth discrimination in the Washington State case are presented in Table 4.4.

Legal debates on the implementation of comparable worth standards continue to this day and involve two primary issues. First, the criteria used to establish job comparability can be endlessly debated. Second, employers can contend that they are merely offering the *prevailing wage* for a particular job. The latter argument, which places the responsibility for inequality outside any particular employer, has sometimes been accepted as part of a successful defense in comparable worth cases. However, it fails to answer the question of how market rates originally got that way (England, 1992). Proponents of comparable worth argue that the market rates embody past discrimination. They argue that comparable worth discrimination should be determined by discrepancies between the measured value or skill of an occupation and its pay, not by disparities between a specific employer's pay scale and prevailing market rates for the work.

The future of comparable worth is uncertain. The courts have not established a clear precedent for ruling in its favor. In addition, there is a movement toward demanding evidence of *intent* to discriminate. This makes defenses based on the concept of prevailing market wages increasingly acceptable. The U.S. Congress seems unlikely to pass additional legislation on this topic, especially given the failure of the Equal Rights Amendment. In their role as employers, local governments have for the most part taken a wait-and-see approach, because it is both easier and less costly than addressing the problem directly. On the other hand, unions representing substantial numbers of women in public employment can be expected to continue as mainstays in the campaign for comparable worth. Lawsuits based on comparable worth discrimination will continue. These lawsuits are at least partly responsible for the improvement in women's relative wages evidenced in recent years (Guthrie and Roth, 1999).

Sexual Harassment

A final problem confronting women in the workplace is their treatment as sexual objects. This problem can range from embarrassing jokes and banter to overt propositions and demands for sex. The Equal Employment Opportunity Commission (EEOC)

T A B L E 4.4 Job Evaluation Points and Gender Differences in Pay (Selected results of the Washington State comparable worth study)

Job Title	Job Evaluation Points	Average Annual Salary, 1983–1984	
		Male Dominated	Female Dominated
Warehouse worker	97	$17,030	
Delivery truck driver	97	19,367	
Laundry worker	105		$12,276
Telephone operator	118		11,770
Data-entry operator	125		13,051
Intermediate clerk typist	129		12,161
Civil engineering technician	133	18,796	
Library technician	152		13,963
Licensed practical nurse	173		14,069
Auto mechanic	175	22,236	
Maintenance carpenter	197	22,870	
Secretary	197		14,857
Chemist	277	25,625	
Civil engineer	287	25,115	
Senior computer systems analyst	324	24,019	
Registered nurse	348		20,954
Librarian	353		21,969

SOURCES: Helen Remick, 1984, "Comparable Worth and Wages: Economic Equity for Women." Manoa, Hawaii: Industrial Relations Center, University of Hawaii at Manoa, Table 3. Reprinted by permission of the Industrial Relations Center, University of Hawaii at Manoa.

defines **sexual harassment** as repeated, unwelcome behavior with a sexual content when submission to such behavior is explicitly or implicitly a condition for the person's hiring or for other employment decisions, or when such behavior creates a hostile, intimidating, or offensive working environment.

How common is sexual harassment in the workplace? Studies of sexual harassment report that, across their careers, as many as a fifth of women have experienced a severe form of sexual harassment, such as sexual blackmail (demanding sex in order to get a job or a promotion). As many as half have experienced moderate forms, such as sexual propositions, verbal innuendoes, or degrading remarks based on gender (Fletcher, 1999). The case of a young file clerk provides a typical example of such "moderate"

forms of harassment. Her supervisor regularly asked her to come into his office "to tell [her] about the intimate details of his marriage and to ask what [she] thought about different sexual positions" (Benokraitis and Feagin, 1986).

Based on her study of an underground coal mine, Yount (1991) developed a typology of three strategies that female coal miners used to confront sexual harassment. "Ladies" sought to cast their co-workers as gentlemen and socially withdrew when they confronted offensive behavior. "Flirts" interacted with men in ways considered to be seductive. This style encouraged come-ons from men and sometimes resulted in more severe harassment if the men perceived that the women were using this strategy to gain preferential treatment. "Tomboys" emphasized their occupational roles as miners and engaged in joking relationships with their male co-workers. Tomboys experienced a great deal of sexual "razzing," but this was often intended to be friendly and inclusionary. Tomboys, who resisted being placed in traditional female roles, seemed to be the most successful at minimizing harassment while being accepted as capable workers. In contrast, the adaptations of Flirt and Lady, which relied on traditional female role behaviors, provoked either active harassment or reduced respect.

Women in low-status occupations who are in regular contact with men in higher-status occupations (such as secretaries in relation to managers) are the most likely targets for sexual harassment. Women in these relatively powerless positions also have the fewest options in responding. They are less likely than women in higher-status jobs to respond assertively and more likely to attempt to placate the person harassing them. Similarly, those women perceived as less powerful are more likely than other women to be harassed (Wilson, 2000). Single and divorced women are more likely to experience harassment than married women. Younger women and minority women are also likely to be targeted (Fletcher, 1999).

Often there are shades of gray between normal, acceptable warmth and sexual harassment. When does an arm around the shoulder cross the line between a show of affection and support and a come-on? When does an off-color joke go beyond harmless banter to become offensive and derogatory? Often the difference lies in a power differential between the parties involved. Greater awareness of the seriousness of sexual harassment and its ramifications for creating a hostile and unproductive work environment may make it possible to enjoy a friendly and supportive workplace in which sexual pressure and innuendo do not make some workers uncomfortable.

MANAGING THE DIVERSE WORKFORCE OF THE 2000s

The workforce of North America in the twenty-first century is more diverse than at any point in history. This diversity represents not only a challenge but also a potential asset. Diverse people bring different contributions to the workplace. In addition, our long history of attempting to absorb and treat fairly people of different cultural backgrounds can become a social foundation for dealing successfully with people of other nations and cultures in an increasingly interconnected world economy. Capitalizing on this potential, however, will not necessarily be easy or automatic and will require sustained effort on the part of all of us (Fernandez and Davis, 1999; Pelled et al., 1999). The existing forms of interaction among classes, races, and genders form a critical challenge to workplace harmony and equity.

SUMMARY

Class, race, and gender are important master social statuses affecting many aspects of everyday life, including work. Good jobs provide not only the means for a comfortable lifestyle but also the means for preparing one's children for their jobs. In this way, social class affects multiple generations. In addition, the opportunity structure, or the relative availability of good jobs, affects the extent to which members of one social class can hope to have their children move to a different class.

Discrimination, by limiting the access that minority groups have to good jobs, can restrict minority group members to the lower classes. Discrimination based on race and gender, while illegal, still exists, although today it may be subtle and difficult to identify. Occupational steering reinforces the segregation of women and minorities into a small set of jobs. Efforts to overcome discrimination, such as affirmative action and comparable worth, are themselves controversial, even though they have contributed to fairer methods of evaluation and promotion to all workers. Finally, because of their social, family, and home roles, women may encounter additional issues of occupational segregation and sexual harassment.

KEY CONCEPTS

master social status

social class

underclass

intragenerational social mobility

intergenerational social mobility

opportunity structure

cultural division of labor

occupational steering

occupational segregation

affirmative action

statistical discrimination

tokenism

glass ceiling

comparable worth

sexual harassment

QUESTIONS FOR THOUGHT

1. Identify the key mechanisms through which discrimination in hiring, pay, and promotions persists. Which of these mechanisms do you think is the most important, and why?

2. Is it possible for employers to practice a sort of statistical discrimination based on social class? How might this happen?

3. Race and gender are social statuses that enjoy some legal protection. Social class does not enjoy such legal protection. What are the reasons for treating social class differently from race and gender?

4. Have you ever experienced, or do you know someone who has experienced, sexual harassment? What changes would be required to reduce sexual harassment in the workplace?

5. Why is employment discrimination an issue for public policy?

MULTIMEDIA RESOURCES

Print

Gary N. Powell (editor). 1999. *Handbook of Gender and Work*. Thousand Oaks, CA: Sage. An extensive compilation on the relations between gender and work.

Barbara Reskin. 1998. *The Realities of Affirmative Action in Employment*. Washington, DC: American Sociological Association. A review of the real, but limited gains of African Americans resulting from affirmative action programs.

Jack Santiago. 1989. *Miles of Smiles, Years of Struggle: Stories of Black Pullman Porters*. Urbana: University of Illinois Press. A compelling history of the black Pullman porters and their union leader, A. Philip Randolph.

Internet

U.S. Equal Employment Opportunity Commission. www.eeoc.gov The official web site of the U.S. Equal Employment Opportunity Commission.

Statistics Canada http://www12.statcan.ca/english/census01/Products/Analytic/companion/paid/contents.cfm An analytic profile of Canada's work force based on the Canadian census.

RECOMMENDED FILM

The Double Burden (1992). A film by Marlene Booth that follows three generations of working mothers, one African American, one Mexican American, and one Polish American. This film shows how social class, race, and gender interact to affect working life.

5

Work and Family

> All the world's a stage,
> And all the men and women merely players.
> They have their exits and their entrances,
> And one man in his time plays many parts,
> His acts being seven ages.

Shakespeare, writing in *As You Like It* (Act II, Scene 7), poetically describes the life cycle, or stages, through which most humans pass. He enumerated the relevant stages in Elizabethan England from the infant and the school child all the way to "second childishness and mere oblivion." Today, although the stages have changed somewhat, we still can identify a distinctive life cycle both for individuals and for families. As both individuals and as family members, we associate some stages with work and others with preparation for work or with leisure.

One important transformation of human work has been its restriction to certain times, places, and ages. For most of human history "work" was what adults did with most of their time. Work shifts, careers, and retirement are relatively recent concepts that describe the packaging of work into hours, days, and years. Commuting came about as work was isolated into places separate from the home. Even people who work at home often designate part of their home as an office. As work became temporally and spatially separate, work time came to be separate from family time (Hochschild, 1997).

This transformation had many consequences. Instead of describing the activity of most adults, work came to mean employment, the activity undertaken for pay or profit. People who worked unpaid at home, especially homemakers, were no longer considered workers. As you saw in Chapter 2, the government does not count homemakers as members of the labor force. For people who were employed, integrating work into their lives and the lives of their families became important issues. This chapter will examine these issues within the context of the individual and family life cycles. It also describes challenges and problems that workers face at different stages in their life cycles, from their preparation for work to their retirement. We will also examine problems of integrating the family life cycle with work, including issues such as child care. This chapter also discusses some possible solutions to these problems.

THE LIFE-CYCLE PERSPECTIVE

The life cycle is the ordering of roles from infancy to death. Many tasks are age-related because they are considered appropriate to different stages of the life cycle. Rules called norms help us understand what is appropriate at each stage. **Norms** are social rules for thinking, feeling, or behaving, and the existence of norms creates expectations for individuals. It is normative in North America for middle-aged men to be steadily employed—it is the expected behavior. Norms are somewhat flexible depending on mediating circumstances. For example, the expectation of steady employment would be relaxed if a man became disabled or if he and his coworkers experienced a layoff.

Norms help define the roles people play. Sociologists understand a **role** as a set of behaviors associated with a particular position in society. All of us play such family roles as daughter, sister, parent, and spouse. Many of us also have work roles within the economy such as employer or employee. The work role may be further differentiated by occupation (cook, barber, machinist, secretary), by hierarchical position (apprentice, supervisor, manager, head nurse), and by permanence of employment (part-time, temporary, tenured). Most of us play multiple roles outside our work and family roles. We are citizens, consumers, and taxpayers; in addition, we may be licensed drivers, church or club members, and tennis players. Modern societies and urban living tend to multiply the number of roles each of us plays.

Individual Life Cycle

In modern industrial societies, as in Shakespeare's society, it is normative that people acquire some roles sequentially. We expect that children will be in school; that adolescents will stay in school, go to work, or join the military; that young adults will seek employment; and that at some point elderly persons will retire. These stages of the **individual life cycle** have economic, social, and psychological significance. Psychologist Erik H. Erikson described the challenges facing people during their life cycle. The resolution of each challenge shapes the personality, which in turn affects how the person faces the next challenge the life cycle brings. Box 5.1 describes how Erikson's analysis of the life cycle applies to work-related challenges.

When applied to work, the normative ordering of roles is sometimes called the **sequential life plan** (Best and Stern, 1977). In the sequential life plan, education comes first as a preparation for work; then come the working years, and finally the leisure years of retirement. The normative ordering is flexible, however. For example, a teenager may work part time while still in school. The sequential life plan has traditionally allowed women some variations to permit time for child rearing or for remaining a full-time homemaker. One challenge in modern work is the search for greater flexibility in the sequential life plan, a topic to which we will return.

B O X 5.1   The Life Cycle, Work, and Personality Development

Erik H. Erikson related the stages of life to the development of personality (Erikson, 1963:247–274). As the life cycle progresses, he argued, everyone encounters important challenges that can be resolved in one of two ways. One resolution leads to good emotional health, but the other resolution leads to poor emotional health and impedes the successful resolution of later challenges. Each challenge, he contended, is related to an important area of social life.

The first two challenges occur in early childhood and have little to do with work. The first challenge, "basic trust versus basic mistrust," refers to infants' eventual belief that others will provide for their needs, especially food, warmth, and comfort. Organized religion is the area of social life related to this stage. The child learning toilet training faces the second challenge, "autonomy versus shame and doubt." Law and order is the social area related to this stage.

The next challenges deal more explicitly with work. The third challenge, "initiative versus guilt," is related to the economy ethos. Erikson wrote of the child at this stage: "He is eager and able to make things cooperatively, to combine with other children for the purpose of constructing and planning. . . . He looks for opportunities where work-identification seems to promise a field of initiative" (1963:258).

The fourth stage, "industry versus inferiority," is also related to eventual work behaviors, and Erikson linked it to the technological ethos of the society. The child learns to win recognition by producing things, developing skills, and performing assigned tasks. Children at this stage learn about the tools and skills of their society, and a sense of inadequacy or inferiority will arise if they despair of learning to work successfully.

The fifth stage, "identity versus role confusion," usually occurs during adolescence. Young people at this stage become preoccupied with their peers' view of them and with the issue of how to connect the skills and tools they have learned with the existing occupational types. Erikson saw explicit work roles becoming important at this time: "In most instances . . . it is the inability to settle on an occupational identity which disturbs individual young people" (1963:262). The social areas related to this stage are what Erikson termed ideology and aristocracy, implying that young people at this age are concerned about whether the best people rule. A sociologist might describe the same issues as part of social stratification.

The sixth stage, "intimacy versus isolation," challenges the young adult to develop close relationships or, alternatively, to remain isolated and alone. The social area to which this stage corresponds is that of sexual selection, cooperation, and competition.

In the seventh stage, "generativity versus stagnation," mature adults discover their need to be needed. *Generativity* refers to the nurturing and guiding of the next generation, either through one's children or through one's work. Generativity includes, but is broader than, productivity or creativity. This challenge is related to child rearing, but also to work.

Erikson referred to the eighth stage as "ego integrity versus despair." This challenge can be met only by the person who has successfully met the previous seven challenges, who "has taken care of things and people and has adapted . . . to the triumphs and disappointments adherent to being the originator of others or the generator of products and ideas" (1963:268). At this stage the older adult assesses both personal relationships and contributions to work. In the integrity that comes from integrating one's life experiences, the human loses fear of death, the final stage of life.

The Career

Sociologists distinguish the individual life cycle from the career. The life cycle refers to a variety of events that occur within one's life, including work events, family-related events, and other age-related events. When sociologists speak of **career**, however, they refer specifically to the sequence of events within a person's work history. Sometimes people use the word career to distinguish a good job from a bad job. "I don't want just a job," they might say. "I want a career." From the sociological perspective, every worker has a career.

Just as we speak of the typical (or normative) life cycle, while admitting that there are many exceptions in the lives of real individuals, we also speak of the typical career. The typical career of a physician, for example, might include medical school, residency, the opening of a private practice, and eventual retirement. Sociologists term this an *orderly career,* in which job changes follow predictable patterns. Many people prefer having orderly careers, with steady advancement in responsibility and compensation.

As a particular physician experiences a career, however, it may have idiosyncratic or disorderly aspects, just as an individual's life cycle may not be normatively ordered. For example, the physician might become a hospital administrator or incur a disabling illness. This physician still has a career, but it might not follow the orderly career pattern. More recently, sociologists have identified the **boundary-less career,** a career that may be pursued with multiple employers and build on the experiences and learning in a variety of work contexts (Baker and Aldrich, 1996). Such careers are becoming more common as the employer-employee relationship comes to be seen as shorter and less secure. The worker does not necessarily choose such a career, but may be forced to adopt it.

The nature of a career appears to affect the worker's personality. Work that remains challenging and interesting stimulates workers and keeps them mentally flexible and optimistic. Some important elements of the challenge appear to be having a variety of things to do, some choice about when and how to perform tasks, and some complexity or conceptual difficulty to the tasks. The tasks do not need to be physically demanding or hard to do, but they do need to present new opportunities for learning and problem solving. *Autonomy,* or the ability to make some decisions about work performance and timing, is perhaps the most significant job characteristic in promoting healthy functioning of the worker (Langfred, 2000).

Nearly every job, even one that is normally challenging and interesting, has its aspects of drudgery and routine. Some jobs, however, offer little else, and workers receive no opportunity to learn or show initiative. Workers in such jobs become more cautious, conservative, and less flexible. One series of studies found that jobs with low requirements for skill and complexity tended to dull the workers' outlook (Kohn and Schooler, 1983; Kohn

et al., 1990). On the other hand, careful design of a job may keep workers interested and committed (Casey, 1996).

Such studies suggest a relationship between career and personality changes. Sociologists have paid particular attention to bureaucratic careers, because more jobs are now located in bureaucratic, hierarchical organizations. Bureaucracy will be discussed in more detail in Chapter 7. For now, it is enough to note that in a bureaucracy, each worker tends to have specific, specialized duties, a detailed set of rules to follow, and a supervisor. Communication in the entire organization is expected to go "through channels," that is, from supervisors to subordinates through several layers of supervision. Most bureaucracies have job ladders so that workers understand the possible promotions and can plan an orderly career in the organization.

Many young bureaucrats begin their jobs with enthusiasm and excitement. Some retain this enthusiasm and are promoted several times. Inevitably, however, there are fewer promotion opportunities than eligible workers, and so some workers eventually feel stuck in their jobs. The "bureaucratic personality" they may develop can lead them to strictly enforce all rules and to avoid any creativity in applying them (Merton, 1968). To an observer they may seem to be going through the motions of their jobs, and their goal changes from getting ahead to playing it safe. In times of downsizing, not even this tactic may be sufficient to ensure job security (Tivendell and Bourbonnais, 2000).

The Family Life Cycle

Besides the life cycle of each individual, there is also a **family life cycle**, which describes the stages of formation, growth, and dissolution of the nuclear family. A nuclear family, which consists of a couple or one or more parents with children, is the most common family structure. Important events in the family life cycle include family formation, birth of the first child, birth of the last child, departure of the last child from the home (the "empty nest"), retirement from work, and the death of one spouse. Divorce and remarriage are additional stages that may occur in the family life cycle (Hill, 1986; Spanier and Glick, 1980). Although there have been changes in the timing of these events—for example, the average age at marriage has risen in recent decades—studies in such diverse countries as Japan, India, Sweden, Belgium, and Canada show remarkable consistency in the phases of the family life cycle (Sussman, Steinmetz, and Peterson, 1999; Hill, 1986).

Three important points need to be made about the family life cycle. First, because of variations in family membership, the model does not describe all families. Childless couples omit the stages involving child rearing, and single-parent families may miss the stages involving an adult partner. Some families experience additional stages as the result of the temporary or permanent addition of family members, such as a grandparent who moves in or an adult child who returns home.

Second, the timing of the family life cycle is highly variable. A divorced parent who remarries and subsequently has children with the second spouse may recycle through earlier stages of the family cycle (Cherlin et al., 1991). The death or the divorce of a spouse may occur at any time after the marriage, regardless of the family "stage" with regard to children.

Third, different social and economic problems accompany various stages of the cycle. Getting married, becoming a parent, or facing the empty nest may all engender crises of adjustment. Each of them is also likely to be accompanied by changes in the financial situation of the family.

An important issue for sociologists of work is the joint sequencing of events in the individual life cycle with those in the family life cycle. Should you get married before completing your schooling? Should you be secure in your job before deciding to have children? How will having children affect a parent's career? If both the husband and wife are employed, how will the family react if one of them receives a job transfer to a distant city? In the remainder of this chapter we will discuss the interaction of the individual's work roles and family roles, the problems people experience in integrating them,

and some of the consequences of these opportunities and difficulties for workers.

SOCIALIZATION AND WORK

Sociologists use the term **socialization** to refer to the process of learning norms, roles, and skills. The teachers in this learning process are called *socializing agents*. Education, training, workplace orientations, and similar arrangements that are specific preparation for a role are called *formal socialization*. In formal socialization the typical socializing agents are educators at all levels and training personnel in the armed services, corporations, and elsewhere. Equally important, however, is the *informal socialization,* a more diffuse preparation for a role that goes on with parents, peers, or the mass media as the socializing agents. Sociologists study both kinds of socialization as a process that prepares children for work and helps workers learn their jobs or adjust to changes in their jobs.

Informal Socialization

Children are first socialized at home, and parents are very important socializing agents. Parents teach children language and basic living skills essential to the socialization that will occur later in school. Parents also model what it means to be a worker, how one goes about working, and the importance of work relative to other activities. Very young children play games of pretending to work, and adults reinforce the significance of work by asking children, "What do you want to be when you grow up?" Little girls may still sometimes receive the message that paid work is not as important for them:

> From the smallest age you don't picture yourself in a job. And if you talk to someone like my husband, he pictured himself in jobs. They were different jobs at different times but that's how you picture yourself. You know, this is what my life is supposed to be like—carry my little briefcase into work. (Dinnerstein, 1992:23)

Different parents teach their children different lessons about work. Sociologists have studied socialization in families of different social classes. Parents with manual jobs tend to have somewhat different child-rearing practices than parents whose jobs involve office work or intellectual work (Kohn et al., 1990; Sears, Maccoby, and Levin, 1957). Parents' own job experience, according to this view, indicates that some behaviors and personality traits are more important than others in being successful, and these are the behaviors the parents emphasize with children (Gecas, 1979; Kohn, 1976). For example, being independent and showing initiative may be very important in professional jobs, but being obedient and taking orders may be more important in manual work.

This informal socialization is most useful when children will take up the same work as the parent's own. When this happens, we say that there has been *occupational inheritance*. Occupational inheritance takes place through the inheritance of capital, as when a child inherits a parent's business, but it can also occur because the parent has prepared the child with the same skills and knowledge as the parent's own. Many children in the advanced industrial societies, however, will occupy jobs in new or rapidly changing industries and require skills their parents never had. For this reason, formal socialization is increasingly important.

Formal Socialization

In school, students' teachers and peers constitute the socializing agents. Schools provide training in literacy, calculation, communication, and other basic skills useful in any job. Even in elementary schools the formal curriculum includes instruction in potential occupational skills such as keyboarding. At advanced levels entire courses of study are devoted to explicit occupational training. Vocational agriculture, business training, and home economics are examples from the high school curriculum. In college some degrees qualify students for specific occupations, such as engineering or accounting, while others, such as the liberal arts, provide general expertise in problem solving and communication.

It would be a mistake, however, to view schools as preparation for work solely in terms of the curriculum. Schooling prepares the young person for life as a worker. For the young child the school represents the first separation of "workplace" from home (Dreeben, 1968). At home the child is accepted more or less unconditionally; failings in terms of one's skill or attitude are often overlooked or offset with compensating virtues. The long-lasting relationships within the family are stressed and nurtured. In school, by contrast, the child is evaluated for performance. Being charming does not compensate for inadequacies in math. The child learns to relate to others in terms of their positions: although the person who is the teacher may change, every teacher will expect similar behavior. In the same way, workers are evaluated principally for their job performance and relate to co-workers and supervisors primarily in terms of positions within the firm. Thus, the school takes children from a setting in which they are valued as individuals and prepares them for a setting in which they will be valued for productivity and competence. This transition may be very gradual in kindergarten and first grade, where the teachers may view themselves as surrogate parents as well as educators. But the process is far advanced by the time the student enters college. Relationships with professors, at least on many college campuses, are likely to be remote and formal, resembling relationships with supervisors in later jobs.

The hidden curriculum of the school conveys many other lessons. Students learn the importance of punctuality, orderliness, and learning and following rules. For example, students learn to raise their hands and wait for recognition before speaking, a behavior that persists in adult gatherings. Students learn to identify and respond appropriately to people with authority, and they come to accept as legitimate the right of authorities to assign tasks and to define how to complete them. Schools reflect the economy and the society that support them. Thus, students will learn an ideology, or way of explaining the existing economic and political patterns. They will also be exposed to any pattern of discrimination that characterizes their society.

Socialization in the Workplace

The workplace is an important socialization site. Supervisors and coworkers can teach many jobs to new workers. Dorinne Kondo describes learning a job in a Japanese bakery:

> The primary morning task was to wrap individual slices of cake with cellophane. Though seemingly simple, this task was hard to do without disturbing the frosting decorations, difficult to accomplish in one smooth, deft movement. My first day on the job, I remember the division chief, Akita-san, coming out from his room after I'd been wrapping for a half hour or so. "Relax your shoulders!" he said, with a concerned look on his face. "Don't worry—you'll get the hang of it before long." But it took a month before he could say to me, "Looks like you've gotten used to it. You're quicker, and you don't seem to get as tired as you did before." (Kondo, 1990:287)

An important difference among jobs is the extent to which the employer is willing to invest the time and cost to provide training. Some companies provide lengthy programs with on-site classroom instruction. For example, banks often offer training programs so that new employees will understand how the federal government regulates the banking system and how bank security is provided. One study indicates that 63 percent of young workers (aged 24 and under) received training provided by their employers. On average, employees received 13.4 hours of formal training and another 31 hours of informal training. Women, college graduates, professional and technical workers, and workers with five to ten years of employment in their current job were most likely to have received employer-provided training. (U.S. Census, 1999:439).

Other employers assume that the worker will know most of what is important before being hired. The employer may rely on a credential from a school or union to certify that the new worker has the required preparation. One observer has argued that the increase in women's labor force participation resulted from increased demand for jobs such as schoolteachers, nurses, social workers, and secretaries, jobs that were stereotypically held by women and for which workers received most of their training before applying for work (Oppenheimer, 1970). The employer's training responsibility for such jobs is limited to a basic orientation concerning the specific rules and procedures at the work site. As a result, the employer's costs are lower.

Apprenticeship is a type of formal socialization widely used in many countries, notably Germany. Canada has also used apprenticeships (Lehmann, 2005), but apprenticeships—typically associated with labor unions—are less common in the United States. Internships, often by students who are still enrolled in school, have become a more common way to gain formal work experience.

Socialization does not end when schooling ends, nor does it end even when formal on-the-job training ends. Coworkers and supervisors continually socialize workers. New electrician apprentices, for example, carry many tools of varying quality and spend time learning specialized terms. They eventually learn from the master electricians to carry only a few well-worn, high-quality tools, and they become completely familiar with the jargon and procedures of the craft (Riemer, 1977). Such socialization prepares workers for eventual promotion or job changes, and it is the way in which they learn new roles: to be a supervisor, to be on strike, to be unemployed, and to retire (Trice, 1993:112–137).

STAGES OF THE COMBINED INDIVIDUAL AND FAMILY LIFE CYCLES

School is the bridge between home and the more impersonal world of work. Young adults face several important and related life tasks. Not everyone will perform each of these tasks, but most will. These tasks include the following: leaving the parents' home; deciding whether to continue schooling; deciding whether to seek permanent, full-time employment; and deciding whether to start a new family. For men the normative order of the last three events is to finish school, find a job, and then get married. There are other, non-normative orderings of these decisions. For example, one may get married, then finish school, then find a job. One study finds young men who follow the normative order receive the highest occupational rewards; those who follow other orderings usually achieve lower positions (Hogan, 1981). Women may follow the same path as men, but it would also be considered normatively acceptable for women to make the decision to bear children instead of continuing school or working.

Entering the Labor Force

Occupational choice is influenced by various factors, including personality and social context. Long before their formal schooling ends, young people acquire a self-image and attitudes toward work involvement and occupations. These personality characteristics play an important part in the jobs they will eventually take and in the success they will experience (Mortimer, Lorence, and Kumka, 1986).

The mix of available occupations also affects the incorporation of new workers into the labor force. In Chapter 1 we mentioned changes in occupational distribution, a topic we will address in greater detail in Part IV. In general there has been a long-term shift from agricultural work to manufacturing work to service work. This process entailed a parallel shift from a labor force that does predominantly manual work to one that does bureaucratic and service tasks. Relatively new occupations tend to attract fairly young workers who have just completed the requisite training. Thus, Internet companies tend to have relatively young employees.

Occupations vary in their *barriers to entry,* or how hard it is to become a member of the occupation. It is relatively easy to become a cashier, in terms of skill, because the duties are thought to be within the capability of most workers. To be a physician, on the other hand, requires long years of specialized training, testing, and licensing, and not as many people are thought to have the ability to master these skills. It is always possible to change occupations, but young workers who choose an occupation with low barriers to entry may eventually regret not having acquired the qualifications for entry into a more desirable occupation. We will discuss in later chapters some of the mechanisms for creating barriers to entry, including craft unions, licensing, and professionalization.

Every young worker enters the labor force in a specific job, which represents the intersection between an occupation and a workplace. Even within the same occupation, jobs can differ a great deal because of the workplace. A nurse, for example, may find that a job in a hospital is very different from working in a physician's office or a nursing home. An entry-level job is a relatively low-level one with minimal job requirements. There are different types of entry-level jobs; one distinction we can make is between so-called dead-end jobs and entry-port jobs.

Dead-End Jobs **Dead-end jobs** require relatively little skill, often have a high turnover, and rarely, if ever, lead to promotions, higher pay, or more responsibility. They often pay only the minimum wage and characteristically provide little security and few fringe benefits. **Fringe benefits** refer to the non-wage compensation and perquisites (or "perks") of workers. Some common fringe benefits include pension coverage, health and other forms of insurance, sick leave, and paid vacation. These benefits may cost the employer 25 percent or more of all salaries or wages. An example of a perquisite would be an employee discount on purchases of the company's product. Many dead-end jobs are part-time or seasonal, or they may be subject to reduced hours or layoffs when business conditions are adverse. Because of the relative unattractiveness of the dead-end job, the employer expects little preparation from the workers, and provides little on-the-job training, as shown in the following account:

> Between four and eight women took three different colors (and lengths) of wire and

inserted them into a small plastic block an inch square and a quarter of an inch deep. This was by far the hardest and most tedious job. Each lead had a square terminal on the end which had to be pushed into a square channel in the plastic block until it locked.... It took a certain amount of force and some finesse as well. If you held the lead too far back, you bent the terminal. If you held the lead too close, you banged your fingers. You could always tell who was new on the job by their bandaged fingers. (Juravich, 1985:47)

From the employer's viewpoint one advantage of the dead-end job is that "anyone can do it." Assuming that labor is plentiful, it will be relatively easy to recruit someone to do the job, because few skills are needed. Recruiting and interviewing prospective workers takes little time. The employer can fire difficult or undisciplined workers and hire new workers. If business is good, the employer can easily create more such jobs. When business lags, the employer can lay off workers or reduce their work schedules.

A dead-end job also may be the only possible arrangement for even a well-meaning employer. For a small employer, or for one whose work flow is seasonal or erratic, such as a subcontractor who does "overflow work" for a larger firm, there may be no reasonable alternative to creating a dead-end job. Such an employer might prefer better trained or more ambitious workers, but those workers will be attracted to larger or better capitalized firms with better working conditions and greater security. Many private householders create what are, in effect, dead-end jobs for housekeepers, gardeners, and child-care workers, simply because there is no realistic way for them to create better jobs.

Dead-end jobs have some disadvantages for the employer, too. Replacing workers is easy if labor is plentiful, but difficult and costly when labor is scarce. Furthermore, because the best workers prefer better jobs, the pool of available workers is likely to have relatively little experience and training. High turnover can disrupt production. In service industries an uncommitted worker can irritate clients and customers and cause them to take their business elsewhere.

It is easy to see the disadvantages of the dead-end job for the workers, yet these jobs do serve some positive functions for young workers. Part-time or summer jobs provide the first working experience for many young people. Although the job may offer no formal training, working itself is new to them, and so they learn some general lessons about the world of work. The employer on a young person's first job is often willing to provide references to colleges, credit bureaus, or other employers. Part-time jobs are attractive to students and young parents who cannot work full time. And for middle-class college students, these jobs often serve as a bridge between their schooling and eventual permanent jobs.

Dead-end jobs are detrimental to the worker who is never able to find better employment. Such a worker may change jobs frequently, looking for better pay or working conditions, only to find that the job is once again a dead end. In this case the worker may never successfully bridge the gap between dead-end jobs and jobs with better prospects. In time the history of job hopping, rather than indicating ambition and a desire for self-improvement, may come to be interpreted as the inability to hold a job or as more generalized instability. Thus, characteristics that originally described the job are used instead to label the worker. And in fact, workers who hold only dead-end jobs learn few new skills and have few incentives to improve their work habits. (Related topics will be discussed in Chapter 14, on marginal jobs.)

Knowing about dead-end jobs helps to explain why the non-normative orderings of life events may lead to lower job rewards. Students who drop out or who have a new family to support may need the income from a job right away, but their lack of education may limit them to dead-end jobs. Although to new workers such jobs may seem to pay well, there will be few annual pay raises and few opportunities to progress to better paid, higher skill jobs.

Entry-Port Jobs   In contrast with the dead-end job, some entry-level jobs are called **entry ports** because they offer the worker the possibility for training, greater responsibility, improved pay and fringe benefits, and promotion. Entry ports are usually filled through formalized recruitment procedures with well-established criteria for hiring. Recruiting and screening are done carefully to ensure that the best available workers are hired. The employer may pay particular attention to education, including degrees earned, the quality of school, and any additional certificates or licenses held. Employers also pay attention to previous work experience, including part-time jobs, summer jobs, and volunteer work.

The conditions of work also differ. Entry ports usually require at least an orientation period (if not a training program) to introduce the new worker to company policies and regulations. There is likely to be a well-defined job description. Many entry ports have a probationary period during which a new worker is carefully watched and evaluated. It is usually easy to dismiss a probationary worker, but afterward the workers are likely to be protected by a job security arrangement and also by formal procedures that prescribe how they may be disciplined. The entry port usually pays more than the minimum wage and provides some fringe benefits. Commonly, there is provision for regular reviews of performance and salary.

Sometimes entry ports offer low initial pay, although above minimum wage. One reason for the relatively low pay, according to economists, is that the workers are subsidizing their own on-the-job training, but their eventual promotions within the firm will compensate them for their low starting salaries. The most distinguishing characteristic of an entry port is that it makes the worker eligible to compete for advancement within the firm. A **job ladder** is the additional position to which one might be promoted within the firm. Job ladders vary among employers in both their height and the intensity of competition for the next rung. Some ladders are very short. A worker hired into an entry port may anticipate only one or two promotions until the ladder ends. In many companies, for example, it is possible to be promoted to supervisory positions, but managerial positions require a different, higher entry port for candidates with a different set of credentials. In other companies

"working your way to the top" may be the company policy, and the job ladder extends, at least in principle, all the way to the top position.

A number of factors determine how competitive promotions are. One factor is business conditions; a company that is expanding can promote its workers, whereas one that is considering layoffs is unlikely to offer promotions. Another factor is the relative steepness of the company hierarchy—that is, whether there are many or few possible levels for promotion. A third factor is the span of control for those higher on the job ladder. They may each supervise only a few workers, in which case more of them will be needed. If, on the other hand, each supervisor has responsibility for a large number of workers, then relatively few supervisors will be needed. Many such factors are invisible or uninteresting to the beginning worker, at least initially, but they can have an effect on how long the worker is willing to remain in the firm.

Most entry-level workers who want permanent jobs would prefer an entry port to a dead-end job. Entry ports generate some advantages for employers by attracting higher quality workers and inducing them to stay with the firm. But employers must also invest more training and compensation in their workers in entry-port jobs.

Regardless of other issues, including the company's financial position, the preferences, beliefs, and strategies of managers affect the characteristics of entry-level jobs. Some companies create entry ports for employees who work closely with the production process and whose growing expertise will be needed by the company in future years. In the same company, other types of work, such as routine maintenance or low-level clerical work, may be packaged as dead-end jobs with little prospect for promotion. Even if dead-end jobs have high turnover, the company is little affected by its ability to produce its major good or service.

Small firms create about two-thirds of the new jobs in the United States, most of these in the service industry (Bednarzik, 2000). New jobs, however, are not necessarily good jobs. Sociologists are concerned that the majority of new jobs will be dead-end jobs, because small firms have few resources for job ladders. Moreover, small firms are most vulnerable to market forces and so are very likely to create short term, seasonal, temporary, or part-time jobs.

There is a possibility that employers may feel pressure to create more attractive entry-level jobs because the birth rates of the industrialized countries have been low since the 1970s. Thus, there will be relatively slow growth in the number of young workers entering the labor market in the late 2000s. Employers may compete for young workers with better wages and working conditions. Moreover, the industrialized countries, which have recently relied on immigrant workers to fill dead-end jobs, have tightened their immigration laws. In 1986, for example, the United States enacted civil and criminal penalties against employers of illegal aliens, and by the mid-2000s there was growing pressure to enforce such laws.

Role Conflict and Role Overload

Besides entering the labor force, the young adult years are associated with many life-cycle tasks, including the completion of schooling, leaving home, and starting a new family. These tasks entail giving up some roles, such as student; reducing the salience of others, such as son or daughter; and learning new roles, such as worker, spouse, and parent. These role shifts may pile up within just a few months, or they may stretch out over a period of years. For example, since the 1970s the age at marriage has been rising, especially among the college-educated, and there is a longer interval between marriage and the birth of the first child. Other people, particularly those who begin work and marry just after high school, may acquire these new roles more quickly.

Multiple life roles can be a source of satisfaction, whereas people with only one life role are at higher risk for depression. Women who play more than one role report greater satisfaction, and married people are more likely than single people to report that they like their jobs (Bersoff and Crosby, 1984). Taking on the many new roles of young adulthood, however, may create **role overload**,

a tension caused by trying to do too much at once. New jobs in many fields are challenging, and employers may expect new workers to put in extra time to master the jobs.

Even worse, the roles may actually be in conflict. **Role conflict** occurs when someone occupies two roles with contradictory expectations of what one should be doing at a certain time. For example, a worker may be expected both to be at work on time and to be caring for her sick infant. Job-related travel, changing work shifts, child care, and unexpected emergencies either at home or on the job are important sources of role conflict. The resolution of role conflict requires setting priorities within both the workplace and the home.

For women, some role conflicts arise from structural ambiguities, in which institutionalized and agreed-upon arrangements for integrating work and family are not yet available (Gerson, 1985:123–124). Men have avoided these ambiguities to some extent because of the widespread assumption that husbands would take principal responsibility for work and that wives and mothers would take principal responsibility for children. The employment of both spouses is often necessary for financial reasons, but the family frequently assumes that the woman's commitment to the home will be undiminished (Hochschild, 1989). Many employed women experience role conflict from the need to maintain both their work tasks and their home tasks.

Another problematic situation for integrating work with family is the one-parent family. Figure 5.1 shows the increase in the number of single-parent families between 1970 and 2005. In about 29 percent of the households with children, there is only one adult present, usually the mother (Census, 2007: Table 69). No division of labor between spouses is possible; the single parent must provide both income and child care. Over 28 percent of female-headed households are below the poverty level (Census, 2007: Table 697). Of the 11.6 million mothers who were custodial parents in 2003, only 7.1 million had been awarded child support, and 1.5 million did not receive any of the support they had been awarded (Census, 2007: Table 554).

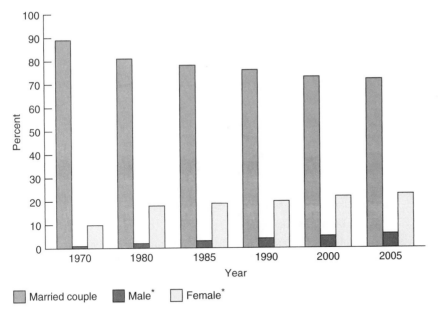

FIGURE 5.1 Percentage Distribution of Family Households with Childeren under 18 years of age, 1970–2005.

SOURCES: U.S. Census.
* no spouse present

Low-paying jobs also contribute to poverty in the single-parent home.

The care of home and children remains important to most young adults, but job responsibilities may appear more urgent. In manual work, especially when the job is covered by a collective-bargaining agreement, overtime and evening work are usually carefully monitored and compensated, although they may be involuntary. In salaried professional and technical jobs, by contrast, the higher pay may be balanced by the sense that the workers have too little control over their time. Full-time workers in 2003 averaged 42.9 hours of work per week (Census, 2007: Table 588). Many workers also bring home bulging briefcases of work for evenings and weekends (Schor, 1991). Superiors expect that workers will routinely make themselves available for additional work whenever a project is nearing completion or a deadline is approaching. At a crucial time in terms of the critical first promotion, a young worker may feel compelled to spend more and more time at work.

Occupations vary in the extent to which workers draw boundaries between their home space and home time and their work space and work time. Sociologist Christena E. Nippert-Eng (1996) found that workers varied in the extent to which they sharply differentiate work from home. Ed, a machinist whom she interviewed, draws a sharp distinction between work and home. His wife has never visited his workspace, and he does not talk about personal things with his coworkers. Nippert-Eng identifies this behavior as "segmentation." John, a scientist, works in the same lab with Ed, but John is married to his scientific collaborator, stays long hours at the lab, works at home, and talks about both work and home in both places. This behavior is called "integration." Table 5.1 shows the criteria that Nippert-Eng used in exploring the continuum from segmentation to integration. Even segmenters, however, may spend longer hours at work if they find home conditions boring or stressful (Hochschild, 1997b).

TABLE 5.1 The Boundary Work of "Home" and "Work" along the Integration-Segmentation Continuum

Integration	Segmentation
Calendars	
Pocket calendar	Two wall calendars, one at home, one at workplace; no overlap in contents
Keys	
Home and work keys on one ring	Home and work keys on two rings; no overlap in contents
Clothes and Appearance	
One all-purpose home and work wardrobe; changing in morning and evening insignificant	Distinct "uniforms" for home and work; changing in morning and evening crucial
Many work- and home-related items in purse/wallet	Few work items in purse/wallet
Eating and Drinking	
Same foodstuffs and drinks consumed in same (un)routinized ways at home and work, throughout day and week	Different foodstuffs and drinks consumed in distinctly different, (un)routinized ways at home and work throughout day and week
Money	
Same monies used for personal and work expenses, incurred at home and workplace	No overlap in accounts or uses of personal and work monies, places where they are spent, or their respective bills, receipts, and IRS forms
Multipurpose bills, receipts, and tax forms	
Talk	
Cross-realm talk within and about both realms	No talk about work at home; no talk about home at work
Same style of talk used in both realms	Realm-specific talk styles
People and Their Representations	
Addresses and phone numbers for all acquaintances kept in one book	Addresses and phone numbers for work and home acquaintances kept on separate lists in separate places
Photographs of coworkers at home; photos of family kept at workplace	Photos of coworkers kept in workplace; photos of family kept at home
Coworkers come to house to socialize with family; family comes to workplace to socialize/work with coworkers	Coworkers socialize together without families, in workplace during workday; family does not come to workplace
Reading	
"Work"- and "home"-related material read and stored anywhere, anytime	"Work" material read and stored only at workplace, during work time; "personal" material read only during "personal" time, away from workspace
Breaks	
No distinction between worktime and personal time during day or year	Distinct pockets of personal time during workday when no wage labor is done; distinct annual vacations when no wage labor is done
Commutes	
"Two-way bridges"; no transformative function	"One-way bridges"; crucial for achieving transformations between realm-specific selves
Phone Calls	
Frequent, random cross-realm calls; intrarealm calls include cross-realm subject matter	No cross-realm calls; intrarealm calls include only realm-specific subject matter

SOURCE: Nippert-Eng, 1996, Table 1, pp. 149–151. Used by permission of the University of Chicago Press.

Work Arrangements among Couples

The setting of priorities between work and home is complicated by the commitments of larger numbers of couples to dual careers and dual earnings. Women's labor force participation has risen in many countries, not just the United States. Although many working women are not married, the large number of wives who work for pay represents a major change within recent decades.

For couples in which both partners work, issues of *time* and *location* are closely related to the issues of *priority*. The time issues can include the possibility that one or both of the partners is expected to work overtime, at nights, or on weekends or holidays. The couple may work different days of the week or different shifts during the day. Such a schedule leaves little time for them to spend together in leisure, in home tasks, or in parenting. When one or both workers is also bringing work home, the home becomes an extension of the workplace, and even the time spent there is not available for sharing. Sharing the housework also presents a time problem for the working couple.

The proximity of the home to work and the availability of transportation are issues involving location. A limited budget for commuting may force one spouse to look for work near home or to use less convenient methods of transportation. If there are children, one parent may also consider working near the children's school or caregiver. Many families develop complex logistics for delivering adults to work and children to school, and for completing the necessary errands during lunch hours or after work.

An even more serious location problem is posed by the *commuter marriage*. Commuter marriages arise when suitable jobs in the same geographic area are unavailable to both spouses. This term is somewhat misleading, for everyone who works outside the home travels a certain distance to work. But in the commuter marriage the distances are so far that returning home every night is not possible. Thus, the commuter marriage maximizes the spatial separation between workplace and home. As an alternative to the commuter marriage, one partner may leave the labor force, accept a job in a different field, or become a part-time worker. These alternatives are costly in terms of both income and the career continuity of the affected spouse.

With some corporations, the military, and certain government agencies, continued progress on the job requires occasional transfers. More than half of all moves in the United States are believed to be work related. Promotions, new job responsibilities, or even just job retention are sometimes attached to geographic moves. While necessary for the individual worker, relocations are difficult for the family. Job transfers have been linked to depression in wives of corporate managers, and they can be stressful to young children and adolescents (Stroh, 1999; Feldman and Bolino, 1998). Transfers can be especially difficult for the couple in which both partners work. A few corporations have begun offering relocation counseling and referrals for the spouses of their transferred employees.

The Arrival of Children

All of the advanced industrial countries have experienced a long-term decline in birth rates and in the average number of children borne by each woman. This trend was briefly reversed during the baby boom following World War II, but today it is firmly reestablished. In 2005, 52 percent of all family households had no children under the age of eighteen, up from 48.7 percent in 1992. The percentage of households with three or more children has declined from 15.3 percent in 1940 to 10 percent in 2005 (Census, 2007: Tables 57, 66; Helmick and Zimmerman, 1984:403). Moreover, roughly 10 percent of all U.S. couples report that they intend to remain childless.

Nevertheless, most couples will become parents and will have children at home for part of their married life. Moreover, many adults will be single parents for at least part of the time that their children are minors. This means that the issue of combining child care with work will still arise for most adults, but it is being handled differently today than it was in the past.

Besides having fewer children, many women are also waiting longer to bear their first child. An even more dramatic change has occurred in the way in which women synchronize their childbearing and work decisions. It was once common for women to leave the labor force at the time their first child was born. In more recent years this practice has nearly disappeared, at least statistically. Figure 5.2 graphs the changes in women's labor force participation rates, by age, from 1950 to 1999. (For a definition of the labor force participation rate, see Chapter 2.) Each successive year's rates are higher, indicating that labor force participation was generally increasing, but the increase shows an interesting variation by age. As the graph shows, women in the prime childbearing years did not participate in the labor force at the same rate in 1950 and 1960 as they did in 1970. Instead, there was a decline between the ages of twenty to thirty-four, followed by a rise after the age of thirty-five. This M-shaped pattern of women's labor force participation can be seen most clearly in the figure by looking at the graph for 1970. By 1980, there was hardly any decline at all in labor force participation in the childbearing years, and by 1985 labor force participation was essentially level at those ages. Today the M-shape has disappeared. The trend in the 2000s has continued this pattern.

More continuous employment has probably helped the careers of many women and has provided more continuous income for their families. It has also challenged the long-held view that mothers should be the primary providers of daytime care for their young children. In 2005, 60 percent of wives with children under age six were working, and 56 percent of mothers with babies one year of age or younger were working (Census, 2007: Table 585). Both proportions have increased dramatically since 1970 (O'Connell and Bloom, 1987). A lack of affordable child care has prevented many more mothers from seeking work (Cattan, 1991).

A couple with children has many options for combining child care with work. One option is for a parent to provide all the child care. One parent (usually the mother) can leave the labor force, or one parent can substitute part-time for full-time work. When one parent is away, the other parent worker cares for the children. About 11.5 million

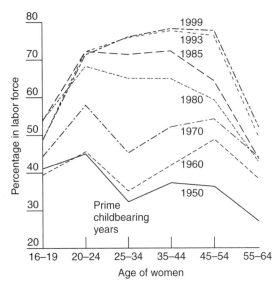

FIGURE 5.2 Labor Force Participation for Women, by Age, 1950–1999

SOURCES: Martin O'Connell and David E. Bloom, 1987, "Juggling Jobs and Babies: America's Child Care Challenge." *Population Trends and Public Policy* 12 (February). Reprinted by permission of the Population Reference Bureau. 1993 data from BLS (1994a). 1999 data from U.S. Department of Labor (2000).

children under the age of fifteen have mothers who are not in the labor force and are cared for only by their parents (U.S. Census, 2007: Table 63). Many women choose part-time work in jobs that accommodate more flexible schedules (Glass and Camarigg, 1992). Another possibility is for the husband and the wife to work separate shifts, so that one parent is always home with the children (Presser, 1988). About 14.8 million workers work on shifts other than regular nine-to-five schedules, and 8 percent of them report that the reason for their shift work is to enable child care (BLS, 2005). Some parents turn to self-employment or to paid work done at home to provide more flexible working conditions. Some preschool children are cared for by their mother while she works at home or at another location.

If the mother is a full-time worker, it is more likely that preschool children will be cared for by a relative (42%), in another home (17%), or in an organized child-care facility or school (24%). Some children receive multiple forms of care, while others do not have a regular child-care arrangement (Johnson, 2005:4). School is the primary source of child care for children aged six to thirteen. Many of these children have after-school care of some sort. Older children are much more likely than younger children to care for themselves after school.

On-site child care at work or nearby has become an important alternative for parents. Many on-site centers permit parents to eat lunch with their children. Cameras linked to secure Internet sites permit parents to check on their children during work hours. Some employers provide child-care vouchers to off-site day-care centers, reserve sites in day-care centers for children of their employees, or offer subsidies for child care as a fringe benefit.

Homemakers and Home Production as a Career

> Q. On the whole, do you prefer raising your children to working outside the home?
> A. Oh, yes. I never plan to go back. I'm too spoiled now. I'm my own boss. I have independence; I have control; I have freedom, as much freedom as anyone is going to have in our society. No job can offer me those things. (Gerson, 1985:129)

Substantial numbers of adults, most of them women, do not work for pay but remain engaged in home production. **Home production** refers to the non-market production of goods and services, usually for the family but occasionally on a volunteer basis for schools, churches, or other groups. What is traditionally known as "housework" is only one aspect of home production, which also includes household budgeting, shopping, care of dependents, and other tasks that go beyond cleaning, cooking, and laundry. If the cooking, sewing, chauffeuring, child care, and so on were monetarily compensated, they would account for billions of dollars each year. Workers engaged in home production without pay are called homemakers. In Chapter 9 we will also discuss high-technology workers who work for pay at home; these workers and others who work for pay, but at home, are called home workers to

distinguish them from homemakers. Home workers are counted in the labor force; homemakers are not.

A major function of homemakers is sometimes called *care work,* which can be defined as care for dependent children, ill people, or elderly, often within the family context and unpaid (Himmelweit, 2000). Care work implies much more than mere custodial care, because it may also involve such functions as home schooling, home nursing care, and emotional and physical nurturing.

Home production offers workers the opportunity to schedule their own work and set their own priorities. Many parents highly value staying at home with their children. On the other hand, home production also has disadvantages, many of which are financial. The homemaker is not covered by pensions, insurance, or Social Security, and is economically dependent on another person (Vanek, 1988). Almost all homemakers are women. About ten years ago observers began to draw attention to the plight of the "displaced homemaker," who when divorced lost her "work" as well as her financial support. Several European countries make provisions for covering homemakers with some benefits and for compelling families to provide homemakers an annual vacation.

Another problem for homemakers is the social devaluation of their work (Oakley, 1975). Housework is not considered a "real" job because many of the tasks can be postponed, and the tasks are thought to be low-skilled and repetitive. Despite evidence to the contrary (Schor, 1991), many people believe that "labor-saving" devices, such as vacuum cleaners and automatic washing machines, do all the housework. Some people consider work outside the home for pay to be more prestigious, perhaps even more "adult." A homemaker who is referred to as "just a housewife" feels devalued because the work has been devalued. Social evaluations of the homemakers appear to be a major factor in their relative happiness; when they feel valued, they are happier (Ferree, 1984).

The Sandwich Generation

In the middle stages of their work career, many workers are well established in their jobs. Compared with the earlier period of role transitions, this stage of work life may be less eventful. Many middle-aged workers find that this is their most productive period, with steady increments in pay and responsibility. Others find that even good jobs, with job ladders and job security, have become boring, that the realistic opportunities for promotion are infrequent, or that economic and organizational changes have made their jobs insecure.

Middle-aged workers are likely to feel some stress from their "sandwich generation" position, with their own growing families relying on them, and often their aging parents as well. In part this stress comes from the economic necessity of providing for young and old dependents. Although young couples may still receive some financial support from their families, by the middle stages of the career there are few outside sources of income, and indeed, elderly parents may begin to need financial support. Not only is the income from the job crucial but also the fringe benefits take on greater significance. With children to care for, the job-related benefits of health insurance, disability insurance, and life insurance become much more important. Also, middle-aged workers become concerned about their pensions. Staying with the same company ensures continuity of fringe benefits, and this may have the effect of "tying" the worker to the company. This is especially true if some fringe benefits or perquisites are pegged to seniority. One reason that layoffs are so disruptive to middle-aged workers is the loss of fringe-benefit coverage for their families (Sullivan, Warren, and Westbrook, 2000).

As children become older, the expense of caring for them also increases. Teenagers eat more than younger children; in addition, providing clothing, transportation, educational expenses, and leisure expenditures for older children is more expensive than it is for babies and preschoolers. College tuition is an even greater expense. This means that many families at this stage need an increase in income. Occupations vary in the extent to which middle-aged workers can expect their incomes to increase. In general, manual occupations are likely to level off in earnings while the workers are relatively young. Although there may be cost-of-living raises thereafter, it is unlikely that

there will be any further spurts in income. In many white-collar occupations, by contrast, merit raises and seniority increases continue well into middle age. For the white-collar worker the increases in income may not slow down or stop until after the children have already left home and the expenses have begun to decline. Manual workers may experience an **income squeeze**, which means that just as the expenses of their families increase, the increments in their annual income stop (Oppenheimer, 1974).

Workers find several ways to deal with the income squeeze. Many borrow money to finance new purchases or to send children to college. Some workers "moonlight"; that is, they take on a second job, often part time or seasonal, to supplement their income. In 1998, 7.9 million workers had second jobs, and over two-thirds of them were between twenty-five and fifty-five (U.S. Census 1999:421). A third alternative is to increase the family labor supply. Homemakers may enter or reenter the labor force to supplement the family's income. In addition, the teenage members of the family may also start to work, usually part time, to help cover some of their own expenses or to contribute to the family's expenses.

Beyond the financial strains on the sandwich generation, there may also be considerable emotional pressure on these workers. Worrying about elderly parents and teenaged children can be a source of strain and tension, and some of this tension may get transferred to the workplace. Box 5.2 discusses some of the ways that the pressures of family life may affect work life.

The "Empty Nest" and Retirement

The "empty nest" refers to the period after the last child has left the parental home. The empty nest may be a time of sadness for the parents, and a homemaker may in particular feel "unemployed" by the departure of her children. This is often a period in which homemakers intensify their volunteer work in the community, or in some cases enter the paid labor force. In addition, the arrival of grandchildren often provides an important new role of "grandparent" to enjoy.

BOX 5.2 The Impact of the Family on Work

Work certainly affects family life, but it is also the case that what is happening at home may affect the workers' performance on the job. All workers who are concerned about the health of a family member or concerned about conflict in family relationships may carry their worries with them into the workplace. Some employers offer Employee Assistance Programs specifically to help their employees deal with health and relationship issues so that they will not lose time or productivity on the job.

Employers sometimes expect that the family's effects on workers will be different for men than for women. For men a stable family life is usually considered an asset to life on the job. Indeed, the family may be more or less incorporated into the job through the expectation that the wife will be available to entertain business guests or to take on volunteer activities that will benefit the company (Pavalko and Elder, 1993). This is sometimes called the *two-person career* (Mortimer and London, 1984:25-26). Children may also be

incorporated through the expectation that workers' families will attend holiday parties and company picnics.

Employers may view women's families, by contrast, as a potential distraction from work. When a woman's children are young, an employer may worry about time lost to care for sick children or to find child care. Even when the children are older, there may be continuing concern that the female worker's attachment to her family will be detrimental to her work. Many workplaces create rules about how often a worker may phone home or receive phone calls from home during the day to limit a conflict between parenting and working. With women workers, the employer may be less likely to expect the husband to be available for business entertainment or social events. One researcher, summarizing the views of managers she had interviewed, writes, "Thus, while men symbolically brought two people to their jobs, women were seen as perhaps bringing less than one full worker" (Kanter, 1977:107).

Another reaction to the empty nest may be an intensified interest in work. Ironically, at the same time the employer may be trying to influence the worker to retire. In the United States laws require employers not to discriminate against workers older than forty and not to make retirement mandatory on the basis of age alone. (A few occupations are exceptions because the physical aging process can be demonstrably linked to poorer job performance. Airline pilots, for example, still have a mandatory retirement age because the gradual deterioration of vision, hearing, and reflexes in older pilots might increase the safety risks of air travel.) Employers may, however, still encourage retirement at or around the customary age of sixty-five, or they may offer special inducements, such as buyouts, to encourage employees to retire.

Retirements are attractive to employers because they provide turnover that allows the hiring of younger workers. In addition, the younger workers are more likely to have acquired the latest technological skills. Older workers are usually bet-

ter paid, especially if wages increase with seniority. Thus, the retirement of an older worker may provide the salary to hire two younger workers, or the opportunity to save on payroll expenses. Finally, although data do not indicate that older workers are any less productive than younger workers, employers may believe them to be less productive and thus encourage them to "move on." There is, nevertheless, concern that the retirement of large numbers of baby boomers starting in 2010 will create staffing shortages in skilled jobs (Dohm, 2000).

Some workers intend to work as long as possible. In particular, observers expect professional workers to remain in the labor force for a relatively long time, while manual workers in harsh environments or doing heavy physical labor are likely to retire earlier. One reason to continue working is financial. Many workers have small or inadequate pension plans, and only half of workers are covered by private pension plans. Legal reforms during the 1970s and 1980s required "vesting" of a worker's pension plan after ten years, meaning that the

TABLE 5.2 Percentage of Workers Covered by Pension Plans, 2005 $

Characteristic	Total[a]	2005 Defined benefit	2005 Defined contribution
Total	50	21	42
White-collar occupations	61	24	53
Blue-collar occupations	51	26	38
Service occupations	22	7	18
Full time	60	25	50
Part time	19	9	14
Union	85	72	43
Nonunion	46	15	41
Establishment characteristics:			
Goods-producing	64	32	50
Service-producing	47	18	39
1 to 99 workers	37	9	32
100 workers or more	67	36	53
Geographic areas:			
Metropolitan areas	52	22	42
Nonmetropolitan areas	44	15	38
New England	51	21	42
Middle Atlantic	55	28	42
East North Central	56	24	46
West North Central	56	22	46
South Atlantic	47	16	41
East South Central	44	13	41
West South Central	45	17	38
Mountain	49	17	42
Pacific	47	23	37

SOURCE: U.S. Census, 2007

[a]Includes defined benefit pension plans and defined contribution retirement plans. The total is less than the sum of the individual items because many employees participated in both types of plans.

worker is assured of benefits at retirement. Subsequent legal maneuvers by corporations have sometimes allowed them to escape this obligation to their workers (Walsh, 2006). In addition, the reforms came too late to help some older workers and workers who changed jobs often, worked in industries without pension coverage, or worked part time.

Table 5.2 shows the variation among workers in pension coverage. A "defined benefit" plan is one in which the employer promises a specific sum of money to the retiree; a "defined contribution" plan is one in which the employer promises to contribute a specific sum to a retirement account in each pay period. Union workers are the most likely to be covered by a pension plan. In general, white-collar occupations, full-time workers, goods-producing workers, and those working for fairly large companies are the most likely to have a private pension.

Women are especially likely to be hard pressed at retirement because they are more likely to have worked intermittently or part time and to have received lower wages. A widow who earned less than her husband is eligible for Social Security benefits based on her husband's earnings. About 92 percent of wives born in 1930–1934 and 82 percent of wives born in 1955–1959 will receive widow's benefits if their husbands die because the benefit will be greater than benefits based on their own earnings (Iams, 1993). A widow may also be entitled to a private pension if her late spouse was covered, but frequently the widow's coverage is a fraction of what the couple would have received in retirement. A divorced woman may be entitled to nothing at all from the estate or pension coverage of her former husband. Such financial concerns may encourage a worker to stay on the job as long as possible.

Many workers, however, prefer to retire early. Between 1980 and 2014 the labor force participation of men over age fifty-five is projected to decrease by 3 percentage points, whereas that of women in the same age group is projected to increase twenty points (U.S. Census, 2007: Table 574). Early retirement is frequently linked to illness or disability. In addition, older workers who are laid off may report themselves retired because they believe that no work is available. Some retirees are interested in pursuing volunteer opportunities (Mutchler, Burr, and Caro, 2003),

There is also an interesting countertrend in the hiring of elderly people. Some older workers seek a new job, often in a different line of work, after they retire. This may be an opportunity for someone to try a new line of work or to pursue a public service interest. Several firms in the fast-food industry have begun to substitute older part-time workers for younger workers, who are becoming less available. Their television advertisements often make a point of showing older workers serving food. Employers report that elderly workers make good employees because they are mature and have good work habits (Pereira, 2000).

Aside from the financial problems associated with retirement, other important family problems arise at this stage of the life cycle. The retiree may feel bored and unwanted, stripped of the productive role that is so important in modern life. Lip service is paid to the significance of volunteer work and home production, but many retirees find little prestige is actually attached to these pursuits, and they miss the activity of their former work environment. Not surprisingly, many retired workers today find that the long-awaited leisure is not what they had anticipated. "Disengagement," or relinquishing their work roles, is difficult. For those with strong work commitments, it is disorienting to have no workplace to report to every morning, and they lose their strong role identity as breadwinners. The retiree may disrupt the nonworking spouse's pattern of daily activities in his or her effort to be useful. In a dual-career family the problem may be that one spouse has retired while the other is still working. Even though the married couple has been looking forward to being together again, they may find that the reality is that they get on each other's nerves. Others find that because of failing health or financial difficulties, the activities they had planned for their retirement years are no longer possible. Inflation may gradually eat away at carefully saved reserves so that the dreams of extensive travel or endless days of golf are never fulfilled.

Historically, preparation for retirement was solely the worker's responsibility. The company might provide the customary dinner and gift of a gold watch to mark the transition to the new status:

> An elaborate ceremony at the time of retirement serves . . . as a rite of separation from both the occupation of firefighting and one's engine company. Its most prominent feature is a verbal "roast" during which the audience and various speakers confront the retiree with many of his misdeeds and indiscretions and generally review his overall performance as a firefighter. . . . These "roasts" are intermingled, however, with compliments and expressions of respect and appreciation. At the end of the dinner, the former firefighter appears publicly in front of his fellow firefighters as a retiree. He is forced into isolation and made to realize the inevitable and irreversible change. (Trice, 1993:123)

As the proportion of the elderly in the population grows, the issue of retirement and its alternatives takes on greater significance. Some companies have begun to consider how they can help their employees transition gracefully into retirement. Significant numbers of male and female workers arrange a transition period of partial retirement between full-time work and complete retirement. Employers may encourage their retirees to consider seasonal or part-time work with the company, and they invite retirees to company social events. Retirees are often eligible to continue participation in the company credit union, athletic teams, and other activities. During the 1980s, some larger employers began retirement-planning seminars for their older workers. These seminars are not yet widespread, but if successful, they could help the workers think realistically about retirement income, activities, and continuing work alternatives.

IMPROVING THE INTEGRATION OF WORK AND FAMILY

A number of changes in the timing of work have been suggested to help improve the fit between work and family. Many of these changes would also have other benefits in terms of increasing the worker's sense of control and productivity.

Repackaging Jobs

One suggestion already in use is **flextime**. Flextime allows workers to set their own hours, within some limits. For example, employers may insist that all employees be on the premises between ten in the morning and two in the afternoon. But workers may design their own schedules for when they will arrive and when they will leave, arranging to work the full work week with flexible hours. Flextime is popular with both workers and employers (Bohen and Viveros-Long, 1981). It allows the parents of small children to help get them off to school in the morning or to meet them in the afternoon. If both parents have flextime, it is much easier to program child care. Flextime also allows workers who are fresher in the morning—or perhaps late in the evening—to work during their most productive hours. When a number of employers in a city use flextime, there may be less traffic congestion during the rush hour, alleviating one more source of worker stress. About 25 percent of all workers are on flextime, and it is most prevalent in financial services and professional and business services (BLS, 2005).

A second change is *block scheduling*, or putting together the traditional forty-hour work week in nontraditional ways. There are several long-standing examples of this scheduling that require the worker to reside briefly at the workplace. For example, many offshore oil crews work one week on duty and the next week off duty. Firefighters often have a "one-day-on-one-day-off" rule. For the twenty-four hours that they are on duty, they sleep and eat in the fire station. Other versions permit the worker to live at home but schedule longer hours for fewer days. Some hospitals now have a forty-hour weekend plan for nurses. The nurse works a full forty hours in one weekend and then has the remainder of the week free for home obligations. Under the 4/10 program the workers work ten-hour days for four days a week. One study shows that male workers on this schedule spend more time with their children and with traditional male household chores, such as mowing the lawn (Maklan, 1977a, 1977b). For employers the principal problem with these plans is that workers' efficiency may decay with the long hours. The principal advantage is in extending the usual business hours.

A third change is **work sharing**. In one version, one full-time job is partitioned into two half-time jobs. These two jobs may be shared by a husband and wife or by two unrelated workers. Especially if it involves a husband and wife, work sharing may allow parents to care for their children themselves. For unrelated workers who, for whatever reason, cannot work full time, work sharing makes a new part-time job available. The principal disadvantage for the workers is reduced income; in addition, many part-time jobs carry no fringe benefits. Moreover, part-time workers may be excluded from the company's job ladder, if there is one. The company saves on fringe benefits, although there are additional bookkeeping costs. There may also be an advantage in keeping workers who might otherwise go elsewhere. It is important for the two part-time workers to communicate with each other about the job; otherwise, the employer will completely lose the continuity that was provided by having only one worker do the tasks.

Family-Related Fringe Benefits

Family-related work benefits can be an important mechanism for helping workers coordinate their families and their jobs (Ferber and O'Farrell, 1991). Maternity, paternity, and family leaves may help ease the transition into parenthood. Canada provides parenting and maternity benefits through unemployment insurance. Most advanced industrial countries have national legislation guaranteeing the right to employment leave and the protection of the mother's job (Ferber and O'Farrell, 1991:161; Kamerman, 1986:60).

The United States has lagged behind these other countries. In 1991, for example, maternity leave benefits were available to only about half of young women workers (BLS, 1993b). In 1993 Congress enacted the U.S. Family and Medical Leave Act of 1993. This law requires employers with fifty or more employees to provide eligible employees up to twelve weeks of unpaid leave for their own serious illness, the birth or adoption of a child, or the care of a seriously ill child, spouse, or parent. To be eligible, an employee must have been employed at least one year and have worked at least 1,250 hours within the previous twelve months. For most families, however, the problem of coordinating work and family is not limited to the actual birth of the babies; rather, it is a long-run problem that needs longer-term solutions.

Many large employers are considering a cafeteria approach to benefits that would give workers the ability to choose among different benefits that best serve their family needs. In the cafeteria approach the worker is given a dollar amount of benefits and asked to choose among the available alternatives. Besides traditional insurance benefits, some employers offer child-care subsidies, college tuition assistance, or elder care assistance.

Alternative Cycles

A more radical set of proposals is to reconsider how major activities are ordered in the lives of workers. In the sequential life plan that we discussed earlier, human work is organized sequentially into a period of education, a period of work, and then a period of leisure (retirement). Our review of these periods, however, indicates problems with the sequential life plan in at least two areas: education and leisure.

Education With rapidly changing technology, the skills even of young workers may become obsolete. There is a need for workers in many fields to update their education periodically through continuing study, sabbaticals, and other means. It is probably cheaper in the long run for employers to make such opportunities available rather than constantly hiring new workers (and losing the expertise of the older workers). These educational needs cannot be met so long as education is relegated solely to the adolescent and early adult years. Employers can provide training benefits periodically to their workers, but they could also consider measures such as sabbaticals that would allow workers a period away from the job to acquire new information and skills.

Leisure Postponing leisure to the retirement days means that many workers will never enjoy it, because of either failing health or low income. Meanwhile, most workers have relatively short vacations that often do not provide sufficient time to refresh themselves for the return to work. Americans, in particular, have shorter annual vacations than workers in most advanced industrial countries, and many Americans pride themselves on not taking all of their available leave time.

With the average age at retirement likely to increase, longer vacations and sabbaticals for workers become increasingly attractive options. The "cyclical life plan" would intersperse periods of education, work, and leisure throughout the life cycle (Best and Stern, 1977). Such a plan might include more internships and work experiences for students, education and leisure sabbaticals for workers, and longer vacations throughout the life-span. Possible advantages of the cyclical life plan include reduced boredom, especially in the middle years of work, and enhanced job productivity. The cyclical life plan could also benefit families by providing them with more time together, especially during longer periods of vacation. The disadvantages of the cyclical life plan include its cost to employers, both to fund the sabbaticals and to provide replacement workers for the absent workers. Workers may also resist these innovations, fearing to miss important workplace developments during a sabbatical.

SUMMARY

The transition from adolescence to adult life is affected both by the personality of the young workers and by the opportunity structure that confronts them. The opportunity structure changes through occupational shifts and through the efforts of employers to create entry ports or dead-end jobs for young workers. Entry ports and career ladders are characteristics of jobs that are more likely to be well compensated, have job security, and have fringe benefits. Such jobs, especially when they offer the worker autonomy in daily tasks, are associated with better mental functioning in middle-aged workers as well as greater financial security.

Socialization for work and work itself affect both the individual life cycle and the family life cycle. The separation of work from home leads to issues of both time and space in reconciling one's career with one's family. Although different work-related challenges must be met at different stages of life, the issues of resolving work and family are always present. There is also a reciprocal, but weaker, effect of family life on work life. Men and women often experience these effects differently because of the continuing assumption that mothers have the primary responsibility for child care.

Many recent changes have affected the ease of integration between family and work. Factors tending to worsen the conflict are the increased employment of working mothers, the increased number of single-parent homes, the number of jobs requiring long hours or travel, difficulty in finding child care, difficulty in finding two jobs in the same geographic area, and the "income squeeze." Factors helping reduce the conflict include novel methods of repackaging jobs into different hours or days and the development of family-related fringe benefits. For the long run, concepts such as the cyclical life plan may provide better ways to combine the demands of work with the needs of families.

KEY CONCEPTS

norm	career	dead-end job	role conflict
role	boundaryless career	fringe benefits	home production
individual life cycle		entry port	income squeeze
sequential life plan	family life cycle	job ladder	flextime
	socialization	role overload	work sharing

QUESTIONS FOR THOUGHT

1. What are the ways in which parents affect their children's work behavior? How does this influence change at different points in the family life cycle and at different points in the children's life cycles? How do children affect their parents' work behavior?

2. Identify the principal problems facing workers in synchronizing their careers with the family life cycle. How are the problems different for men and women workers?

3. What role, if any, do you believe employers should play in helping workers accommodate

family and career? What role, if any, should the government play?

4. Take a proposal for change that was mentioned in this chapter, such as flextime. Suppose that all employers adopted this change. Trace the likely effects, both positive and negative, for the labor force and for the economy.

5. Why is housework devalued? What, if anything, could be done to change the devaluing of housework?

MULTIMEDIA RESOURCES

Print

Cynthia Fuchs Epstein and Arne L. Kalleberg (editors). 2004. *Fighting for Time: Shifting Boundaries of Work and Social Life*. New York: Russell Sage Foundation. This book examines how dividing time among work, family, and personal obligations has become an important contested area in American life. Various chapters examine the length of the workweek, non-standard schedules, and how time affects workers.

Arlie Hochschild and Anne Machung. 1997. *The Second Shift: Working Parents and the Revolution at Home*. Avon Books, reproduced edition. An excellent account of women who work their "second shift" when they care for their families.

Leslie Perlow, 1997. *Finding Time: How Corporations, Individuals, and Families Can Benefit from New Work Practices*. Cornell University Press. A collection of pragmatic suggestions for improving the fit between home and work.

Internet

Social Security Online. *http://www.ssa.gov* The official Web site of the Social Security Administration. Information on disability support, Supplementary Security Income, Medicare, and retirement planning.

Work-Life and Human Capital Solutions. *http://www.workfamily.com* An information clearinghouse for work-life professionals. Contains a great deal of information on child care and a list of best practices.

Shiftworker.com *http://www.shiftworker.com* Information on the quality of life for workers who work shifts. This site includes information on nutrition, relationships, and safety, and includes links to research reports on shift work.

RECOMMENDED FILM

Mrs. Doubtfire (1993). Directed by Chris Columbus. Robin Williams plays a voiceover actor who can't keep a steady job; he and his career-oriented wife eventually divorce. Dressed as "Mrs. Doubtfire," he comes back to his family to do housework and child care. A funny but provocative portrayal of work, divorce, housework, and child care. Rated PG-13.

6

Collective Responses to Work

When the Union's inspiration through the workers' blood shall run,
There can be no power greater anywhere beneath the sun.
Yet what force on earth is weaker than the feeble strength of one?
But the Union makes us strong.

Solidarity forever!
Solidarity forever!
Solidarity forever!
For the Union makes us strong.

Is there aught we hold in common with the greedy parasite
Who would lash us into serfdom and would crush us with his might?
Is there anything left to us but to organize and fight?
For the Union makes us strong. [chorus]

It is we who plowed the prairies; build the cities where they trade;
Dug the mines and built the workshops; endless miles of railroad laid.
Now we stand outcast and starving, 'midst the wonders we have made;
But the Union makes us strong. [chorus]

All the world that's owned by idle drones is ours and ours alone.
We have laid the wide foundations; built it skyward stone by stone.
It is ours, not to slave in, but to master and to own,
While the Union makes us strong. [chorus]

They have taken untold millions that they never toiled to earn,
But without our brain and muscle not a single wheel can turn.
We can break their haughty power; gain our freedom when we learn
That the Union makes us strong. [chorus]

In our hands is placed a power greater than their hoarded gold;
Greater than the might of armies, magnified a thousand-fold.
We can bring to birth a new world from the ashes of the old.
For the Union makes us strong. [chorus]

("SOLIDARITY FOREVER," LYRICS BY RALPH CHAPLIN, FROM *SONGS OF THE WORKERS*, 34TH ED., PP.4–5. REPRINTED BY PERMISSION OF THE INDUSTRIAL WORKERS OF THE WORLD, 3435 N. SHEFFIELD, SUITE 202, CHICAGO, IL 60657.)

"**S**olidarity Forever," sung to the tune of the "Battle Hymn of the Republic," is a widely recognized union inspirational song. It was first popularized by the Industrial Workers of the World, a North American union active at the beginning of the twentieth century. Its verses evoke images of workers' collective power to control the industrial world in which they toil.

People who work together often come to see that their personal troubles are not unique but are shared by others in their workplace. Such shared problems provide the basis for collective actions, ranging from bargaining over wages, hours, and working conditions to organizing workers' cooperatives to overthrowing the existing political and economic system through mass revolt.

This chapter explores the reasons behind workers' collective actions. We provide a brief history of working-class collective activity, focusing on trade union activity. In the second half of the chapter, we examine the roles of labor unions and professional associations in the twenty-first century. Workers also pursue their interests through political action and electoral politics. These topics, however, are largely outside the focus on this book and are better addressed in courses and texts on political sociology and related fields.

WHY DO PEOPLE NEED LABOR ORGANIZATIONS?

Workers organize themselves into unions to bargain collectively with employers over specific grievances in the workplace. These grievances include inadequate wages and benefits, work that is too rapidly paced, unfair retention and promotion practices, and exclusion from decision making. When a problem first arises, workers often blame themselves for their frustrations. They discuss the problems only with a few trusted coworkers, if at all. As you learned in Chapter 3, the responses at this stage are typically individualistic and may include apathy, withdrawal, and quitting. In the second stage, the informal work group begins to frame complaints in collective terms. Solutions, however, tend to remain largely personal. Sometimes a specific event stimulates the move from this second stage to organized collective action in the workplace. The transition frequently requires leadership from the more outspoken workers. Collective action will occur, however, only if the workers are committed enough to their jobs to forego the option of leaving and if a significant core of workers sufficiently overcomes fears of management reprisal to start organizing their coworkers (Morris, 2005).

Union Membership

What kinds of workers belong to unions? As Table 6.1 shows, membership varies by industry, occupation, age, and other characteristics. Workers in *transportation, government, manufacturing and mining,* and *construction* are more likely to be union members than workers in trade and services. *Manual workers* are more likely to be union members than

TABLE 6.1   Who Belongs to Unions?

Criterion	Percent Represented
Total	14%
Industry	
Manufacturing	14
Mining	10
Construction	17
Transportation	27
Trade	7
Service industries	8
Government	42
Age	
16–24	6
25–64	17
Gender	
Men	16
Women	13
Race	
White	14
Hispanic	12
Black	18
Asian	13
Hours	
Full-time	16
Part-time	7

SOURCE: U.S. Department of Commerce, Bureau of the Census, *Statistical Abstract of the United States, 2004*. Washington, DC: U.S. Government Printing Office.

are other workers. Because of greater commitment to their jobs, workers *twenty-five and over* are more likely to be members than younger workers. *Men* are more likely than women to be members. *Black workers* are more likely to be members than are whites, Hispanics, or Asians. *Full-time workers* are also more likely to be union members than are part-time workers.

Most local unions are part of a larger national union. Table 6.2 lists the twenty-five largest national unions in the United States. The largest is a professional association, the National Education Association (NEA), with just over two million members. Professional associations such as the NEA take on some union roles, such as lobbying in the professions' interests, but do not necessarily bargain directly with employers over wages and benefits. The next largest union is the International Brotherhood of Teamsters (www.teamster.org), a union with almost 1.3 million members. As the table reveals, workers have organized into labor unions across a wide range of industries and occupations. Unions in the United States are not confined to blue-collar jobs in the manufacturing sector. In recent years the white-collar, professional, and service sectors have been important growth areas for unions.

Sociologists study unions because these are the most important mechanism through which workers give voice to their grievances and demands. Unions have thus had an impact on the nature of work in modern society far beyond their own membership.

AN OUTLINE OF NORTH AMERICAN LABOR HISTORY

As background for understanding how unions have shaped work in advanced industrial societies, we briefly overview the history of workers' collective movements, focusing on the United States and Canada.

Local Craft Unions

Skilled workers formed the earliest unions in the United States. These **craft unions** first developed shortly after the Revolutionary War and included associations of shoemakers (Philadelphia, 1792), carpenters (Boston, 1793), and painters (New York, 1794) (Borjas, 2005). These associations had much in common with the medieval craft guilds of which they were direct descendants (see Chapter 1). A local craft union drew its membership from only one trade and from only the local area where direct contact between members was possible. The primary goal of these unions was to provide members with higher wages and a measure of economic security based on

TABLE 6.2 The 25 Largest Labor Unions in the United States

Labor Organization	Membership (in thousands)
National Education Association (NEA)	2,530
Teamsters (IBT)	1,402
Food and Commercial Workers (UFCW)	1,381
Service Employees (SEIU)	1,374
State, County and Municipal (AFSCME)	1,300
Laborers (LIUNA)	818
Electrical Workers (IBEW)	741
American Federation of Teachers (AFT)	736
Machinists and Aerospace Workers (IAM)	731
Automobile Workers (UAW)	672
Communication Workers (CWA)	618
Steelworkers (USW)	612
Hotel and Restaurant Employees and Neddletrades (UNITE HERE)	479
Mail Handlers (NPMHU)	398
Operating Engineers (IUOE)	379
Plumbers and Pipefitters (UA)	319
Postal Workers (APWU)	312
Paper and Chemical Employees (PACE)	311
Letter Carriers (NALC)	303
Firefighters (IAFF)	242
Government Employees (AFGE)	197
Transit Workers (ATU)	175
Sheet Metal Workers (SMWIA)	146
Iron Workers	134
Office and Professional Employees (OPEIU)	134

SOURCE: Court Gifford (editor), 2002, *Directory of U.S. Labor Organizations.* Washington, DC: Bureau of National Affairs.

such benefits as accident and sickness relief and aid to the widows and orphaned children of deceased members. In addition, they performed important social functions such as providing meeting rooms for literacy classes and social gatherings.

Solidarity Group **solidarity** based on common interests is essential for the success of any union. Solidarity includes mutual defense and support in times of crisis or challenge. Common interests among members of a single craft in a local area, however, did not provide an adequate basis for collective action on issues of wider importance that crossed regional or craft lines. In a world increasingly dominated by regional and even world economies, wider issues such as trade policy and extending rights of free speech became increasingly important. The transition from local craft unions to workers' organizations based on broader interests provided the motivation for much of the ensuing history of labor in the United States and Canada.

Workers' Political Parties

Many early efforts by workers to organize around broader class interests took the form of political parties. By the 1820s workers had established such parties in over a dozen states. For a brief time these parties held the balance of power between the major political parties. The interests that they represented, however, were often still divided along regional and craft lines, preventing a unified program with broad appeal across the working class (Lipset and Marks, 2000).

Many reasons contributed to the failure of working-class political parties in the United States. These parties did not, as in Europe, have to lead the working class in a struggle to overthrow feudal society and secure political democracy. Also important was the diffusion of interests resulting from "the spreading of the population across the continent, and persistent infusions of new immigrants from abroad" (Dulles and Dubofsky, 2004:50). Working-class parties also failed because the United States political system makes it difficult for third parties to succeed. In this winner-take-all system, which mandates a general election for President, third-party views often go unrepresented. In parliamentary systems, as in England and Canada, the legislative members elect the prime minister. The process of negotiation that goes into this vote allows greater room for third-party influence and can even make third parties pivotal in forming governments. A final important factor was violent repression of working-class movements by the U.S. government throughout the eighteenth and nineteenth centuries.

These forces combined to keep American workers, and the movements that they built, weak and internally divided. Workers in the United States were unsuccessful in organizing as a unified political party to pursue their collective interests. Instead, they went forward with a more fragmented posture, continually struggling to reconcile the competing interests of different occupations with the more general interests of the working class as a whole.

Early National Unions

The earliest and most significant national union seeking to coordinate the interests of workers from different trades and regions was the National Labor Union, founded in 1866 by Bill Sylvis of the Iron Molders Union. The principal goal of the NLU was the establishment of an eight-hour workday, which it helped win for federal government workers and for workers in six states. This victory, unfortunately, was largely symbolic. The eight-hour day was impossible to enforce because most workplaces were small and widely dispersed, and the government had only a limited ability and willingness to intervene. The withdrawal of the NLU candidate in the presidential election of 1872 resulted in the collapse of the union and provides an example of the difficulties unions experienced in identifying issues to mobilize all workers. New technological developments resulting in larger and more centralized workplaces would, however, help workers in their efforts to organize.

A Central Role for Railroads The railroad industry, the largest and most concentrated industry in the United States during the nineteenth century, gave birth to the first large national unions, with memberships in the tens of thousands. During the 1870s and 1880s these unions played a significant role in the tumultuous origins of the modern American economy. This role was facilitated by the increased size of the railroad companies and by the opportunities for contact and communication among workers that such large enterprises provided. Widespread railroad strikes over wages, hours, and conditions occurred across the country in 1877. In Baltimore the government called out two army regiments. These soldiers opened fire on the strikers, killing ten and wounding more than twenty. In Pittsburgh when officials called out the local militia, the militiamen joined the strikers and marched to the rail yards, which they helped burn down after a pitched battle with hired company guards (Brecher, 1997).

Such strikes took on an increasingly general nature in the 1880s as more and more workers joined in the strike activity. Strikes involving several trades simultaneously are called **general strikes**. When civic participation occurs in the form of demonstrations, marches, and insurrections, these general strikes become **mass strikes**. Mass strikes are a revolt by the general population against some major aspect of the social order. General strikes leading to mass strikes were crucial events in the nation's history in the 1880s and again in the 1930s.

May Day, 1886 One of the most significant mass strikes in United States history was the May Day strike of 1886. It was organized by a consortium of unions headed by the *Eight-Hour League* under the leadership of Albert Parsons. The organizers planned a nationwide mass strike for May 1, 1886, to demand the eight-hour day in order to humanize the conditions of labor and alleviate unemployment. Over 190,000 workers went on strike, and 340,000 workers and citizens paraded in more than a dozen cities. The strike centered in Chicago because of the city's significance as a hub of rail transportation. Over 80,000 workers struck in Chicago alone (Brecher, 1997).

Near the close of the day's events, the police fired into a crowd of picketers who were attacking strikebreakers as they left the McCormick Harvester plant, one of the few businesses in Chicago that had attempted to remain open. Four picketers were killed. Three days later, people crowded into Haymarket Square to hear speakers protest the killings at the McCormick plant. As the speeches ended and the crowd began to thin, a column of police arrived and ordered the crowd to disperse. A bomb was thrown among the police, killing seven, and injuring sixty-seven others. The police opened fire on the crowd, killing four, and wounding fifty or more. Subsequently, the police went on a months-long rampage, breaking into union offices, burning pamphlets, smashing printing presses, arresting union leaders, and seeking the deportation of union members who were recent immigrants (Dulles and Dubofsky, 2004). Eight men who had given speeches at Haymarket Square were arrested and tried for murder. Four of them, including Parsons, were subsequently hanged. The American union movement suffered severely in the aftermath of the Haymarket affair. The accusation that unionists were "foreign-born, bomb-throwing anarchists" was used to legitimate widespread repression against labor unions. The difficulty of recruiting members also increased. To this day, workers around the world remember this pivotal moment in the history of labor through May Day marches and demonstrations. The United States is one of the few nations that does not commemorate these events and instead celebrates Labor Day at the end of the summer.

The Pullman Strike   The late nineteenth century continued as a period of widespread labor unrest. In 1894 a strike in Chicago against the Pullman Company rekindled the flames of mass strike and civic protest. The company manufactured passenger cars for the busy railroads of the time. The Pullman strike was brought about by a long series of accumulated grievances against the company and was precipitated by wage cuts. To work for Pullman, workers were forced to live in the company-owned "Pullman Town," where rents were twice as high as in neighboring areas, water and gas were quadruple their price elsewhere, and the newspaper editor and church preacher were hired by Pullman himself (Dulles and Dubofsky, 2004). Given the vastly superior resources of the company, a strike by these workers would have

had little chance for success. However, the strike was supported by the newly formed American Railway Union (ARU) under the leadership of Eugene Debs. The ARU agreed to sidetrack any trains that included Pullman cars. Such sympathy strikes in which one union supports another by refusing to handle struck goods are called **secondary boycotts**. Secondary boycotts are illegal under today's labor laws.

The Pullman strike was eventually broken when a federal judge found the ARU guilty of interference with the mail. The company had transferred the mail to Pullman cars to encourage federal action against the strikers. The hiring of five thousand armed men, who were deputized for the occasion by the state of Illinois but paid by the Pullman Company, the deployment of six thousand U.S. troops and three thousand Chicago "temporary" police against the strikers, and the fatal shooting of more than thirty men and women provided the final blows against the strike (Boyer and Morais, 1970). Debs was subsequently convicted in federal court for "restriction of trade" as the first case tried under the Sherman Antitrust Act.

The May Day and Pullman strikes were important battles in the struggle for free trade unions, as well as in the struggle for many other rights and benefits that we take for granted today. The 1980s witnessed a similar wave of general and mass strikes in Poland over such issues as the price of basic commodities, the right to form independent unions, the right to strike, and the five-day work week. These strikes started with work stoppages in the Gdansk shipyards but rapidly evolved into mass strikes involving participants from virtually every sector of society. The Gdansk strikes were one of the pivotal actions bringing about an end to communism in Poland and throughout the Soviet bloc.

General Unions: The Knights and the Wobblies

In the late 1800s and the early 1900s, two **general unions** emerged in the United States that attempted to enlist workers from all walks of life: **Knights of Labor** and the **Industrial Workers of the World**.

BOX 6.1   Fighting for Literacy, Education, and Workers' Rights

During the first third of this century, lectors—people who read to cigar-factory workers—came to Ybor City, Florida, from the cigar factories of Havana and Key West. They were educated men who had great acting abilities and a vast capacity to entertain and educate....

They were paid 25 cents per week by each worker. As many as four hundred workers contributed, so the lector was among the highest-paid employees in the cigar industry....

Lectors usually read the *Tampa Tribune* in the opening hours, which meant that they had to be up early to translate the news to Spanish. Many liked to read from two different types of books. The first

literary reading would be from the classics: Cervantes, Hugo, Shakespeare, or Moliere. The second would be a popular dime novel. The workers voted on what popular novel was to be read....

Over the years, the cigar makers were transformed into the best-educated workforce in the world. Since the lectors were their teachers, the workers looked to them for leadership. In time, the lectors also read from political tracts, which were often of a socialist nature, and argued for workers' rights. They supported unionization, better working hours, higher wages, medical benefits, and pension funds.

SOURCE: Excerpted from Ferdie Pacheco, 1997, *Pacheco's Art of Ybor City*. Gainesville, Fl.: University of Florida Press, p.10.

Although neither survives today as an active union, both had a strong influence on the direction of the American labor movement.

The Knights of Labor   The Knights emerged as a national organization in 1878 under the leadership of Terence Powderly. The Knights enlisted workers of all kinds, including craft workers, unskilled laborers, farmers, small-business owners, immigrants, and women. The only occupations barred were saloon keepers, professional gamblers, lawyers, and bankers. The goal of the Knights was to improve the position of the "direct producers" through education and social reform. The fervor of workers' desire for literacy and education is illustrated by the history of Florida cigar makers as reported in Box 6.1.

The Knights called for government sanctions against the monopolies and encouraged the growth of producers' cooperatives. The union grew rapidly in the 1880s in response to a series of depressions that both impoverished workers and thinned the ranks of craft unions. It had grown to a membership of 50,000 by 1883, only five years after its emergence. The depression of 1884 and 1885 sent large numbers of unemployed trade unionists into the open ranks of the Knights, and by 1888 membership had jumped to 700,000. In Canada, at their height, the Knights had

252 locals in eighty-three different cities, with their greatest strength in the industrial towns of Toronto, Hamilton, and Ottawa (Palmer, 1992).

The Knights officially opposed strikes, favoring political and social reform over work stoppages. They believed that shop-floor actions tended to address only the specific needs of each craft and that political and social reform could better serve the general interests of the working class. However, much of their membership was taken from the struggling trade union movement and favored direct action on the shop floor as well as longer-range social reform strategies. Because of these divergent membership goals, the Knights became unwillingly embroiled in a spectacular but unsuccessful strike against the Southwest Railroad System in Texas and Louisiana in 1886. The negative political reaction from this defeat, as well as from the Haymarket affair that same year, fell heavily on the Knights as the most visible national labor organization of the times and was an important cause of their demise. The Knights' failure, however, must ultimately be traced to their inability to devise a program behind which a broad-based working-class coalition could stand. An agenda of political and social reform without a strong activist trade union movement on the shop floor did not provide a sufficient organizational basis for sustained working-class collective action.

BOX 6.2 Rallying Cry of the IWW

Preamble of the Industrial Workers of the World
The working class and the employing class have nothing in common. There can be no peace so long as hunger and want are found among millions of working people and the few, who make up the employing class, have all the good things in life.

Between these two classes a struggle must go on until the workers of the world organize as a class, take possession of the earth and the machinery of production, and abolish the wage system.

We find that the centering of the management of industries into fewer and fewer hands makes the trade unions unable to cope with the ever growing power of the employing class. The trade unions foster a state of affairs which allows one set of workers to be pitted against another set of workers in the same industry, thereby helping defeat one another in wage wars. Moreover, the trade unions aid the employing class to mislead the workers into the belief that the working class have interests in common with their employers.

These conditions can be changed and the interests of the working class upheld only by an organization formed in such a way that all its members in any one industry, or in all industries if necessary, cease work whenever a strike or lockout is on in any department thereof, thus making an injury to one an injury to all.

Instead of the conservative motto, "A fair day's wage for a fair day's work," we must inscribe on our banner the revolutionary watchword, "Abolition of the wage system."

It is the historic mission of the working class to do away with capitalism. The army of production must be organized, not only for the everyday struggle with capitalists, but also to carry on production when capitalism shall have been overthrown. By organizing industrially we are forming the structure of the new society within the shell of the old.

SOURCE: *Songs of the Workers*, Thirty-Fourth Edition, 1984. Reprinted by permission of Industrial Workers of the World, 3435 Sheffield, Chicago, IL 60657.

In spite of their rapid decline the Knights of Labor left many important legacies. Their vision of worker-owned cooperatives remains one of their lasting contributions. Equally important is the legacy of including women in their locals. Previously, women had largely been excluded from the labor movement. The Knights organized several locals among laundresses and seamstresses, as well as in other predominantly female trades. Both in ideology and in practice they did much to advance the efforts of women to improve their working conditions.

The Industrial Workers of the World The **Industrial Workers of the World** (IWW, www.iww.org), also called the Wobblies, was founded in 1905 and included unskilled factory workers, miners, lumberjacks, dock workers, and even cowboys. The IWW organized workers not only in the industrial East but also in the mining towns of the Rocky Mountain West and in the lumber towns and ports of the Pacific Northwest. It called for an overthrow of capitalism and its replacement by committees of workers in each enterprise. The IWW believed that "an injury to one is an injury to all" and held that it was the responsibility of all workers, no matter what their occupation, to put down their work and assist their fellow workers whenever and wherever they were in conflict with the capitalist class. Many of the goals of the IWW are included in its preamble, which is reprinted in Box 6.2. Because of its goal of establishing a general union for all workers in order to radically transform society, the IWW was often in competition with the craft unions of the time, which sought more limited goals.

A textile strike in Lawrence, Massachusetts, set up one of the IWW's most significant victories and marked the height of its power. The strike was foreshadowed by increased national concern with working conditions in the textile and apparel industries. On March 25, 1911, 146 women burned to death or jumped to their deaths in the infamous Triangle fire (www.ilr.cornell.edu/trianglefire). This tragedy resulted from the Triangle Company's refusal to provide safety measures of any sort. The workers were trapped in the burning building because the company had locked the doors on each floor of the tall apparel factory from the *outside* to prevent workers from shirking or stealing.

In January 1913 poor working conditions and wage cuts in Lawrence precipitated a strike by more than 20,000 textile workers, many of them women and children. Police brutality against the strikers and extensive press coverage of the strike ironically transformed a desperate struggle by the poorest of workers into one of the most significant victories in American labor history. As a result of the strike, not only were wages raised and conditions improved in the textile industry as a whole but important legislation was also enacted that restricted the exploitation of child and female labor (Borjas, 2005).

The IWW was also very active in Canada, leading numerous strikes and protest movements. The most famous of these was the Winnipeg general strike of 1919. Workers throughout Canada were agitating for change in the face of harsh working conditions, low wages, and deteriorating living conditions resulting from soaring wartime inflation. A crisis was reached when local employers in Winnipeg refused to recognize and bargain with the building and metal trades workers. The Winnipeg Trades and Labour Council called a general strike to support union recognition for the workers. In response, local employers hired armed vigilantes and convinced the government to use the Royal Northwest Mounted Police to charge through lines of picketers. The general strike was broken, strikers were arrested and jailed, and the workers' demands were left unmet (Krahn and Lowe, 1998).

Repeated episodes of violent repression, including the long-term imprisonment or deportation of more than one hundred IWW leaders on sedition charges, brought about the eventual decline of the union. In their wake, however, the Wobblies left the labor movement with many cultural heroes, such as leaders Bill Haywood and Mother Jones and songwriter Joe Hill. They also created a lasting image of a distinctly North American version of radical mass unionism.

Employer Resistance   Fierce resistance from employers often limited the success of early labor unions. Companies were reluctant to give workers a share in either the profits or the decisions. Employers' strategies of resistance included using *blacklists* to identify and refuse employment to union sympathizers. Employers also used *yellow-dog contracts* in which workers had to sign a promise that they would not join a union or engage in any collective action. (The unions argued that anyone who would sign such an agreement was a "cowardly yellow dog.")

Companies also employed spies, *scabs* (replacements for striking workers), professional strikebreakers, and armed guards to undermine unions and intimidate workers. The well-known Pinkerton detective agency first came into national prominence through providing such services. One particularly infamous incident occurred in the mining town of Ludlow, Colorado, during a strike by the United Mine Workers (www.umwa.org) against the Colorado Fuel and Iron Company, owned by John D. Rockefeller. The workers had been evicted from their company housing and were living with their families in tents. On Easter night of 1914, while the men were at a meeting, company gunmen set fire to the tents. Thirteen women and children died in the fire or were shot to death as they were running out of the tents. Five other strikers were shot to death as they tried to help the women and children escape (Boyer and Morais, 1970). This massacre of innocents sparked widespread public outrage. In the end, however, the miners failed to win their demands, and none of the company guards were ever prosecuted.

Employers also enlisted the government to intervene against workers. In the early 1800s unions in the United States were considered conspiracies for the purpose of limiting owners' free use of their property. Unions were forced to exist as secret societies. By the mid-1800s the conspiracy doctrine had been relaxed, and unions were no longer seen as inherently a conspiracy against the rights of property. Instead, unions were legal unless their *objectives* or the *means* used to secure these objectives were *conspiratorial* (of possible injury to others). In plain language, this meant that unions were legal, but strikes were still illegal.

As the conspiracy doctrine dwindled in importance, however, unions faced yet another obstacle. This was the **injunction** (court order) to cease and desist a specific action, such as a strike (Borjas, 2005). Injunctions were used to block strikes throughout the nineteenth century and into the twentieth century. Violating an injunction put strikers in contempt of court. It could also evoke the entire arsenal of state power, including the use of fines against unions, the imprisonment of union leaders, the deployment of the police and the military, and, ultimately, the use of direct violence against workers. Following the passage of the Sherman Antitrust Act in 1890, workers could also be found guilty of monopolistic practices if they attempted to bargain wages through collectively withholding their labor. In 1914 the Clayton Act explicitly exempted labor unions from prosecution under the Sherman Antitrust Act. In 1932 the Norris-LaGuardia Act limited and regulated the use of injunctions in labor disputes. But well into the twentieth century government intervention in favor of owners represented a major barrier for unions.

The AFL and Craft Unionism

Samuel Gompers, the president of the Cigar Makers' Union, was the first head of the **American Federation of Labor (AFL)**, founded in 1886. The AFL is the oldest existing labor organization in North America and helped set a successful and enduring pattern for what labor organizations and industrial relations look like in the United States and Canada to this day.

Gompers's vision was based on an effort to reconcile the differing needs of each craft or trade. The AFL was organized as a *decentralized* federation of unions; that is, each member union (the carpenters, the cigar makers, the boilermakers, and so on) had autonomy over its own affairs. The AFL served as an umbrella organization responsible for coordinating and supporting these efforts and pursuing the general interests of the working class at the regional and national levels.

Strikes and Collective Bargaining Gompers emphasized collective bargaining by each trade on specific issues rather than general and mass strikes over broader issues. In his vision strikes by each trade were the crucial tool in the struggle for better working conditions:

> A strike on the part of workmen is to close production and compel better terms and more rights to be acceded to the producers. The economic results of strikes to workers have been advantageous. Without strikes their rights would not have been considered. It is not that workmen or organized labor desires the strike, but it will tenaciously hold to the right to strike. We recognize that peaceful industry is necessary to civilized life, but the right to strike and the preparation to strike is the greatest preventive to strikes. If the workmen were to make up their minds tomorrow that they would under no circumstances strike, the employers would do all the striking for them in the way of lesser wages and longer hours of labor. (U.S. Congress, 1901:606)

With workers in each trade firmly in charge of their own affairs, the AFL could pursue the goals of political and social change through its activities as a pressure group.

By 1890, only four years after its founding, the AFL was the largest labor organization in the United States. It used strict business principles to organize the collection of dues, the raising of strike funds, and the creation of old-age, sickness, and burial funds for its members. From 1890 to 1914 the AFL pursued policies that increased average weekly wages for unionized workers in manufacturing from $17.57 to $23.98 and reduced average hours from 54.4 to 48.9 per week. The AFL achieved these gains by institutionalizing conflict through the use of collective bargaining and abandoning the more volatile policies of mass unionism and social upheaval. By 1914 the AFL had a "membership of 2,021,000 workers for whom it had won higher wages, shorter hours, and increased security" (Boyer and Morais, 1970:181).

Many of the early craft unions organized under the umbrella of the AFL excluded women and minority workers in an effort to keep wages from being underbid by these cheaper sources of labor. Other AFL-affiliated unions, however, actively organized female and minority workers. Between 1910 and 1930 the International Ladies' Garment Workers Union (ILGWU), composed largely of female workers, was one of the most rapidly growing unions in the United States (Milkman, 1985). Similarly, in 1933 the all-black Brotherhood of Sleeping Car Porters was given an international charter by the AFL to organize railroad porters. On August 25, 1937, the Porters signed a contract with the Pullman Company. This was the first labor contract ever negotiated between an African-American union and a major United States corporation (Foner, 1982).

The CIO and Industrial Unionism

The AFL served as the key organizing and coordinating umbrella for trade unions until the Great Depression of the 1930s. By the mid-1930s unemployment stood at over 25 percent, and AFL membership was hard hit. The federation was also racked with internal dissent. Some members believed that the AFL should include only skilled craft workers. Others believed that it should attempt to organize the large numbers of semiskilled workers in the new mass-production industries of automobiles, rubber, steel, and glass. In 1935 John L. Lewis, head of the United Mine Workers and chair of the AFL's Committee for Industrial Organization, withdrew eight unions loyal to his vision of industrial unionism and formed the **Congress of Industrial Organizations (CIO)**. In Lewis's vision of **industrial unionism**, all workers in an industry, regardless of their particular craft or their level of skill, would be in the union.

Sit-down Strikes in Mass Production Wages had fallen dramatically in the mass production industries due to the Depression. Employers had also sped up the pace of work on the new assembly lines, and workers in these industries were eager to be unionized. Withholding their labor in a strike, however, would have been ineffective because, unlike skilled craft-workers, they were easily replaceable. In response, mass-production workers developed a new form of collective action, the **sit-down strike**. In a sit-down strike workers stayed in their places but stopped working. Sit-down strikes were staged as protests against speed-ups or against specific abuses by foremen, such as unfair firings or disciplinary actions. They also were used to pressure a company to recognize and bargain with the union. They were generally of short duration, lasting from a few minutes to a few hours. But even short sit-down strikes could effectively disrupt production, especially in the highly coordinated mass production industries. Most importantly, they prevented companies from replacing workers with strikebreakers in order to continue production.

The Flint Sit-down Strike Sit-down strikes spread in the 1930s, becoming both longer and more frequent. Lewis and the CIO capitalized on these strikes to organize workers into the CIO. The most famous sit-down strike was staged against the General Motors Corporation in Flint, Michigan, in January and February of 1937. The sit-down started at Fisher Body, Buildings I and II. Police attacked the striking workers by breaking windows and firing tear gas shells inside. The workers doused the shells and held their ground against three police assaults by using fire hoses that had been installed as safety devices. Police gunfire wounded fourteen unarmed workers, but the workers refused to leave the plant. The next day the state militia massed outside the plant. Meanwhile, workers at the Fleetwood plant had also gone on strike, and the governor of Michigan refused to escalate the volatile situation any further. A siege lasting forty-four days ensued. The siege included the famous "Battle of the Running Bulls," in which the police (bulls) attacked the workers' wives, who were trying to smuggle food into the plant. The police were forced to retreat in front of the workers' fire hoses (Brecher, 1997).

General Motors eventually capitulated to the CIO. Within a three-week period in early 1937 the CIO organized not only the automobile industry but also the steel industry. This brought into being two of the largest and most important unions in American history, the United Auto Workers (UAW, www.uaw.org) and the United Steelworkers of America (USWA, www.uswa.org). Sit-down strikes became the craze and even spread to such occupations as soda-fountain clerks, waiters and waitresses, stenographers, dry-goods clerks, and teachers (Boyer and Morais, 1970).

The Fisher Body Plant in Flint produced car bodies for fifty more years after the famous sit-down strike until December 1987, when it was finally closed. Although the sit-down movement gradually faded, it left a lasting mark on American history through the unions it helped organize and the legacy it left of direct action. This legacy helped set the groundwork for similar sit-down tactics in the civil rights and antiwar movements of the 1960s and 1970s, the nuclear-freeze movement of the 1980s, and the environmental movement of the 2000s.

Because of high unemployment, the Great Depression may at first seem an unlikely period to have witnessed rapid gains in union membership. However, the social and economic situation in the Depression was so desperate that there was a perceived need for a radical transformation of the social order. In this context many people considered labor unions an important part of the solution to the problems of unregulated capitalism that had given rise to the Depression.

Legislative Gains Important federal legislation passed during the 1930s solidified workers' hard-won gains into lasting structures. Legislation gave concrete form to workers' demands and also sparked additional union growth by granting legitimacy to the labor movement. The most significant of these acts was the **National Labor Relations Act** of 1935 (NLRA, www.nlrb.gov), commonly known as the Wagner Act. The NLRA calls for secret-ballot elections in which workers can choose whether or not to be represented by a union. If the union wins the election, it becomes the sole bargaining agent for the workers, and the owners must bargain in good faith with the union. The NLRA does not spell out wages and conditions. Rather, unions and companies are required only to meet, bargain in good faith, and put their final agreement in writing. The NLRA thus endorses the process of collective bargaining but leaves the results of the bargaining completely open to the parties involved. It also provides certain safeguards against unfair

labor practices. The most important of these is the stipulation that workers cannot be fired for trying to organize a union at their workplace. The NLRA is the legal basis for modern American unionism.

Canadian labor law, though similar in content, came into being in a more piecemeal fashion over a longer period of time. The traditions of English and European labor law influenced Canada more strongly than the United States; this influence helped establish the legal basis of unionism in Canada at an earlier date. As a result, a greater diversity of union organizations developed in Canada than in the United States, including a powerful presence of Catholic labor unions in French-speaking Quebec (Gunderson and Ponak, 2000).

Postwar Challenges and Opportunities

The organizing efforts of the AFL and the CIO, operating under the new protection of the NLRA, resulted in a fivefold growth of union membership between 1933 and 1945. By 1945 the combined membership of the AFL and the CIO numbered fifteen million (Brody, 2005). The postwar period, however, saw political setbacks for the union movement, as well as continuing change, challenges, and opportunities.

The Taft-Hartley Amendments In 1947 Congress passed the **Taft-Hartley Amendments** to the NLRA. These amendments contained setbacks for labor in three major areas. First, they outlawed secondary boycotts and sympathy strikes. Secondary boycotts and sympathy strikes had enabled workers in one union to strike or otherwise pressure their employer not to do business with another company because of its labor practices. This exclusion took away an important tactical weapon, weakening union's bargaining power and eliminating an opportunity for cooperation between workers in different trades.

Second, the Taft-Hartley Amendments allowed states to outlaw contracts requiring union membership as a condition of employment, commonly called *union shop* contracts (Dixon, 2005). The NLRA requires unions to share the wage-and-benefits package they win with all workers in their unit, including the full procedure for settling grievances. As a result, most union workers believe that once a union has been elected by majority vote, all workers should be required to be members of the union and should pay dues. Union workers see nonunion workers as getting a *free ride* on benefits secured by the union. Twenty-one states, however, have enacted so called *right-to-work* laws, which prohibit contracts including such compulsory membership clauses. In these states, workers do not have to join a union at their workplace, even if one has been elected by majority vote to represent the workers. Such workplaces are sometimes called "open shops" in contrast to "union shops."

Third, the Taft-Hartley Amendments allow the President to force striking workers back to work in cases of national emergency. Unions feared that this would reintroduce the extensive use of injunctions against strikes and labeled the Taft-Hartley Amendments the "slave labor act." The national emergency clause, however, has been invoked only a handful of times and in the final analysis has represented a much less significant setback to labor than the exclusion of secondary boycotts and sympathy strikes and the granting of states' rights to prohibit union shops.

Union Retrenchment To confront these and other restrictions on unions, the labor movement closed ranks and carried out a series of internal purges. In 1949 and 1950 the CIO expelled eleven unions that had substantial communist membership. In 1955 the AFL and the CIO, under the leadership of George Meany and Walter Reuther, merged to form the AFL-CIO (www.aflcio.org). The merger resulted from concerns about a leveling off of membership growth and also from a narrowing of ideological differences between the two organizations. With the continued growth of mass production industries, both the AFL and the CIO unions had begun to move toward a hybrid form of unionism that included both semiskilled workers and skilled workers in the same unions.

In 1957 the AFL-CIO expelled the graft-ridden Teamsters, led by Jimmy Hoffa, in an effort to clean up labor's image. In 1959 Congress passed its final major piece of legislation dealing with unions, the Landrum-Griffin Act. The act regulates the internal practices of unions, including the election of officers, and encourages political democracy within unions. Illegal use (and mishandling) of dues and retirement funds still sometimes occurs and sullies the reputation of a union, but it is increasingly uncommon. In the 1990s, after a series of reforms, the International Brotherhood of Teamsters was readmitted to the AFL-CIO, although it would again split from the AFL-CIO in 2005 (see section below on "Divisions in the AFL-CIO").

Racial Equality Well into the period after the Second World War, some labor unions, especially those in the building trades, kept out minorities or restricted them to jobs in specific neighborhoods or to lower-paying positions within the trade. Practices of this type occurred as late as the 1960s among such unions as the Plumbers, Electricians, and Sheet Metal Workers. However, these practices were greatly reduced by the mid-1960s due to pressure from the government, the public, and other, integrated unions in which African-American and other minorities made up a substantial share of membership. By the early 1960s the AFL-CIO had became a strong supporter of the Civil Rights movement, and it played an influential role in getting the equal employment opportunity section, Title VII, included in the landmark Civil Rights Act of 1964 (Borjas, 2005). Today, blacks are more likely to be union members than are whites (Hunt and Rayside, 2000), and blacks and other minorities have also secured leadership positions in many racially integrated unions.

During the 1960s the labor movement and the Civil Rights movement often worked together to improve the position of black people. For example, on February 12, 1968, black sanitation workers went on strike in Memphis, Tennessee. The strike had the active support of the local African-American community, the AFL-CIO, and civil rights organizations such as the Southern Christian Leadership Conference. The Reverend Martin Luther King, Jr., went to Memphis three times, speaking at rallies and leading marches in support of the strikers. He was assassinated in Memphis on April 4, 1968, during his third visit in support of the strike (Foner, 1982).

The union practice of promoting seniority as the basis for allocating layoffs, however, has sometimes had inadvertent negative effects on minority employment. Seniority-based protection against layoffs has tended to favor whites over more recently hired women and minorities. However, court settlements in favor of racial quotas in layoff procedures, and the aggressive organizing of new female and minority workers have tended to reduce these negative consequences (Colgan and Ledwith, 2002). Asian Americans are the only minority group significantly less represented in the labor movement today than majority whites. And active organizing drives among garment workers, hotel and restaurant workers, and medical workers are attempting to increase Asian-American participation in the labor movement (Chen, 1993).

Women in Unions Starting in the 1970s, the AFL-CIO began to be strongly influenced by the women's movement, and a group of female trade unionists formed the Coalition of Labor Union Women (CLUW) (Green, 2000). The CLUW brought women's issues increased attention in the union movement. These issues have included the elimination of the restriction of women from certain occupations because of circumstantial factors such as height or weight, the need for pregnancy leave, and proposals for comprehensive child-care programs (Morris, 2005). The increased visibility of women in the union movement has met with resistance from some men. It has been supported, however, by many others. As early as 1979, for instance, 1,400 members of the International Woodworkers of America, most of whom were men, went on a successful strike over the unfair firing of a female worker at a plywood plant. The worker's firing had resulted from her efforts to file a discrimination suit against the company because she was restricted from preferred jobs and shifts for which she was qualified (Kauffman, 1979). Such examples of solidarity between male and female workers have had an important role in easing tensions as women expand into previously male occupational domains. As a result of aggressive organizing in heavily female occupations and industries, female membership in unions has grown significantly at a time when male membership is still declining. Unions have been particularly successful in organizing women in the fast-growing telecommunications industry. Women's wages in unionized jobs are higher than in nonunion jobs, and the disparity between men and women's wages is significantly less. Box 6.3 presents some of the issues highlighted in an organizing drive among largely female clerical workers at Indiana University by the Communication Workers of America (CWA, www.cwa-union.org).

Public Sector Unions In recent decades the public sector has been the fastest-growing area of union organizing. In 1960 less than a third of federal employees belonged to labor organizations; by 2000 almost two-thirds were union members. In this same period union membership among state, county, and local employees increased from one million to five million, bringing the unionization rate among eligible government employees above 40 percent, higher than any private-sector industry. A majority of states now utilize final and binding arbitration for some or all state employees. This mechanism allows collective bargaining in the public sector without the use of strikes that might disrupt important public services. Most states now have either statutory or tacit recognition of state workers' rights to join unions and negotiate their conditions of employment (Borjas, 2005).

Professional Workers Professional workers have also begun to unionize to improve their bargaining position relative to the large bureaucratic organizations in which they are increasingly employed. In the past professional organizations focused their activities on membership training, the defense of members' legal rights, and legislative lobbying. In recent years many such organizations have become increasingly eager to represent their members in collective bargaining as well. This organizational transformation has resulted in large leaps in union membership (Borjas, 2005). For instance, the combined membership of the National Education

BOX 6.3   Organizing Clerical Workers at Indiana University

IU/CWA Clerical and Technical Organizing Committee
WHAT ARE SOME OF OUR CONCERNS?

SALARY EQUITY: Are you being compensated fairly for all the hard work you do? Did reclassification raise your grade but not your salary? We believe CWA can make a difference. With a union we can address the issue of salary equity—no full-time clerical or technical should qualify for state or federal assistance.

JOB SECURITY: You should not have to work fearing that the next day your job could be eliminated. Job security is part of CWA's strength. Putting in many years of hard work should count for something—we will work together to address the issue of job security.

HEALTH INSURANCE: Everyone knows health costs are going up, but is it always necessary that employees absorb the rising costs? Who can afford a higher deductible? Why don't we have vision care? This is another area where CWA can draw upon their past experience for a fresh approach to the problem.

PENSIONS: Everyone should be able to live comfortably when they retire. Who wants to do with less?

Administrators and faculty live comfortably after retirement, why shouldn't we? CWA will help us to address the issue of pensions and work towards a happier retirement for everyone.

CHILD & ELDER CARE: Why don't IU employees have affordable child care? Presently facilities are expensive and available space is limited. The majority of the clerical/technical workforce at IU is female—working women need good child care. Elder care is also becoming an increasing responsibility for many—together with CWA we can work for improved conditions in both child & elder care.

We are the backbone of the university and we deserve respect. We work hard for IU and for our pay. We believe that together with CWA we can change things for the better. All clericals and technicals need to join together and let the university know our concerns so we may work toward bettering our jobs, our lives and our future.

SOURCE: Communications Workers of America.

Association and the American Federation of Teachers is over 2.7 million, making teachers the largest group of organized workers in the United States.

The largest strike ever in the United States among white-collar workers occurred in 2000. Over 23,000 engineers and technicians at Boeing Aircraft went on strike over such classic trade union issues as pay, benefits, and health insurance. A favorable settlement was reached after thirty-seven days off the job.

Even such high-status professions as university professors have begun to organize unions. In California, faculty in many of the state colleges are organized by the American Federation of Teachers (AFT, www.aft.org). Among college and university faculty nationwide, 170,000 of 400,000 full-time and 300,000 part-time faculty are organized into unions, and labor activists expect additional organizing gains in the future (Morris, 2005). In response, many locals of the American Association

of University Professors (AAUP, www.aaup.org) have felt competition from the more aggressive AFT and have started to engage in collective bargaining in addition to more traditional lobbying and professional development activities.

Farm Workers   Farm workers have also been actively organizing in recent decades. California became the first state to include farm workers under labor laws with the passage of the Agricultural Labor Relations Act in 1975. Since then, other workers in heavily agricultural states such as Florida and Texas have also been actively organizing, though they have met stiff resistance from entrenched agricultural interests in state legislatures. Nevertheless, their efforts have had some success. In the 1980s, for example, the United Farm Workers (www.ufw.org) was successful in getting farm workers in Texas covered for the first time under the state unemployment compensation system.

T A B L E 6.3 **International Comparisons in Union Membership**

Year	U.S.	Canada	Australia	Japan	Denmark	Germany	Italy	Sweden	United Kingdom
1955	33%	31%	64%	36%	59%	44%	57%	62%	46%
1960	32	30	61	33	63	40	34	62	45
1965	28	28	46	36	63	38	33	68	45
1970	27	31	43	35	64	37	43	75	50
1975	22	34	48	35	72	39	56	83	53
1980	22	35	47	31	86	40	62	88	56
1985	17	36	47	29	92	40	61	95	51
1990	16	36	43	25	88	39	65	95	46
1995	15	37	35	24	80	29	44	91	33
2000	14	36	34	23	81	27	46	88	32
2003	14	36	33	23	81	25	47	86	31

Percent of Total Civilian Wage and Salary Employees

SOURCE: Clara Chang and Constance Sorrentino, 1991, "Union membership statistics in 12 countries," *Monthly Labor Review* (December), p.48; *Human Development Report*, New York: Oxford University Press, 1998; *www.eiro.eurofound.eu.int/2004*

International Comparisons The United States currently has the lowest rate of unionization among all the Western industrialized nations. Workforce shortages, a tighter labor market, and new organizing initiatives, however, all suggest the stabilization or reversal of this trend (Stewart, 2005). Perhaps most importantly, there is a large reserve of workers in the United States who want to be unionized but who do not have a union in their workplace. Among workers who do not have a union, fully 33 percent say they would vote in favor of union representation (Freeman and Rogers, 1999).

Unionization rates in some of the major industrialized nations are displayed in Table 6.3. Not only are these rates uniformly higher than in the United States, ranging from 23 percent in Japan to 86 percent in Sweden, but several are on the increase during the last half century, including the unionization rate for Canada. These trends suggest as yet unrealized possibilities for unionization in the United States.

Recent decades have also seen dramatic, if localized, upsurges in union activity in Italy, Poland, and Brazil (Turner, 2005). Union membership numbers are difficult to ascertain for indust-

rially developing nations because of the nature of union activities in these settings. In developing nations, unions are more key players in promoting broad social justice agendas. As was the case for North American workers in the nineteenth and early twentieth centuries, the goals sought are diverse and include such issues as the right to form unions and to strike, freedom of speech, and basic public education and welfare benefits. Union membership and activities in many developing nations continue to be illegal, dangerous, or both.

Lessons from Labor's History

We have seen that the North American labor movement is not monolithic. Rather, its history is one of deep internal schisms as organizations representing different workers and different strategies for articulating and realizing their goals contend for power (Clawson, 2003; Fantasia and Voss, 2004). The labor movement today embraces a variety of often divergent viewpoints and needs, ranging from those of doctors and skilled craft workers to clerical workers and janitors.

Labor unions in North America have often appeared to *react* to circumstances rather than to take a leadership role in shaping events in the economy. In many ways this is not surprising. Unions have suffered severe repression at various points in American history. The organizations that have survived have generally been conservative and cautious in their behavior. Conservative leadership, however, may be more of a liability than an asset in times of rapid technological change and increased foreign competition such as the present. American labor unions will need to take innovative stands if they are to survive and prosper. History suggests that unions *will* adapt successfully to change. In their two-hundred-year history in the United States, unions have repeatedly had to adapt to changing circumstances. The emergence of the CIO and industrial unionism in the 1930s provides one of the clearest examples of such creative adjustments to altered circumstances. Unions were not created just to bedevil the current generation of managers. They are an enduring part of the social, economic, and political landscape of industrial societies. In the next section, we explore the functioning of contemporary North American unions as well as the various creative innovations they are developing to address the changing needs of workers.

TWENTY-FIRST CENTURY
LABOR UNIONS

The struggles of the labor movement have produced many gains for working people in the United States and Canada. These gains include higher wages, shorter hours, safer working conditions, and increased bilateral governance of the workplace by negotiated rules rather than management fiat. The legacy of the union movement, however, is more than just these specific gains. It is also a vision and a program of how these gains can be protected and extended. That vision includes the central roles of organizing and collective bargaining, the maintenance of effective grievance procedures, job security, and the targeted use of political lobbying.

How do labor unions fulfill their many current roles? Which unions are declining, which ones are growing, and why? How are unions confronting rapid technological change and increased international competition? What innovative programs are unions developing to confront the challenges of the twenty-first century? In this section we explore the answers to these questions organized around the three key union roles of *bargaining, organizing,* and *lobbying.*

Collective Bargaining

Negotiating wages and conditions and enforcing the contract thus agreed on are key roles of North American labor unions. After a union wins a certification election, it tries to negotiate a contract with the employer, who is required only to bargain with the union in good faith. About 150,000 labor contracts cover the sixteen million unionized workers in the United States, with over half of these workers covered by the two thousand largest contracts. Labor contracts, and the precedents they set for workplace relations, represent significant extensions of the property rights of workers over their jobs and over their conditions of employment.

How high are union wages? Higher than those of nonunion workers but lower than popular belief frequently portrays them. Comparisons are difficult to make, because union workers are typically more highly skilled and are employed in higher-paying industries and larger organizations than nonunion workers. In 2003 the average weekly pay of *nonunion workers* was $599. The average weekly pay of *union workers* was about 20 percent higher, at $730 (Census, 2004). If the skills of workers, the sizes of workplaces, and similar differences are statistically controlled, then union workers on average earn about 10 to 15 percent more than nonunion workers (Wallace, Leicht, and Raffalovich, 1999).

Union wages are relatively compressed between the lowest- and the highest-paid workers relative to nonunion establishments. A reduction in pay differentials helps to maintain community of interests among workers. Bringing up the wages of the lowest-paid employees also protects all workers by reducing employer incentives to replace highly paid workers with poorly paid workers.

Fringe benefits, such as health insurance, retirement, and paid holidays, make up an increasing share of total payments to workers. Unions have played an important role in extending these benefits from salaried employees to all workers. Union employees receive 24 percent of their total compensation in fringe benefits, as compared with 18 percent for nonunion workers (Mishel and Voos, 1992). This difference further widens the gap in total compensation between union and nonunion employees.

Collective bargaining agreements are open contracts that can include almost anything to which both parties agree. Besides wages and benefits, contracts also typically include overtime and shift premiums, promotion and layoff procedures, provisions for due process in discharge cases, and a formal procedure for handling workers' grievances. In addition, elected union officials, such as the president or vice president of the local, are often given a certain amount of *release time* from their jobs to attend to union business, such as the handling of grievances. In small units or units with weak contracts, such a clause may be absent or may allow only one afternoon a week, even for the union president. In larger shops release time is also bargained for a number of **shop stewards**, whose role is informing workers of their contractual rights and handling their grievances about violations of these rights.

Provisions concerning work rules are less common. Managers generally regard the setting of work rules as their inalienable right, but some work rules may be negotiable, especially in hazardous work settings. A union contract for workers at an electrical power plant, for example, may specify that two workers (not one) must be sent to work on remote power stations and that both must be qualified electricians. The NLRA explicitly forbids *featherbedding* clauses, which mandate unnecessary workers.

Most union contracts include provisions for due process in discharge cases. The procedures usually include a statement of "just cause" and may even permit a third-party review of the discharge on union request. **Seniority** clauses are also common. Seniority may be used to allocate protection from layoffs, offer first choice of shifts or job transfers, and provide access to apprenticeship training programs based on years of service. Seniority clauses are typically strongest in relation to layoffs. They are often weakest in relation to promotions. Managers generally prefer promotion criteria based on a supervisory evaluation of merit.

A crucial section in a union contract concerns **grievance procedures**—the set of procedures for handling of workers' complaints about violations of their rights under the contract. The shop steward or other appropriate union official has the right and duty to represent workers when they feel their rights have been violated. If a worker has been docked pay for what he or she feels should have been a legitimate paid sick day, for example, the shop steward will try to resolve the issue with the worker's supervisor. If the issue cannot be resolved at this level, the shop steward will advance the grievance, in writing, to the next highest organizational level, perhaps to the plant manager and the union president. If they cannot reach an agreement, most grievance clauses mandate final and binding arbitration by a neutral third party acceptable to both union and management. Most arbitrators are members of the American Arbitration Association. They typically act as arbitrators on a part-time basis; many are also employed as lawyers or university professors.

Providing a fair and effective system for handling workers' grievances is often a union's most important contribution to a workplace. Individually, workers are relatively powerless in relation to managers. When disputes arise, this powerlessness can produce bitterness and resentment. The grievance procedure allows workers the right to an impartial hearing if a dispute cannot be settled by direct negotiation. Many observers argue that due process is also in the interests of the company because it channels potentially bitter personal disputes into more peaceful mechanisms of conflict resolution. The successful resolution of such conflicts helps clear the air and prevents the emergence of lingering resentments.

Strikes Strikes and the threat of strikes are the most important mechanisms through which unions press their claims. However, strikes themselves are quite infrequent and are not always successful (Green, 2000). In 2003 there were only fourteen strikes in the United States that involved over a thousand workers. A total of 129,000 workers were involved in these strikes. Lost time due to strikes represented 0.01 percent of the total time worked by the labor force (Census, 2004). The peak of strike activity since the Second World War occurred in 1971, when 2.5 million workers were involved. Throughout the postwar period an average of less than 0.05 percent of work time has been lost due to strikes. This amounts to less than one day lost per thousand days worked, a tiny fraction of the amount of time lost to the common cold or to industrial accidents.

The benefits won by union members directly influence the wages and working conditions of nonunionized workers. Union gains in one plant or industry generally bring up the prevailing wage in related plants and industries. In addition, many large nonunion companies keep their wages and benefits comparable with those in union plants to help undercut union organizing efforts at their plants. The workers in such nonunion plants are indirect beneficiaries of the struggles and sacrifices of union workers, though they pay no dues and have never had to risk their jobs in a strike. Even the benefit packages of professional and supervisory workers in unionized plants are strongly influenced by the benefits won by union workers (Borjas, 2005).

In recent years workers have been increasingly reluctant to strike because of fears that management will permanently move their jobs elsewhere. This fear has forced unions to look for alternatives to strikes such as political lobbying and worker ownership. (See the section below on "Innovative Organizing and Bargaining Strategies.")

Promoting Safety and Health   Worker safety and health have gained increasing attention as union priorities. This concern results from both increased use of hazardous chemicals at the workplace and increased public awareness of health issues. Unions strongly support *right-to-know legislation,* which requires the labeling of chemicals at the workplace. Unions also support more stringent limits on industrial chemicals proven to be hazardous and increased information and education for workers on workplace hazards. Unions have also taken a leading role in developing programs to limit the hazards of video display terminals in the workplace and associated joint problems from prolonged sitting and repetitive wrist movements (Morris, 2005).

A Broader Role in the Manufacturing Sector   In the manufacturing sector, union goals focus on increasing the participation of workers in managing work and setting organizational goals (Appelbaum et al., 2000). The UAW has been a leader in this regard. Current UAW contracts with the major automakers call for extensive training programs jointly managed by the union and the company. These programs help workers learn the skills needed to make the most efficient use of new technologies. Innovative programs include daily meetings of senior union representatives and plant managers about the operation of the plants, as well as an active role for thousands of teams of workers and supervisors trying to increase efficiency at every level of operation.

Union goals in the manufacturing sector also include improving the quality of work life, as well as improving product quality. Issues of concern include ergonomics, flexible work arrangements to accommodate child care, and family leave policies. Worker initiative in identifying areas of concern has a proven history for generating important and often cost-effective changes in the workplace (Biagi, 2002).

Unions have a strong interest in programs of worker participation, but they have reservations stemming primarily from concerns for job security. Many large corporations follow antiunion strategies at the top levels of strategic planning (Katz, Batt, and Keefe, 2003). Such strategies move plants to regions of the country less hospitable to unions or even move work overseas to avoid unions. In this context, unions may find that local plant management is a willing partner in participation programs to fulfill productivity objectives but that top management is

Milk Parker 1985, *Inside the Circle: A Union Guide to QWL*. Boston: South End Press, p.56. Used by permission of the artist, Mike Konopacki.

simultaneously moving jobs overseas. Unions are thus sometimes suspicious that participation programs may be a short-run strategy to get more work out of workers prior to layoffs. Quality control circles and other workplace participation programs have also been used to encourage workers to spy on each other and report union organizers and sympathizers to management (Grenier, 1988). Unions are suspicious because quality circles can be a carrot with a stick attached. On the other hand, increased productivity and increased quality are essential for the long-term survival of high-paying manufacturing jobs in the United States.

Because of these reservations about the integrity and range of management-sponsored participation programs, unions have also sought to increase the rights of workers over investment decisions. Such rights may be manifest in contractually negotiated job security clauses. Investment rights can also be realized through various forms of worker ownership. Worker ownership is an increasingly common strategy as workers confront plant closures and other forms of disinvestment. Such strategies by workers are common both in North America and throughout the world. For example, in the UK a manufacturer of commercial vehicle bodies wished to sell off its repair business to focus on the more profitable selling of new commercial vehicles. The twenty-four employees in the repair unit invested about five thousand

dollars each to purchase the unit. The process took over a year to complete and concluded with the new worker-owners renting the premises and paying for the plant and machinery under long-term contract. Stable relationships have been secured with both vendors and clients, and profits have increased in each of the three years since the conversion (Employee Ownership Options, 2005).

Unmet Membership Concerns   Unions have also had many failures in the postwar United States. They have not always been able to address some of the pressing goals and concerns of their members. They have had limited effectiveness, for example, in negotiating working conditions and work rules. Management has retained control over determining work rules in almost all settings. Although unions have been successful in establishing seniority as the *basis for layoffs* they have had little success in actually *reducing layoffs*.

Organizing

In addition to collective bargaining, unions must also constantly recruit and organize new members in order to keep up with the rapidly changing industrial and occupational structure. This challenge is further complicated by international competition for jobs and increased company resistance to unions.

Running Fast to Stay in Place   Union membership in the United States grew rapidly during the Great Depression and steadily, though at a slower pace, during the 1950s, 1960s, and into the 1970s. Figure 6.1 displays trends since 1900 in union membership. Membership reached a peak in 1978 with 22,757,000 members enrolled and perhaps 10 percent more covered by collective bargaining agreements but not enrolled as dues-paying union members. This figure represented 22.2 percent of the total labor force and 26.2 percent of nonagricultural employees (Borjas, 2005). Union membership as a *proportion* of nonagricultural employment reached its peak during the Second World War at 35.5 percent. During the 1970s and 1980s union membership declined to less than 20 percent of the labor force. In the 1990s these declines ended, and union

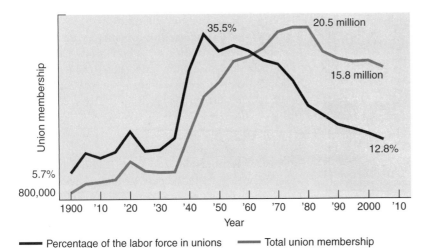

FIGURE 6.1 Union Membership Trends in the United States

SOURCE: Larry J. Griffin, Philip J. O'Connell, and Holly J. McCammon, 1989, "National Variation in the Context of Struggle." *Canadian Review of Sociology and Anthropology* 26,1 (February): 37–68; Census, 2005. Reprinted by permission of the Larry J.Griffin.

membership in the United States actually increased slightly. Growth areas include both reasonably well-paid professional workers on the one hand, and low-wage service workers on the other. The rate of union membership in Canada, by contrast, is over double that in the United States and has been stable or growing during the last several decades. In part these differences reflect a less hostile legal environment for unions in Canada.

What accounts for the decline of union membership in the 1980s and 1990s in the United States and its return to a more stable position in the early 2000s? Ironically, one factor that has limited new union membership is the union's very success in securing benefits for workers through collective bargaining, social welfare legislation, and spillover effects on non-unionized jobs. It is easy for workers to forget how hard it was to win these benefits. It is relatively more difficult to remain active in the union movement in order to secure and extend these benefits. The erosion of job benefits and job security in the 2000s has promoted a resurgence of union membership.

Industrial Shifts   Shifts in the composition of industries and occupations have tended to undermine traditional union strongholds. Employment has declined in industries in which unions became

strong in the 1930s, such as mining and steel, because of technologically based automation. And it has increased in white-collar occupations and service industries in which unions had historically been weak. The union movement has had to organize aggressively in areas of employment growth just to maintain membership at current levels. These efforts have been partially successful. Unionization among white-collar workers increased from 12 percent in 1970 to 16 percent by the early 2000s and now exceeds membership in blue-collar occupations. Similarly, the proportion of union membership among women workers has increased from 10 percent to 13 percent. In the most recent decade, well over half of new union members were women.

International Competition   The decline of union membership also results from additional factors acting to depress manufacturing employment and discourage unionization. High levels of unemployment in manufacturing have resulted at least partially from increased international competition and the movement of American factories and corporations overseas in the search for cheaper labor, lower taxes, and less stringent environmental and safety and health regulations. Unionized steelworkers declined from a peak of 1,062,000 in 1975 to just over 600,000 by 2002 (Gifford, 2002).

Based on the declining competitiveness of many American-manufactured goods and services in the world economy, many corporations have sought either contractual concessions from labor unions or the elimination of labor unions altogether. In the 2000s, many automobile plants in the United States and Canada have closed as market share has been lost to Japanese and European automakers and as U.S. corporations have transferred their production overseas. Chronically high unemployment has produced a new bargaining environment that encourages a much more aggressive stance on the part of management (Milkman, 1997).

Increased Company Resistance More direct attacks on unions have also increased. Paid spies and sophisticated antiunion consultants are increasingly used by companies to ferret out and eliminate pro-union workers (Perkins, 2004). It has been estimated that 75 percent of employers who experience union organizing drives hire antiunion consultants to orchestrate their efforts to avoid unionization (Tilly, 2004). Illegal management tactics during NLRB elections have become an important—perhaps the most important—reason for the failure of many such campaigns. The odds that a worker exercising his or her legal rights will be fired for union activity have become extraordinarily high (Freeman and Medoff, 1984).

In recent years workers have attempted to organize at the world's largest retailer—Wal-Mart, which has over 1.2 million employees. Workers' concerns have included a company policy of locking in overnight employees who are restocking shelves. This policy is supposedly to protect workers in high crime areas, but others argue it is to prevent any theft by employees. Whatever the reason, it has resulted in difficult and dangerous situations for workers, including two whose wives went into childbirth labor while they were locked in and at least one other who suffered a heart attack. To date, Wal-Mart has blocked all unionization drives by extensive use of anti-labor consultants, outsourcing tasks being unionized (e.g., switching to prepackaged meat when butchers seek to organize), and flooding units being organized with new employees screened for anti-union attitudes (Greenhouse, 2005).

Why do employers increasingly attack their workers' unions? Are unions responsible for the decline in the United States' competitive position in the world market? The available evidence suggests that union workers are *more* productive than non-union workers. Their greater productivity results from better training programs, lower turnover, greater tenure, increased commitment to the job, and more professional management in unionized firms (Gunderson and Ponak, 2000). Greater productivity in unionized firms results from management having been forced to develop high productivity routes to competitiveness because unions resist low-wage routes to competitiveness (Belman and Voos, 2004). Unionization does tend to lower the *profit* rate of firms, because union workers expect higher pay for their more skilled labor. On average, unions reduce the profit rate of companies by about 18 percent (Freeman and Medoff, 1984:183). The resulting paradox between increased productivity and decreased profitability can create pressure on management to resist unions. Eliminating unions, however, can undermine productivity over the long run.

Growth Areas Even in the face of industrial shifts, international competition, and increased company resistance, many unions have grown. Among the fastest-growing unions are the American Federation of State, County, and Municipal Employees (AFSCME, www.afscme.org), the United Food and Commercial Workers (UFCW, www.ufcw.org), and the Service Employees International Union (SEIU, www.seiu.org), all of which represent service workers. Similarly, the American Federation of Teachers more than quadrupled its membership in the last three decades, moving from 165,000 to 736,000.

Workers in large public institutions such as hospitals and schools have had increasing organizational successes in recent years. In the public sector, 42 percent of workers are unionized, while in the private sector, only 13 percent are members. Part of the reason is that public-sector employers are not free to manipulate and intimidate their workers to the same extent as private-sector employers, because

---

BOX 6.4 **Organizing Topless Dancers**

$

Topless dancers at one of [San Diego's] oldest and best-known bars voted ... on becoming California's first unionized nude nightclub.

The dancers—joined by bouncers, bartenders and disc jockeys—are angry about pay and working conditions.... Dancers earn $4.35 an hour, plus $100 or more in tips per 6–8 hour shift. But they say club policy forces them to:

- Pay the club $5 an hour for each hour worked.

- Give 15 percent of their tips to bartenders and disc jockeys, who say they give 40 percent of all tips to the club.

- Buy costume accessories from the club at inflated prices. Dancers claim cheap garters are sold for $3.50

and last only a night. They say they buy stockings at $10.50 a pair, which last two or three nights.

Mary, 23, a college student and dancer [at the bar], who won't give her name for fear of being fired, says some nights dancers don't make anything for their work, but owe the club....

Dancers also charge sexual harassment from management, safety violations and lack of job security.... Organizer Robert Fisher of Hotel and Restaurant Employees Union Local 30 is confident a majority of the ninety dancers and twenty other employees will vote to unionize.

SOURCE: Ross, Bob. 1993. "Bottom line for topless dancers." *USA Today*, June 4, p.2. Reprinted with permission.

---

government officials must ultimately answer to the public through elections.

Organizing Low-Wage Workers The growth of low-wage industries and occupations has increasingly moved the United States toward becoming a *two-tier society*. The upper tier consists of reasonably well-paid jobs, and the other, growing tier, consists of marginal employment (see Chapter 17 for a fuller discussion of the movement toward a two-tier society). The lower tier of jobs provides a major focus for current union organizing activities. Poor wages and virtually nonexistent fringe benefits in lower-tier jobs have resulted in unions having more success in organizing workers in smaller firms than in larger corporate giants (Bronfenbrenner, 1998). Unions promote the theme of social justice for these workers who may feel excluded from the mainstream of American society.

Service workers, including health care workers, hotel workers, and janitors are a growth area. The expansion of unionization into the service sector represents a revival of union effort to organize and serve the poorest segments of society. An affiliate of the SEIU called "Justice for Janitors" (www.seiu.org/property/janitors) has made widespread organizing gains among janitors. In some large cities, such as Washington, DC and Los Angeles, 40 percent to 90 percent of janitors are organized. The group uses '60s-style sit-downs and demonstrations to pressure building owners into negotiating. In 2004 a large janitors' strike in San Francisco succeeded in making lasting gains in wages and health benefits for janitors in the city.

Unions have made important advances in organizing low-wage service workers in the health care industry. New members include hospital employees as well as workers in residential homes for the elderly, the mentally handicapped, and the physically disabled. The fast-food industry is also being targeted by union organizers as a ready pool of poorly paid and poorly treated workers ripe for organizing (Morris, 2005). It seems likely that the future of organized labor in North America significantly rests on the success of such organizing drives among low-wage service workers in recreation, health care, and related industries (Lopez, 2004). Box 6.4 reports on a union movement among topless dancers, a seemingly unlikely occupation to organize, but one with serious demands and objectives similar to those of other workers.

Lobbying

Labor unions exert political power by registering voters, especially low-income and working-class voters, and by encouraging them to vote. They mobilize noncash union resources, including staff time and volunteer efforts, for the political campaigns of pro-labor candidates. Finally, they lobby for legislation favorable to workers. Because unions directly represent less than 20 percent of the labor force, they have had limited success in securing legislation directly favorable to organized labor, such as liberalization of picketing laws or elimination of states' rights to outlaw union shop contracts. They have enjoyed greater success in coalitions with other groups supporting issues such as public education, anti-poverty legislation, civil rights, voting rights, health insurance, public housing, and occupational safety and health.

Union leaders believe that progress is possible on some labor issues, but that sweeping advances are unlikely. Consequently, the union movement puts much of its effort into backing legislation that favors the working class and the working poor more generally, rather than labor unions specifically. First and foremost, this means supporting laws that promote full employment through tax incentives, public works programs, and whatever other means are available.

Improving the Image of Unions

To revitalize the union movement, unions seek to improve their image with the public and with their own members. The prosperity of the second half of the twentieth century led many people to feel that unions were no longer necessary. More recent economic stagnation has reawakened concerns about job security and wages. However, this awareness came at a time when union membership was declining because of layoffs, and employers were increasingly aggressive against unions. In these attacks management blamed unions for the country's economic problems. To counteract these attacks, unions have sought to improve their image by addressing issues of general interest. As you have seen, these issues include safety and health, plant closings, trade policies, and education.

Unions are mounting a more active counterattack to defend and bolster their public image (Martin, 2005). Current union policy priorities of safety and health and child care are selected partly with public appeal in mind. The AFL-CIO also initiated a multimedia project called "Union, YES!" that has probably had its greatest visibility through car bumper stickers. Even the Boy Scouts have been targeted, and the AFL-CIO has been successful in getting a labor badge included in the merit badge system of the Boy Scouts. Similarly, the California Federation of Teachers developed a set of lesson plans for teaching about the role of trade unions in resolving workplace conflicts. The lessons involve a role playing exercise in which students work in a fictional company, "Yummy Pizza," and encounter various management restrictions, such as rules against talking and limited restroom breaks. Unions are also taking on increasingly active roles in promoting such community service programs as United Way (Ness and Eimer, 2001). Such strategies are potentially effective—recent national polls suggest that a strong majority of Americans approve of unions and that unions' approval ratings have been improving steadily since bottoming out in the early 1980s (Freeman and Rogers, 1999).

Innovative Organizing and Bargaining Strategies

A union movement that had become conservative and cautious in the prosperity of the second half of the twentieth century has had to become bolder and more innovative in the twenty-first century. Unions have developed new organizing and bargaining strategies, such as the **corporate campaign**, which expands union activity outside the workplace by targeting financial backers of the company, consumers, and the public (Martin, 2004). Public boycotts of a company and related negative publicity can be very effective, especially in customer-oriented service industries (Jensen and Hammerback, 2002). A union might also target the principal banks lending money to a company and distribute literature linking the banks with wage cuts, unsafe work practices, layoffs, shutdowns, or other problems the workers are experiencing with their employer. Other unions sometimes support such actions by closing their accounts with the banks, including large retirement accounts. The bank in turn—to protect its own image and interests—may pressure the company to act more responsibly toward its employees. The Hotel Employees and Restaurant Employees Union (HERE, www.unitehere.org) has been a leader in using such pressure tactics to force large hotel chains to the bargaining table.

Such innovative bargaining strategies return to the social movements roots of modern unions and may be more effective bargaining tools than traditional organizing and bargaining strategies that rely on oversight by the National Labor Relations Board (NLRB). The NLRB is headed by political appointees, and often these officials have been unable or unwilling to ensure that union recognition elections and subsequent contract negotiations proceed within a reasonable time frame. When companies are able to delay elections and negotiations for years, their power over employment and investment decisions can circumvent and undermine a union, even one initially strongly supported by workers. Tactics based outside a single workplace are also effective and increasingly needed where workers are dispersed in small establishments, as in many service industries.

Unions and other activist groups are also developing new hybrid organizational forms that transcend boundaries between people acting as employees and as citizens. In Central Florida, for instance, current and former employees of Wal-Mart have formed a voluntary association to put pressure on the state of Florida to protect the interests of Wal-Mart workers. They complain that Wal-Mart cuts hours and makes employees choose between paying the rent and having food on the table. Wal-Mart's health care plan has such high premiums and deductibles that many workers cannot afford to enroll in it and must rely on federal or state aid for whatever health care they receive. "We are building something that's never been seen; it's neither fish nor fowl [says the association's chief organizer]. . . . We're focusing on Wal-Mart because it is the largest employer in the area—and in the whole nation—and is setting standards that affect

communities and employment relations across the nation" (Greenhouse, 2005b).

Divisions in the AFL-CIO

The tension between traditional bargaining to service members' needs and aggressive new organizing strategies produced a major rift in the AFL-CIO in 2005. In the largest reorganization in organized labor in 70 years—since the creation of the CIO in 1935—seven large unions pulled out of the AFL-CIO and formed a new group, "Change to Win" (www.changetowin. org). *These* unions include some of the largest and fastest-growing unions in the nation—the Teamsters (IBT), the Service Employees (SEIU), and the combined needle workers and hotel and restaurant employees (UNITE-HERE)—and take with them nearly half of the AFL-CIO's former membership.

Opinions vary on what happens next. Some observers fear that a period of raiding of other unions' members will follow in which resources are wasted on defaming other unions and encouraging their members to vote out their old union and vote in a new one. Other observers believe that competition between unions will invigorate the union movement and create pressures for accountability. Members with choices may feel entitled to hold their unions more accountable for improving wages and working conditions. Such optimistic scenarios note that the decades between 1935 and 1955—a period in which the AFL and CIO competed with each other for members—saw fastest union growth in the nation's history. Observers also note that some of the most vigorous and successful union movements exist in nations with divided labor organizations, such as Canada, France, Korea, and Argentina (Cutler, 2005).

Transnational Strategies

National unions are today confronted with the important reality that the production of goods and services is increasingly international while their own organizing, bargaining, and lobbying activities remain national or even local in orientation. Without the ability to engage in strategic action at the international level, unions are increasingly disadvantaged in bargaining with corporations that can move facilities and resources around the globe at will. How can unions learn to act effectively on a global stage?

A number of strategies are being developed and utilized by unions today to bargain, pressure, and otherwise represent their members' needs on a global stage. Unions in industrialized nations are increasing their support for unions in developing nations as these unions struggle for recognition and for workers' rights (Sandoval, 2005). Support includes providing trainers, workshops, and legal and other resources. The success of such efforts is essential not only for the well-being of workers in newly industrializing nations but also for those in already industrialized nations.

A second major strategy is for unions to pressure their own governments to place restrictions on imported goods from nations that violate basic worker rights. Unions argue that global standards for workers' rights should include prohibition of child labor, the right to organize collectively, and basic health and safety standards (Dreiling, 2000). They argue that such standards are essential as a basis for **fair trade** between workers and companies in different nations. (See Chapter 16 on the global economy for further discussion of the "free trade" versus "fair trade" debate.)

Unions actively seek to pressure world economic organizations, such as the World Trade Organization, the International Monetary Fund, and the various regional trading blocks to include recognition and support for workers' rights as a prerequisite for opening markets to foreign goods. These efforts have met with resistance but also with some incremental success (Kay, 2005). Perhaps most importantly, the emergence of transnational economic bodies has provided a platform for labor unions to speak on issues of general importance to the working people of the world and to become a part of broader coalitions seeking redress for poverty and injustice on a global scale.

SUMMARY

Workers organize themselves into unions to seek redress for grievances at work. On some issues there will always be a divergence between workers and managers. Modern labor unions emerged as the consequence of a long struggle by workers to improve their situation. In the process of this struggle, many alternative forms of organization were explored, including workers' parties, producers' cooperatives, and mass unions. In the United States and Canada the most lasting and significant form of worker organization has been the trade union, which focuses on collective bargaining over wages and conditions. Today, trade unions organize both skilled craft workers and less skilled workers in mass production industries, as well as increasing numbers of government, service, and professional employees. Unions have been able to coordinate the potentially divergent interests of these groups of workers by combining into national federations—chiefly the AFL-CIO in the United States and the Canadian Labour Congress. An important key to the success of unions in the United States and Canada is the simultaneous achievement of trade union autonomy within these federated structures and the presentation of a united front for labor that can speak with a single voice on important national issues.

The primary legacies of the labor movement are heightened job security, the provision of a grievance system, and collective bilateral bargaining with managers over the conditions of work. These rights are secured and protected through the use of strikes. Although strikes are infrequent, they play a vital role in making the collective bargaining system work.

Union membership in the United States declined in the last forty years of the twentieth century because of employment shifts from manufacturing into white-collar and service jobs, the movement of jobs overseas or into lower wage areas, and increased management attacks on unions. These losses have been partially offset by membership gains in the service industries, particularly in the hotel, restaurant, and health industries. Union priorities for the future include continued efforts to promote full employment, increased participation of workers in setting organizational goals in the manufacturing sector, increased attention to organizing lower-tier jobs in the service sector, and efforts to improve the image of labor as an essential actor in advanced industrial society.

In some ways the decline of membership in labor unions in the 1980s and 1990s is reminiscent of similar declines in the 1920s. On the other hand, "the labor movement today numbers over 23,000,000 members, compared to 3,000,000 in 1929, and today's unions include millions of women, nonwhites, and profes-

sional employees absent from the labor movement of the 1920s" (Dulles and Dubofsky, 2004:400). It is impossible to predict whether the next decades will witness a boom in union membership comparable to the 1930s. Further dramatic declines, however, seem unlikely. The bulk of membership losses in the manufacturing sector have probably already occurred, and organizing drives in service and white-collar jobs have been increasingly successful. How successfully unions will respond to the challenges and opportunities of the future remains an open question.

KEY CONCEPTS

craft union

solidarity

general strike

mass strike

secondary boycott

general union

Knights of Labor

Industrial Workers of the World

injunction

American Federation of Labor

Congress of Industrial Organizations

industrial unionism

sit-down strike

National Labor Relations Act

Taft-Hartley Amendments

collective bargaining

shop steward

seniority

grievance procedure

corporate campaign

fair trade

QUESTIONS FOR THOUGHT

1. Describe some of the different ways in which workers have organized themselves and the reasons for the successes and failures of different types of organizations.

2. Why was the Great Depression the pivotal turning point for labor unions in North America?

3. What are the major roles of labor unions in the workplace today?

4. What problems are labor unions facing today? Which of these challenges do you think unions

will be able to meet? Which will continue to be problems?

5. Describe current areas of union growth and the reasons that these are growth areas.

6. What strategies are labor unions developing to confront the changing economic and political situation of the 2000s? Do you think these strategies will be successful in helping to revitalize the union movement?

MULTIMEDIA RESOURCES

Print

Foster Dulles and Melvyn Dubofsky. 2004. Labor in America, 7th ed. Arlington Heights, IL: Harlan Davidson. Perhaps the best comprehensive history of the American labor movement.

Richard B. Freeman and James L. Medoff. 1984. *What Do Unions Do?* New York: Basic. The best source book available on the role of labor unions in the contemporary economy.

Richard B. Freeman and Joel Rogers. 1999. *What Workers Want.* Ithaca, N.Y.: Industrial and Labor Relations Press. Reports a comprehensive new survey on the desires of workers for unions and other forms of representative participation.

Daniel B. Cornfield and Holly J. McCammon (editors). 2003. *Labor Revitalization: Global Perspectives and New Initiatives.* Amsterdam: Elsevier-JAI. Articles from around the world on the efforts of unions to adapt to the challenges of the twenty-first century.

John Pencavel. 2002. *Worker Participation: Lessons from the Worker Co-ops of the Pacific Northwest.* New York: Russell Sage. Key insights on the human and economic successes of worker ownership in the plywood industry of Washington State.

John Steinbeck. 1939. *In Dubious Battle.* New York: Modern Library. A compelling fictional account of the rise and defeat of an early unionization effort among California's agricultural workers.

John Perkins. 2004. *Confessions of an Economic Hit Man.* New York: Penguin. A disturbing look at how an operative for U.S. multinational companies bribed and coerced foreign governments into economic and labor policies against the interests of their own citizens.

Internet

Labor Notes news magazine. www.labornotes.org *A quarter-century-old independent workers' news magazine operating from Detroit.*

International labor news. www.labourstart.org *Labor news from around the world.*

Canadian Auto Workers. www.caw.ca.

Walter Reuther Labor Archives. www.reuther.wayne. edu *The most important archives of labor information and history in North America.*

Institute for Global Communications. www.igc.org *Progressive causes network.*

Ohio Employee Ownership Center. dept.kent.edu/oeoc *Useful resources for both academics and workers on building employee ownership.*

RECOMMENDED FILM

Norma Rae (1979). Sally Fields and Beau Bridges. The true-life story of a poor female textile worker who helps organize a southern mill.

PART III

Industries and Technologies

In the first two parts of this book you learned about the history of work, the problems and satisfactions of work, how people enter a particular career, and how people respond individually and collectively to work. In this section we examine in greater detail the nature of work in specific settings.

Companies that produce the same product or service are collectively called an *industry*. Each enterprise in an industry is likely to have a similar, though not necessarily identical, technology and organizational structure. Industries are thus bundles of products or services with attendant technologies and organizational structures. The industry in which one works determines the nature of one's work: different industries require different skills and provide more or less stable employment opportunities. In this section we discuss how these differences influence the nature of work.

Why do employment levels in different industries change across time? Many employment changes result from changing technologies that allow more goods or services to be produced with less labor. The most dramatic example of such technologically driven changes in employment is provided by agriculture. In 1790 nearly 90 percent of the labor force was required to produce food for Americans (Chandler, Amatori and Kikino, 1997). By 2000, less than 2 percent of the labor force was needed to produce an even greater quality and variety of food (see Table A). Employment changes also result from older goods being replaced by new goods and services. For example, few blacksmiths are employed to shoe horses in a modern economy. However, a great many mechanics are employed to maintain and repair automobile engines.

Some industries have grown dramatically in their share of employment. Educational services have grown as a result of the greatly expanded need and desire for literacy and higher education. Professional services, including medical,

T A B L E A Percentage Distribution of American Workers by Industry, 1870–2012

Industry	1870	1880	1890	1900	1910	1920	1930	1940	1950	1960	1970	1980	1990	2000	2012[a]
Agriculture	50.8	50.6	43.2	38.1	32.1	27.6	22.9	19.2	12.7	7.0	3.7	3.4	2.7	1.8	1.7
Mining	1.6	1.8	2.0	2.6	2.9	3.0	2.5	2.1	1.7	1.1	0.8	1.0	0.6	0.3	0.3
Construction	5.9	4.8	6.1	5.8	6.4	5.3	6.5	4.7	6.2	6.2	5.8	6.2	6.5	7.2	5.9
Manufacturing	17.6	18.4	20.2	22.1	22.8	26.4	24.4	23.9	26.3	28.3	25.9	22.1	18.0	14.4	12.0
Transportation and utilities	5.0	5.0	6.5	7.3	8.8	10.2	7.6	7.0	7.9	7.1	6.8	6.6	6.9	6.4	6.7
Trade	6.5	7.1	8.4	9.6	9.3	9.8	12.6	14.5	15.8	16.1	16.9	20.3	20.6	18.6	18.1
Finance and real estate	–	–	–	–	1.4	1.9	3.0	3.4	3.5	4.3	5.4	6.0	6.8	6.7	6.0
Domestic service	7.4	6.3	6.5	6.1	6.0	4.1	6.5	5.3	3.2	3.1	1.7	1.3	0.9	0.5	0.6
Other personal services	2.0	2.1	2.7	3.4	4.2	4.0	4.7	8.7	8.9	8.2	8.3	8.0	10.8	12.2	13.3
Education	1.5	1.9	2.2	2.3	2.5	2.8	3.1	3.5	3.8	5.4	8.6	7.7	7.3	8.1	8.4
Other professional services	1.1	1.1	1.5	1.7	2.1	2.6	3.5	4.4	5.4	8.0	10.5	12.0	14.2	19.4	22.3
Government	0.8	0.8	0.8	1.0	1.5	2.2	2.8	3.3	4.5	5.2	5.6	5.4	4.8	4.4	4.7

[a]Projection based on assumption of moderate economic growth.

SOURCES: U.S. Department of Commerce, Bureau of the Census, 2004, *Statistical Abstract of the United States;* and Bureau of the Census, 1975, *Historical Statistics of the United States,* Washington, DC: U.S. Government Printing Office.

engineering, accounting, and legal services, have grown rapidly and are now the largest single sector in the economy. Similarly, government (public administration) expanded through the 1960s because of the growth of government-provided services. Recent growth has occurred only at the state, county, and municipal levels with the federal government actually decreasing in size in recent decades.

Retail and wholesale trade is also a large employer. This sector has grown because of the increasing availability of consumer goods and the need to sell these goods. Personal services have also grown over time. Personal services include such industries as hotels, eating and drinking places, day care for children, repair, laundry, and entertainment. Housewives and domestic servants working at home provided many of these services in the past. Thus, the growth of personal service parallels a decline in domestic service and the movement of women out of home work and into the paid labor force. The combined growth of domestic and personal service from about 9 percent to 14 percent of the labor force indicates a steady increase in the demand for personal services coupled with a significant transformation in the way personal services are delivered.

Other industries have experienced more moderate changes. Transportation and utilities reached a peak level of employment in 1920 but have today declined to the level they occupied in the 1890s. Despite the steadily increasing demand for transportation, communication, and utility services, significant labor-saving technologies in these industries from the 1930s onward have resulted in fewer jobs. These advances include motorized transportation, more efficient electric generation and distribution systems, and the development of highly efficient telecommunications systems. Construction has also maintained a relatively stable percentage of employment. This stability results from moderate growth in the demand for physical structures such as houses, buildings, and roads, coupled with a relatively slow pace of technological innovation in the construction industry.

Manufacturing employs only about 14 percent of the labor force—a level similar to that in 1870. However, this stability at the endpoints of our time line masks the rise and subsequent decline of manufacturing employment in the United States. In 1960 manufacturing employed more than 28 percent of the labor force. This peak percentage represents the dominance of American manufactured goods in the post–World War II world economy. The subsequent decline in manufacturing jobs is due to increased international competition and technological innovations that have reduced labor requirements in manufacturing.

These employment changes and the concepts necessary to understand them are the subject of the four chapters in Part III of this book. Chapter 7 explores the roles of technology and organization in determining the nature of work. Chapter 8 examines changes in the extractive and manufacturing sectors. Chapter 9 explores changes in the nature of work across a range of industries and occupations resulting from the rapid spread of microchip technology—the "Third Industrial Revolution." Chapter 10 examines the nature of work in the expanding service sector.

7

Technology and Organization

Any industrial labor force is characterized by (1) the separation of the workplace from the household, (2) a distinction between the worker as a person and the position he occupies, (3) widespread employment in large-scale organizations with both bureaucratic and professional forms of authority, (4) individual accountability for the performance of tasks judged according to standards of competence, and (5) by the affiliation of individuals to organizations through contractual agreements (DREEBEN, 1968:114–115).

In industrial societies most work takes place inside large, complex organizations that use highly specific production technologies. The above quotation identifies some of the key characteristics of work in such organizations. No longer do most people work alongside family and community members to produce goods for local consumption. Instead, they enter into contractual employment agreements and work according to the dictates of bureaucratic or professional standards to produce goods and services for mass markets.

Technology and organization are, in essence, two sides of the same coin. That coin, called the *social relations of production,* comprises all the material and nonmaterial means and techniques used to produce goods and services. Thus, social relations of production include the tools and machines used, the skills needed, the formal and informal group structure utilized, and the structure of the larger organization and its relation to other organizations in society. Although technology and organization are thus inseparable, it will be useful to discuss them separately for purposes of presentation. It should always be remembered, however, that they are intricately related.

In this chapter we develop the ideas of technology and organization and examine how they jointly determine the nature of work. We also examine the nature and some of the limitations of the modern bureaucratic organization of work.

TECHNOLOGY

Technology is the application of knowledge and skills for the achievement of practical purposes. It includes both physical apparatus, such as tools and machines, and the knowledge required to build and use them and to solve problems in their application to the production of goods and services. Technology is commonly defined as having three components: operations technology, materials, and knowledge (Hughes, 2004).

Operations Technology

The people and machinery that produce a good or service, along with the rules and procedures that pattern their use, are called *operations technology*. For example, the operations technology of college education includes professors scheduled to teach specific subjects at certain times, classrooms, libraries, and living and recreational accommodations for students. The operations technology of a chicken-processing factory includes a receiving dock for crates of live chickens, an assembly line where the chickens are slaughtered and dressed, and a packaging area where the dressed chickens are prepared for shipment.

Materials

The *materials* to be used in producing a good or service are also part of the technology of its production. In an automobile assembly plant, engine blocks, chassis, radiators, transmissions, windshields, and other components are the materials. These materials and previously manufactured subcomponents are part of the essential technology of the manufacture of automobiles. In a chicken-processing plant, live chickens are the essential material. However, scalding water, plastic wrap, waste bins, refrigeration packs for the dressed chickens, and shipping crates also are required as raw materials. These examples also illustrate how the technology in any one industry (for example, automobile assembly or chicken processing) is integrated with the technology used in other parts of the economy to produce the needed materials and subcomponents.

Materials can be more or less uniform, and this has important implications for the technology that can be used. The materials used to produce automobiles and packaged chicken are either very uniform or somewhat uniform, respectively, and this uniformity allows significant standardization and even automation in the case of automobiles. The materials in design, service, and repair industries are less uniform and include significant variability from case to case, which limits technological routinization of production.

Knowledge

The final aspect of technology is *knowledge*. Knowledge is required to operate the various machines, deal with the exceptional cases that are always characteristic of production, and coordinate production activities. Production always involves variability and uncertainty. Unpredictable aspects of production include access to materials, variable quality in these materials, wear and breakdown of tools and machinery, and variations in weather and staffing. Production workers must have a thorough knowledge of the technology and materials in order to anticipate and accommodate the unexpected. For example, the chicken snatchers in a chicken-processing plant must have knowledge of chicken behavior and movement in order to seize and hang their prey on a moving assembly line. Variability in chicken behavior makes these jobs difficult to automate. Box 7.1 provides an example from nursing of both formal and informal knowledge in the application of technology.

ORGANIZATION

Organizational structure is the established pattern of relationships among the various parts of an organization and among the various employees in the organization. Organizational structure is not

---

B O X 7.1 **Formal and Informal Knowledge in Nursing**

The following episode involves a hospital nurse briefing her replacement at shift change on the needs of a patient with a tube draining a wound. This brief episode illustrates the extensive informal knowledge behind even seemingly simple procedures. The segments in brackets are reports by the ethnographer of the actions being modeled by the nurse.

I have had trouble with his leg—see [she touches the patient on the right arm, and the left leg spasms and moves sharply] and that has affected this [points to the

tube draining a wound], so I found that if I put this around here like this [demonstrates the way that the tube has been stretched around the leg of the bed], then if he moves involuntarily it doesn't kink. But you have to pull this here like this [she demonstrates the way that the loop needs to be pulled] before you touch him so that he doesn't wrench it out.

SOURCE: Excerpted from Annette Fay Street, 1992, *Inside Nursing*, Albany: State University of New York Press, p. 178.

---

visible in the same sense as the structure of a material object, such as a bridge or archway, but its consequences are just as real. Sociologists use the related but more general concept of **social structure** to describe such diverse phenomena as the family and the state. Families have structures made up of relations between different members of the family. For instance, the family structure of a single-parent family with two teenage girls consists of the relationship between each daughter and the parent and the relationship between the two sisters. Different families have different structures, just as do different governments and different organizations. These structures specify the patterns of obligations and responsibilities that the incumbents of different roles have in relation to one another. These structural relationships also influence the sympathies, affections, and animosities that different members of the group are likely to experience toward one another.

Different organizations, such as government agencies, economic organizations, religious organizations, and political parties, have different aims; what they share in common is having identifiable structures for the attainment of these aims. It is these organizational structures that we seek to understand in this chapter. In formal organizations, such as large corporations, specific job positions and the hierarchical and lateral relations

among these positions are the key parts of these structures.

According to the organizational behaviorists, French, Kast, and Rosenzweig (1985), the structure of economic organizations consists of the following parts:

1. the pattern of formal relationships and duties (the organizational chart plus job descriptions)

2. the way various activities or tasks are assigned to different departments and people in the organization (differentiation)

3. the way these separate activities or tasks are coordinated (integration)

4. the power, status, and hierarchical relationships within the organization (authority structure)

5. the planned and formalized policies and procedures that guide and control the activities and relationships of people in the organization (administrative structure).

Researchers have developed a number of concepts to describe variations in these structural characteristics. These concepts include *complexity, formalization,* and *centralization* (Hall and Tolbert, 2004). *Complexity* includes a vertical dimension, a horizontal dimension, and a geographic dimension. Or more simply, complexity means how many *layers,* how many *divisions or specialties,* and how *spread out* the organization is geographically. A locally owned corner grocery is obviously much less complex than Wal-Mart—on all three dimensions. *Formalization* means the extent to which actions must follow pre-set rules or are allowed to emerge depending on the unique needs of each situation. In settings with highly standardized production, formalization is possible and often implemented. In production situations with greater variability, formalization may be either impossible or counterproductive. *Centralization* refers to the concentration of decision making at the top of the organization. Where production is routine, centralization of decision making is possible and perhaps even preferable, but where specialized knowledge is needed to resolve ongoing problems, centralized decision making tends to be counterproductive.

TECHNOLOGICAL DETERMINISM?

Do forms of technology determine resulting organizational structures? Is our future **technologically determined?** Technology influences which organizational structures are viable and most effective, and this influence has been the object of much study and analysis. Seminal studies by British organizational analyst Joan Woodward (1980) demonstrated a connection between type of production and resulting organizational structures. Her studies showed that *small batch or unit* production, such as in shipbuilding or aircraft manufacturing, is characterized by a relatively low ratio of managers to workers (low vertical complexity). *Large batch or mass production,* such as in automobiles and many types of light manufacturing, is characterized by front-line supervisors being responsible for many workers and a high centralization of power. *Continuous process production,* such as in chemical or petroleum manufacture, is characterized by more decentralized power and decision making. Woodward also observed that the fit between technologies and organizational structures is important for organizational effectiveness.

Is the nature and future of work tightly determined by technological imperatives? Probably not. Subsequent studies have shown that the match between technologies and organizational structures is far from precise and much slippage occurs (see Hall and Tolbert, 2004). Technology poses certain limits on organizational forms, but significant choices remain. Certain technologies cannot be used, or cannot be used effectively, except in moderately large enterprises. For instance, coal-powered electrical generation plants must reach a certain size to be economical. Thus, this technology demands a certain organizational structure—in this case, a minimum size—to be efficient. However, the minimum size required for the effective use of most technologies is quite modest, and, in most cases it is well under that of today's large corporations (Blau, Ferber and Winkler, 1998).

A key reason that the core parts of the economy are dominated by large, centralized firms is that these firms have the power to do so. There is no technological reason that this has to be the case. In recent decades international competition and technological change have reintroduced a high degree of uncertainty into the environment of many organizations. In this situation large bureaucratic organizations face many challenges. Diverse organizational forms are competing effectively in the world markets of the twenty-first century, and this has introduced new opportunities and challenges for economic organizations and those who work within them.

As you can see, there is clearly substantial overlap between the concepts of *technology* and *organization.* For example, are the division of labor and the assignment of workers to different tasks part of the organizational structure, or part of the technology of production? Obviously they are both. When examining the nature of work in upcoming chapters, we avoid entering into debates about whether some aspect of the social relations of production is a characteristic of technology or of organizational structure. The key insight is that technology and organization are closely interrelated and both influence the nature of work. It is to these influences that we now turn.

HOW DOES TECHNOLOGY INFLUENCE WORK?

In this section we present a brief history of technology and its influence on work. We also discuss the concept of skill and how skills are changing because of changing technologies.

Changing Technologies

Technology has developed in distinctive stages, with each stage retaining aspects of the earlier stages and adding to them. Many technologies were discussed in the more general history of work presented in Chapter 1. Here we organize the presentation in terms of distinct stages.

Simple Tool Technology The earliest technologies were used by primitive humans to wrest material sustenance from the environment. These technologies included hand tools, made from stone and bone, and woven baskets. Early technologies also included the knowledge needed to make and use these tools and materials. People had to know where to find stones that could be flaked or ground to a sharp edge, how to fashion these stones into usable tools, where to find and how to work with clay or reeds to make bowls or baskets, how to locate edible plants, and how to locate and kill small game. We refer to this level of technology as *simple tool technology.*

Craft Technology The second level of technological development entails the use of more advanced tools, which require greater effort to manufacture. For instance, the production of ovens for firing clay pots requires a technology to build the ovens in such a way that desired temperatures can be reached and maintained. Similarly, mining and rudimentary smelting of copper or iron ores requires more advanced technology to mine and transport ore and to build kilns. The metals produced can then be used to make a variety of tools and final products. This level of technological development is associated with a division of labor into different crafts, such as ironworking, glass blowing, and boot-making. Accordingly, it is called **craft technology**. The skills needed for craft technology are typically acquired through apprenticeship.

Mass-Production Technology The third level of technological development is that required for the mass production of commodities. **Mass-production technology** depends on the availability of a variety of specialized goods and services and on the integration of these into a mechanized production operation (Noble, 1997). Because tools and products are more specialized under mass-production technology, there is a great need for specialized knowledge. People acquire such knowledge through apprenticeships or through college training. Additional knowledge is also required to coordinate diverse production operations. The term "Fordism" is sometimes used as a short-hand expression for mass-production technology because of the important role of the Ford Motor Company in developing assembly-line technologies in the early years of the twentieth century.

Microchip Technology The latest stage of technological development entails the use of numerical control processes and electronic microprocessors. These technological innovations allow the automation of many production processes and the almost instantaneous collection and tabulation of data about production. Computer-assisted technologies have had far-reaching consequences for the nature of the skills and knowledge needed in production. This new stage of advanced **microchip technology** will be our focus in Chapter 9.

Even before the widespread use of microprocessors, American sociologist Daniel Bell (1976) argued in an influential book, *The Coming of Post-Industrial Society,* that new knowledge-based technologies would transform the nature of society. Bell maintained that knowledge would become the central resource in society. He predicted that the economy would become *knowledge-intensive* rather than *capital-intensive.* He also forecast that the economy would shift from the production of goods to the production of services and that women would have more job opportunities. Some of these predictions have come true, others have not. For instance, women have gained new job opportunities. However, these opportunities have come less in the new knowledge-intensive industries, such as engineering, than in more labor-intensive industries, such as personal services. The expanded opportunities of women have thus occurred in those industries that have been *least* influenced by technological change (Epstein and Kalleberg, 2004; Florida, 2004).

In the workplace of the late twentieth century, workers produce more goods, as well as a greater variety of goods, than was ever before possible. The development of new technologies has brought with it increased productivity and the possibility of shorter work hours, increased safety, greater employment security, and material abundance for all members of society. In the following sections we will see how organizational structures have both facilitated technologically based increases in productivity and shaped the distribution of rewards in such a way that many people in society are excluded from the benefits of increased productivity.

What Exactly is Skill?

Changing technologies have important consequences for the types of skills needed by workers. Sometimes these skills are more complex than previous skills; sometimes they are simpler. Skills may involve *physical, mental, or interpersonal* aspects of work, or a combination of these. Social scientists generally measure the skill level of a job in terms of complexity, diversity, and autonomy. **Job complexity** refers to the level, scope, and integration of physical, mental, or interpersonal tasks on a job. **Job diversity** refers to the number of different tasks and responsibilities required by a job. **Job autonomy** implies self-direction and the potential for creative improvisation (Spell, 2001).

The amount of *training time* required to qualify for a job is often taken as an indicator of the level of skill required. Many *tacit skills* essential for a job, however, can only be learned by long experience

on the job. Tacit skills are those bits of information and knowledge that are not easily expressed as formal knowledge but are nevertheless essential for doing work correctly and efficiently. Tacit skills vary with each job but generally involve the ability to evaluate a range of options in determining how to proceed with a given task. Such tacit skills combine with formal knowledge to define the *working knowledge* required on a job. Thus, formal training provides only some of the actual skills required on a job. An over-reliance on formal training as opposed to on-the-job experience as an indicator of skill tends to exaggerate the skills of those with formal credentials over those with experiential knowledge (Vallas, 2003). Box 7.2 describes some of the tacit skills acquired over time and used in the daily tasks of a machinist.

Some researchers argue that skills have historically declined because of automation (Braverman, 1974). This phenomenon is called **deskilling**. Evidence based on case studies of specific industries, such as printing or meatpacking, often supports such concerns (Brueggemann and Brown, 2003). Studies based on broader samples of occupations, however, generally find that skills have been relatively stable since the mid-twentieth century. That is, technological changes have had offsetting effects, with some jobs becoming more skilled and others becoming less skilled, so that the average level of skill demanded has remained fairly stable (Leigh and Gifford, 1999). (See Chapter 9 for a more complete consideration of skill upgrading and deskilling in the context of computer-assisted production.)

Acquiring New Skills

Workers enter the labor market with a certain level of education and experience. Between 1940 and 2005 the average level of formal education of adult members of the labor force (those twenty-five years old and older) increased dramatically from 8.6 years to 13.4 years. Because of changing skill requirements, workers increasingly need additional training after completing high school to qualify for a job. This training is acquired in various ways. Of the workers who needed more than high school

BOX 7.2 The Working Knowledge of a Machinist

When you were tempering something you get it to what they call a cherry red. One piece of steel you might need to get to a cherry red, maybe another one a little redder. You cool it in certain ways as you go along. It draws the temper into the steel. Makes it harder. But if you cool it too quickly it gets tempered so hard it's just like glass—you can break it.

They have what they call flame temper, an oil temper, or a water temper. Like if you sharpen a pick—you hammer the point out on a pick and then you want to temper it so it won't burr over when you hit a stone—that's a cherry temper. But if you temper it *too* hard and you hit a stone, it'll pop the end right off.

You dip it in the water slow. And it'll turn a bluish color as the temper works out into it. And your coal temper—a temper out of a coal forge—is a lot better than your gas temper. See, they use gas forges now. Or I can temper with a torch, but you've got to be very careful with it. When you're using gas, you're only heating one side at a time. When you're using coal, you're poking the metal right into the hot ashes. It heats more evenly, all the way through and around. Where with your gas you don't get that. And you only heat one side with the torch, and it's not as good.

SOURCE: Excerpted from Douglas Harper, 1987, *Working Knowledge.* Chicago: University of Chicago Press, pp. 32–33.

---

training for their current job, 52.1 percent reported they had attended a training school or college. In addition, 50.1 percent said they acquired needed skills through informal on-the-job training, either at a previous job or on their current job. More than 17.5 percent said they had learned the necessary skills through a formal company training program. Other significant sources of skill training included friends and relatives or activities not related to work, such as hobbies or avocations (5.9%), armed forces training (3.5%), and correspondence courses (1.4%) (BLS, 2005a). These percentages add up to more than 100 percent because some workers report making use of more than one source.

Community Colleges and Vocational Training The sources workers use for acquiring job skills vary by occupation. Enrollments in four-year colleges are steadily increasing. Teachers, engineers, social workers, accountants, and many other professionals earn their degrees through four-year colleges. Advanced degrees, such as those for doctors and lawyers, build further on these four-year degrees as a base.

Community college and vocational school enrollments, however, have grown even faster than, and now exceed enrollments in, four-year colleges. Two-year associate degrees cover a range of fields including medical, dental and biological technicians, computer and electronics technicians, aircraft and power plant maintenance, computer programming, business management, and hotel management. Vocational training credentials are offered in a range of fields including hairdressing, upholstery, automobile mechanics, refrigeration technology, and truck driving. Such additional formal schooling, as provided by both colleges and vocational schools, is thus widely used across white-collar, blue-collar, and service occupations.

Apprenticeship Programs Formal **apprenticeship** training programs are more likely to be used by skilled craft workers. Training programs in the skilled construction trades must adhere to national standards and involve a combination of classroom instruction and supervised on-the-job training that typically lasts from two to six years. Skilled factory workers, police officers, and telephone installers and repairers are also likely to learn their skills in company-sponsored training programs. Many of these programs are operated as joint union-management apprenticeship programs. However, there are no national standards for these programs, and completion of such programs may not produce credentials that are universally acknowledged. (See Chapter 16 on the world economy for a discussion of more developed apprenticeship programs in some of the European countries and the contributions of these programs to national development.)

On-the-Job Training   Skills are also acquired and upgraded through both formal and informal **on-the-job training**. In the skilled trades workers often attend company-sponsored training classes as well as honing their skills through experiential knowledge and through interaction with peers on the job. Managers and supervisors also learn needed skills through both formal training classes and informal on-the-job learning.

The armed services are also an important source of training. Occupations whose incumbents receive substantial training in the armed forces include aircraft engine mechanics and electricians and electronics technicians. Correspondence courses and paid workshops are used by repairers of electronics and industrial equipment, and by workers in securities, financial sales, and service occupations. Occupations that rely significantly on friends or experiences outside of work for their training include dressmakers, musicians, farmers, carpenters, and painters.

Training in computer skills has become increasingly widespread across a wide range of occupations. Computers are used for inventory control, communication and correspondence, and word processing. The widespread use of computers on the job is illustrated in Table 7.1.

HOW DO ORGANIZATIONS INFLUENCE WORK?

The skills required at work are determined by the way tasks are divided and subdivided and by organizational structure, as well as by the technology used. In this section we discuss the increasing division of labor and the ways in which the social

TABLE 7.1   Workers Using Computers on the Job

Characteristics	Percent using a computer at work
Total	53.5
Male	47.9
Female	59.9
Managerial	80.4
Professional	78.9
Technical	75.0
Sales	57.7
Clerical	73.6
Service	23.3
Skilled	29.9
Operators and laborers	19.5
Farming	19.1

SOURCE: U.S. Department of Commerce, Bureau of the Census, 2004, *Statistical Abstract of the United States*, Washington, DC: U.S. Government Printing Office, Table 617.

organization of work allows a few people to control the work of many others.

The Division of Labor

The division of labor was initially discussed in Chapter 1. Here we examine its implications for organizational structure. The earliest form of the division of labor is that between men and women and between children and adults. This stage is called the *division of labor by gender and age*. In this earliest division of labor, men specialized in hunting and women specialized in gathering. Children tagged along as adults went about their activities helping as needed and as their abilities allowed. This activity helped them acquire adult skills. At this stage organizational structure was identical with family and group structure.

The Social Division of Labor   The social division of labor—the division into different crafts or trades—was typical of work in feudal society. The specialization of workers in crafts results in more efficient production and in better-quality goods. At this stage of development, the main organizational structures of work were feudal relations in the rural areas and guild structures in the cities.

The Manufacturing Division of Labor Most work in modern industrial society is organized in terms of the manufacturing division of labor. In this stage the different activities involved in each craft are separated. Instead of making the soles for a pair of shoes, then making the tops, and then stitching each set together, a shoemaker might make a number of soles and a number of tops and then stitch them together. This breaking down of a task into parts is called the *analysis of labor*. The analysis of labor produces efficiencies in handling materials and the rhythm of work. The manufacturing division of labor may also involve the assignment of the different parts of the job to *different workers*. Thus, one worker might make soles, another tops, and a third stitch them together. The division and reassignment of tasks allows more efficient mass-production techniques to be used, such as those based on the assembly line. The price of finished goods may be further lowered because some workers are paid less for doing parts of the task that require less skill. Starting with this stage, work is increasingly carried out in large complex organizations.

The manufacturing division of labor also creates the preconditions for a further heightening of productivity through mechanization. Once tasks have been separated into parts, it is possible to develop machinery to do some of the work previously done by hand. In 1776 Adam Smith argued that the division of labor produces efficiencies because of the increased dexterity of the worker as he or she specializes in one task, because of the time saved "in passing from one sort of work to another," and because of the introduction of machinery. In contemporary views the introduction of machinery is seen as the most significant of these developments. The other two aspects that Smith mentions, increased manual dexterity and reduced time lost in transferring

from one aspect of work to another, are frequently negated by the increased alienation of workers and by their lost enthusiasm, factors he did not fully consider. Such factors are especially important today with more sophisticated production systems that require great worker involvement and commitment.

Organizational Structure as Control

The division of labor produces not only specialized positions but also a vertical differentiation based on power. The workers whose tasks are finely subdivided suffer a loss of skill, a loss of power, and, finally, a loss of wages. Meanwhile, the power and income of those who organize the labor of others increases. The analysis of craft work into parts and the development of machinery to help with some of the tasks clearly increase productivity. However, some researchers argue that the key goal in assigning minute tasks to different workers is to deskill labor and cheapen its price (Jacoby, 2004). Thus, the organization of work also becomes a means through which employees are controlled and manipulated.

Direct Personal Control To understand how this takes place, the American economist Richard Edwards (1993) categorized the control of the labor process into stages. The first stage is that of **direct personal control**. The owner shows the workers how to do the work and indicates the appropriate pace. The owner also evaluates the workers and rewards or punishes them according to their performance. Early in the industrial revolution, punishments included verbal and physical coercion and the threat of firing. Today punishments are mainly restricted to firings, and a greater number and diversity of positive rewards are used, such as promotions and raises. Direct personal control works reasonably well for employers. It typically gets the work out on schedule and according to plan. Employers still use it in small enterprises today. However, the increasing prevalence of large enterprises undermined the ability of employers to use direct personal control. As a result, it is not how most work is organized in industrially advanced economies.

Foreman's Control   With the growth of large enterprises, direct personal control is replaced by foreman's control. Under this system the owner hires foremen who take on the duties of recruiting and supervising workers. This system of labor control prevailed in steel manufacturing, railway construction, textiles, and many other industries until about 1900. Foreman's control still exists in such industries as agriculture, construction, and landscaping. But it was largely abandoned in most industries because it proved inadequate for organizing work under mass-production systems that needed more coordination and more standardization. Each foreman had tremendous latitude in organizing tasks. In mass-production industries, however, large numbers of interchangeable components are needed. Therefore, it is essential to use *identical* techniques throughout the operations and to coordinate activities in order to ensure the quality and consistency of the final product.

Scientific Management   New systems of control were thus developed in an effort to implement more standardized procedures in mass-production industries. The American industrial engineer Frederick Taylor developed a system that he called **scientific management**. Taylor believed there was "one best way" to do every task. This way could be discovered by first carefully observing how the workers did the task and then devising a more efficient way to do it. For instance, he experimented with different ways of holding cutting tools on metal lathes in order to produce machine parts as quickly as possible.

Taylor was also concerned with the problem of workers who resisted working as fast as possible; he saw this problem as a major impediment to efficiency. Taylor observed that many workers engage in **soldiering;** that is, they intentionally work well below their capacity. He concluded:

The greatest part of systematic soldiering ... is done by the men with the deliberate object of keeping their employers ignorant of how fast work can be done.

So universal is soldiering for this purpose, that hardly a competent workman can be found in a large establishment ... who does not devote a considerable part of his time to studying just how slowly he can work and still convince his employer that he is going at a good pace (Taylor, 1911:32–33).

Taylor's solution was to fire skilled workers and hire less skilled workers to replace them. The new workers could then be trained to do the work Taylor's way. They would receive a higher wage than they had received for less skilled work but would be paid less than the skilled workers whom they replaced. Box 7.3 presents Taylor's outline of the principles of scientific management.

Technical Control   Like scientific management, technical control was developed to standardize work procedures in the mass-production industries. Under technical control the worker is controlled and paced by the machinery. The classic example of this is the assembly line. The activity and pace of work are directly controlled by the assembly line, which delivers materials to the worker. In many ways the technical control of work is the most rigid form yet developed.

Worker Resistance   Scientific management and the technical control of work appear to offer great efficiencies because they specify precisely how the work is to be executed and how quickly it is to be done. However, they also have serious drawbacks and limitations. Most importantly, they make limited use of workers' skills. Production rarely occurs exactly as planned, and machinery and parts often fail. When the workers have been robbed of the skills necessary to understand their work and their tools, or have never been given the opportunity to learn these skills, they are not in a position to handle unexpected situations nor to facilitate production. Scientific management and technical control also rob workers of their *enthusiasm for work.* Employees who have been alienated from their

---

BOX 7.3 The Gospel of Work According to Taylor

Frederick Taylor is considered the father of scientific management. The following passages express his outline for organizing work:

The managers assume ... the burden of gathering together all of the traditional knowledge which in the past has been possessed by the workmen and then of classifying, tabulating, and reducing this knowledge to rules, laws, and formulae ...

All possible brain work should be removed from the shop and centered in the planning or laying-out department ...

The work of every workman is fully planned out by the management at least one day in advance, and each man receives in most cases complete written instructions, describing in detail the task which he is to accomplish, as well as the means to be used in doing the work ... This task specifies not only what is to be done, but how it is to be done and the exact time allowed for doing it ... Scientific management consists very largely in preparing for and carrying out these tasks.

SOURCE: Frederick W. Taylor, 1911, *The Principles of Scientific Management.* New York: Harper and Row, pp. 39, 63, 98–99.

---

work in this way may engage in **worker resistance,** whereby they make a science of finding ways of allowing production to lag, or at least they take little initiative in increasing productivity. Scientific management and technical control, because they treat the worker like a machine or an animal, deny the humanity of workers. These systems of organizing work thus do not reap the full benefit of workers' enthusiastic participation and may generate many unanticipated consequences as well. In particular, workers may organize to resist scientific management and technical control. The greatest growth in the American labor movement, for instance, occurred immediately after the advent of these techniques.

Rediscovering the Worker

In 1927, as mentioned in Chapter 2, a series of studies was begun at the Western Electric Company's Hawthorne plant near Chicago under the direction of Elton Mayo (Roethlisberger and Dickson, 1939). These studies were motivated by a desire to increase productivity. However, they yielded many surprising findings that gave birth to what has come to be called the **human relations school** of industrial relations. Initially, experiments were done with various combinations of

lighting levels and break schedules to find the optimum conditions for maximum productivity. The researchers observed that productivity increased in both the experimental group and the control group whenever *any* change in procedures occurred. In an effort to understand this surprising result, additional studies were undertaken in a relay assembly test room and a bank wiring room.

In the relay assembly test room the experiments continued for two years. The women in this group were placed in a separate room for observation. In addition, they were given regular physical examinations at which cake and ice cream were served. Two women were also removed from this group during the two-year period for "talking too much" and for having generally "bad attitudes" toward the experiment. Again, as in the case with the lighting experiments, productivity tended to increase no matter what the experimental manipulation. This kind of result became known as the **"Hawthorne effect"**: added social attention, regardless of its content, increases productivity (see Chapter 2). The discovery of the Hawthorne effect was taken as a refutation of the vision of "economic man" proposed by scientific management. Economic man could be manipulated by wage rates and incentive plans. In the place of economic man, the human relations theorists proposed a vision of "social man." This theory argued that workers needed a positive social context within which to achieve maximum productivity (Grint, 1991).

In the bank wiring room the researchers used a different strategy. Here, instead of using experimental manipulation, they kept detailed observations of workers' behavior in order to understand the causes of increased productivity. They observed that there was a prevailing norm among workers for how many switches should be wired in a day. Workers who exceeded that norm would be ridiculed and ostracized by their fellow workers. Workers who exceeded these informal quotas were given such caustic nicknames as "Shrimp," "Rate Breaker," "Slave," or "Speed King." This finding further reinforced the idea that understanding "social man" was of central importance for increasing productivity.

In more recent reflections on the Hawthorne studies, the observations in the bank wiring room have also been interpreted as indicating that workers are capable of resisting productivity drives that they define as exploitive (Jones, 1992). Similarly, productivity increases in the relay assembly test room have been reinterpreted in light of the firing of the two recalcitrant women during the test period. These reservations highlight the fact that management coercion and worker resistance do not automatically disappear from the workplace just because supervisors give greater attention to workers' social needs.

THE GROWTH OF BUREAUCRACY

At the dawn of the twenty-first century, most work takes place inside large, complex enterprises. Modern enterprises are complex in their level of horizontal, vertical, and spatial differentiation. *Horizontal differentiation* refers to the division of labor into component tasks. Thus, in automobile manufacturing some workers are responsible for welding the body of the car and others for installing the windshield. *Vertical differentiation* refers to the creation of multiple levels of hierarchy in complex organizations. In a university, for instance, professors report to depart-

mental heads, who report to deans of colleges, who report to the president of the university, who reports to either a system-wide president or a governing board. *Spatial differentiation* refers to the geographic dispersion of different aspects of production or different product lines among different plants, sometimes located in different regions, or even in different countries.

Because modern organizations are so large and complex, they have had to develop bureaucratic structures to coordinate their activities. In this section we define bureaucracy, discuss variations in the basic bureaucratic form, and describe the informal workplace cultures that often emerge within formal bureaucracies.

Defining Bureaucracy

The German sociologist Max Weber initiated the systematic study of **bureaucracy** as a form of social organization at the beginning of the twentieth century. He enumerated six characteristics distinguishing bureaucratic organizations from other forms of social organization:

1. There is the principle of fixed and official jurisdictional areas.

2. The principles of office hierarchy and of levels of graded authority mean a firmly ordered system of super- and subordination in which there is a supervision of the lower offices by the higher ones.

3. The management of the modern office is based upon written documents ("the files").

4. Office management, at least all specialized office management—and such management is distinctly modern—usually presupposes thorough and expert training.

5. When the office is fully developed, official activity demands the full working capacity of the official.

6. The management of the office follows general rules, which are more or less stable, more or less exhaustive, and which can be learned. (Weber, 1946:196–198)

Although other forms of social organization may share some of these characteristics, such as hierarchy, only organizations that have all or most of these characteristics are appropriately characterized as *bureaucratic*. Weber felt that increasing bureaucracy was the fate of modern society, because bureaucracies are so efficient in achieving their stated objectives. He was not completely favorable toward this future, however, because he also felt that bureaucracies stifled human initiative and creativity. He famously referred to bureaucracy as the "iron cage of the future."

Bureaucracies are indeed more efficient than, for example, the administrative structures of feudal society. Feudal administrative structures, such as those used for collecting taxes, relied on bonds of loyalty and obligation. Graft, corruption, and inefficiencies of all kinds were commonplace. Bureaucracy, however, is not necessarily the final word in efficiency. As you will see, bureaucracies generate their own sorts of inefficiencies and may eventually be replaced or modified to allow for greater initiative and innovation by those who work within them.

Staff and Line Management   Contemporary bureaucracies may also involve recent innovations beyond Weber's model. A relatively early innovation was specialized staff positions. In addition to the arrangement of authority in a hierarchy, many modern bureaucracies also include staff positions that are outside the linear chain of command. Positions on the linear chain of command are called **line positions**. Ancillary support positions are called **staff positions**. Staff positions are filled by specialized workers trained in some specific profession, such as safety and health, law, accounting, personnel relations, or other important functions that support the main activity of the organization. These staff personnel report directly to someone in a line position at a given level of the organizational hierarchy. However, they have no direct relationship to those higher up in the hierarchy or to those in subordinate positions. In effect, they are supplementary experts needed at specific levels of the organization, but they are not included in the formal chain of command. Promotion opportunities may be less available

for staff workers than for line workers, because they have less clearly defined job ladders.

Matrix Organization   A second bureaucratic innovation is the **matrix organization** of authority. Under this system each worker must report to two different supervisors. For instance, suppose that an aircraft manufacturing plant has five different functional divisions: research and development, engineering, manufacturing, finance and accounting, and marketing. In addition, the plant is working on three different types of airplanes. Workers in the plant report both to the supervisor of their functional division and to the supervisor of their particular aircraft project. Matrix organizations are popular when there is a need to share technical information and coordinate activities between different projects but also a need to allow each project to develop independently (Harris and Raviv, 2002).

Bureaucratic Control

We discussed direct personal control, foreman's control, scientific management, and technical control as techniques for controlling labor and getting the maximum amount of work out of employees. Bureaucracy can be seen as the most recent version of efforts to control labor (Jacoby, 2004). In a bureaucracy, procedures are no longer determined by how the boss or foreman says the job should be done, or dictated by the machinery, or by an industrial engineer. Instead, organizational rules spell out the procedures. These procedures generally include both the specific techniques to be used for a given task and the criteria used for evaluating and promoting workers.

Internal Labor Markets   Bureaucracies rely heavily on rewards and inducements to control workers. For instance, they fill most positions internally from within the organization. Internal recruitment helps create the expectation of advancement (Ospina, 1996). This expectation provides an important source of motivation for employees. Internal recruitment also reduces the costs of training new employees in company procedures and company-specific skills.

The job ladders organized in this way, called **internal labor markets,** are an important way for bureaucracies to motivate and control workers. Workers enter the organization only at certain jobs that serve as entry ports. Access to higher jobs is mainly through the job ladders starting at these entry ports. For instance, the usual way to become a full partner in a law firm is to begin as a junior associate. Internal labor markets are an important mechanism for cultivating and retaining experienced employees within a firm. Internal labor markets help to increase the value and effectiveness of **human capital** within an organization and thus can be an important component of organizational effectiveness (Elliott and Turnbull, 2005).

The spread of such internal labor markets in the latter half of the twentieth century increased average job tenure among workers. Nearly 40 percent of men aged thirty or older have jobs they will hold for twenty or more years. Similarly, though women are more likely than men to change jobs, 15 percent of women thirty or older can expect to hold their current job for twenty or more years (Osterman, 1999). However, as we discussed in Chapter 14 on marginality and in Chapter 15 on large corporations, long-term commitments between employers and their employees appear to be weakening, and strong internal labor markets may be increasingly displaced by more short-term and contingent employment arrangements (Epstein and Kalleberg, 2004).

Customizing Bureaucracies

Although almost all large, complex organizations are organized and controlled bureaucratically, important variations occur because different types of organizations have different goals. For example, the goals of a department of natural resources may be to preserve these resources and provide recreational facilities for the public. The goal of a manufacturing establishment may be to produce the maximum amount of goods at the minimum price. A goal of an electronics research company may be to foster innovation among its professional staff. The successful attainment of these goals may

require different organizational structures. For instance, a military unit or a bank may function best with a strong vertical hierarchy. However, a research, educational, advertising, or architectural firm requiring high levels of creativity may function better with a more dispersed system of authority and greater individual or group autonomy.

Contingency Theory Sociologists also point out that organizational structure depends on the environment that the organization faces. This theory of organizational structure is called **contingency theory** (Donaldson, 2001). If the market for the product of a company is highly variable, the organization will fare better if it has a relatively loose organizational structure. If the market is highly predictable, more rigid structures are viable and may increase efficiency. If the environment is so harsh that the survival of the organization is in question, centralized control is generally the most effective form of organization. Armies are good examples of organizations that face very harsh environments and have highly centralized control. Technologies generate additional contingencies: they place demands and constraints on organizational structure. When standardized technologies are used, highly rigid systems of authority are possible. When there is great uncertainty in materials or in the available technologies or when production systems are highly complex, more flexible systems of authority and organization may be more efficient.

Informal Work Cultures

Although bureaucracies are known for the painstaking detail with which they spell out procedures, **informal work cultures** also typically emerge. Informal cultures are based on social relationships that emerge among the people who work in an organization. For instance, two workers in an engineering firm may become friends and help each other with their work. A third worker in their section may be excluded from this relationship, and his or her satisfaction in work and productivity may both suffer as a result. These informal cultures have a tremendous influence on how

---

BOX 7.4   The Dark Side of Informal Relations on the Job

In a blue-collar setting in the American Midwest, millwrights are secretive with their knowledge and unwilling to share it with apprentices.

Steve was convinced the millwright was prejudiced against him because he was young, and because he was Serbian, and that he was keeping him from learning anything about the job.... Steve was beginning to boil over, and one morning when the millwright was thumbing over a blueprint, holding it purposely out of Steve's view, as though it were personal and confidential, he popped off, "Look here, Polak, I'm gonna' learn everything there is to know about this [expletive] millwright job, no matter what you think. Cut this [expletive]. I lost four [expletive] years in the army, while you were sittin' on your [expletive], making easy money, paying off on a house, stashin' it away in the bank."

Relations in a bank are no better. Although the style and presenting issues appear different, the underlying issues—fears about differential privilege and access to opportunity—are much the same.

Individuals try to fashion other, more personal identifications by criticizing others, thus indirectly asserting their own individual worth and even superiority.... Each individual knows that she herself may become an object of criticism. This creates an enervating apprehensiveness of others' judgments. An analysis clerk, when asked if she was staying for a branch party, expresses a widely felt anxiety: "I stay because if I don't, they'll talk about me." [Another worker reports:] "Everyone is two-faced, and you have to watch out for yourself" (p.122).

A clerk discusses social divisions in her office and describes her personal feelings towards the workers who form the antithesis, and the defining point, of her own group:

> There is a division between us and [another group of workers].... They go around with their noses in the air.... [They're] *female, very female*. Like when a male walks in the office, they're like a bunch of high school girls. They're always giggling. They're a bunch of phonies. There's one who drives me crazy the way she walks. She is shaped like a seahorse, and the way she walks is like she's saying, "I'm beautiful." They are snotty. I feel awful if I'm around any of them.

SOURCE: Excerpted from Charles Spencer, 1977, *Blue Collar*, Chicago: Lakeside Charter Books, pp. 63–64; and Robert Jackall, 1978, *Workers in a Labyrinth: Jobs and Survival in a Bank Bureaucracy*, Montclair, NJ: Allanheld and Osmun, pp. 121–122.

---

the bureaucracy actually operates, in contrast to how it operates "on paper." Informal work groups may facilitate the attainment of stated organizational goals, or they may develop their own goals, which can be either tangential or oppositional to the stated organizational goals.

Sociologists sometimes refer to informal cultures as "negotiated orders" in recognition of their emergence from the ongoing informal negotiations among different members of the organization (Fine, 1999). "The negotiated order on any given day could be conceived of as the sum total of the organization's rules and policies, along with whatever agreements, understandings, pacts, contracts, and other working arrangements were currently obtained. These include agreements at every level of organization, of every clique and coalition, and include covert as well as overt agreements" (Strauss, 1978:5–6). Workers refer to such negotiated orders simply as "office poli-

tics." In Chapter 3, Box 3.1, we reported on the mutual support and solidarity that can characterize informal work groups. In Box 7.4 we report on the meanness and pettiness that can also characterize informal coworker relations.

Making Out   Within the context of formal and informal rules, employees attempt to make a livable niche for themselves in which they can survive and even prosper—they attempt to "make out." **Making out** refers to workers' efforts to meet the requirements of the job without completely exhausting and alienating themselves in the process. In such ways, workers come to terms with the organizations in which they work. The nature of work is thus not determined solely by organizational rules; rather, it is determined by a process of negotiation and sometimes conflict among the various actors in the workplace. Box 7.5 describes the process of making out in a machine tool factory.

BOX 7.5 "Making Out" on the Shop Floor

Sociologist Michael Burawoy coined the phrase mak-ing out to describe the process through which work-ers devise ways to reach desired rates of production without exhausting themselves in the process. He describes the process of making out in the following excerpt:

After the first piece has been OK'd, the operator engages in a battle with the clock and the machine.... The question is: Can I make out? It may be necessary to figure some angles, some shortcuts, to speed up the machine, make a special tool, etc. In these undertakings there is always an element of risk—for example, the possibility of turning out scrap or of breaking tools. If it becomes apparent that making out is impossible or quite unlikely, operators slacken off and take it easy. Since they are guaran-teed their base earnings, there is little point in wearing themselves out unless they can make more than base earnings—that is, more than 100 percent. That is what Roy refers to as goldbricking. The other form of "output restriction" to which he refers—quota restriction—entails putting a ceiling on how much an operator may turn in—that is, on how much he may record on the production card. In 1945 the ceiling was $10.00 a day or $1.25 an hour, though this did vary somewhat between machines. In 1975

the ceiling was defined as 140 percent of all opera-tions on all machines. It was presumed that turning out more than 140 percent led to "price cuts" (rate increases)....

In 1975 quota restriction was not necessarily a form of restriction of *output*, because operators *regu-larly* turned *out* more than 140 percent, but turned *in* only 140 percent, keeping the remainder as a "kitty" for those operations on which they could not make out. Indeed, operators would "bust their ass" for entire shifts, when they had a gravy job, so as to build up a kitty for the following day(s). Experienced operators on the more sophisticated machines could easily build up a kitty of a week's work. There was always some discrepancy, therefore, between what was registered in the books as completed and what was actually completed on the shop floor. Shop management was more concerned with the latter and let the books take care of themselves. Both the 140 percent ceiling and the practice of banking (keeping a kitty) were recog-nized and accepted by everyone on the shop floor, even if they didn't meet with the approval of higher management.

SOURCE: Excerpt from Michael Burawoy, 1979, *Manufacturing Consent*, Chicago: University of Chicago Press, pp. 57–58. Copyright 1979 by the University of Chicago. Reprinted by permission of the publisher.

Indulgency Pattern The violation of rules by employees is sometimes overlooked by supervisors as part of an implicit bargain for greater effort on other fronts. This arrangement has been labeled by sociolo-gists as the **"indulgency pattern."** Thus, for example, in an ethnography of underground mining, Gouldner (1964) observed that mine foremen overlooked the men taking unscheduled days off for personal reasons in exchange for the heroic efforts that they often under-took while on the job. A key limitation of the indul-gency pattern is that it can be withdrawn at any time, and privileges that workers had assumed can become a basis for sanctioning and even firings.

It is frequently argued that women and minorities receive fairer treatment in large bureaucracies than they do in smaller organizations. This argument is based on

the observation that formal rules make discrimination more difficult. Because of the importance of informal cultures in large organizations, however, women and minorities may still be excluded from the "insider" situations and "old boy" networks in which important coalitions emerge and in which key decisions are made (Ashkanasy, Wilderom and Peterson, 2000).

LIMITATIONS OF BUREAUCRACY

Bureaucracies may be efficient, but they also have many drawbacks and limitations. These problems are discussed in this section. In the next section we discuss some possible alternatives to these problems.

Top-Heavy Management

Perhaps the most important consequence of the rise of bureaucratic organizations has been a significant increase in the proportion of employees engaged in administrative work. In contemporary society people's lives seem dominated by bureaucrats and red tape. This rise in the number of administrative workers has also reduced the proportion of workers directly engaged in production. Between 1900 and 2000, the proportion of non-production employment rose from less than 10 percent to more than 20 percent (Census, 2000). Although bureaucracies may be highly productive, they take on extra administrative baggage. Administrative costs may take up too large a share of the budget, even overshadowing the direct costs of production. This top-heavy nature of modern bureaucracies has become a focus of increased concern in the 2000s as U.S. firms attempted to regain their competitive position in the world economy. In recent years U.S. firms have increasingly sought to reduce middle management and create "lean" managerial structures with fewer intermediate levels and fewer managers at each level (Ferraro, Pfeffer and Sutton, 2005).

The Centralization of Control in the Economy

Not only does bureaucracy centralize control in a large managerial structure, but it also contributes to the **centralization of control** among companies. As firms have grown larger, a greater share of economic activity has come to be controlled by the top managers of a few large firms. In the 1920s and 1930s this process occurred through the growth of *multidivisional firms* (Chandler et al., 1997). In multidivisional firms the central office coordinates the activities of a number of operating divisions or product lines through allocating funds, personnel, and other resources. For example, the General Electric Company (www.ge.com) operates seven major divisions: technical research, services and materials, power plant systems, industrial products, consumer products, personal finance, and aircraft engines.

Conglomerates In the 2000s many large firms have come to control even larger shares of the economy by acquiring companies in unrelated product lines. Such *mergers and acquisitions* expanded the control of large companies into product lines that in many cases are outside their core expertise. In 2000, the 500 largest firms in the United States controlled 82 percent of corporate assets. These 500 corporations represented less than 0.10 percent of the firms in the United States. Control of such huge market shares by a few firms can lead to higher profits for these firms, but also to higher prices for consumers and to reduced employment (Borjas, 2005). (See also the discussions of concentration ratios in Chapter 2 and corporate mergers in Chapter 15.)

A Ruling Elite The centralized control of the economy by a few large firms also translates into the existence of a powerful upper class. Control of corporations rests on the ownership of large blocks of stock. The richest 1 percent of the United States population owns over 50 percent of corporate stock. The richest 1 percent also owns over 28 percent of the *total wealth* in the country. This class constitutes a relatively closed social group with tremendous power. Its members attend exclusive preparatory schools, colleges, clubs, and resorts, and intermarry with other members of the ruling class (Domhoff, 2002). There are, of course, self-made tycoons, whose lives parallel the famous rags-to-riches novels by Horatio Alger. But in reality, the heads of the largest corporations are almost exclusively born into their position (Braun, 1997).

Reduced Creativity

The centralization of control in large organizations can also have negative consequences for the employees of the organization and for the organization itself. These drawbacks may limit the continued expansion of bureaucratic control. Centralization of control increases alienation among members of the organization, because they have little say in the decisions that influence their lives. Decisions may also be slow or faulty, because it is difficult for a few people to receive and process all the relevant information essential for good decision making.

Bureaucratic Rigidity Innovation and creativity may also be lessened by over-conformity to bureaucratic standards. Such bureaucratic rigidity is often a reaction to excessive control by top management:

> An analysis of instances of extreme rigidity in hierarchical organizations reveals that they are usually associated with fear of superiors. . . . Bureaucratic superiors cannot generally censure a subordinate for following official regulations exactly, regardless of how inefficient or ridiculous such action may be in a particular case. . . . Feelings of dependency on superiors and anxiety over their reactions engender ritualistic tendencies. (Blau and Meyer, 1971:104)

Terkel reports the following account of the damaging effects of bureaucratic rigidity:

> I'll run into one administrator and try to institute a change and then I'll go to someone else and connive to get the change. Gradually your effectiveness wears down. Pretty soon you no longer identify as the bright guy with the ideas. You become the fly in the ointment. You're criticized by your supervisors and subordinates. Not in a direct manner. Indirectly, by being ignored. They say I'm unrealistic. . . .
>
> It took me six months to convince my boss to make one obvious administrative change. It took her two days to deny that she had ever opposed the change. (Terkel, 1974:448–449)

In contrast, the conditions that have been found to promote innovation and change include the decentralization of power, low levels of formalization, equity of rewards, low emphasis on volume, low emphasis on cost cutting, and high levels of job satisfaction (Hage and Aiken, 1970; Hall and Tolbert, 2004: 169). In brief, excessive hierarchy and bureaucracy may interfere with productivity rather than promote it. At some point, excessive rationality becomes irrational (Ritzer, 2004).

Corporate Accountability

A final problem that we confront in the modern economy is that of establishing **corporate accountability** for the huge organizations that have come to produce most of the goods and services in the world. Such organizations may come to feel entitled to break or bend the law on a regular basis because of their immense size and power. Common illegalities include price fixing, manipulation of stock prices, illegal environmental pollution, false and misleading advertising, bribery, tax evasion, political payoffs, and the production and sale of unsafe products (Gobert and Punch, 2003). Box 7.6 describes the Firestone tire scandal as an example of corporate misconduct at its worst.

Externalizing Costs The ability of large enterprises to lower the costs of their economic activity through unlawfully dumping dangerous chemical waste into the environment, instituting dangerous work practices, and producing unsafe products has become a major concern in the 2000s. Individuals and communities absorb the **externalization of costs** in terms of lost health, destroyed communities, and degraded environments. For example, on March 24, 1989, the largest oil spill in North American history occurred when the tanker *Exxon Valdez* ran aground in Prince William Sound, Alaska (www.evostc.state.ak.us). Although Exxon was sued for over $5 billion by the State of Alaska and by other individuals affected by the spill, full payment for the economic costs of the disaster is unlikely ever to be recovered (Picou, Marshall and Gill, 2004). Smaller scale repeated abuses, which are unlikely to be recognized or publicized, may cumulatively be even more damaging.

The Price Tag Given the magnitude of the problems resulting from corporate irresponsibility, many social analysts argue that the issue of corporate accountability should receive greater public attention (Ermann and Lundman, 2001). The Senate Judiciary

BOX 7.6 The Firestone Tire Scandal

In the summer of 2000, Firestone announced a recall of several of its tire models for the popular Ford Explorer sport utility vehicle. As the news unfolded over the summer the public and Congress became outraged to learn that over 100 people had already died because of the tires, with several hundred more seriously injured and the death toll steadily rising.

It was supposed to be a mission of mercy. Victor Rodriguez piled the family into his Ford Explorer over Labor Day weekend to visit a sick aunt at a Laredo, Texas, hospital. But as Rodriguez cruised down Interstate 35, he was startled by a thump and looked back to see the tread shredding off a Firestone Wilderness AT tire on his Explorer. The 53-year-old father was unable to control the vehicle, which flipped, ejecting five of its passengers. Among them: his 10-year-old son Mark Anthony, who died instantly....

Mark Anthony Rodriguez had just become the latest victim of a rapidly widening safety crisis that has driven fear into the hearts of motorists who viewed the sport utility vehicle as the ultimate family car. Even as the Rodriguez family planned Mark Anthony's funeral, top executives from Ford Motor Co. and Bridgestone/Firestone were summoned before congressional panels....

Documents show that Firestone was chronicling a pattern of tire failures for the last three years, while Ford was also getting an early warning about safety problems from its own warranty data. And an internal Ford memo shows suspicions between the business partners about the safety record of the tires.

Firestone executive vice president Gary Crigger said that the data were compiled to assess only the company's financial liability. They were not shared, he said, with the company's safety engineers.... For their part, lawmakers have rejected the companies' claims of ignorance about the lethal problem....

The families of the victims are not in a forgiving mood. To them, no apology or excuse will ever make up for the fact that the companies could have warned consumers earlier. "If they would have told people, maybe they wouldn't have made as much money, but it would have saved dozens of lives," says Sara Romero, 12, who survived an Explorer rollover in Florida last December that killed her 37-year-old mother. "My mom didn't have to die. They could have told us."

SOURCE: Naughton, Keith, and Mark Hossenball, *Newsweek*, September 11, 2000, *www.msnbc.com*

Subcommittee on Antitrust and Monopoly estimates that corporate crime costs the public between $174 billion and $231 billion annually. In contrast, "the yearly losses from street crimes [are estimated] at about $4 billion—less than 5 percent of the estimated losses from corporate crime" (Coleman, 1998:6). Additional casualties due to corporate neglect include 14,000 deaths a year from industrial accidents and 30,000 deaths a year from unsafe consumer products. By contrast, about 14,000 people are also murdered each year in the United States, and this is considered a major social epidemic.

Whistle-Blowing   One potential way for employees to fight back against corporate malfeasance is by exposing clearly unlawful corporate actions. Such activities are called "whistle-blowing" (Johnson, 2003). Federal laws protect employees from retaliation by the company in situations where employees expose illegal company activities that involve violation of federal laws such as the Clean Air Act or other environmental laws. The reporting of company fraud against the government is also protected. Such laws, however, are difficult to enforce and the negative consequences for employees can be devastating to their lives and careers. Corporations have tremendous power and financial resources—they have deep pockets when it comes to hiring lawyers and pursuing law suits against employees who they claim have unfairly damaged their reputation. A company can also sometimes succeed in shifting the blame for the misconduct to the worker who reports the problem. Whistle-blowing theoretically has great potential for limiting corporate misconduct, but employees' fears concerning retaliation greatly limit its practical impact (Rothschild, 2000). Whistle-blowing often occurs only in extreme cases, and even then only if someone has the courage to withstand the resulting hailstorm of accusations and counterattacks.

DIRECT WORKER PARTICIPATION

Early theories of industrial relations, such as those of Frederick Taylor, assumed that workers needed to be coerced (by threat of firing) or bribed (by promises of pay raises) into working harder. These ideas are sometimes referred to as **Theory X** of organizational motivation. You saw how these theories were replaced by human relations theories, which assumed that workers would be more productive if they received more humane consideration and attention. This latter view is sometimes referred to as **Theory Y** (McGregor, 1985). It still portrays the worker as a passive object to be manipulated by management, though in this vision socially oriented techniques of manipulation replace economically oriented techniques.

Efficiency through Participation

In recent years, inefficiencies and rigidities associated with bureaucratic management structures have called both of these theoretical traditions into question. The new theories emerging to replace Theories X and Y include a more active role for workers and are sometimes referred to collectively as **Theory Z** (Ouchi, 1981). These theories view productivity as embedded in workers, in their skills, and in their attitudes rather than in specific procedures.

In Europe these theories have taken form in the widespread use of autonomous work groups. Under the group organization of work, teams of workers have extensive control over decision making about their day-to-day operations and activities (Hodson, 2002). Increased employee participation is also typical of enterprises that are owned by workers. Theories X, Y, and Z are discussed further in Chapter 12 on management. Box 7.7 describes the high level of participation and cooperation in collectively owned plywood companies in the Pacific Northwest of the United States.

Lifetime Employment   In Japan such theories of productivity have been implemented through **lifetime employment**. Guaranteeing employment for workers insures their job security and, thus, their trust and enthusiasm. It also retains their presence in the organization as a repository of skills and knowledge. This benefit is amplified by having workers rotate through a variety of jobs during their careers, thus building their knowledge base. This knowledge is invaluable for coordinating activities among different parts of the organization and is a more efficient mechanism for integrating production than more cumbersome bureaucratic structures that may undermine enthusiasm and commitment. Greater reliance on workers' knowledge and decision-making abilities also reduces the need for first-line supervisors and for middle-level managers, thus producing additional significant cost savings (Pfeffer, 1998).

Although participatory systems of worker involvement and commitment have become identified in the public mind with the "Japanese system," in many ways Japanese industrial relations provide a very incomplete model for worker participation. The Japanese system of lifetime employment is available largely for male workers in large firms in the manufacturing sector—less than a third of the labor force. Women and workers in secondary parts of the economy do not have much security. Even employees supposedly covered by lifetime employment may, at mid-career, find themselves transferred to "client organizations" of their company, such as suppliers or wholesalers, as their high-paying jobs are taken over by younger and more aggressive workers. In Chapter 16 we discuss other aspects of the Japanese model of industrial relations. In Chapter 17 we discuss in more depth the possibility of increased worker participation in the workplace of the twenty-first century.

BOX 7.7 Direct Participation in Worker-Owned Plywood Companies

The plywood producer cooperatives are among the most fully developed and enduring democratic industrial enterprises in the United States.... The Pacific Northwest plywood cooperatives, most of which have been in existence for twenty-five to thirty years, not only have persisted through the good and bad times characteristic of that industry but have remained enterprises in which those who work also decide policy....

Although the worker-shareholders may choose to delegate many day-to-day concerns and responsibilities to the general manager and the elected board of directors, in a formal sense they are responsible for the total governance and direction of the enterprise. Worker-shareholders may decide to fire the manager and hire a new one, alter hourly wage rates, build a new plant, or reach any other decision about the enterprise that seems appropriate to them. In membership meetings, each shareholder is entitled to only a single vote....

Central to life in the cooperatives is the sense that the worker-shareholders are in charge, that they run the enterprise, are responsible for what goes on in it, and have the opportunity, within certain boundaries, to make of their environment what they will.

> If things get too bad, the stockholders can just say, "Wait a minute ... we are going to change this." I think that's great because there's a lot of companies that take advantage of the workers, and there's nothing that can be done about it....

Ownership and participation in the co-ops also fosters an extremely strong sense of collective responsibility and mutuality.

> You just find it's kind of a big family attitude.... Here you get in and do anything to help. Everybody pitches in and helps. The people stick together, that's the reason we've gone so far and production is so high, cuz everybody works together.

SOURCE: Edward S. Greenberg, 1986, *Workplace Democracy: The Political Effects of Participation.* Ithaca, NY: Cornell University Press, pp.28–43.

SUMMARY

As technology has advanced from simple hand tools to sophisticated microchip technology, the social organization of production has become more complex. The earliest division of labor, based on gender and age, has been replaced by a detailed division of labor in which tasks are subdivided into many minute processes assigned to different workers. This detailed division of labor and the subsequent hierarchical and bureaucratic organization of the workplace create their own limits because of the inefficiencies introduced by top-heavy managerial structures, lack of organizational flexibility, and lost worker initiative. An additional problem is establishing accountability in a world dominated by large corporations. Increased worker participation has possibilities for ameliorating some of the problems associated with hierarchical and bureaucratic control.

KEY CONCEPTS

technology

organizational structure

social structure

technological determinism

craft technology

mass-production technology

microchip technology

job complexity

job diversity

job autonomy

deskilling	worker resistance	human capital	corporate accountability
apprenticeship	human relations school	contingency theory	externalization of costs
on-the-job training	Hawthorne effect	informal work culture	Theory X
direct personal control	bureaucracy	making out	Theory Y
foreman's control	line position	indulgency pattern	Theory Z
scientific management	staff position	centralization of control	lifetime employment
soldiering	matrix organization		
technical control	internal labor markets		

QUESTIONS FOR THOUGHT

1. What are the major trends in employment in different industries, and why have these occurred?

2. Define technology; define organizational structure. In what ways do these concepts overlap, and in what ways are they distinct? Apply these concepts to an industry in which you have worked. What aspects of the job were reflections of the technology, and what aspects were reflections of the organizational structure?

3. How do sociologists define and measure skill? Select one job each that would commonly be considered highly skilled, moderately skilled, and unskilled. Apply the definition to skill to these jobs. In what ways does the definition of skill fit or fail to fit each job?

4. What are the major types of employee control in the workplace? Which of these would you most/least like to work under?

5. Define bureaucracy. What are some of the limitations of bureaucracies? Do you think bureaucracy will be the ultimate form of social organization? Why or why not? What other forms of social organization might replace bureaucracy?

6. Describe some informal work cultures in which you have been involved that encouraged behaviors other than those organizationally prescribed.

7. How might increased corporate accountability be achieved?

MULTIMEDIA RESOURCES

Print

Richard C. Edwards. 1979. *Contested Terrain.* New York: Basic Books. A useful framework for understanding worker-management conflicts as these have varied across history and across different settings in the contemporary economy.

George Ritzer. 2004. *The McDonaldization of Society.* Thousand Oaks, CA: Pine Forge. A compelling discussion of the over-rationalized nature of modern society.

Hernando De Soto. 2000. *The Mystery of Capital.* New York: Basic Books. An articulate statement of that idea that small producers have the potential to help developing nations rise above the constraints of corruption and cronyism among ruling economic elites.

M. David Ermann and Richard J. Lundman. 2001. *Corporate and Governmental Deviance,* 7th edition. New York: Oxford University Press. The best current survey of white-collar and corporate crime.

Internet

London School of Economics, Employment Relations and Organisational Behaviour Group. *www.lse.ac. uk/collections/EROB* One of the most renowned centers of learning on economic and social issues.

Tavistock Institute for the Study of Human Relations. *www.tavinstitute.org* Origin of the Human Relations school of industrial relations.

Labor and Employment Relations Association. *www.lera. uiuc.edu* The key American association supporting academic work to understand and improve employment relations.

Canadian Industrial Relations Association. *www.cira-acri.ca* Highlights research on workplace change and innovation that foster efficient and equitable workplace practices.

Economic Sociology and Organizational Studies at Pennsylvania University. pesos.wharton.upenn.edu Reports latest research on issues related to organizations.

The *Economist* news magazine. *www.economist.com* The premier news magazine dedicated to international economic issue.

Multinational Monitor. *www.multinationalmonitor.org* Dedicated to uncovering and exposing corporate scandals and abuses of power.

RECOMMENDED FILM

The Corporation (2004). A troubling documentary that starts with the legal status of corporations as individuals and shows how this status facilitates actions by some of the largest and most powerful entities in the world that in true individuals would be labeled sociopathic—actions based on single-minded pursuit of their own interests without consideration of injury to others or any standards of morality.

8

From Field, Mine, and Factory

Looking around the shop, first day on the job, I sorted things out. . . . Metal chips were everywhere on the pitted cement floor, scattered in piles formed as they flew off machines. Squinting, intent milling machine operators turned feed cranks with one hand and brushed on cutting oil with the other, pushing metal stock against spinning milling cutters. Lathe workers watched closely while long, spindly chips curled away from their whirling workpieces.

In this little neighborhood nonunion shop we were making parts for the military, thousands of them; many were subcontracted from larger corporate defense contractors. The casual, mom-and-pop character of the place didn't seem to fit the deadly nature of its products. . . .

No one in the 30-man shop wore safety glasses. Open buckets of naphtha, a highly flammable cleaning fluid, stood around the floor. Parts were being sloshed in them to get the oil off. . . .

My actual job assignment was far less interesting than the surrounding scene. I was to countersink a particular pair of holes in rotor mounts, aluminum holders for electrical parts in missiles. This meant using a drill press to put a rounded bevel on the top edge of each hole, so that a tap, or screw-thread cutter, could enter it properly. The foreman led me to a new, lightweight drill press made in Taiwan. He showed me how deep to make the countersink, gave me a couple of plug gauges to test its size, and walked off. I countersunk about a hundred pieces, or 200 holes. Then I moved on to tapping the holes. After that I "deburred" the part by filing off the sharp edges. Time crawled. There were two other guys on drills beside me, a young Haitian and another guy my age who I later found was from Barbados. They said very little to me during my first days.

(TULIN, 1984:1–2)

This passage recounts the first day on the job for a young worker in manufacturing. Manufacturing activity continues to be a crucial economic foundation for industrially advanced societies. Employment in manufacturing, however, has declined because of automation even as productivity has risen. Manufacturing, unlike services which are often spatially tied to a specific location, is highly mobile and can be easily moved around the world in the search for cheaper labor and more lenient environmental regulations. In this chapter we look at the concerns of manufacturing workers and workers in related industries.

In Chapter 7 we discussed the general concepts of *technology* and *organization*. More concretely, economic activity occurs in three principal sectors: extractive, manufacturing, and services. In **extractive industries**, such as agriculture, forestry, fishing, and mining, a product is removed from the environment. The process of extraction may use sophisticated equipment, but the final product retains much the same form as it has in the natural environment. The raw products are further processed in **manufacturing industries** into more usable forms. For instance, vegetables are canned or frozen. Iron ore and coal are used to make steel, which is fabricated into such things as girders and cables. In this way products are manufactured that may be quite far removed from their origins as raw materials. Cars, refrigerators, airplanes, and computers are examples of such products.

Service industries are even more diverse than manufacturing industries. They include financial and accounting services, medicine, entertainment and recreation, education, government administration, and social welfare services. Services are distinct from goods in that they cannot, in general, be stockpiled, stored, or transported. Many services, such as restaurant meals, must be produced at specific times of the day and in specific places. Goods are transferable from buyer to buyer, so that it is possible to have long chains of intermediaries involved in their production. Services, on the other hand, tend to be nontransferable and must be delivered by the producer directly to the final consumer. Because of these factors the production of services tends to be more geographically dispersed than the production of goods. Partly as a result, automation and the global movement of jobs has had less of an impact on the production of services, although that may be changing. Service work is the focus of Chapter 10.

POSTINDUSTRIAL SOCIETY?

A hundred years ago, even in what are today industrialized nations, most people were involved in agriculture. Over time, because of increasing agricultural productivity, fewer workers were required to meet society's agricultural needs. The result was a period of employment growth in manufacturing. Later, increasing productivity in manufacturing allowed more workers to be engaged in the delivery of services. The often-used phrase **postindustrial society**, however, is a misnomer for today's society. The current economy, with a large share of the labor force in services, has not gone *beyond* being an industrial society. Rather, our economy is based on a highly productive industrial base in which it is possible to produce more goods with fewer workers. Because the demand for manufactured goods can be expanded to a greater extent than the need for agricultural products, the economy is unlikely ever to reach a state in which only a tiny share of employment is in manufacturing.

American society is better characterized as an **advanced industrial society**. This label implies a small but highly productive extractive sector, a larger and also highly productive manufacturing sector, and a growing labor-intensive service sector. This service sector can be expected to continue growing because the demand for services, such as education, health, and recreation, is even more insatiable than the demand for goods.

OCCUPATIONS AND INDUSTRIES

Many different occupations may be involved in the production of a specific good or service. Some of these occupations are common to many different industries. For instance, all industries employ clerical workers and managers. Because these occupations are not unique to any one industry, we consider them in separate chapters in Part III of this book. In this chapter and in Chapters 9 and 10 we focus on the jobs of those employees who are the direct producers of specific goods or services. Thus, in manufacturing industries we focus on blue-collar workers. In the next chapter on high-tech work, we focus on professional and production workers engaged in the design and manufacture of high-tech products such as computers and computer applications. In Chapter 10 on service industries, we focus on the employees directly engaged in providing services, such as fry cooks, waitresses, orderlies, janitors, garbage collectors, and taxi drivers.

RAW MATERIALS: AGRICULTURE, FORESTRY, AND FISHING

Agriculture, forestry, and fishing are the oldest extractive industries, and they still utilize the most traditional technologies and organizations of work. Many aspects of production still rely heavily on time-honored traditions.

Agriculture

We used to work early, about four o'clock in the morning. We'd pick the harvest until about six. Then we'd run home and get into our supposedly clean clothes and run all the way to school because we'd be late. By the time we got to school, we'd be all tuckered out. Around maybe eleven o'clock, we'd be dozing off. Our teachers would send notes to the house telling Mom that we were inattentive. The only thing I'd make fairly good grades on was spelling. I couldn't do anything else. Many times we never did our homework, because we were out in the fields. The teachers couldn't understand that. . . .

The hardest work would be thinning and hoeing with a short-handled hoe. The fields would be about a half a mile long. You would be bending and stooping all day. Sometimes you would have hard ground and by the time you got home, your hands would be full of calluses. And you'd have a backache. Sometimes I wouldn't have dinner or anything. I'd just go home and fall asleep and wake up just in time to go out to the fields again.

If people could see—in the winter, ice on the fields. We'd be on our knees all day long. We'd build fires and warm up real fast and go back onto the ice. We'd be picking watermelons in 105 degrees all day long. When people have melons or cucumber or carrots or lettuce, they don't know how they got on their table and the consequences to the people who picked it. If I had enough money, I would take busloads of people out to the fields and into the labor camps. Then they'd know how that fine salad got on their table. (Terkel, 1974:34,38 [Roberto Acuna, farm worker])

Rising Productivity Agriculture still involves much hand labor, as described eloquently above by farm worker Roberto Acuna. Mechanization and productivity in general, however, have increased dramatically over the last century in agriculture. This was particularly true during the latter half of the twentieth century because of the expanded use of machinery, fertilizers, and improved varieties of grains and other crops. Because of this increased productivity, fewer people are required to produce agricultural goods. From the standpoint of increased productivity, this is an important step forward into industrial society. From the standpoint of the individual farmer, technology has lightened many physical burdens. But it has also contributed to the chronic treadmill of falling farm prices. As farm productivity has increased, the prices of farm products have fallen. Fewer and fewer family farmers are able to make a living from working the land because they have to compete with large, heavily mechanized corporate farms. In response to falling prices, family farmers may take on second jobs in neighboring towns to continue farming, or they may be forced to mortgage and eventually sell their farms and abandon farming altogether (Lobao and Meyer, 2001).

For many small farmers the problem may simply be having too small a farm to support themselves and their families. Small black-owned farms in the South, for example, historically had *higher* yields per acre than commercial farms as a whole, but many were so small that they could not produce sufficient income to sustain a family. As a result, the amount of farmland owned by blacks in the South fell from fifteen million acres in the early 1900s to five million acres by the 1970s (McGee and Boone, 1979).

Large corporate farms contribute to the displacement of family farms by further driving down farm prices based on economies of scale. Some of the largest corporate farms have developed huge factory-like production facilities for hogs and chickens. These farms drive down prices for pork and poultry but in so doing undercut small farmers who had been able to sustain themselves with these activities (Falk, Schulman, and Tickamyer, 2003). In addition, in the Carolinas and other parts of the country, large corporate farms have been associated with significant environmental problems resulting from the huge tonnage of animal waste produced at single locations.

It is somewhat ironic that at the same time small farms are declining in the largest capitalist nations, they are making a comeback in the formally socialist countries of Eastern Europe. In Hungary, for example, which had a highly collectivized agricultural system of large cooperatives, small privately owned farms now produce nearly half the fresh vegetables and fruits and farm animals (Schimmelfenning and Sedelmeier, 2005). In the United States some farmers are attempting to maintain their economic viability by producing specialty products that do not compete directly with those produced on larger farms. Such crops include organically grown produce and meats and other products for local "farmers' markets."

Rising Costs and Falling Prices Farmers have also been burdened with rising fuel prices and high interest rates, which increase the costs of production while prices for farm commodities continue to fall. In 1998, for example, the average cost of growing a bushel of corn was $2.42. The average price received for a bushel of corn was $1.95, *24 percent less* than the average cost of its production (Census, 2000:687). Obviously, only farmers with below-average production costs can survive for long in such a market. What occurs in this context is a *shaking-out* process in which smaller or less-efficient farms can no longer produce a living income. This process can be very traumatic for the farmers and farm families involved, resulting in anxiety, depression, and even increased risk of suicide. In 1975 there were 2.8 million farms in the United States. By 2005 almost a third of these had folded. Ownership of the remaining farmland is extremely concentrated. The U.S. Department of Agriculture reports that more than 60 percent of the nation's food supply is produced by just sixty thousand farms (Census, 2005).

Technological Advances   Given their financial constraints, technological advances in agriculture may come as a curse to small farmers. They may not be able to afford the expensive new technology and thus may fall even further behind larger producers. For instance, the use of growth hormones and antibiotics in animal feed has greatly increased yields for farmers who can afford these purchases, but it has also lowered prices for beef and other meats, putting additional pressure on small producers. The development of expensive new genetically engineered seeds has further heightened these pressures.

Technological advances in agriculture may also be a mixed blessing for Third World nations struggling to feed their growing populations. In the 1980s U.S. hogs were imported into Haiti under a joint U.S.-Haitian program, largely in response to a fear that African swine fever among the Haitian hogs would spread to the United States. However, the lifestyle of the U.S. pigs may have been too rich for desperately poor Haiti: "Haitian pigs roamed freely in the countryside and lived off orange peels, mango seeds, and garbage. Their American replacements are supposed to live on a concrete floor, eat imported food and be given expensive vaccines. Even worse, unlike the now-defunct Haitian pig, which had long legs and could be led to the market by a string, the American pig refuses to walk" (Oppenheimer, 1986).

Adding to the difficulties faced by farmers is instability in agricultural prices. Prices for farm products are highly dependent on how bountiful the harvest has been. Thus, in good years the farmer has a lot to sell, but prices tend to be low. Annual variations in weather patterns, insect infestations, and diseases all produce dramatic swings in farm productivity. In addition, the *fixed costs* of farming are very high, averaging more than 50 percent of total costs (North, 2005). Fixed costs are those costs that farmers must pay regardless of whether they sell their crops or at what price. They include such things as depreciation on buildings

and equipment, property taxes, and interest on land and equipment loans. Fixed costs are a large expense, and the farmer must have at least some income to pay them. Thus, it is often rational for the farmer to continue producing even when the prices received do not cover the full costs of production. The money earned at least helps pay the fixed costs of production. The residual is made up by taking on additional debt, which then demands a greater interest payment the next year. The cycle finally comes to an end when the farmer can no longer take out any more credit or cannot survive on earned income. At this point the farmer must sell the family farm, pay the proceeds to creditors, and look for nonagricultural employment.

Federal Price Supports   Federal farm policy in the United States attempts to slow the decline of the small farmer by supporting prices for selected farm products and providing credit to farmers. Such policies establish a floor for farm commodity prices, thus ensuring the farmer a certain minimum price for products. Unfortunately, such price supports and credit arrangements have benefited mainly large farmers who have the acreage to take advantage of the programs. Agricultural price supports have also been reduced in recent years in response to international trade agreements that limit such subsidies and encourage market prices for commodities (Crowley and Roscigno, 2004).

Farm incomes have also been reduced by increased world competition in agricultural commodities. For example, rapid expansion of Third World countries into sugar production in the late twentieth century resulted in a fall in sugar prices in the 2000s. This decline almost eliminated sugar production in Louisiana and Hawaii (sugar cane) and Colorado (sugar beets), and it had even more severe consequences for the economics of countries, such as Cuba and the Philippines, that depended heavily on cash income from sugar production.

In response to the threat of foreclosure and rising poverty rates in rural areas, the members of farm families have increasingly taken jobs away from home. Often this means that the husband looks for paid employment off the farm. Such arrangements may not be very lucrative, however, because traditionally male jobs in rural areas tend to pay low wages. Farm women are often able to secure higher status employment off the farm (such as clerical work) than are farm men (Lobao and Meyer, 2001). High unemployment in rural areas unfortunately often cuts off even this "safety valve" for rural distress. There are often long lines of unemployed workers waiting for any job openings that become available. The children of farm families, many of whom would prefer to remain in farming, often face disheartening odds. Agriculture in the United States, in spite of high and rising productivity, thus operates in a state of chronic crisis.

Farm Laborers   In the United States an additional three million people work as farm laborers at some time during the year. These workers are almost as numerous as farmers and unpaid family members who work on farms. Over half a million farm laborers work full-time (250 or more days per year). Another half million are long-term seasonal workers who are employed at least two to three months per year in farm work. The remainder—short-term seasonal workers—are primarily students and housewives or nonfarm workers who take occasional second jobs in agriculture. Full-time farm laborers earn an average of about $27,500 annually. Long-term seasonal workers earn much less—about $8,000 annually. And short-term seasonal workers average less than $3,000 annually (Census, 2006).

In contrast to the decline in the number of small farmers, the employment of farm labor is stable and likely to remain so. The employment of farm labor is concentrated on the largest farms, which are growing in size and number. Fewer than 2 percent of farms hire more than a third of all farm labor. About a third of American farms, however, employ at least some hired labor during the year. Farm work is regionally concentrated in California, Texas, and Florida. As is also the case for farmers and their families, most seasonal farm laborers do not depend on agriculture as their sole source of income.

Farm workers, especially full-time farm workers, have an intense interest in securing the advantages available to other workers, such as regular employment, health insurance, accident insurance, retirement benefits, and union representation (Dalton, 2003). Farm workers, however, have traditionally been excluded from laws allowing workers to organize into unions and engage in collective bargaining. Cesar Chavez championed the cause of the farm worker from the 1960s until his death at age sixty-six in 1993. He organized boycotts, hunger strikes, union organizing drives, workers' cooperatives, and marches to highlight the plight of the farm worker. His efforts culminated in the emergence of the United Farm Workers (UFW, www.ufw.org), which continues his efforts into the 2000s. Chavez's innovative approach to organizing farm workers is described in Box 8.1.

To reduce the problems of unstable employment and low earnings faced by farm workers, the U.S. Department of Agriculture has recommended educational programs for employers, labor contractors, workers, and their respective organizations. Such programs would seek to improve the skills of workers and to encourage modern labor-management practices by employers. Such practices would include the provision of benefits, such as health care and retirement, to provide a more stable employment situation. Increased skills for workers and improved labor-management practices would increase the productivity of farm labor and allow a general increase in farm wages. The Department of Agriculture also recommends further development and equitable enforcement of labor-related laws in the agricultural sector (Wells, 1996).

Forestry

A real artist minimizes the heavy labour for himself by dropping the trees within inches of where he wants them. And that is no mean trick, any novice of the game will find. By dropping the tree mid-way across his skid-pile, the real artist can cut into lengths, strip the branches and pile neatly with little more than a twist of the wrists or the leverage allowed by a handy pike pole. (Radforth, 1986:251)

BOX 8.1   Cesar Chavez and the United Farm Workers

To organize farm workers [Chavez] had to tackle four problems at once. He needed to organize them, to get the leverage so that organizing meant something, to unite bitterly divided regions and nationalities, and to simultaneously sustain all these long enough for growers to pay attention. Chavez organized [farm workers] in fraternal service organizations. He got the leverage through boycotts fueled by feverish enthusiasm for liberating America's poorest workers. He united races, nationalities, religions, and languages in the fields and cities across the nation by speaking and living a simple life dedicated to justice. And he sustained all of it by his personal example.

"You can't organize on money," he used to tell us. "There isn't enough money to organize now. There never was, and there never will be. Once you depend on money, you're finished." He lived on the UFW stipend of room, board, and $5 a week....

He invented a type of organizing that synthesized the tactics of the civil rights and anti-war movements, cooperatives, labor unions, community organizing, and religious ministries. Everyone in the union was uncomfortable with at least one aspect of this medley. But he got us all to live with each other because the union needed every one of these tactics and constituencies. No where else then or since could you find nuns working beside immigrant laborers and college students, a routine combination in the UFW. He didn't care whether he had to march 1,000 miles or set up a shrine outside a vineyard—if it worked, we were going to do it.... "There's a moment," he would tell us, "when the growers, politicians, corporations and sheriff's deputies believe we want it more, stronger, and longer than they want to keep it from us. That's the moment we win."

SOURCE: Excerpted from John Gardner, "Community Organizing: Seeds of Justice," *In These Times*, May 17, 1993, pp. 18–19. Reprinted by permission of the publisher, www.inthesetimes.com

---

Forestry employed 79,000 workers in 2005, about 4 percent of the total employed in agriculture, forestry, and fishing and less than 0.1 percent of the total civilian labor force (Census, 2005). Forestry is like agriculture in that it requires a high degree of skill. These skills, however, often go unrecognized and unrewarded because they are acquired informally during childhood and adolescence in rural areas.

Skilled work remains the norm in the logging industry. Mechanization in the harvesting of pulpwood, however, has dramatically changed the nature of the skills needed in this part of the industry. Pulpwood production is the most rapidly growing part of the forestry industry. Pulpwood, made from small trees not suitable for lumber, is used to make paper and compressed wood products. Workers use a machine called a wood harvester to cut 150 to 180 small trees per hour (about three trees per minute). The large machine dwarfs its single operator, who "by controlling numerous levers, gadgets, and joy sticks, guides the harvester through the forest, using the machine's huge hydraulic shears to fell the trees, which are then automatically processed into logs and stored at the rear of the vehicle" (Radforth, 1986:261).

Forestry shares with other extractive industries extreme vulnerability to price fluctuations. In British Columbia and in the Pacific Northwest of the United States, the lumber industry periodically collapses because of downturns in housing construction. The economy of this lush region has been described as a "vulnerable hinterland economy based on a rich but limited resource" (Marchak, 1996). When a region is tied to a single raw product, such as lumber or pulpwood, it is highly vulnerable to economic downturns and price fluctuations, no matter how rich its natural resources.

Fishing

The accommodations aboard the [tuna] boats are very comfortable. Most are air-conditioned throughout the main deckhouse and bridge. All have hot showers and nice, even fancy, head (toilet) facilities. The galleys are modern and well-equipped. Many boats have a crew lounge with a card table, color television, and stereo system. . . . Each bunk is large—approximately four feet by seven feet—and has a reading and a night light, extra space for books, and a curtain which closes it off from the rest of the cabin. . . .

Being at sea is dangerous. Three boats in the fleet—big boats—have sunk in the last year and a half. . . . Working around heavy equipment, especially heavy running rigging, is also dangerous. . . . The seine net on the tuna boats is primarily simply a way of putting up a "corral" around the tuna, and then slowly tightening the boundaries until the fish are swimming in the net right next to the boat. . . . The net is simply too large (three-quarters of a mile long, 350 feet deep, weighing by itself seven tons) to be controllable after it has been set, and the fish weight contained in a good set (50 to 100 tons) is too great for one to depend on even the strongest of equipment to hold it in the kind of weather in which the boats often have to set. This lack of control and consequent dependence on precise timing make the entire process dangerous for all those involved. (Orbach, 1977:19–20, 28-9)

In the year 2000 the catch of fish from the world's oceans was over 130 million tons (Census, 2005). As a source of protein for human beings, fish is twice as important as poultry and more than half as important as cattle, pigs, sheep, horses, and goats combined. Between 1950 and 2000 the catch expanded at about 5 percent annually, significantly faster than the 2.5 percent annual growth in cereals and the 2 percent growth in meat.

The largest exporter of fish in the world is Canada, with a yearly catch of two million tons, most of which is exported. The United States has a yearly catch of about five million tons, with most of this being consumed domestically. Japan both imports the largest amount of fish and has the largest catch in the world (seven million tons, or about 15 percent of the world total).

The Ocean's Limitless Bounty? At one time ocean fishing was expected to provide relief for the starving millions in the world. In the last decade, however, the world catch has grown at only 1 percent annually. The reason for this reduced rate of growth is mainly overfishing of coastal waters, which forces fishing boats to move farther offshore. In addition, many fisheries have been systematically overfished. Deep-sea fishing today can involve thousands of baited hooks on lines stretching eighty miles across the ocean. Trawlers line up abreast to sweep life from swaths of ocean with nets large enough to cover a small city. In North America, both East Coast cod and haddock fisheries and West Coast salmon fisheries have declined dramatically. In the Mediterranean, the bluefin tuna catch has come to a virtual end because of worldwide overfishing of this highly prized delicacy. In coastal waters these problems are sometimes compounded by pollution. Human-generated environmental pollution can greatly reduce or render unfit for consumption harvests of shellfish and other species dependent on coastal habitats.

Overfishing and coastal pollution, combined with rising fuel prices, has limited the expansion of the catch and has even resulted in dramatic downturns for many prized species (Apostle et al., 1998). Ocean fishing may have already peaked and be on the decline unless significant conservation measures are implemented—measures that would entail multinational agreements that are difficult to negotiate and enforce.

Technology and Organization in Fishing The technology and organization of fishing varies from small shore boats with crews of two to three, which are at sea for a day; to larger boats with crews of nine to eighteen, which are gone for up to ten days; to the largest trawlers, which may have crews of twenty or more and be gone from port for up to three weeks or longer. Traditionally, fishing has been organized around independently owned boats, each staffed by a crew that shares in the catch. This organization has been favored over wage-labor systems because of the need for cooperation and teamwork to secure the catch and to face the dangers and challenges of the sea.

Ocean fishing has supported fishing communities around the world in the Mediterranean, Canadian Maritime, Pacific Northwest of Canada and the United States, South Asia, the coastal waters of South America, and elsewhere. Many of these communities today face chronic hardship because of the declining catches. In the Gulf Coast of Texas, shrimp and crab fishing have been mainstays. Gulf Coast communities, however, have faced hard times because of increasing regulations on the season and the catch and because of imports of foreign and farm-raised shrimp. In recent years these communities have also had to adjust to Vietnamese refugees who migrated to the Gulf Coast as familiar terrain and who have competed successfully with native shrimpers because of industriousness and a forced willingness to live on meager means (Maril, 1995).

In recent years the increased capital investment required in fishing has often made it more difficult for individuals to purchase the necessary boats and equipment. Absentee ownership of boats and extension of wage-labor relationships into fishing may erode the cooperative basis of the work and the working conditions of fishermen. Risks are minimized when those in the boat are free to arrive at a consensual decision about when to fish and when to return to port. When this decision is made by an owner back on shore or according to a set of bureaucratic rules, lost opportunities and dangerous situations are more likely to occur (Doeringer, Evans-Klock, and Terkla, 2002). In addition, unlike large corporations, family-owned fishing enterprises are much less likely to quit fishing and move to more profitable sectors during economic downturns (Norr and Norr, 1997).

Aquaculture   A bright spot in the world fishing outlook is the growth of **aquaculture**, also known as *fish farming*. Under this technology fish are raised in ponds or in *fish ranches* (large floating cages in the sea).

In 2005 ten million tons of fish were raised in this way, and the potential for expansion is considerable. Nearly five million tons were raised in China alone with only a limited investment of capital and technology. With improved technology even the existing fish ponds in China are projected to have a potential yield of fifty million tons a year, almost half the current annual catch from the sea.

MINING

"First time I got hurt, I got covered by a rib roll (a collapsing wall) and got three vertebrae busted and a busted pelvis. I got a 25 percent disability for that, but I came back to work anyway.

"[Eight years later] I slipped in some grease and broke my neck and got my back messed up. It was my fault. I was in too much of a rush.

"Oh, yeah. I've also got first-stage rock dust—silicosis. I got paid a flat $3,100, but my lungs are so bad I can't hardly get around."

His buddies half-listen to Raymond's story. Most of them can match it, fracture by fracture, wheeze by wheeze. . . .

Over the years the men build up friendships with each other that are not equaled in many marriages or families—although most of them would be embarrassed and would deny it if the subject were verbalized. Men call each other "buddy" with an open affection that seems quite foreign to a visitor from the cold, impersonal urban world. (Vecsey, 1974:19, 124)

Down in the Mine   Mining is dangerous and demanding work. But it generates a sense of purpose and a collective identity. This identity is built on the twin pillars of shared group responsibility for the work below ground and geographically isolated work communities above ground. Miners' work ethic prescribes that the work be done according to certain standards of safety and efficiency. This ethic also stresses competition between work crews in the effort to produce the most tonnage. Teams of miners are responsible for organizing their own activities below ground and resent external control. The miners' sense of group solidarity is intensified by the many dangers they face. These hazards include poor illumination, inadequate ventilation, dangerous gases, use of explosives, poorly supported roofs, unsafe tunnels, flooding, working with high-voltage electrical equipment, and chronic inhalation of dust (Warren, 2000). Mine accidents are still a common occurrence, often killing a dozen or more miners in a single accident and continuing to make mining one of the most dangerous occupations.

Occupational Solidarity Above ground, this collective identity is reinforced by the frequently isolated location of mining communities. In mining regions a distinct culture often emerges that provides its members with a shared identity. Sociologists refer to such subcultures as **occupational communities**. They are typical not only of miners but also of lumber workers, longshoring workers, and other trades that work in isolation from the mainstream of society. Occupational communities may even be said to distinguish some professionals, such as doctors, who live their lives in relative social isolation if not in geographic isolation.

In the United States, miners have often been a force in national politics. For instance, coal miners had a key role in promoting a more even-handed government approach to trade unions. In the coal strike of 1902 and 1903, President Theodore Roosevelt threatened to nationalize the industry (eliminating the owners' profits) and have the army mine the coal (eliminating the workers' jobs) unless the owners and workers bargained collectively. This episode is frequently cited as the first instance of the federal government taking a neutral stand between capital and labor. Previously, the government had sided with capital by jailing strike leaders and supporters and using troops to protect hired strikebreakers and company property.

American miners, however, have never played as pivotal a role in national politics as have British miners. Their relative weakness has resulted partly from their geographic isolation in Appalachia (the

center of the American coal mining industry) and in the West (the center of much of the metal mining industry). These chronically depressed, semirural regions have provided a weak springboard for miners' demands. In times of economic downturn, miners have often reverted to rural and agricultural pursuits such as gardening, hunting, and producing eggs and butter for local markets. Even in the best of times American mine workers have kept strong ties to the rural economy. The availability of these alternative pursuits, even if they are only minimally productive, has been an important factor limiting the intransigence of miners in demanding redress for the problems of their industry.

Strip Mining In recent years the coal mining industry has shifted from Appalachia westward toward the Rocky Mountains. In the West coal seams are thicker and closer to the surface. As a result, labor-intensive deep mining technologies have been replaced by more capital-intensive strip-mining technologies, long used in the West for mining metals such as copper and aluminum. As with agriculture and other extractive industries, improvements in technology have reduced employment in mining even as output has increased (Berger, 2002).

CONSTRUCTION

The difficulty is not in running a crane. Anyone can run it. But making it do what it is supposed to do, that's the big thing. It only comes with experience. Some people learn it quicker, and there's some people can never learn it. . . .

This is a boom crane. It goes anywhere from 80 feet to 240 feet. You're setting iron. Maybe you're picking fifty, sixty ton, and maybe you have ironworkers up there 100, 110 feet. You have to be real careful that you don't bump one of these persons, where they would be apt to fall off. . . .

There's a certain amount of pride—I don't care how little you did. You drive down the road and say, "I worked on this road." If there's a bridge, you say, "I worked on this bridge." Or you drive by a building and you say, "I worked on this building." Maybe it don't mean anything to anybody else, but there's a certain pride knowing you did your bit. (Terkel, 1974:49–50, 54 [Hub Dillard, crane operator])

Pride in Skilled Work Like miners, construction workers take great pride in their work. They often have a high level of skill and exercise a good deal of autonomy on the job. Construction is difficult to supervise, because it is spatially dispersed and because much of it requires significant judgment. Often, only the construction worker knows how best to do a given job. Because of these factors, construction workers are not closely supervised. Rather, their work is inspected, either by their employer or by a government building code inspector, *after* it has been completed. Their skill and the resulting autonomy give construction workers a sense of power and pride in work (Applebaum, 1999).

Many construction workers belong to craft unions, and the wages and conditions for other workers in construction are strongly influenced by these workers' contracts. Union construction workers are generally more qualified than nonunion workers. Unions work with building contractors and state governments to design apprenticeship programs that train craftsmen who are highly skilled in all aspects of a trade—not just in a narrowly defined set of tasks (Allen, 2001). It has been estimated that union craft workers are between 7 percent and 11 percent more productive than their nonunion counterparts, even after adjusting for their higher wages (Freeman, 1994). Craft unions also run hiring halls where contractors can secure large numbers of skilled workers on relatively short notice, thus reducing the costs of recruiting, screening, and supervising workers.

Nonunion Workers Even though they are on average less productive, nonunion workers do an increasing share of construction work. The reason is that, in many settings, price competition is more important than quality. In housing markets, price competition is a particularly important factor. Union construction workers have responded by making wage concessions and stressing their ability to produce high-quality work on schedule, factors that still have to figure prominently in the decision making of construction contractors.

Because construction work is highly regional, the demand for construction workers can vary dramatically between areas. Due to population shifts, construction activity has been higher in the West and South than in the Northeast and Midwest. In spite of these short-term problems, the long-term employment outlook in construction is relatively stable. Technological innovations in construction have been largely incremental and, given the nature of construction work, few major employment displacing innovations are expected in the future.

In spite of generally stable employment in construction, women have increasingly gained entrance into this previously all-male field. As new entrants into the construction trades, women have had to confront resistance from some of their male coworkers. Box 8.2 reports on the frustration that can be caused by such resistance. It also illustrates that resistance can, in some cases, be overcome.

MANUFACTURING

Manufacturing industries produce a tremendous variety of goods in industrial societies. They thus employ a wide range of occupations. Manufacturing workers can be broadly classified as craft workers (skilled workers), machine operators and assemblers (semiskilled workers), and laborers (unskilled workers). In this section we discuss these three types of manufacturing jobs and look at current developments in three key manufacturing industries. Manufacturing also employs many professional, managerial, and clerical

BOX 8.2 Gaining Acceptance as a Female Carpenter

For a woman to survive in the trades . . . you have to be tactful, not be hostile, not alienate people. You really have to learn professional survival skills, because men's masculinity is threatened by you being there. Society recognizes construction workers as being very macho and virile. When a woman comes along who's five foot three and a hundred and twenty pounds and can get in there and do their type of work, it's a blow to their ego, a real shock. So the men are threatened by it.

The men show that in different ways. If a foreman comes up to a group, he'll look at everyone except you. He'll delegate jobs to everyone but you, and then you have to go up and ask him. It's sort of uncomfortable things like that. Or the carpenters will pair up, and you're the one left out. Or men you're working with just won't talk to you. I used to not push in those situations, but lately I've felt more self-confident, and sometimes I play games with those people. Like, for example, today on the job site there is another fourth-year apprentice who is sort of the foreman's pet. He just came over and picked up some wood and started doing some form work on top of the footing another carpenter and I had built. And I yelled at him, "What's going on?" Not in a hostile way, just curious. We'd

finished that footing, and I wanted to know what changes had been ordered. He wouldn't answer, so I got a bit flustered and asked again. He still wouldn't answer. I kept probing, and he kept not answering. Finally I said, "I'm talking to you. I want a response." I was just very aggressive, which I'd never been before. But I've gotten to the point where I'm tired of not getting recognition. Finally he answered me and explained there was a change in plans, not any mistake. I could see he was a little taken aback, and I felt so good. I had won something. What was more interesting was the reaction of the other man. Right in the middle of all this he looked up at me and smiled.

The crew I'm working with now—it's never been better. I'm accepted. They kid me like one of the guys. . . . Like if a carpenter I've been working with is talking with another man about something we built, he'll mention my name, say, "Elaine and I were working on this. . . ." Even a week ago . . . I wouldn't have gotten that recognition. So they include me, acknowledge me.

SOURCE: Jean Reith Schroedel, *Alone in a Crowd: Women in the Trades Tell Their Story*, pp. 38–39. Copyright 1985 by Temple University. Reprinted by permission of Temple University Press.

---

workers in support positions. We discuss those occupations in Chapters 11, 12, and 13.

Craft Workers

The U.S. Bureau of the Census classified over fifteen million workers as **skilled craft workers** in 2000. About four million craft workers are employed as precision production operators. These occupations range from tool and die makers to power plant operators. An additional five million craft workers are mechanics and repairers, including vehicle and equipment mechanics, telephone line installers and repairers, heating, air conditioning, and refrigeration mechanics, and heavy equipment mechanics (Census, 2005). The remaining six million workers in the skilled trades are employed in construction.

Craft Apprenticeships Skilled workers typically learn their trade through an apprenticeship program. At any given time about 250,000 workers are enrolled in apprenticeship programs in the United States. Most of these programs are jointly administered by a company and a craft union. Apprentices receive about half pay during the years they are in the program. Depending on the trade, workers spend one hundred to eight hundred hours a year in classroom education for two to four years. The rest of their time is spent on the job working under the supervision of more senior members of their craft.

Machinists, for example, are required to complete about 570 hours of classroom training and about 8,000 hours of on-the-job training during a four-year program (Wright, 2000; www.iamaw. org). Machinists manufacture specialized parts for the machines that mass produce other products. (The workers who operate these machines are classified as "machine operators" and are discussed below.) At the end of their apprenticeship, machinists are expected to be able to work with a variety of metals and machinery, including complex electronic and computer-driven machining tools. Besides knowing how to cut, drill, and shape metal, they must also know how to program their machines and how to make precise measurements, sometimes down to a millionth of an inch, so that the parts they produce will work perfectly.

As a result of having skills that are difficult and time consuming to acquire, craft workers generally have more security against layoffs than do semiskilled machine operators. In addition, automation has affected craft workers less than it has affected less skilled workers (Kimeldorf, 1999). The tasks required of a skilled craft worker are too diverse to be easily automated. The printing industry, however, provides an exception to this pattern. In the past this industry employed a large number of skilled craft workers to set type for newspapers and magazines. With the development of electronic and computer-assisted printing and layout, many of these operations have been eliminated or are now performed automatically by computers. The extent to which other crafts are being affected by computer automation is explored further in Chapter 9.

Exclusionary Practices   In the past women and minorities were strongly discriminated against in the skilled trades. Exclusionary practices by employers and efforts by craft unions to protect the jobs of their predominantly white, male members were both to blame. In recent years most of these practices have been greatly reduced. As you saw in Chapter 4, however, such practices are deeply ingrained in established prejudices and procedures and are difficult to eradicate completely. In 2005, 14 percent of white and Hispanic male workers were in the skilled trades. However, only 9 percent of black workers and only 2 percent of women held jobs in the skilled trades (Census, 2005). The strong occupational segregation of women out of the skilled trades continues because of different socialization of men and women, continuing discrimination, and a lack of informal job contacts in the skilled trades for female workers.

Machine Operators and Assemblers

I start the automobile, the first welds. From there it goes to another line, where the floor's put on, the roof, the trunk hood, the doors. Then it's put on a frame. There is hundreds of lines.

The welding gun's got a square handle, with a button on the top for high voltage and a button on the button for low. The first is to clamp the metal together. The second is to fuse it.

The gun hangs from a ceiling, over tables that ride on a track. It travels in a circle, oblong, like an egg. You stand on a cement platform, maybe six inches from the ground.

I stand in one spot, about two- or three-feet area, all night. The only time a person stops is when the line stops. We do about thirty-two jobs per car, per unit. Forty-eight units an hour, eight hours a day. Thirty-two times forty-eight times eight. Figure it out. That's how many times I push that button.

The noise, oh it's tremendous. You open your mouth and you're liable to get a mouthful of sparks. (Shows his arms.) That's a burn, these are burns. You don't compete against the noise. You go to yell and at the same time you're straining to maneuver the gun to where you have to weld. . . .

I don't like the pressure, the intimidation. How would you like to go up to someone and say, "I would like to go to the bathroom?" If the foreman doesn't like you, he'll make you hold it, just ignore you. Should I leave this job to go to the bathroom, I risk being fired. The line moves all the time. (Terkel, 1974:221–222 [Phil Stallings, spot welder])

In 2005 fourteen million workers were employed as machine operators and assemblers in the United States. Most of these workers are

---

BOX 8.3   Bored to a Stupor

Based on over four years of full-time employment as a production worker in the beer bottling industry, Sociologist Clark Molstad reports the following experiences of time spent in truly boring work.

When [working on the bottling line] I experienced strong feelings of mental regression. My fantasies became progressively more childlike, until I was actually holding imaginary conversations with the beer cans in my hands. I worried about their dents or streaked labels as though they were animate objects. I wondered where they would be shipped and how they would like it when they got there. I wondered how a can felt as it was being crushed and shredded. It was only with some difficulty and effort that I could

muster my consciousness to return to normal after hours of this work. Yet interruptions were troublesome because they required that I return to an adult mode of thought and take up the burden of conscious life in the brewery....

For that reason, I resisted interruptions and resented problems that forced me to return my attention to the work and the brewery. At times I even regretted going to the lunchroom on breaks because it required focusing my attention on the here and now. I preferred to stay in my fantasies.

SOURCE: Clark Molstad, 1986, "Choosing and Coping with Boring Work," *Urban Life* 15, 2 (July):221–22.

---

considered **"semiskilled."** The Bureau of Labor Statistics defines **semiskilled work** as requiring *less than two weeks of training.* The period required to become fully proficient at many of these jobs, however, may be much longer.

The largest group of machine operators and assemblers, numbering over seven million, operate stationary machines. The largest subgroup of these work in textiles and apparel. Machine operators and assemblers also include punching and stamping machine operators, lathe operators, molding and casting machine operators, assembly-line welders, printing machine operators, and laundry and dry cleaning machine operators. The second largest group, just over five million, operate transportation equipment. These occupations include three million truck drivers, as well as bus drivers, forklift drivers, and other mobile equipment operators. The final group, just under two million, are assemblers, testers, and graders. These workers do hand assembly and sorting that does not require the regular use of machinery (Census, 2005).

Repetitive Work   Many semiskilled jobs involve work on mechanically paced lines. Such work is extremely repetitious. It requires a high degree of

surface mental attention without corresponding mental absorption (Appelbaum, Bernhardt, and Murnane, 2003). Repetitive work, especially if it is time pressured, can be stressful, and it can lead to mental distress and breakdown. Other semiskilled jobs involve sitting at a bench and assembling a subunit. Bench workers take parts from one pile, assemble them, and place them in another pile. Bench work has the advantage of not being machine paced, but it is still highly repetitive, and workers have to meet production quotas.

Workers in semiskilled occupations are also often stressed out because the assembly line or the production quota pushes them so fast that they cannot do quality work or take pride in their work. Jobs that demand sloppy work are demoralizing—maintaining pride in one's work is difficult or impossible in such situations (Fink, 1998). Box 8.3 describes the experience of doing boring work in a bottling factory.

Working Ahead   In response to job pressures, semiskilled workers devise a variety of creative ways to give themselves at least limited control over the pace of their work. One common technique is to *work ahead*. This may mean either working *up the line* on the assembly line, by moving ahead to work on parts before they arrive at one's station to secure a brief break later, or "building a bank" of parts, by pushing oneself in spurts in order to slack off later.

Another strategy is *doubling up,* in which one worker temporarily takes on two jobs while a second worker takes a break. A new worker in a Toyota plant in Japan reports help from a more experienced worker who regularly engages in working ahead:

> Those of us who remain have to find some pleasure somehow. Shimoyama, for instance, works ahead and then comes to help me. He takes over my position, puts the six bolts into the lock, and tightens them all at once with a nut runner—which takes some skill. When he succeeds, he yells excitedly. I do too, whenever I can manage it. Completing the task in two or three seconds, hearing the bolts slide in with a nice click, gives me real pleasure. Even in this kind of detailed, boring work, you need some sort of satisfaction, or you can't go on. (Kamata, 1982:89)

Managers generally oppose such practices, but workers contend that they make fewer errors by working in intense pushes and then slacking off. Workers argue that a continuous grinding pace creates greater boredom, poorer concentration, and more errors (Graham, 1995).

The jobs of machine operators and assemblers can be quite different in core and peripheral parts of the economy. In the *core* economy, made up of large manufacturing companies with a significant share of their market, wages and benefits are generally good and provide at least partial compensation for the repetitive nature of the work. In *peripheral* parts of the economy, dominated by small and financially vulnerable firms, wages may be at or near the legal minimum, and benefit packages may be nonexistent. In these situations the problems of repetitive, alienating work are compounded by problems arising from poverty-level income. Female and minority workers tend to be overrepresented in these latter jobs.

Assembly line work in many industries is undergoing significant change. As we will see later in this chapter, as well as in Chapters 9 and 17, automation has eliminated many assembly jobs while placing greater demands on other jobs. For instance, assembly workers today are often expected to participate more actively in improving quality. It is unclear, however, if these jobs are any less stressful today than in the past.

Unskilled Labor

Poultry must be handled, initially, as live birds. Workers (almost invariably male because of the weight handled) must snatch live birds from cages unloaded from tractor trailer trucks and hang them, upside down, on shackles attached to moving conveyor lines. The "hanging" job may even involve 30–40 pound turkeys. The "hangers" are subjected to wing battering by the dirty, squawking birds who not infrequently urinate and/or defecate on the workers handling them. As the flopping, noisy birds move down the line, they undergo an electric shock intended to relax all muscles for a thorough bleeding after the throat is cut. This step also results in additional excretory discharges from the birds. All five senses of the workers are assaulted. One "hanger" who was interviewed revealed, on weekends, he took six to eight showers trying to rid himself of the stench.

After being shocked, the birds are slaughtered by having their throats cut, either by hand with a knife or by a machine with a worker standing by to kill birds where the machine fails to do so.....

Poultry processing of necessity involves the commodious use of water, and thus there is an inevitable high degree of dampness and frequency of standing water constituent to this industrial process. The water used may have to be scalding, to remove feathers, or near-freezing, in order to cool the carcasses of the birds.... Thus, employees may have to work in damp, cold, or hot rooms, literally standing in water (and sometimes blood, such as in the "killing room"), handle blood, gore, offal, and visceral organs and materials. (Bryant and Perkins, 1986:158)

In 2000 about five million workers in the United States were employed as laborers, materials handlers, equipment cleaners, and helpers—jobs all classified as **unskilled labor** (Census, 2005). These jobs include stock handlers, garbage collectors, hand packagers, and machine feeders and offbearers. Some **unskilled work**, such as garbage collection, is relatively autonomous and is not closely supervised. A greater share of unskilled work is closely supervised and involves cleaning, loading, unloading, or preparing for some more complicated step in manufacturing. The loading of chickens onto the "disassembly line" described above is a good example of such work.

Laboring jobs are often closely supervised under the supposition that workers will not do the tasks right unless forced to do so. Whatever the merit of such considerations, close monitoring serves as an additional negative factor in much unskilled work and acts to undermine whatever autonomous motivation workers have.

"No Experience Needed"   Most laboring jobs require little, if any, training—in effect, people can be hired "off the street" to do the work. No training is required beyond a brief explanation of the job and those skills picked up while performing the task. Much of the work is physically demanding and is often done under harsh physical conditions. The work of female laborers is no exception to this rule. In describing her job as a "craw puller" (remover of the throat pouch from the dead chicken) on the poultry processing line described above, one woman felt that men could not physically do the job: "Take a man, he won't stick to some of those harder jobs like craw pullin'. A man will stick his finger in there and gets a sore finger and walks the floor all night long—he will quit that job, he won't stick to it" (Bryant and Perkins, 1986:162).

Minorities in Unskilled Jobs   Blacks and Hispanics are *over*represented in unskilled work because of a historical pattern of discrimination in better paying jobs. Women are *under*represented in unskilled work largely because many of them have sought employment opportunities in other fields, such as clerical work, as preferred options. Blacks make up 12 percent of the labor force but 16 percent of unskilled workers. Hispanics make up 10 percent of the labor force but 18 percent of unskilled workers. Conversely, women make up 46 percent of the labor force but only 20 percent of unskilled workers. Blacks are well represented in all types of unskilled labor but most strongly as garbage collectors and as machine feeders and offbearers, where they make up 38 percent and 23 percent of workers, respectively. Women are most likely to have unskilled jobs as machine feeders and offbearers, where they make up one-third of workers, or as hand packers, where they make up two-thirds of workers (Census, 2005).

Career Mobility   The occupational outlook for people in laboring jobs is quite restricted. Of the male workers at the poultry processing plant discussed above, about 25 percent said they had no particular career aspirations, and about 25 percent said they wanted to be mechanics. Of the women, 37 percent said they had no particular occupational aspirations, but those who did have aspirations had somewhat higher ones than their male counterparts. Many aspired, or had aspired, to professional, semiprofessional, or white-collar occupations, such as nursing, teaching, or secretarial work (Bryant and Perkins, 1986:160). Reduced aspirations are a common response to the problems that plague unskilled workers. Among these are the limited job opportunities in local areas, as well as family and other local and regional ties that make workers reluctant to leave home in the search for better work elsewhere. There are only limited opportunities for on-the-job advancement in unskilled work. From the viewpoint of the unskilled worker, dreams of occupational advancement are difficult to fulfill. The work provides little extra money to save for continuing one's education, and about the only skills the worker can demonstrate or perfect on the job are speed and endurance.

THREE KEY MANUFACTURING INDUSTRIES

Having discussed the three major job categories in manufacturing, we now take a closer look at some of the major manufacturing industries and trends in these industries. We briefly consider the automobile industry, the steel industry, and the textile industry. All these industries have faced automation and increased world competition in recent decades.

Automobiles

Automobile manufacturing has been *the* key manufacturing industry since the middle of the twentieth century. Worldwide, over twenty million workers—roughly equivalent to the entire population of Australia—are involved. The production of automobile components and subassemblies involves one of the most globally dispersed networks of any industry.

The world production of automobiles peaked temporarily in 1978 at thirty-two million annually, faltered in the 1980s, and resumed growing at a reduced rate in the 1990s, reaching a peak of over forty million by the early 2000s. Why has this pattern of irregular growth occurred? One reason is that the market for cars in the industrialized nations has become saturated. The market for cars in the less industrialized world is growing, but only slowly, because relatively fewer people in these nations can afford an automobile. These factors are compounded by rising oil prices, which have greatly increased the costs of owning and operating an automobile.

Increased World Competition   The problems of a sluggish world market have been amplified for American automobile manufacturers by increased competition from Japan, Korea, and Western Europe and also from countries in Eastern Europe and Latin America that have begun to build cars to supply their own markets. In 1960 the United States produced over half the automobiles in the world; by 2000 its share had fallen to less than a quarter. An additional reason for this decline is that American consumers have increasingly turned to smaller, better engineered, and more fuel-efficient cars. Japan, for instance, exports almost four million more cars than it imports. The United States imports almost three million more cars than it exports—and a large share of these imports come from Japan. In addition, 35 percent of the production of General Motors cars and 62 percent of the production of Ford cars take place outside the United States (primarily but not exclusively in the area of subassembly manufacturing). Thus, many so-called "American cars" contain a substantial proportion, or even a majority, of parts made or assembled overseas (Milkman, 1997).

Why has the United States lost out so dramatically in the market for automobiles? One reason is that American management has been slow to respond to market changes, preferring to specialize in higher-priced cars and light trucks with greater profit margins. Unfortunately, this decision has spelled disaster for American automobile workers. Another reason is that American manufacturing technology and work practices have failed to respond to the Japanese challenge. Even when Americans have tried to build more competitive cars, they have been less efficient in doing so than their Japanese counterparts. The American automobile industry has long had the highest rates of layoffs, turnovers, and absenteeism of any major industry. Such organizational practices and worker-management relations have provided an insufficient base from which to compete with Japanese automakers who utilize more efficient group production techniques (see Chapter 16).

Starting in the 1990s Japanese automakers have increasingly opened assembly plants in North America to increase their access to the American market. Two of the largest of these are the joint General Motors-Toyota plant at Fremont, California, and the Nissan plant at Smyrna, Tennessee (Besser, 1996). Other plants are located in Ohio, Kentucky, Indiana, and Michigan. American automobile manufacturers have learned innovative production strategies from these plants. The contrast with work systems in traditional plants is readily apparent. At the California plant, for example, workers have the right to stop the assembly line if problems arise—something strictly forbidden in traditional American manufacturing facilities.

Workers at the Japanese transplant facilities report mixed reactions to the new work systems. Although they like many of the new forms of worker involvement, they also report high levels of pressure, stress, and resulting injuries (Graham, 1995). Canadian workers similarly report resentment about erosions of union power and worker solidarity that are sometimes experienced in Japanese automobile assembly plants (Rinehart, Huxley and Robertson, 1997). At a Japanese plant in Indiana, workers are organized into "teams" but often balk at cheering for the company, Japanese-style, each morning (Graham, 1995).

The challenge of world competition has nevertheless promoted long needed changes in the American automobile industry. A new contract between Buick and the United Auto Workers (UAW), for instance, stipulates a "pay-for-knowledge" plan in which workers are compensated according to the number of jobs for which they are qualified, rather than solely on the basis of their current job. Chapter 17 considers in greater detail some of the job redesign programs that are gaining increasing acceptance in American industry. Many of these programs have emerged out of the struggles of the American automobile industry.

Steel

The steel industry has met with hard times in the United States and Canada. Once again, stagnant demand in industrialized countries is partly to blame. A great deal of steel is used to build industrial plants and facilities, but because of excess manufacturing capacity in much of the industrialized world, there is little demand for steel for constructing new plants. Delayed and deferred reconstruction of bridges and other large structures in these countries has further reduced the demand for steel. The demand for steel continues to rise in the industrializing nations, such as China and Korea. Because of the costs of transportation, however, steel is generally produced near the area where it is to be used. Steel has not been heavily involved in international trade. In addition, its production is strongly *vertically integrated*—that is, closely linked to the mining of iron ore and the production of coke. The production of these raw components, the production of steel, and the production of finished steel products often occur at nearby sites. These factors make it difficult to subdivide the process of making steel and retain selective activities in older locations.

Outdated Equipment   The problem of a diminishing market for American steel is compounded by the lack of investment by North American steel companies in new technology. With few competitors after the Second World War, American steel companies used their profits to buy companies in unrelated industries rather than to upgrade and expand their capacity for manufacturing steel. USX, formerly United States Steel and once the largest steel company in the world, now makes only 11 percent of its operating profits in the steel industry. It has systematically shifted its operations to other areas, including oil exploration, chemicals, and real estate. As a result, the American steel industry operates with outdated technology and equipment. Such "vintage capital" and technology perform poorly in competition with the new capital and technologies used in the German and Japanese steel industries, which have been rebuilt from the bottom up since the Second World War. In addition, just as in automobiles, the Japanese and Europeans have outpaced North American manufacturers in new technological advances. As late as 1970 most American steel was still being manufactured with open-hearth furnaces. By that time 80 percent of Japanese steel was being made with more advanced basic oxygen process technology (Warren, 2001). Although American firms have since moved into this technology, they have been unable to regain the competitive advantage against the Japanese, who also benefit from greater use of automation at all stages of production.

Employment in the American steel industry declined from 450,000 in the 1970s to only about 130,000 by 2005. This reduction in employment has entailed both permanent layoffs and wholesale plant closings. By 2005 the largest steel producer in the world was Nippon Steel, headquartered in Tokyo. This company produces two and a half times as much steel as its nearest U.S. competitor. The leading steel-producing nations and their share of world production are listed in Table 8.1.

Specialty Steel Products One bright spot in the American steel industry is the recent emergence of smaller companies called **mini-mills**, which use electric furnaces to turn scrap steel into basic steel products, such as concrete reinforcing bars and light construction products. These mills are both productive and profitable and are capturing a significant share of the domestic market.

Advances have also occurred in the employment of women in the steel industry in the 2000s. Women have made these strides even in the face of declining employment in the industry as a whole. The women who have found work in the steel industry report less sexual harassment than women working in such traditional female occupations as clerical work (Leach, 2005). The reason is that women in steel plants are working with men as peers. Conversely, the majority of female clerical workers are supervised by male managers, and subordination in work roles encourages sexual harassment. On the other hand, 20 percent of female workers in the steel industry are employed as janitors, whereas only 2 percent of male workers are so employed. Thus, even though the gender barrier has been broken in the steel industry as a whole, gender-based occupational segregation within the industry remains a problem.

TABLE 8.1 **World's 10 Leading Steel Producers**

Country	Share of World Production
China	25.8%
Japan	10.7
United States	9.3
Russia	6.2
South Korea	4.5
Germany	4.4
Ukraine	3.7
Brazil	3.1
India	3.1
Italy	2.7
Top 10 share of total	**73.5**

SOURCE: International Iron and Steel Institute, Brussels, 2005, (worldsteel.org).

Textiles

Textile and apparel manufacturing is a huge worldwide industry. It employs twenty-five million workers as well as additional millions of unregistered workers laboring in sweatshops and in home production. Worldwide production of textiles increased fourfold from 1950 to 2000 (Census, 2005). During this time the nature of fabrics also changed, with synthetics growing to almost half the total production. Synthetics include fibers derived from wood products, such as rayon and acetate, and petrochemical-based fibers, such as polyester, acrylics, and nylon.

The United States is the largest producer of synthetic fibers in the world, with a 25 percent share of production. This share, however, has dropped from 32 percent in 1970. Much of the growth in textile employment has occurred in Third World nations, such as Indonesia, Bangladesh, and Sri Lanka because of cheaper labor costs in these nations, as well as in industrializing nations, such as China and Korea. Additional employment shares have been captured by technologically advanced European competitors.

Mill Workers Only a small portion of textile and apparel manufacturing requires skilled labor. The unskilled nature of much of the work has greatly facilitated the movement of the textile and apparel industries to less industrialized countries. Only about 12 percent of textile workers in the United States are skilled craft workers. Unskilled textile workers have typically been viewed as acquiescing to the demands of their work with little resistance:

> The textile worker's powerlessness is expressed in constant work pressure, an inability to control the pace and rhythm of his work activity, a lack of choice of work techniques, and the absence of free physical movement. In his lack of freedom and control, the textile hand's situation is virtually the polar opposite of that of the free craftsman. (Blauner, 1964:66)

Waves of unionization among textile workers occurred in the United States in the 1930s but turned out to be short lived (Roscigno and Danaher, 2001). Today, textiles remain one of the least unionized American manufacturing industries. Partly, this is because much of the industry remains in the South. In this region white workers often receive preference in jobs, treatment, and pay over black workers. As a result white workers perceive that their self-interest lies in the maintenance of the existing system. The black workers, who are short-changed in this system, have historically been powerless to demand changes, especially in the face of resistance from white workers (Minchin, 1999).

As a result of a declining share of the world market, workers in the American textile and garment industry have experienced layoffs and permanent job losses in recent years. These employment losses result from the importation of less costly foreign-made apparel and from the export of jobs by American textile and apparel companies to areas with cheaper labor, such as border plants in Mexico and production facilities in Pacific Rim nations (Gereffi, Spener, and Bair, 2002).

Sweatshops Return Sweatshop conditions, entailing long hours, no fringe benefits, harsh and unsafe working environments, and the use of child labor have reemerged in the textile and garment industry in North America and elsewhere in the world as a result of increased competitive pressures:

> The subcontracting system allows manufacturers and retailers to slash the cost of labor and facilities, and—since subcontractors, not manufacturers, are legally responsible for any labor law violations in their shops—leave subcontractors with the burden of ensuring decent working conditions. Manufacturers and retailers reap huge benefits from the sweatshop system. Garment workers in Los Angeles, for example, each produce about $100,000 worth of goods a year, but are paid less than 2 percent of the total value. For a dress that retails for $100, $1.72 goes to the sewer, $15 to the contractor, and $50 goes to the manufacturer. (Louie, 2001:5)

Sweatshops have also sprung up on the West Coast, where they employ large numbers of Asian and Latin American immigrants (Bonacich, 2001). College students have recently become sensitized to these issues because of increased awareness of the use of sweatshop labor in the manufacture of popular clothing brands. Students have organized into a group called United Students against Sweatshops (www.studentsagainstsweatshops.org) to protest the conditions under which these workers labor.

Specialty Textiles Not all of the American market share in textiles has been lost to countries offering cheap labor. A significant market share has also been lost to Germany and Italy, which pay wages comparable to, or higher than, those in the United States. These countries have been able to increase their market share by a steady stream of incremental innovations in manufacturing design. The innovations have allowed textile and apparel manufacturers in West Germany and Italy to customize their products, making significant inroads against mass-produced goods. Such innovations also put these firms in a better position to take quick advantage of new market trends and developments. This technological progress has been encouraged by trade associations, cooperative banks, and apprenticeship programs under the joint sponsorship of unions and governments (North, 2005). In the United States, with its tradition of more individualistic forms of competition, such practices would be viewed as collusive and as an unfair constraint of trade. In the European nations they are seen as cooperation for the sake of economic development.

GLOBAL COMPETITION AND THE NEW WORLD ORDER

Many manufacturing industries in North America and in other industrially advanced nations have stagnated in recent decades. What are the reasons? One is increased international competition. Japan and Europe have been able to rebuild their industrial capacity, which was destroyed during the Second World War. In addition, many Third World nations have been pursuing the path of industrialization. Some nations have met with only limited success in this regard. Many others, however, such as China, South Korea, and Brazil, have secured sizeable market shares for themselves in the world economy.

The Wrong Policies at the Wrong Time

American manufacturing firms have responded inadequately to this heightened competition. They have pursued three main strategies: (1) exporting jobs overseas in the search for cheaper labor, (2) attempting to drive down wages at home, and (3) manipulating balance sheets and profit margins through what has come to be known as **paper entrepreneurialism** (Reich, 2002). These strategies all rest on the increased worldwide mobility of capital allowed by telecommunications and jet transport. Rapid transportation and virtually instantaneous communications allow the coordination of engineering and marketing among diverse facilities scattered around the world. In addition, the control of capital has become more concentrated over time

within larger and larger corporations. Large corporations have the financial power and political leverage to take advantage of these new worldwide networks of production and marketing (see Chapter 15). At the dawn of the twenty-first century, capital has simply become more mobile than labor. And this **capital mobility** has allowed American capital to increasingly leave the United States in the search for cheaper labor costs and less restrictive environmental regulations.

The strategy of driving down the wages of American workers has been carried out in part through demands for wage and benefit concessions from trade unions. Often threats of bankruptcy and plant closings have backed up these demands. As a result there has been strong downward pressure on the wages and benefits of American workers. This strategy has also been pursued through attacks on Social Security, education, and other social programs that were secured by middle-and working-class Americans during the latter half of the twentieth century (Reich, 2002).

Although the dual strategy of exporting jobs and reducing wages at home has secured profits for some firms, this strategy has been costly for the American economy as a whole. It has resulted in the stagnation of real income for average Americans as well as tremendous costs to local communities in the wake of large-scale layoffs and plant closings (Brady and Wallace, 2004). New jobs that have been created in the service sector have generally been much lower-paying than those lost in manufacturing. Clearly, a more rational policy, focusing on heightening productivity rather than exporting jobs and driving down wages, would be preferable. In this section we examine current manufacturing policies and their consequences and look at some policy alternatives.

Financial Shell Games Some American manufacturers can be characterized as *hollow corporations*. Instead of investing in productivity-enhancing activities at home, such as technological innovation and job redesign, they subcontract production outside the United States. As we discussed in the section on automobiles, many of the goods we

---

B O X 8.4 **Arguments for and against Plant Closing Legislation**

Recent debates on plant closings suggest the following arguments for and against legislation to regulate such closings.

The principal arguments in *favor* of legislation:

1. Advance notice and income guarantees are necessary to mitigate the economic and psychological burden of closure on employees.

2. Businesses get substantial economic incentives and rewards from government; workers deserve the same protection.

3. Communities also deserve advance notice, because they lose a major source of tax revenue when plants shut down and face the added burden of social welfare payments to displaced workers.

4. Plant closure has a multiplier effect within a community, causing a greater loss of jobs than those in the plant itself.

5. Because businesses usually plan shutdowns far in advance, prior notification requirements pose no great burden.

The principal arguments *against* legislation:

1. State plant closure laws would place unconstitutional restraints on interstate commerce.

2. State plant closure laws would create more, not less, local unemployment, because large firms would establish or increase operations in other states to avoid penalties.

3. Workers receive adequate economic protection through state unemployment benefits and job search services.

4. Businesses need to be free to close inefficient plants with obsolete equipment and replace them with new facilities.

5. Based on their educational qualifications and skills, manufacturing workers receive generous wages; as a quid pro quo they should assume the risk of closure.

SOURCE: Adapted from Paul D. Staudohar and Holly E. Brown, 1987, *Deindustrialization and Plant Closure*. Lexington, MA: D.C. Heath, pp. 275–276.

---

think of as being "American made" are actually made in foreign nations, with the American company providing little more than the packaging and the labeling.

Plant Closings Many of the jobs lost in the United States were exported overseas by American corporations. For instance, in 2005 General Motors closed four assembly plants and cut its North American workforce by 25,000. General Motors, which employed 600,000 workers in North America in 1979, now employs only 125,000. At the same time General Motors has opened new plants overseas, thus retaining profits but not domestic employment (Hakim, 2005). These **plant closings** cost workers in terms of lost livelihoods, but they also cost communities and states in terms of lost tax revenues and increased social welfare expenses. Box 8.4 presents arguments for and against legislation regulating plant closings.

Externalization Plant closings are encouraged by the fact that large companies pay only part of the total costs entailed in moving from location to location. Other costs are **externalized** and are paid by taxpayers or by individuals. Such costs include tax write-offs negotiated by companies with the new localities in which they locate, accelerated depreciation allowances, and utility discounts. These costs reduce local revenues. Local services must then be curtailed or other taxes increased, such as personal property taxes or sales taxes. These adjustments create additional difficulties for communities trying to attract skilled labor and professional workers. Attracting and retaining skilled labor is made more difficult by the underfunding of social services, such as educations, parks, and police.

From the company's standpoint, this pitting of communities and regions against one another offers some attractive advantages. By complaining about pollution standards, taxes, and other regulations in one area, a company can induce a community into offering concessions on these issues to attract jobs. To entice a company to relocate, a city may offer tax breaks or offer to build an industrial park on city-owned property. The costs are absorbed by the city and by taxpayers and are *real costs* of relocation. Even moves that are *economically irrational* for the nation as a whole may thus be made because they are *profitable* to the individual company, which receives a subsidy from the new host community. Additionally, the costs of moving families, shutting down schools and rebuilding them in new locations, and building new streets and sewers are all real costs of relocation that are not paid by the company but by the taxpayers or by the individuals affected. In addition, federal tax codes allow much of the value of the physical plant and equipment of closed factories to be counted as depreciation against income from new plants located elsewhere or even against earnings from newly purchased subsidiary companies operating in entirely different industries. The loss of jobs and the loss of social safety nets because of plant closings have contributed to the increased problems of poverty, homelessness, and marginality discussed in Chapter 14.

Downsizing and Flexibility   A key corporate strategy used in recent decades to adjust to increased competition is to **downsize**. Workforce downsizing has become increasingly common in the 2000s. By downsizing core employment and subcontracting many functions, corporations believe they can cut costs through increased **flexibility** in matching their resources to market needs. Displaced automobile workers report losing an average of 44 percent of their previous earnings during the first two years after being laid off. Workers in steel, meatpacking, and aerospace report similar losses. A worker periodically laid off from General Motors reports: "People have to be taught about the things displaced workers face every day. I'm very discouraged to find out the number of people who are ignorant on this matter or just don't care. I can't wait to retire so I can stop working for one of the most inept, unpatriotic companies in recent memory" (Phillips, 1998:iii).

A Declining Middle   Heightened pressures on jobs and wages have helped create a trend toward a **declining middle class** (Braun, 1997). Traditionally, in manufacturing and construction, wages clustered around the middle range of income. Employment in these sectors thus helped create a prosperous working class that was a bulwark of middle-class society. Today, real wages in these jobs are stagnant or declining. Average earnings have stagnated since 1980 for married couples with children in the bottom 60 percent of the income distribution but have risen by 30 percent and 60 percent respectively for families in the top two quintiles (Mishel et al., 2005:104). Families have compensated for stagnant earnings by increasing their hours of work for both husbands and wives (Wallace, 1998). Unmarried individuals have fared less well.

The economic strategies pursued by the largest American corporations have resulted in the loss of millions of jobs and have slowed growth in the gross domestic product. The reason is not that the heads of these corporations are ignorant or stupid. It is because what is good for large corporations in a global economy can be disastrous for local and national economies.

Current federal policies do little to ameliorate these tendencies and nothing to prevent them. The Trade Adjustment Act provides some retraining benefits to workers losing their jobs to international competition. But it has been interpreted very narrowly by federal administrative bodies and excludes workers displaced because of American corporations moving overseas. Similarly, Title 3 of the Job Training Partnership Act is earmarked for retraining redundant workers. However, only about 2 percent of the agency's budget has been allocated for this purpose (Borjas, 2005).

Unexplored Alternatives

A potentially important set of policy options that could address these issues center on establishing "fair trade" between nations instead of unregulated "free trade." The goal of fair trade policies is to establish common standards for the treatment of workers worldwide that will allow nations to compete with each other without doing so on the basis of lowering wages, working conditions, and environmental standards to the lowest common denominator.

Recent years have also witnessed increased public concern about the actions of the World Trade Organization (WTO, www.wto.org), World Bank (www.worldbank.org), and the International Monetary Fund (www.imf.org). These non-elected global institutions have acted to open markets and to promote the free flow of goods between nations.

These actions have increased trade and development in some areas. These global institutions, however, have been much more reluctant to implement policies that provide protections for workers or for the environment. As a result, their actions in support of free trade have accelerated the competition of workers in different nations against each other without providing any framework for establishing minimum standards of child labor, safety, or workers' collective rights to organize.

SUMMARY

Advanced industrial societies include a diversity of industries, which can be broadly classified into extractive, manufacturing, and service sectors. Productivity in extractive industries has increased to such an extent that only a relatively small part of the labor force is still employed in this sector. Similarly, a declining share of workers is employed in manufacturing. The decline in manufacturing jobs is based on rising productivity, renewed world competition, and the movement of American corporations overseas in the search for cheaper labor and more lenient safety, health, and environmental regulations.

Some jobs in manufacturing are highly skilled. Workers in these jobs experience a great deal of autonomy and pride in their work. Such jobs are similar to professional jobs in their training requirements and the autonomy they afford the worker.

Other jobs in manufacturing are boring and repetitive.

Employment growth in advanced industrial economies is occurring mainly in the service sector. Some researchers argue that the growth of the service sector foretells an increasingly unequal society as middle-level positions in manufacturing decline in number and are replaced by poorer paying service jobs.

As manufacturing productivity increases, a decline in its relative share of employment is inevitable. However, the sharp decline of employment in American manufacturing due to a loss of competitiveness is neither inevitable nor desirable. Investments in technology and in worker training can improve productivity and competitiveness and help keep manufacturing jobs in the United States.

KEY CONCEPTS

extractive industries

manufacturing industries

postindustrial society

advanced industrial society

aquaculture

occupational communities

skilled craft work

semiskilled work

unskilled work

mini-mills

paper entrepreneurialism

capital mobility

plant closings

externalization

downsizing

flexibility

declining middle class

QUESTIONS FOR THOUGHT

1. Describe some of the problems facing farmers in North America today. What policies might help address these problems?

2. Why has employment in manufacturing in North America declined in recent years? Construct arguments for and against the following proposition: "The decline of manufacturing is an acceptable and desirable aspect of industrial growth and development in advanced economies."

3. Why do skilled craft workers generally resent being closely supervised? Is the autonomy of skilled workers a roadblock or a potential asset for economic growth in advanced economies?

4. If you were working as a machine operator in a factory, what would you like most about the job? What would you like least?

5. Identify three policies that you believe would both increase the competitiveness of North American manufactured goods in world markets and increase employment opportunities for North American workers. Rank order these three policies by their probable impact and defend your ranking.

MULTIMEDIA RESOURCES

Print

Timothy C. Lloyd and Patrick B. Mullen. 1990. *Lake Erie Fishermen: Work Identity and Tradition*. Urbana, IL: University of Illinois Press. Oral histories of work and community life in Lake Erie's commercial fishing community.

William M. Adler. 2000. *Molly's Job*. New York: Scribner. The story of a metal stamping machine, and the job attached to it, as it moves from New England, to Mississippi, to Mexico.

James W. Rinehart, Christopher Huxley, and David Robertson. 1997. *Just Another Car Factory? Lean Production and Its Discontents*. Ithaca, N.Y.: Cornell University Press. An analysis of the lives of workers as they chose between early retirement and new work relations in the automobile industry.

Deborah Fink. 1998. *Cutting into the Meatpacking Line: Workers and Change in the Rural Midwest*. Chapel Hill: University of North Carolina Press. An ethnographer's experiences as a worker in an Iowa pork meatpacking plant.

Internet

Government publications. *infomine.ucr.edu/search/govpub search.phtml* A comprehensive guide to online government information.

Corporate Agribusiness Research Project. *http://www. electricarrow.com/carp/* Reports and links on the centralization of agricultural production.

Upjohn Institute for Employment Research. *www.upjohninst.org* Innovative research on workplace issues.

Economic Policy Institute *www.epinet.org* Insightful reports on contemporary economic and workplace issues—a truly comprehensive site.

Labor Posters from the Second World War. *american history.si.edu/victory/victory4.htm*

Thomas Register of American Manufacturers. *thomasregister. com* A core resource for finding any part, gadget, or service a manufacturer could ever need.

RECOMMENDED FILM

The Perfect Storm (2000). Gripping true-life story of a North Atlantic fishing vessel caught in the storm of the century.

9

The High-Technology Revolution

With the availability of small computers that can be plugged into immense data banks, word-processing equipment, and other devices, more and more people are setting up what Alvin Toffler calls the "electronic cottage." . . .

Pat Lee, a consultant and special projects manager, likes the flexibility: "It means if you want to work on a rainy day for 12 hours, and spend 12 hours in the sun the next day, you can. If you feel like sitting in your chenille bathrobe at midnight writing an employee handbook, and you really get into it and want to stay up till 4:00 in the morning doing it, that's terrific." . . .

Author Jessica Lipnak, who also works at home with her husband, likes the opportunity for parenting: "We both really enjoy it in relation to our small children, because we've had a lot of access to the kids—they're seven months and three years. I'm nursing the baby, so it's made it possible for me to be one of those rare women who can work and nurse. That's a big issue for a lot of people with infants. So that's been wonderful. Being able to see them on and off all day and make choices about spending time with them has been great."

(APPLEGATH, 1982:46)

Maria, a twenty-six-year-old political refugee from Argentina, who chooses not to be known by her real name, found work in the Silicon Valley, but she did not strike gold. Maria quit her $4.10 an hour production job at Memorex to have her first baby. For two years, she illegally stuffed and soldered thousands of printed circuit (PC) boards in her home. Her employer, a middle-aged woman she calls "Lady," subcontracted assembly work from big firms—so Maria was told—like Apple and Memorex.

> Maria gladly accepted the low piece-rate work because child care would
> have eaten up most of her after-tax earnings at a full-time job. She quit,
> however, when Lady asked her to wash her assembled boards by dipping
> them into a panful of solvent, heated on her kitchen stove. Maria, unlike
> most Silicon Valley cottage workers, had studied chemistry before
> immigrating to the United States, and she knew that the hydrocarbon
> fumes could make her young son, crawling around on the kitchen floor,
> seriously ill.
>
> (SIEGEL AND MARKOFF, 1985:138–139)

These workers' experiences illustrate the potentially diverse consequences of high-technology production. The effects of high technology on the workplace are still not fully understood despite the widespread attention often focused on technology's potential for spurring economic growth.

In this chapter we examine the consequences of new technologies for workers and for organizations. We present competing views of the influence of high technology on work. We examine the mix of job opportunities and the changing skill content of jobs. Finally, we look at how advanced technology affects the experience of work and the functioning of organizations. The complex and differentiated effects of high technology on work and workers highlight the importance of examining issues of power and the social organization of work simultaneously with the examination of technological change.

COMPETING VIEWS OF HIGH TECHNOLOGY

What exactly is high-technology production? At the heart of high technology is the tiny silicon computer chip. With the widespread application of the microchip to a variety of work settings beginning in the late twentieth century, entirely new industries, such as robotics, have sprung up, and others, such as computer manufacturing, have expanded dramatically. Electronic technology has also altered the nature of production across a wide range of settings. These settings include both traditional blue-collar manufacturing and white-collar clerical and professional work. Our investigation of the high-technology workplace thus focuses both on high-technology industries and on more traditional jobs that have been transformed by microprocessor technology.

The U.S. Bureau of Labor Statistics defines **high-technology industries** as those employing 1.5 times the average proportion of technology-oriented workers (engineers, life and physical scientists, mathematical specialists, engineering and scientific technicians, and computer specialists); having research and development expenditures twice the average for all industries; or being above average on both these criteria (Neumark and Reed, 2004). This definition identifies forty-eight high-technology industries, including electronics, machinery, ordnance, chemicals, instrumentation, pharmaceuticals, aerospace, genetic engineering, and communications equipment.

There are competing perspectives on the influence of advanced technologies on work. One perspective emphasizes the benefits resulting from technological advances. The work of American economist Faye Duchin (1998) is representative of this tradition. Duchin views technology as liberating individuals from the necessity of performing undesirable work and as a source of higher living standards. A key characteristic of this optimistic view is its conception of technology as a neutral and inevitable feature of human existence (Mowshowitz, 2002). Technologically based potentials for improved working conditions and skill upgrading also figure prominently in this view. In this vision, the benefits of new technology far outweigh any disadvantages and lead the way to increased job opportunities and greater prosperity.

Other analysts, however, describe the impact of technology in more pessimistic terms. These social scientists identify technological growth more as a cause of current problems, such as unemployment, than as a solution to these problems. A leading spokesman in this more pessimistic view of technology is sociologist David Noble (1997), who argues that there is nothing automatic about either the development or the consequences of technology. In his view technology is neither neutral nor inevitable but is, instead, a tool to increase management's leverage in bargaining with workers. Noble argues that when options exist, employers systematically select the technologies that weaken workers' autonomy and solidarity. Systems that rely on expensive automated machine tools, for example, are selected over less expensive machine tools requiring more skill to operate. In this view the selection of new technologies reflects and reinforces the unequal distribution of economic and social power.

Another sociologist, Mike Cooley (1999), faults the new technologies for reducing workers' skills and dissipating their motivation. Computers and other forms of automation in the workplace can have negative effects on the experience of work. Two negative changes may occur when control is removed from workers and placed in computers and other advanced forms of technology. The first is that the work may become less diverse and less rewarding. The second is that auto-

mation may actually *decrease* efficiency and quality because achieving high quality requires skilled human input, which is excluded in automated systems (Shaiken et al., 1997). Without monitoring by a skilled and knowledgeable worker, machines may turn out junk rather than precision manufactured products.

We have seen that there are competing views about the consequences of new technologies for work and for workers. However, there is a reasonable degree of consensus on at least two aspects of high technology: (1) the major effect of new technologies is to transform existing jobs rather than to create new ones; and (2) technological innovations can increase productivity, product quality, and the ability to customize products, thus improving the competitive position of those organizations that use them effectively (Ozaki, 1999).

In the following sections we look at specific areas of agreement and disagreement about the consequences of high technology. Our discussion revolves around three core issues: (1) What is the balance between job displacement and job creation resulting from high-technology production, and, more specifically, what new occupations are created, and what jobs are eliminated? (2) How are new technologies transforming the skill requirements of different occupations? (3) How are working conditions being modified by new technologies, and how are new technologies affecting the nature and meaning of work?

JOB DISPLACEMENT AND JOB CREATION

Some occupations are eliminated by new technologies while others are created or expanded. Are more jobs *displaced* or *created* by high-technology production? There is a range of views on this important issue. Many researchers focus on the creation of high-prestige jobs such as systems analyst and ignore the displacement of other workers. Others focus only on the potential displacement effects of new technologies

---

B O X 9.1   Futuristic Automation?

In 1952 Kurt Vonnegut, Jr., wrote Player Piano. The novel portrays a future society in which automation has eliminated all jobs except those of a few select engineers. Unemployment is pervasive in spite of public works projects and a chronic state of manufactured war with an illusive enemy. Vonnegut's fictional scenario rests on the futuristic projection of data tape automation first used in the 1950s to automate machine milling work. This type of automation is almost identical to that used in a player piano.

In Vonnegut's story engineers have captured the skills of master machinist, Rudy Hertz, by attaching recording instruments to his lathe. His movements are then reproduced, with variations, ad infinitum.

Rudy hadn't understood quite what the recording instruments were all about, but what he had understood, he'd liked: that he, out of thousands of machinists, had been chosen to have his motions immortalized on tape.

And here, now, this little loop in the box before Paul, here was Rudy as Rudy had been to his machine that afternoon—Rudy, the turner-on of power, the setter of speeds, the controller of the cutting tool. This was the essence of Rudy as far as his machine was concerned, as far as the economy was concerned, as far as the war effort had been concerned. The tape was the essence distilled from the small, polite man with the big hands and black fingernails....

Now, by switching in lathes on a master panel and feeding them signals from the tape, Paul could make the essence of Rudy Hertz produce one, ten, a hundred, or a thousand of the shafts.

SOURCE: Excerpt from Kurt Vonnegut, Jr., 1952, *Player Piano.* New York: Dell, pp. 9–10.

---

and ignore the jobs that are being created. Many offsetting forces are involved, but their effects are difficult to estimate, and so the overall effect of high technology on employment remains an open question. The possibility of further significant unanticipated developments is also high. For example, the declining growth rate in clerical occupations because of word processing was not anticipated prior to the rapid expansion of microcomputers. Conversely, increased productivity due to new technology has led to a growing economy and increased employment opportunities for many workers (North, 2005).

Job Displacement

Many researchers believe there has been a net displacement of jobs and increased job insecurity resulting from the introduction of new technologies, particularly microprocessor-based technologies and robotics. Such an elimination of jobs is called **technological displacement**. Clerical workers, especially typists, stenographers, and office-machine operators, are often identified as the group most affected (Hartmann, 2000). In addition, robotics may make a wide variety of manufacturing labor redundant. Finally, there is likely to be some displacement of machinists, tool and die makers, and metal workers resulting from the introduction of computer-controlled machine tooling systems (Wright, 2000). Box 9.1 presents Kurt Vonnegut's futuristic vision of an automated society in which numeric-controlled machine-tooling systems have made workers obsolete.

Is the High-Technology Revolution Really Different?   Two critical differences between the current wave of technological change and the wave of continuous-process automation that occurred in the mid-twentieth century may amplify current negative effects on employment. First, microchip applications have the capability of transforming the production of a much wider range of products than continuous-process automation. Second, this wave of automation is occurring at a time when economic growth can no longer be taken for granted

(Mowshowitz, 2002). A key difference between electronic technology and the automation of the 1950s is that the consequences of microcomputers are at least as significant for white-collar work as for blue-collar work. The potentially negative impact of computer technology on employment can be glimpsed by examining some of the industries in which its impact has been, or is being, felt.

The telephone industry employed 213,000 operators in 1960 but, in spite of tremendous expansion in the telecommunications industry, the occupation of telephone operator had virtually disappeared by 2005, due largely to the introduction of electronic switching. Railroads employed over 1 million workers in 1955 but only 215,000 in 2005 as a result of automated traffic switching, sophisticated machines for track maintenance, and mechanized warehouse loading operations (Census, 2005). Simultaneously, the tonnage of freight moved on railroads has actually increased. The printing industry has also experienced advances in productivity coupled with dramatic declines in employment. Retail trade, hotel management, libraries, and many other white-collar, trade, and service industries are undergoing similar transformations as their operations and accounting systems are increasingly automated (Mowshowitz, 2002).

Employment Losses in White-Collar and Management Occupations The decline in employment growth in white-collar occupations has occurred at a time when these jobs were being counted on to absorb manufacturing labor displaced by automation (North, 2005). Rough times ahead are also projected for middle managers as more information processing is built into technological systems and centralized under the control of top management. Currently popular management theories pose further threats to middle managers. For instance, sociologist and management consultant Rosabeth Moss Kanter (2001), in her best-selling book, *Evolve!,* argues in favor of "lean" managerial structures with few levels and with few supervisory personnel. Such management theories both reflect and add to technologically based threats to employment opportunities for middle managers (see Chapter 12).

Robotics The growth of robots and robotics applications in manufacturing has been dramatic and has had dramatic effects on job displacement. It is estimated that by 2000 robots had displaced 35 percent of welders and 60 percent of production painters in the automobile industry.

In 1980 there were about 30,000 robots in use worldwide. This represented only about one robot for every 10,000 manufacturing workers. By 2000 the number of robots was estimated to have increased to about 200,000, but this still represents only about one robot for every 1,500 manufacturing workers. About 90,000 new jobs were created in the robotics industry during this twenty-year period. But robots displaced about 300,000 production workers (Colestock, 2005).

The cost of a programmable (servo) robot ranges from $30,000 to $120,000, and the purchase price does not include installation, site preparation, or maintenance, all of which can be very expensive. These costs slow the deployment of robots and narrow their displacement effects. However, more versatile robot-assisted devices are increasingly being used across a range of manufacturing settings. It is also possible that the use of robot-like devices can be extended to nonmanufacturing settings. Mechanized warehousing and inventory control provide examples of these possibilities. Such computer-related mechanization of support services may result in further employment losses.

One of the newest frontiers in robotics is in medical surgery, where miniaturized optical and cutting devices remotely operated by surgeons are being developed to allow, for example, surgical repair of tiny arteries in the brain. Such devices can even be used to allow specialized surgeons in one location to participate in a surgery at a remote location in another state or even in another nation.

Job Creation

In spite of highly visible displacement effects from robotics, many researchers argue that relatively few employees are laid off as a direct result of technological change. They argue that such displacement is often moderated by techniques used to prevent layoffs, such as providing notice and retraining and reassigning displaced employees to new jobs. In addition, computer technology has introduced entirely new occupations such as systems analyst, programmer, and web master. Expanded job opportunities are also occurring for engineers and research-and-development scientists. Acknowledging that some workers are hurt by the introduction of new technologies, these researchers argue that technology is not the culprit. Instead, they see displacement problems as resulting from limits in adjustment, retraining, and relocation programs for workers.

Many possible effects of advanced technology may offset short-term job losses. New products and industries are being created; demand may be stimulated by cheaper, better products; and economic growth may occur through increased investment in new technologies. Increased demand for machine maintenance required by ever-greater investment in technology will also tend to increase employment.

Internet   The potential for computer applications to expand employment has been demonstrated through economic growth brought about by the rapid expansion of the Internet. In spite of the "dot-com bust" of the early 2000s, most analysts believe that the potential for Internet-stimulated economic growth is considerable. The financial success of Google "has propelled its stock market value to $112 billion, more than the combined value of Disney, Ford, General Motors, Amazon.com, and the media companies that own *The New York Times,* the *Wall Street Journal,* and the *Washington Post"* (Vise, 2005).

The economic impact of the Internet is not just that it makes access to information easier but that it has the potential to lower the costs of marketing and sales and also to increase competition among sellers, thus lowering prices and stimulating demand (Freund and Weinhold, 2004). Savings include reduced costs for transactions and increased efficiency through better management of supply chains and through better communication within the firm and with customers and partners (Litan and Rivlin, 2001). The significance of such savings is

that they are not unique to computer or Internet companies but can also accrue to "old economy" industries that take advantage of the Internet for managing information.

Long-Term and Indirect Effects   Drawing final conclusions about job displacement versus job creation is difficult. The problem in determining the bottom line effect on employment arises from the difficulty of calculating the **long-term and indirect effects** of new technologies. When the direct effects alone are examined, job displacement clearly outweighs job creation. But when indirect effects are considered, the analysis quickly becomes clouded. For instance, if the introduction of robotics leads to higher productivity and lower prices, it may stimulate demand and create a significant number of jobs throughout the economy. Similarly, Internet-based informational efficiencies can spark an increase in sales and trade. These potentially important consequences, however, are difficult to verify. Conclusions about the effects of new technologies on job creation and job displacement must thus be made with considerable caution.

Increasing Segmentation?

Sociologists are also concerned about the consequences of advanced technology for the relative prevalence of desirable and less desirable jobs. At first glance, job structures in many high-technology settings appear to suggest an increasingly middle-class occupational distribution. For example, professional and technical occupations make up 10 percent of jobs in manufacturing as a whole but 33 percent in the electronics industry (Naisbitt, 1999).

Examining the labor force as a whole, Levine (1995) concludes that positions requiring the highest and lowest level of skill have decreased, resulting in a broadening of the middle. Nevertheless, he is cautious in interpreting these results because they rely solely on an analysis of general educational development while ignoring more specific vocational training, on-the-job training, and earnings.

Other researchers have found that new technologies sometimes create dual occupational structures in high-technology settings (Tolbert, Lyson and Irwin, 1998). The increasing use of educational credentials as a screening device can also heighten barriers to mobility between nonconnecting career lines (Burris, 1993). For example, secretarial and accounting career paths are typically segregated on the basis of a degree in accounting and certification as a certified public accountant (CPA). Such limits on mobility can lessen motivation for all employees (Glass, 2000).

Even in high-technology industries, many of the newly created jobs will be in traditional "low-tech" occupations with less than average earnings. Only 25 percent of the jobs in high-technology industries can be classified as truly high-technology occupations. In Santa Clara County, California (popularly known as Silicon Valley), 28 percent of jobs in the semiconductor industry are in the top third of the national earnings distribution. But the middle third is almost empty, with only 9 percent of the jobs, and 63 percent of the jobs are in the *bottom third* of the income distribution (Appelbaum et al., 2000).

This earnings distribution is unlikely to improve with further advances in technology, because the profit imperative demands that highly paid labor be automated before poorly paid labor. For example, one of the most successful areas of automation in computer manufacturing is in computer-aided design. This technological breakthrough has displaced large numbers of reasonably well-paid engineering drafters. Meanwhile, the development of automatic equipment for inserting microchips into printed circuit boards has advanced much more slowly, partly because much of the work is done by poorly paid workers in the Third World.

The International Division of Labor When examining the consequences of high technology for the occupational distribution, we need to look at the full range of jobs created by high-technology companies, including jobs in the company's foreign operations. The six major high-technology companies operating in Massachusetts, for example,

employ 28 percent of their workforce overseas. Hourly wages for these workers range from an average of $1.26 in Mexico, to as low as twenty-one cents in India (United Nations, 2000). If these workers are brought into the equation, the resulting distribution of jobs is much more negatively skewed.

The available evidence thus suggests a strong possibility that advanced technology may contribute to an increasingly unequal occupational distribution made up of a few highly paid jobs and an increasing share of poorly paid, relatively alienating jobs. We discuss this possibility, and the opportunities for avoiding such a future, in Chapter 17.

Public Policy and Employment

Because of concerns about technological unemployment, various types of employment programs have gained increasing public attention. The programs include such diverse proposals as advance notice of plant closings, more extensive retraining, reductions in the work week, early retirement, and expansion of public service jobs, especially for young workers. If technological unemployment in the face of advancing productivity is indeed to be the fate of the industrially advanced nations, then systematic planning and adjustment needs to take place.

Local Impact of High-Technology Development One important aspect of technology-related employment changes is their impact on local communities. Many community planners believe that development policies stressing the recruitment of high-technology industries are ill-conceived. For example, even after an aggressive (and expensive) state program to attract high-technology firms, fewer than 7 percent of all jobs in North Carolina are in high-technology businesses (Luger, 1996). Programs narrowly targeted to attract high-technology businesses may actually be detrimental to states' long-run economic interests. Many high-technology industries have limited employment-creating potential, and a substantial outlay of public funds is often required to entice them to move to a specific location (Kenney and Florida, 2004).

It is likely that development funds could more effectively be targeted to help *existing* businesses become more technologically advanced (Colclough and Tolbert, 1992). Such development plans would be comprehensive rather than targeted toward high-technology businesses per se, and they would stress integrated economic development. Without broader and more comprehensive plans, strategies for high-technology development may become policies of social abandonment of regions hoping for development and employment (Malecki, 2004). An important component of more integrated development programs is the requirement that companies receiving publicly subsidized financing give advance notice of plant closures or major lay-offs and agree to refund public investments made to attract the company.

CHANGING JOB CONTENT

To better understand the consequences of high technology for the future of work, let us first look at how it has affected the overall mix of occupations. Significant changes in the distribution of occupations occurred at the end of the twentieth century and the beginning of the twenty-first century. Relative to the distribution in 1980, a significant increase occurred in professional jobs (from 15.6% of the labor force to 20.3%) and service workers (12.4% to 16.0%). A moderate increase also occurred for managers (from 9.5% to 10.5%). Declines for clerical workers (17.8% to 14.2%) and machine operatives (15.7% to 10.4%) offset these increases. Sales workers, craft workers, and laborers roughly maintained their relative share of employment (Census, 2005).

These figures may come as a surprise to those who believe that in the future everyone will be a high-technology worker. This misconception arises from paying too much attention to which occupations will have the *fastest growth rates* and too little attention to which occupations will actu-

ally provide the *largest number of jobs*. Database managers, systems analysts, and computer engineers are expected to be among the fastest-growing occupations, all with growth rates of 50 percent or more in the first decades of the 2000s. However, the *total number of jobs* created in these occupations will be fewer than the number of new positions created for cashiers, home care aides, truck drivers, or teachers (Census, 2005).

Large numbers of new jobs will *not* be created in the high-technology industries themselves. However, microprocessor-based change has dramatically transformed and will continue to transform many occupations. In this section we examine recent changes in the nature of work for some of the occupations most affected by advanced technology. Such occupations include engineering, assembly work, machine maintenance, service work, clerical work, management, technical work, and work in the home.

Engineering

Engineering has always been the glamour occupation of advanced technology. Engineers frequently report that one of their most important motivators at work is the opportunity to work on intellectually challenging projects (Meiksins and Whalley, 2002). It is highly unlikely that the creative aspects of engineering will ever be automated away. Box 9.2 describes the work of a young computer programmer involved in video game production.

Engineers make up 10 percent to 15 percent of the workforce in electronics, compared with about 3.5 percent in the U.S. economy as a whole. In robotics manufacturing, this figure runs as high as 23.7 percent, with an additional 15.7 percent of the workforce employed as engineering technicians. Well over 50 percent of the jobs in robotics require two or more years of college training, compared to fewer than 20 percent in the rest of manufacturing. Salaries for computer programmers averaged $36,000 in 2000—compared to about $29,000 for the labor force as a whole (Wright, 2000). These employment figures suggest a

B O X 9.2   You Are Shooting a Lot—You Are Killing a Lot

I started making computer games when I was ten years old. My dad had a small PC business. . . . I was just lucky. I went to MIT, took a lot of science and liked it. . . . I first worked on *Mech Warrior 2* and then did an add-on to that—*Ghost Bear's Legacy.* . . . That add-on was probably the funnest thing I've ever done. It took two and a half months and was very low budget, and we just had a great time. It was such a small team, and all of us really dug the design we were working with. We put hundred-hour weeks into it, sleeping at the office a lot, and I was like a kid again, you know? It was just a joy to work on. . . . And it sold three hundred thousand units, which the company loved. We were all golden boys for a while. . . .

The company I work for now, it's a pretty small company. We just do development of games. We don't actually sell the game afterwards. My old company Activision does that for us. They're kinda like our client—they have to like the game we make, which means usually they pick the topic. . . . They give us the main idea, they distribute it, we do the fun stuff. . . .

This is a dream job for me. It's the best job in the world. It doesn't change the world for the better, but it's at least giving people some enjoyment

for a couple of hours a day. And it's only going to get bigger in the future. Five years from now a lot more people are going to have games. You're going to have the game system just built into your television. The PlayStation is pretty close to that now, and WebTV is starting with that direction, too. So many more people are going to have games. Which means the audience is gonna get much broader. . . . When I started in this, we were all geeks who played games, so you could make games for geeks. Now you have to make games that have big action stars in them, or have cool music—like the Beastie Boys might be in one of the games. Something that really draws in a large crowd. . . .

There is definitely a violent aspect of it all. You are shooting a lot. You are killing a lot. But, that's kind of obviously the point, right? And it's more than that, too. . . . It's an experience. It's a release. . . . You're watching someone play, and you see their eyes get wide. "Wow!" They're hurting their thumb playing the game. It's very good to see.

SOURCE: John Bowe, Marisa Bowe, and Sabin Streeter. 2000. *Gig: Americans Talk About Their Jobs,* pp. 305–309. New York: Crown Publishers. Reprinted with permission.

---

positive job outlook for many engineering and computer specialties.

Assembly Jobs

At the other extreme are assembly jobs in high-technology industries. Frequently, these jobs are dull, repetitive, poorly paying, and hazardous. Many of them are exported overseas to countries with large pools of labor willing to work for what would be considered subminimum wages in North America and Europe. Much of the economic "success" of these industrializing countries comes at the expense of their citizens who labor under harsh conditions to manufacture goods for the world market:

A 48-year-old woman and her husband who labor in a camera factory in Shenzhen, China,

[report being] ashamed of their meager income and their dependence on relatives in Hong Kong and their native Shunde. "I never buy new clothes now. All the things I wear I get from my Hong Kong relatives. . . . My relatives saw that we could not even afford a television, so they gave us one as a present. When my husband got sick last year, they also lent us money, but who can take care of long-term poverty? Our relatives in our home village are doing much better than us." (Lee, 1999:51)

Because partially assembled electronic components are easy to export by jet, North American and European workers employed in such jobs are placed in direct competition with lower-paid workers around the world. Working conditions in electronics assembly in the advanced nations are not as bad as those in the Third World, but the availability of cheap labor places strong downward pressure on the wages and conditions of electronics assembly workers in the industrially advanced nations.

Some assembly work can be automated through the use of robots or robot-like devices. However, productivity gains are not immediate or assured because of the expense and inflexibility of robotics applications relative to human workers. Most assembly workers have somewhat ambivalent, though generally positive, attitudes toward robots. In a study made two and one-half months *before* robots were introduced into a manufacturing plant, 87 percent of the workers thought that the robots would make their company more competitive. However, 50 percent also worried that the robots would displace workers. Five months later (two and one-half months *after* the introduction of robots) attitudes were decidedly less positive. Workers complained that robots had increased costs, increased accidents, and lowered the quality of the product. They were also no longer convinced that robots increased productivity (Argote, 1999).

Machine Work

The work of skilled craft workers is also being affected by computer technology. Skilled machinists provide one of the clearest examples. These workers make the tools, patterns, molds, and machine parts that make modern industrial production possible. Today, these jobs are being transformed by the use of **numeric control (NC)** and **computer-aided numeric control (CNC)**. Under such systems, the metalworking lathes, drills, and cutting tools operate automatically. The machinist is left with the residual job of feeding and unloading the machine, while the computer controls the cutting tools, their speeds, and their depths. An example of NC-controlled machining is provided in Box 9.1 based on Kurt Vonnegut's *Player Piano*.

In CNC production systems the right to program the control device becomes a central issue. Is this work to be done by machinists after they receive training in programming, or is it to be done by computer programmers after they receive training in machine tooling? Shaiken et al. (1997) argue that in most cases the best programmers are machinists, because the programming software is not hard to learn. The machinists' skills, however, are very complex, requiring a long training period and much experience to master, and this knowledge is essential for designing computer programs that will actually work.

Microprocessor-based technology also influences the work of those who maintain and repair machinery. Because of the rapid pace of technological change, maintenance workers will no longer be able to specialize in one narrow craft or type of application (Greenbaum, 2004). Instead, the trend will be toward the multi-role, polyvalent craft worker who is simultaneously machinist, electrician, and computer programmer (Mort, 2000).

Clerical Work

Some of the most dramatic effects of high technology are being felt in clerical work. Clerical jobs have been dramatically transformed because their basic task is handling information, and it is this task for which computers are most suited. For instance, the introduction of word processing is estimated to have reduced the cost of producing a letter by a secretary from seven dollars to two dollars.

The development of computerized office automation can be divided into two stages. In the first stage individual workstations are automated through the introduction of personal computers. This stage dramatically increases individual productivity. In the second stage, as yet incompletely realized, individual workstations are eliminated, and the entire information accounting system becomes fully automated. Jobs in this stage may be more highly skilled, but they will be largely programming and machine maintenance rather than clerical, and the reduction in clerical jobs is projected to be even more dramatic than in the first stage (Wright, 2000). The development of voice recognition software may further displace clerical and data entry workers as information they once

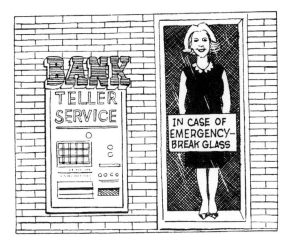

SOURCE: Mark Belanger. 1983. *The Facts* 5, 7 (September): 53. Reprinted by permission of The Canadian Labour Congress.

typed into a computer is now recognized as voice input from a front-line worker.

Capital outlay per worker will increase with the development of the automated office. Office work has traditionally been badly undercapitalized, with an average investment of only about $2,000 per worker. By contrast, factory work has an average investment of about $25,000 per worker. In automated office work investment typically rises by fivefold or more to about $10,000 per worker (Wright, 2000). Office automation will remove much repetition and drudgery from clerical work, such as retyping multiple drafts of documents. It is also likely, however, to intensify the stress and pace of clerical work.

Middle Management

Middle management is also being affected by computer technology. Traditionally, the job of middle managers has been to search out, compile, and digest production and marketing information and then pass this information on to top management. These are precisely the activities that are most easily automated through computerized **management information systems** (**MIS**) that tally inventory, handle payroll, and print schedules. Where computerized accounting systems are used, middle man-

agers may face displacement and more pressured working conditions because they are in direct competition with the more cost-effective systems. Such systems also allow closer monitoring, not only of workers, but also of middle managers, thus eroding important aspects of their traditional power and autonomy. Researchers have observed increasing resistance to automated production systems among some middle managers, a group normally identified with eager compliance to organizational goals (Smith and Walter, 2003).

Technical Workers

In addition to affecting existing occupations, new technologies are creating entirely new occupational specialties and greatly expanding others. These occupations include computer programmers, health technologists, and engineering and science technicians. Many of these occupations require two-year college degrees, and they are some of the fastest growing professional and semi-professional occupations. Many of the jobs being created are in rapidly growing high-technology industries. For example, chemical and biological technicians work in laboratories evaluating medical and scientific specimens for the rapidly growing medical and biotechnical industries. Other technical workers provide support functions to more traditional professions that have experienced rapid technological transformations in the nature of their work (see Chapter 11). Nurse anesthetists, radiologists, dental hygienists, and biological and chemical laboratory technicians are examples of such rapidly growing technical support occupations (Rabinow, 2003).

Telecommuting

Because of the falling costs of personal computers, it has become economically feasible for many employees to do an increasing share of their work away from the office (North, 2005). **Telecommuting** is doing work that would normally be done in the office at home on a personal computer or remote terminal. It is estimated that by 2005, 10 percent of the labor force was telecommuting at least part time. Telecommuting can have both positive and negative consequences for workers.

Positive consequences of telecommuting include increased flexibility of working hours and reduced commuting time. The greatest benefits are likely to occur for professional workers who use personal computers to do parts of their work at home during the evenings or on weekends. Telecommuting allows significantly increased flexibility in work schedules for these workers (Ellison, 2004). However, telecommuting also potentially allows work to invade the home twenty-four hours a day, seven days a week.

Other negative consequences include reduced personal contact with coworkers and clients and exclusion from informal sources of feedback that may be essential for doing a job correctly and efficiently. Isolation and diminished visibility can also reduce an employee's opportunities for making contacts essential for promotion and upward mobility. Blue Cross Blue Shield provides an example of the negative side of telecommuting. "The company pays clerical home workers piece rates, offers no paid vacation or benefits, and charges $2,400 a year in equipment-rental charges. Their 'cottage keyers,' as the workers are called, process more than 200 medical claims a day and net only about $100 a week" (Moore and Marsis, 1984:13). Processing the average claim requires about two and one-half minutes, and the home workers net about ten cents for the task.

Isolation is inherent in telecommuting and is a core problem, but working on a piece-rate basis is also problematic for many employees. According to a developmental editor in the publishing industry:

> I find in my freelance work that most employers require an estimate, which allows little flexibility. Though they offer an hourly wage, they give me a ceiling for the entire job. Thus, I must either stay confined to the number of hours allotted to the job—even if the job needs more hours than estimated—or not get paid for the extra hours worked. And some employers work on a piece-rate

basis alone. The rate offered is almost invariably based on fewer hours than the work requires (anonymous personal communication).

Clerical workers who are forced to choose home-based work because of lack of affordable child care may find that their problems are compounded by trying to work at home and take care of children at the same time. These problems may be further intensified by difficulties in finding adequate work space in already cramped quarters.

"Offshore" Telecommuting

Perhaps the most ominous trend for the North American labor force is the growth of "offshore" telecommuting (Nickson, 2004). Most large companies now do at least some of their routine clerical work overseas. The telecommuted work includes keypunching magazine and newspaper subscription lists, airline reservations, and survey questionnaires, and preparing book manuscripts for electronic typesetting. Data can be electronically transmitted between countries even more readily than partially assembled electronic components, and the threat to the jobs and conditions of North American workers is at least as immediate. As the United Nations' *Human Development Report* (2001:27) notes, "Today's technological transformations are intertwined with another transformation—globalization—and together they are creating the network age."

More recently, many companies have moved all or parts of their telephone-based services overseas. India, particularly the Bangalore region, has been one of the leading recipients of this movement of jobs. The success of the Bangalore region is no accident. India has pursued a public policy of providing education for information technology. "India's English-language technical colleges turn out more than 73,000 graduates a year.... And India has invested in infrastructure, especially high-speed links and informational gateways" (United Nations, 2001:37). Box 9.3 describes the nature of work in India's booming "call centers."

BOX 9.3 The Call Center Down the Street: Bangalore, India

There are currently about 245,000 Indians answering phones from all over the world or dialing out to solicit people for credit cards or cell phone bargains or overdue bills. These call center jobs are low-wage, low-prestige jobs in America, but when shifted to India, they become high-wage, high-prestige jobs....

The call center [where I interviewed] is a cross between a coed college frat house and a phone bank raising money for the local public TV station. There are several floors with rooms full of twenty-somethings—some twenty-five hundred in all—working the phones. Some are known as "outbound" operators, selling everything from credit cards to phone minutes. Others deal with "inbound" calls—everything from tracing lost luggage for U.S. and European airline passengers to solving computer problems for confused American consumers. The calls are transferred here by satellite and undersea fiber-optic cable. Each vast floor of a call center consists of clusters of cubicles. The young people work in little teams under the banner of the company whose phone support they are providing. So one corner might be Dell group, another might be flying the flag of Microsoft....

Most of the young people ... give all or part of their salary to their parents. In fact, many of them have starting salaries that are higher than their parents' retiring salaries.... Most of the young people start their work day [at six p.m.] to coincide with dawn in America....

Male operator: "Merchant services, this is Jerry, may I help you?" (The Indian call center operators adopt Western names of their own choosing. The idea, of course, is to make their American or European customers feel more comfortable.)

Woman operator getting an American's identity number: "May I have the last four digits of your Social Security?"

Woman operator giving directions as though she were in Manhattan and looking out her window: "Yes, we have a branch on Seventy-fourth and Second Avenue, a branch at Fifty-fourth and Lexington."

Male operator selling a credit card he could never afford himself: "This card comes to you with one of the lowest APRs."

SOURCE: Excerpt from Thomas L. Friedman, 2005, *The World is Flat: A Brief History of the Twenty-First Century.* New York: Farrar, Straus, and Giroux, pp. 21–24.

MICROPROCESSOR TECHNOLOGIES AND SKILL REQUIREMENTS

Besides displacement and changing job content, a second major issue surrounding high-technology work is its effects on the skill level of jobs. To address this question, we consider three divergent positions in the debate about the effects of advanced technologies on skill. Recall that we first discussed skill upgrading and deskilling in Chapter 7.

The Skill-Upgrading Thesis

The most optimistic position argues that new technologies have increased skills. Among the oldest and most frequently cited empirical studies dealing with technology and skill upgrading is Blauner's (1964) study of

continuous-process automation in the chemical industry. Continuous-process systems involve a continuous flow of the product (for example, chemicals) through a production system. Human intervention is required only to monitor pressures, temperatures, flows, and so on. Blauner found that continuous-process automation requires a greater proportion of skilled maintenance workers than do less automated manufacturing systems. In addition, machine operators in the chemical industry have greater responsibility for the care and proper functioning of expensive capital equipment than do machine operators in mass-production settings.

More recently, in a Communications Workers of America (CWA, www.cwa-union.org) membership poll, 78 percent of the respondents indicated that technological change had increased the skill requirements of their jobs. Automated systems often require workers to utilize both high levels of technical knowledge and skills acquired only through lengthy experience (Zuboff and Maxmin, 2004). It is easy to underestimate the depth of knowledge required by the technicians who press buttons on automated equipment or who click boxes on computer screens. Trained and experienced workers are necessary for the effective utilization of technologically advanced automated systems, even where managers devise systems that they believe are foolproof.

Based on a case study of banks and bank tellers, Adler (2005) arrives at similar conclusions about skill upgrading. Focusing on worker responsibility, cognitive learning, and job interdependence (rather than mastery of a fixed set of tasks), Adler finds that the least skilled jobs are the ones most affected by automation within banks. Many of these jobs are rapidly being eliminated. Therefore, the overall effect of automation is to upgrade the average skill requirements of the remaining jobs. Similarly, in a study of companies that made significant investments in microprocessor production techniques—computer-aided numeric control machine tools, computer-aided design, and management information systems—Francis (1994) finds little evidence of deskilling. He concludes that increased use of microprocessors produces no discernible reduction in skill requirements.

Several studies have emphasized the new skills that workers must acquire to operate technologically advanced production systems. Based on analysis of work in automated paper mills, Penn, Rose, and Rubery (1994) find that high-level maintenance skills requiring autonomous choices increase with advances in technology. In a review of technological change in three industries (printing, banking, and metalworking) Ozaki (1999) finds that the increasing use of high technology results in greater requirements for formal knowledge, precision, and autonomous decision making. Box 9.4 describes some of the potentially empowering aspects of working with sophisticated computer-assisted production systems.

The Deskilling Thesis

Substantial evidence, however, also suggests that deskilling often accompanies new technology. Greenbaum (2004), although noting that automa-

tion produces increasing levels of responsibility for some workers, concludes that, overall, it creates a tendency toward declining skill requirements. She argues that as mechanization progresses, initial changes that demand increased skills give way under automation to a progressive loss of skill resulting in an inverted U-shaped skill curve. According to this thesis, as automation progresses, skills first rise and later decline. Moreover, Greenbaum questions whether the growing demand for trained technicians to operate automated technology is not just a form of credential inflation rather than a true upgrading of skills. That is, degrees and other credentials may be required because so many workers have them, not because the skills they represent are actually required on the job.

Two case studies describe in detail how new technologies have deskilled workers (Boddy and Buchanan, 1981). The first study involves the transformation of copy typists into video display typists. Deskilling occurs because with automated equipment there is less need to type text correctly the first time. With the new equipment corrections are easier, and the printer positions the paper and takes over other functions formerly performed by copy typists. However, some new skills are required of video display typists. These include increased concentration and familiarity with codes for formatting and editing text.

The other case documented by Boddy and Buchanan involves the introduction of automated mixing equipment at a large cookie factory. The major consequence of technology was the transformation of the "doughman" into a mixer operator. Formerly, a master baker had held the position of doughman. But as the computer replaced the need for human intervention in the mixing process, the doughman suffered a loss of craft skills. The new automated equipment left the doughman with the residual responsibility of pressing a button to start the mixing cycle. Skilled maintenance positions at the factory also declined in skill as repair work became merely a matter of running a series of simple tests and replacing defective parts. No aspects of the automatic technology demanded that the mixer operators or maintenance workers

---

B O X 9.4 Computer Technology and the Flow of Information

Based on her study of a variety of facilities undergoing computer-based automation, Shoshana Zuboff believes that the new technologies tend to disperse information to workers at the front lines of production.

Some believe that organizations in the future will achieve competitive advantage based on their ability to better understand their own businesses and apply imagination to newly available data in order to generate higher levels of innovation. A corporate vice president, reflecting on the emerging manufacturing environment, struggled to formulate such an alternative:

There has been a fear of letting [management information] out of our hands—that is why information is so carefully guarded. It could be misused or misinterpreted in a way that cannot be managed. Traditionally, we have thought that such data can only be managed by certain people with certain accountabilities and, I hesitate to say, endowed with certain skills or capabilities. But with the new technology it seems there is an almost inevitable kind of development if you have as a goal maximizing all business variables and maximizing the entire organization's ability to contribute to that effort. I don't think you can choose not to distribute information and authority in a new way if you want to achieve that. If you

do, you will give up an important component of being competitive.

Judgment means the capacity to ask questions, to say no when things are not right. It also creates the possibility of asking "why?" or "why not?" One of the managers most respected for his willingness to "pass the knowledge down," described what he called "the developmental learning process" that operators must go through if they are to become critically competent at the data interface.

At the first level, of course, people need to know how to keep the equipment running. But the next step is to ask, "Why am I doing what I am doing?" Only if people understand why, will they be able to make sense of the unknowns. The third step is process optimization and diagnostic problem solving. At that point, they can hone in on the real issues. . . .

According to an operator at the factory:

We need to know the whys of this process. I can't just punch this button because I was told to. I have to do it because I know why and what happens. That's the only way I can run it better.

SOURCE: Excerpt from Shoshana Zuboff, 1988, *In the Age of the Smart Machine.* New York: Basic Books, pp. 288–290.

---

acquire new skills or knowledge to perform effectively. Workers became bored, apathetic, and careless. They rejected responsibility for breakdowns of the new system. Further, they developed few new skills that would have made them promotable. The factory managers thus lost a valuable source of potential recruitment into supervisory positions.

Based on her analysis of several insurance companies in New York, California, and Pennsylvania, Appelbaum (1984) identifies deskilling as the main pattern associated with the introduction of new technology in insurance underwriting. Her work indicates that in 80 percent of the cases computers make the underwriting decision on personal insurance lines. In short, as their work became increasingly standardized, insurance underwriters retained their skill in title only.

In a review of problems posed by the use of automated methods for industrial and office workers, Mowshowitz (2002) finds skill requirements for workers mainly declining after the introduction of new technologies. For office workers computerization makes work more standardized and formally defined. Workers also have less autonomy as more of the decision making involved in the production process is assumed by higher level managers or built directly into automated information systems.

The Mixed-Effects Position

Finally, many analysts view the impact of new technologies on skill requirements as a dynamic process in which some skill requirements are increased and others are reduced. Milkman and Pullman (1991) argue that the introduction of robotics and other automated processes into manufacturing industries produces both deskilling and skill upgrading. They conclude that robotics affect workers' skills in a positive way because the jobs created in robotics and robotics maintenance require more technical background than did the manufacturing jobs they replace. A solid understanding of post-high school math and science is essential for robotics technicians. For the most part this training can be obtained through technical programs offered at community colleges. However, they also warn that many of those displaced by robots may end up in low-skill, low-wage service jobs.

Automation also has certain inherent limits that necessitate the continuation of skilled positions. In the machine tool industry, for example, as lathes and other tools age and settle, tolerances spontaneously get out of whack. When machine tools are computerized or numerically controlled, problems arise in achieving precision cuts. Without a skilled craft worker in charge of the machine, such problems can be difficult to fix (Shaiken et al., 1997). Many computer-driven systems may simply be so fragile and vulnerable to disruption that they inevitably require close maintenance by a skilled craft worker to be utilized effectively (Hirschhorn, 2002). To the extent that this is true, craft skills can never be eliminated from many manufacturing operations.

Related problems can occur in automated engineering and drafting systems in which the computer generates generic parts and fits them into a design. Although such systems facilitate the work of the drafting engineer, they also eliminate less demanding parts of the task that provided essential "mulling" time for the engineer to survey and comprehend the broader goals of the project. Such limitations suggest that computers will never completely replace skilled workers.

Sociologist Kenneth Spenner (1990) and Industrial Relations expert Jeffrey Keefe (1999) argue that part of the problem in drawing firm conclusions about the effects of technology on skills is that skill is often a vague and undefined concept.

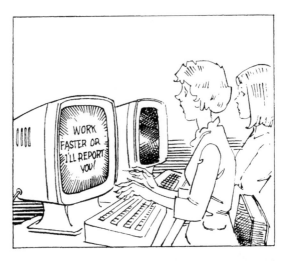

SOURCE: Mark Belanger. 1983. *The Facts* 5, 7 (September): 36. Reprinted by permission of The Canadian Labour Congress.

What is meant by skill appears to change as technologies evolve. Thus, it may be difficult to determine whether skills have increased or declined as a result of technological change. The observed general stability in skill levels may thus be a result of offsetting changes. Upgrading of skills appears to be occurring through the creation of jobs that require more training. Simultaneously, deskilling appears to be occurring through the downgrading of job content over the lifetime of many jobs.

Variability in High-Technology Effects It is difficult to avoid the conclusion that the effects of high technology are highly diverse and vary from one setting to another. In some situations, automation demands that workers learn new skills to monitor sophisticated equipment. Automation may also increase the responsibility of workers for complicated and expensive integrated production systems. In other situations manufacturing workers may be reduced from operating equipment that required advanced skills to simply loading and unloading an automated version of the same equipment. The latitude of clerical workers to make decisions about their work may be similarly restricted as they labor under increasingly routinized systems. There are also situations in which workers find that their old skills are obsolete but that they need new skills in mathematics, electronics, science, or programming to handle their new responsibilities. Rather than universally upgrading skills or deskilling jobs, advanced technologies appear to have mixed and contradictory effects. Some jobs are upgraded, some are deskilled, and some experience changing skill requirements (Liker, Fruin and Adler, 1999).

One important mechanism through which this transformation occurs is that older workers with more vintage skills are laid off, and younger workers with newer training are hired. This works well for the companies involved, which avoid the costs of in-house training to update the skills of their existing employees. Plus, they may also be able to pay the new workers less than existing workers with seniority. This solution to upgrading the skills of a company's workforce, however, does not work well for middle-aged workers who are left without jobs and with few employment options. Many employers have also recruited technically well-trained immigrants from countries with lower wage structures, such as Ireland and India. Some employers even subcontract work such as software production to off-shore sites in the search for cheaper labor. These strategies further weaken the position of North American workers, especially older workers.

Skills appear to increase in settings where workers have the power to insist that new technology be introduced in a manner that upgrades their skills. Workers' power to demand skill upgrading may rest on their organization into a union or on their professional expertise. Skills also appear to be upgraded when managers perceive that deskilling the workforce is counterproductive because of the complexity of the production process. When worker power is absent and managers perceive no need to maintain or cultivate workers' skills, high technology has resulted in the deskilling of work.

The insights of researchers in the mixed-effects position allow us to see that the effects of high technology on skills are not determined solely by technological imperatives. They are often determined by the social context in which the technologies are introduced and by the relative power of the actors involved.

Training for Changing Skill Requirements

New technologies profoundly affect workers' needs for training. Some researchers maintain that existing workers can be retrained to fill the new jobs. In practice, however, this process can be very difficult. Who pays for a year's leave for a worker in mid-career to learn a new skill? Retraining the workers directly displaced by robots for new jobs in robotics may not always be realistic. For example, substantial training would be required to teach an assembly-line welder to repair and maintain the welding robot that will be doing the welder's job in the future (Bills, 2004). On the other hand, retraining skilled plant maintenance workers to maintain industrial robots is relatively simple.

Continuing Education   Many researchers argue that the educational demands for new high-technology jobs will be substantial but that they can best be met by continuing-education programs and on-the-job training rather than by additional training for new workers before they enter the labor market. They argue that the new technology changes so rapidly that it often requires highly specific training, which can most effectively be acquired on the job or through continuing education classes (Ozaki, 1999). As a result, training for high technology jobs will become a lifelong endeavor, with new training being required as new technologies emerge.

Liberal Arts Education   The increased focus on job-specific training also includes a continuing role for broader education in the liberal arts. Much of the skill upgrading associated with new technologies involves cognitive and interactive skills, exactly the kind of skills taught in traditional liberal arts curriculums (Wolff, 2000). In addition, job seekers flooding highly focused technology training programs may create an overabundance of computer technicians and related specialists. Unfortunately, the actual number of new jobs in these fields may be relatively small. For these reasons, a liberal arts education with a broad background continues to make sense as a starting point for lifelong learning in a high-technology economy.

Training Options   How then is training for high-technology jobs best provided? One option is formal, degree-granting programs. For example, twenty-seven degree-granting programs in robotics maintenance exist in the United States. In addition, 343 robotics courses are taught at various other institutions in the United States (Lapan et al., 2000). The non-degree courses may be at least as important as the degree-granting programs, because they can be combined into flexible training programs to meet rapidly changing needs.

Another alternative is to institute such training programs as part of collective bargaining agreements. Starting in the 1980s, the CWA succeeded in bargaining for training programs to upgrade members' skills in working with sophisticated telecommunications equipment. In the initial contract $36 million was allocated for training programs for current employees (Noble, 1986:10). Today the CWA, in cooperation with the regional telephone companies, runs a wide range of lengthy training programs and shorter courses targeted to meeting and anticipating specific skill needs in telecommunications. This commitment to training is an important part of the CWA's and the regional telephone companies' plans to remain competitive in the rapidly changing telecommunications industry.

WORKING IN HIGH TECHNOLOGY

Technological changes are capable of increasing both productivity and job satisfaction. To what extent are these possibilities being realized? In this section we explore the consequences of high technology for the meaning of work and for organizational dynamics. We also examine union responses to the challenges of technological change.

Computer Technology and the Meaning of Work

There is considerable evidence that new technologies often increase job satisfaction (Lincoln and Kalleberg, 1996). Reviewing the literature on job satisfaction and computer technology, Danziger (2004) argues that workers tend to experience technological change as mildly benign. Japanese workers appear to be even more favorably disposed toward technological change. The majority welcome such change, including robotics, and only small numbers have negative attitudes.

In contrast, other observers see advanced technology as having less than ideal consequences for job satisfaction. Examining the work of machinists, Shaiken et al. (1997) find that NC systems create heightened alienation and stress. Specifically, they see increased alienation stemming from workers' loss of control over the production process. This heightened stress is partly due to growing isolation at work, as proportionately more machines and fewer workers are involved in production. Shaiken et al. are not completely negative in their evaluation of NC, however, noting that it can also mean less noise, more accuracy, and more cleanliness. In particular, they note that NC and CNC in the machine tool industry have created a situation in which workers are less frequently confronted with chronically "cranky" machines that are difficult or impossible to operate within prescribed tolerances.

Computer technologies can also have negative effects for white-collar workers. These include lessened variety, reduced physical mobility, computer-paced pressure to produce, reduced interaction with other workers, and "feelings of being zombie-like" because of prolonged, intense interaction with a computer while being physically and psychologically isolated from other workers (Buchanan, 1997). For example, one of the most significant negative effects faced by copy typists when their work is automated is the loss of the personal, informal working relationship between authors and typists. More characteristic of the literature as a whole is Dyer-Witheford (1999), who observes that isolation and constant monitoring can create stress but that, simultaneously, new technologies are eliminating much tedious and dangerous work.

Computer Technology and Organizational Dynamics

How are organizations being influenced by computers, and what will high-technology organizations be like in the future? Many researchers argue that the introduction of computer technologies is fundamentally altering the nature of organizations. Some argue that the new technologies are increasing the availability of information and thus dispersing power throughout the organization. Researchers find that alienation is lessened under these circumstances. Others argue that computers are leading to a centralization of control.

Dispersion of Information A key component of the thesis that technological change leads to improved working conditions is the argument that advanced technology tends to **disperse information** and authority more broadly throughout the organization. This thesis has gained popular recognition in Naisbitt's (1999) slogan that "computers destroy hierarchy" and in Cleveland and Anderson's (1999) thesis concerning the "twilight of hierarchy." This thesis, ultimately based on technological determinism, is given further support by current management schemes that attempt to promote productivity through the involvement of all employees (Drucker, 2003).

Many researchers also argue that technological advances have resulted in greater interdependence among tasks and have made it easier, and more important, for groups to determine the nature and pace of their work. Because high-technology industries make a large capital investment per worker, they depend heavily on workers' understanding of the job and on their good will, motivation, and commitment. Technological advances may thus empower workers and improve the experience of work for many. In addition, quality-control functions are sometimes reintegrated into pro-

duction rather than being allocated to separate divisions. This reintegration results partly from management decisions about how best to pursue quality control. But it also rests on the increased ability of the workers who use advanced technology to monitor their own work and produce consistently high-quality products (Hirschhorn, 2002).

Centralization of Control In direct contrast to those who believe that advanced technology lessens organizational inequalities, other researchers observe a connection between technology and the centralization of authority. These researchers argue that the natural tendency of automation is to concentrate the functions of control and decision making in the upper levels of management (Daday and Burris, 2001). Organizational analysts have long noted a connection between advances in technology and organizational centralization (Pfeffer, 1998). It is important to be aware that the flexibility of computer-aided production systems is primarily flexibility in *information retrieval* and *product design*. This flexibility does not necessarily translate into greater organizational, task, or interpersonal flexibility.

Electronic Surveillance Advances in technology have also created the specter of **electronic surveillance** at work. Electronic devices are routinely used to monitor the work of many employees, especially in large companies. This monitoring may include regular printouts for managers on keystroke rates, error rates, and break times. Even the work of grocery checkout clerks has become closely monitored with the advent of electronic scanning of grocery items (Marx, 2003). In combination with chemical surveillance through urine tests, electronic surveillance significantly increases the ability of corporations to intrude into the lives of their employees both on and off the job.

In Canada electronic monitoring of work is considered an invasion of privacy and a violation of the Canadian Charter of Rights. Even more stringent restrictions on the electronic monitoring of individual workers have been implemented in many European countries (Ozaki, 1999). Box 9.5

BOX 9.5   Internet Surveillance in China

Hearings before the U.S. House of Representatives have accused four American Internet and technology companies—Google, Yahoo, Microsoft, and Cisco—with collaboration with the Chinese government in monitoring and silencing the voice of dissidents there.

Among the chief issues is the alternation of online products in the Chinese market—from search engines to blogging tools—to conform with the repressive requirements of the government there.

Also of concern is the sale to China of Internet hardware that the Chinese government has been able to deploy in the surveillance of its online population, as well as the role American companies are being forced to play in the undemocratic imprisonment of Chinese citizens for online behavior that in the West would be considered simple free speech....

Representative Tom Lantos, a California Democrat whose own Congressional Human Rights Caucus

was snubbed by all four companies when it invited them to speak two weeks ago, had sharp words for the executives. "I don't understand how your corporate leadership sleeps at night," Mr. Lantos said....

"Many, if not most, of you here know that one of Google's corporate mantras is 'Don't be evil,'" Mr. Schrage of Google said in his statement. "Some of our critics—and even a few of our friends—think that phrase arrogant, or naïve, or both. It's not. It's an admonition that reminds us to consider the moral and ethical implications of every single business decision we make," the statement continued. "We believe that our current approach to China is consistent with this mantra."

SOURCE: Excerpt from Tom Zeller, Jr., 2006, "House Member Criticizes Internet Companies for Practices in China," *The New York Times*, February 15.

describes the controversial activities of some major American companies that have adapted their Internet products to facilitate electronic eavesdropping in China.

Management from the Rear   Various other organizational changes besides increasing hierarchy and the possibility of electronic surveillance also suggest at least the possibility of more stressful working conditions under high-technology systems. One commonly cited problem concerns the stresses caused by working on accelerated project schedules (Rabinow, 2003). Perhaps more importantly, many high-technology companies struggle to find competent managers. This problem arises for two reasons. First, even if managers are promoted from within the production staff, rapid product changes may quickly make their knowledge obsolete. Without continuing hands-on involvement in design and production, even engineering managers rapidly begin to lose touch with the new technology and with its problems, possibilities, and limitations. Second, managerial orientations toward short-run profit

may come into direct conflict with the efficient operation of new production systems. Managers may rely excessively on cost-cutting techniques or may make poorly considered demands for getting the product out too fast or on too tight a schedule. Whatever its cause, working under chronically incompetent managers can devastate worker morale and long-run productivity (Roscigno and Hodson, 2004). The popular Dilbert cartoon series uses the issue of management incompetence in a high-tech engineering setting as its primary source of comic tension.

The High-Technology Life Cycle   A rapid cycle of corporate birth and death is characteristic of high-technology industries. High-technology companies frequently produce a wide variety of spin-off companies, often started by engineers leaving the parent company. Spin-off companies generally specialize in products either competing directly with or complementing the products of the parent company. For example, Massachusetts Computer is an offshoot of Digital, Stratus is an offshoot of Data General,

Automatix is an offshoot of Computervision, and Precision Robotics is an offshoot of Teledyne.

Just as high-technology companies come into existence quickly, so do they quickly pass out of existence, often through corporate mergers. These mergers may occur when larger corporations seek to buy out smaller firms with compatible product lines. Acquisitions may also occur when highly diversified conglomerates attempt to move into high-technology fields. The nature and consequences of the accelerating pace of corporate mergers are discussed further in Chapter 15.

A final characteristic of high-technology industries is a quickened cycle of boom and bust. Tens of thousands of workers in the electronics industry were laid off in the early 1980s and again in the early 2000s. Such booms and busts occur because of the rapid movement of new firms into high-technology fields, which produces periodic gluts in the market. These problems are increased by rapid changes in product lines as new products quickly eclipse older ones.

Union Responses

Unions are concerned about a range of issues associated with the introduction of new technologies (Ozaki, 1999). These concerns include the following issues: (1) Advance notice of technological changes is necessary to give unions time to study their impacts and develop reasonable strategies for accommodating these changes. (2) Unions need to be included on a consultation basis from the very earliest planning stages so that they can have a role in determining which new technologies will be selected. (3) It is important to initiate technological changes on a trial basis so that their unintended effects can be examined. (4) Workers need to be protected from reclassification to lower grades or pay scales. (5) Training programs for teaching existing workers the new skills required are strongly preferred over hiring new workers to displace existing ones. (6) Job security against technological layoffs is a particularly important issue and may be based on reduction of the work week, voluntary early retirements, or redeployment to other facilities. (7) Workers need to be protected from potential health hazards associated with new technologies and from electronic surveillance at work.

The above list of contractual provisions is what unions would like, but what have they actually been able to get? Fewer than 5 percent of workers in the U.S. electronics industry are unionized (Mort, 2000). The American Electronics Association reports only ninety collective bargaining contracts among its more than 1,900 member companies. Even when a high-technology company has a union, contractual provisions dealing explicitly with technological change are not common. For example, the AFL-CIO reports that only about 14 percent of agreements among its member unions contain provisions for worker retraining after the introduction of new technology.

New Technologies and Union Power One reason that unions have not been more successful in negotiations over technological change is that union power is often reduced in the situations where technological change is occurring. Automation of production makes it increasingly possible for managers to run production operations without workers, at least until repair and maintenance problems mount. As a result, the effectiveness of labor's ultimate bargaining weapon, the strike, is reduced. In addition, there are well-grounded fears that too many demands from North American workers may encourage companies to move their production operations overseas.

In spite of these adverse conditions, union organizing and bargaining activity does occur in high-technology industries. In Massachusetts the CWA, the United Auto Workers (UAW), the Machinists, and the Electrical Workers have joined in a coalition with several other unions and are seeking to organize workers at the scores of plants and industrial laboratories along Boston's Route 128. Probably the greatest union successes have been in the area of demanding training for existing workers so that they can successfully utilize the new technologies (Cutcher-Gershenfeld, 2005).

International Agreements In Canada unionized workers have been able to negotiate stronger agreements than those in the United States. Advance notice and consultation are required in 30 percent of contracts. Provisions for retraining are specified in 19 percent of contracts. Even in Canada, however, 60 percent of unionized workers and 85 percent of the labor force as a whole have no contractual rights concerning technological change (Gunderson and Ponak, 2000).

European unions have been actively engaged in negotiating technological change longer than North American unions and generally bargain on at least two additional issues. First, they frequently bargain for advance training for workers' representatives so that selected shop stewards can be fully trained in all aspects of the new technology before its introduction. Second, they generally specify that electronic devices cannot be used to collect personal or production data on individual employees.

The greatest strides in agreements in Great Britain have been made in white-collar industries, particularly banking, where strong job security provisions are the norm. These agreements typically contain provisions concerning safety and health, as well as job security. Britain has also been a leader in worker retraining with its state-financed Training Opportunities Programs (Casey and Gold, 2000).

The most comprehensive technological agreements have been reached in the Scandinavian countries. Sweden's largest and most dynamic sector is advanced machinery and machinery parts, an industry requiring the very latest technology to maintain its competitive international position. Sweden has the largest number of industrial robots per capita in the world. Many of the Swedish provisions concerning technological change are legislative rather than contractual, including a complete prohibition against the electronic monitoring of individual workers' output. Additional provisions are included in collective bargaining agreements. In Norway the primary innovation is the creation of **data stewards**. Data stewards keep abreast of the latest technology being considered by the company, consult and negotiate with the company concerning its deployment, and protect workers' legislative and contractual rights.

Because of the ability of high-technology companies to move production jobs around the world, improved conditions for electronics assembly workers may increasingly depend on organizing workers in newly industrializing countries and on the ability of workers in different countries to coordinate their demands for better conditions. We develop these themes further in Chapter 16 on the global economy.

NEW FRONTIER IN HIGH TECHNOLOGY

Scientific advances, many of which have been facilitated by the microprocessor-based information revolution, are currently spurring new developments in biotechnology and nanotechnology. **Biotechnology** is "the application of biological knowledge and techniques pertaining to molecular, cellular, and genetic processes to develop new products and services" (Cortright and Mayer, 2002:6). Biotechnological science has leaped into prominence in recent decades because of dramatic advances in genetic engineering that allow the modification of plants, animals, and organic processes. Biotechnology has also allowed rapid advances in the treatment of human diseases, including many previously untreatable diseases and conditions. The implications of biotechnology for both new products and increased productivity are highly significant with over 5,000 patents currently being issued per year in biotechnology fields. Employment in the life sciences exceeds 200,000, including large numbers of research scientists and technicians.

Nanotechnology is broadly defined as chemical, electrical, and metallurgical techniques involving microscopic processes. Nanotechnology is *not* miniature robots in the conventional sense. Rather, nanotechnology involves the positioning and alignment of chemically reactive molecules and electronically reactive atoms in order to build new products or improve old ones. Nanoscaled products and devices bear a closer resemblance to nature's nanodevices—proteins, DNA, and membranes—than to conventional robots (Zucker and Darby, 2005). For example, nanotechnology has been used to develop new synthetic fabrics with molecular structures that repel stains. Other products can potentially be made that are lighter, stronger, and cheaper to manufacture than current products. Uses include thin films, chemical sensors, fluid dynamics, adhesives, and health applications. The uses of nanotechnology are probably even broader than those of biotechnology since nanotechnology has the potential to modify the properties of virtually all organic and inorganic materials.

In the next few decades biotechnology and nanotechnology are likely to produce important new products and provide employment for large numbers of research scientists and technicians. People will benefit from these products but, unlike the micro-processor based information, which has had a truly widespread influence on the nature of work, only the jobs and working conditions of those directly involved are likely to be dramatically affected. In the next *hundred years,* however, the possibility that biotechnology and nanotechnology will pervasively affect the lives and work of large sectors of society, in ways as yet unforeseen, remains a very real possibility.

SUMMARY

New technologies create new occupations, destroy old ones, and transform skill requirements of existing occupations. They also produce changes in working conditions and organizational dynamics. There are few consistent, undisputed findings concerning the consequences of technological advances.

Nowhere are these contradictory outcomes more apparent than in the area of skill requirements. In some cases new technologies destroy requirements for traditional craft skills; in others they create new skill requirements. Often the prevalence of deskilling or skill upgrading has little to do with the nature of the new technology itself. Instead, it may depend on the power of workers to demand the preservation and extension of their skills and on managers' perception of the importance of preserving a skilled workforce.

Technological displacement is a significant problem resulting from the utilization of new technologies. Older workers with now outdated skills are especially vulnerable. The majority of the studies favor the notion that a dual occupational structure emerges after the introduction of new technologies. A dual occupational structure appears to be a prominent feature in the electronics industry. In fields such as the manufacturing of microprocessors, computers, and scientific instrumentation, large numbers of lower-level production workers are required along with smaller numbers of highly trained engineers and technicians.

There is substantial agreement that advanced technologies are producing new sources of alienation and stress for workers. Increasingly, control and discretion are removed from workers and placed directly in automated equipment or in computer-assisted management information systems.

Many analysts cite the role of new technologies in eliminating dirty, tedious, and dangerous jobs as an important element in improved working conditions. But this point tends to ignore the fact that displaced workers may not have any job after the new technologies are introduced. Because new technologies create few jobs in contrast to the number they eliminate, acquiring new skills may be of little benefit to these workers. Long-term and indirect employment generating consequences of high technology may help offset these employment losses, but these effects are hard to verify.

Union responses to technological change focus primarily on the issue of job security for workers threatened by technological displacement and on the availability of new jobs for displaced workers. Unions also encourage the development of programs that provide training for workers in new skills and that protect workers from electronic monitoring.

KEY CONCEPTS

high-technology industry

technological displacement

long-term and indirect effects

numeric control (NC)

computer-aided numeric control (CNC)

management information systems (MIS)

telecommuting

dispersion of information

electronic surveillance

data stewards

biotechnology

nanotechnology

QUESTIONS FOR THOUGHT

1. What aspects of the current computer-based technological revolution generate the greatest potential for job displacement? Do you think other factors will be able to offset this job displacement and, if so, what will these factors be?

2. Do you think the jobs in the year 2025 will be more skilled or less skilled as a result of widespread use of computer technology?

3. Describe the potential benefits and problems of telecommuting. Do you think you will be telecommuting at least part time in your future career? What about this experience will be positive? What will be negative?

4. Computers can allow both greater access to information and greater centralization of control. Which of these do you think will be their dominant effect? Consider how some people might benefit from greater access to information while others might suffer from additional control and surveillance.

5. Can you envision a future working in biotechnology or with nanotechnology? How might these emerging technologies affect your employment, earnings, and meaning in work?

MULTIMEDIA RESOURCES

Print

Devinatz, Victor G. 1999. *High Tech Betrayal: Working and Organizing on the Shop Floor.* East Lansing: Michigan University Press. A first-hand account of the drudgery of doing assembly work in a modern high-technology biomedical factory.

David Noble. 1997. *The Religion of Technology: The Divinity of Man and the Spirit of Invention.* New York: Knopf. An analysis of the causes and consequences of technological change and its role in human development.

Shoshana Zuboff. 1988. *In the Age of the Smart Machine.* New York: Basic. Presents the argument that the essential character of computer technologies is their ability not only to automate work but also to provide information about the production process, which then serves as a source of potential power for workers.

Internet

U.S. Department of Education. *www.ed.gov* A valuable site for information on high-technology training opportunities.

National Center for Education Statistics. *nces.ed.gov* Comprehensive information on educational programs and educational attainment of the labor force.

Journal of Industrial Teacher Education. *scholar.lib.vt. edu/ejournals/JITE* The best site for information on vocational-technical training.

Public Broadcast System (PBS) Online. *www.pbs.org* High-quality information (and entertainment) on a wide range of social issues.

Internet Resources for Sociologists. *www.umsl.edu/~ sociolog/ resource.htm* Plentiful links to information and research on a wide range of topics of sociological interest.

RECOMMENDED FILM

Matrix (1999) and its sequels. A thought-provoking window on the future of technology.

10

Services

"A post-industrial society is based on services. Hence, it is a game
between persons. What counts is not raw muscle power, or energy,
but information."
(BELL, 1976:127)

" . . . [T]he reasons for the rapid growth of service occupations in both
the corporate and government sectors of the economy [are]: the
completion by capital of the conquest of the goods-producing activities;
the displacement of labor from those industries, corresponding to the
accumulation of capital in them, and the juncture of these reserves of
labor and capital on the ground of new industries; and the inexorable
growth of service needs as the new shape of society destroys the older
forms of social, community, and family cooperation and self-aid."
(BRAVERMAN, 1974:359)

The great majority of the North American labor force already works in service
jobs, and most new jobs are being created in the service sector. Students
reading this book are likely to be employed in service occupations or industries
for most of their working lives. As the two quotations above indicate, however,
observers disagree on the implications of the transformation of the economy from
goods-producing to service-producing. On the one hand, the concept "information society" conveys the idea of a productive economy in which most people
can be freed from agriculture and manufacturing to pursue occupations that
emphasize ideas, images, and creativity. On the other hand, low-skilled service
jobs may merely replace with wage labor some tasks that are better done by
families and communities. In this chapter we will examine the nature of service
work, the problems that face service workers, and the challenges that lie ahead for
individuals and firms that produce services.

WHAT ARE SERVICES?

Members of the labor force provide either goods (tangible items) or services to be sold. In Chapters 7, 8, and 9 we discussed the production of goods such as agricultural products and manufactured products. **Services** are acts provided in return for payment. The distinction between goods and services reflects how close in time or space the work is to consumption by the user. Pressers in a clothing factory are counted as manufacturing workers, but pressers doing exactly the same job in a dry-cleaning establishment are considered service workers (Braverman, 1974:360–361). A worker in a frozen-food factory who prepares and freezes a casserole for later sale is performing a manufacturing task. The restaurant cook who microwaves the casserole for a seated customer would be considered a service worker—so would the supermarket cashier who rings up the frozen casserole for purchase by a shopper. But the consumer who microwaves the casserole at home performs this behavior outside the market economy, and so this food preparation would be considered home production. The difference is that the frozen-food worker is preparing a *product*. The cook and the salesperson perform acts (cooking, selling) that make the product accessible to the consumer, and their actions are considered to be *services*. As you learned in Chapter 2, any actions undertaken by the final consumer are not officially counted as work.

Service workers may be self-employed, employees in private enterprises, or employees of governments. Their customers may be individuals, organizations, or governments. Some service workers receive direct payment based on performance—for example, automobile salespeople typically receive a commission for each car they sell. Some service workers receive payment for the services they are *ready* to provide during a certain period of time, even if there are no customers. For example, a cook in a restaurant would be paid even on an evening when no diners came. We pay for some services that we hope never to need, such as emergency medical services or national military defense. The insurance industry is such a service: the policyholder pays for the protection of the policy but hopes that the policy is not needed (Zelizer, 1979).

Characteristics of Services

Bound in Time and Place The many types of services share some common characteristics. Most services are **time-bound** because there are culturally defined times for providing the services. Some services, such as electrical utilities, hospitals, or police protection, are offered twenty-four hours a day. Other services are offered at times most convenient to customers or providers. Restaurants do most of their business, and must have their workers present, at the traditional mealtimes. Health clubs are open evenings and weekends when their members have leisure time. Some services must be offered according to regularly occurring deadlines. Accounting firms, for example, require longer hours from workers as the April 15 tax deadline approaches. Youth camps employ counselors only during the summer, and ski resorts employ workers only during the winter.

Services are also time-bound in the sense that their production and consumption occur nearly simultaneously. Services cannot usually be stockpiled, as goods can be produced and stockpiled in a warehouse. An airlines reservations clerk may have no calls during one hour of her shift and then dozens of calls during a brief period of time. The earlier idle time, however, cannot be put to use later when she is busier. Fast-food restaurants may "bank" hamburgers, which means cooking additional hamburgers in anticipation of a rush at lunchtime, but this tactic is limited by the needs to forecast accurately the customers' orders and to provide fresh, hot food. The limited ability to stockpile services reinforces the need to match workers' schedules to the customers' convenience.

Many services are also limited by place for the convenience of their customers. It does not matter to the hungry customers in this city how many hamburgers have been banked at a fast-food restaurant in another city; the demand for food is place-bound. Cities have multiple establishments performing the same services for the convenience of customers who live in different neighborhoods.

Urgency, convenience, and the travel habits of the customers affect the location of services establishments. Today, in our hyper-busy society, there seems to be a coffee shop and a fast-food restaurant on every corner.

The Internet and improved telecommunication have relaxed the constraints of time and place for some services, leading to the creation of new work sites such as call centers (Wickham and Collins, 2004). Call centers provide catalog and Internet sales, software consulting, and various other types of customer service that can be delivered even if the service provider is far away from the customer. This fact lets service industries take advantage of the fact that when it is night in North America, it is daytime in Asia. **Outsourcing**, or the relocation of jobs overseas, takes advantage of time differences by hiring people in other continents to provide twenty-four-hour customer service. Most outsourced workers are paid much less than standard North American wage rates.

Low Productivity   Services have traditionally been characterized as having *low productivity,* although there is much variation among the different types of services (Fuchs, 1968). There is also concern that measures of productivity appropriate for goods production is less meaningful for services. Suppose that a child-care worker who has been caring for four infants must now care for five. From one perspective the worker's productivity has increased. Others, however, might conclude instead that the quality of the service provided to the babies will decline (Malambre and Clark, 1992).

Compared with manufacturing jobs, technology enhances service productivity in a more limited way. Low productivity derives in part from the need to work with clients or customers individually. The most efficient airlines reservations clerk can handle only one call at a time and cannot be productive when there are no calls. Telephone technology may aid the clerk's efficiency—for example, call-waiting and automated messages may pacify the customers waiting for service—but the productivity gains are limited by the need for customized service. In a hospital, electronic monitors may substitute for a nurse's or an aide's direct observation

of a patient, but other tasks require direct interaction with one patient at a time, such as giving baths, administering medication, or assisting with physical therapy.

Restaurants provide a second example of the difficulties in increasing productivity. Technology may reduce the time required to prepare food in a restaurant, but a waitperson is still needed to bring the dishes to the diners' tables. In the fast-food industry, innovations in organization have been used to increase productivity while keeping prices low. Salad bars, buffets, and cafeterias reduce costs by substituting the customer's labor for the waitperson's.

Important consequences flow from the lower relative productivity of service work. As agriculture and manufacturing become more productive, they require fewer workers to maintain equivalent levels of output. The displaced workers seek employment in the service industries. Managers in the service industries seek higher productivity to increase their profits; to the extent that they are successful in raising productivity, there may be a further movement of workers within the service industries from more productive to less productive jobs. Economists identify low productivity as one reason that compensation in service jobs is lower than in manufacturing jobs. Other factors contributing to lower compensation for service workers are their lower levels of unionization, the greater likelihood that they work in small firms, and the greater probability their work is part time or seasonal.

Insatiable Demand   Finally, there may be an *insatiable demand* for services, in the sense that people can always consume more services, or more expensive versions of services they now consume. The demand for agricultural and manufactured products is more limited. There are only so many bowls of cereal or television sets that a consumer will buy. For services, by contrast, the consumer with more disposable income will find increasingly more attractive services to purchase. As long as consumers can afford to pay for services, service employment will grow.

Sources of the Demand for Services

The demand for services can grow in several ways. First, service industries can arise from *new manufactured products*. The manufacturing of computers eventually generated a large number of service jobs. Wholesalers and retailers sell computers, peripheral equipment, and software. Consulting firms advise companies about purchasing computers; other firms maintain and repair the equipment; and employee training firms develop programs to help employees use both hardware and software. In a similar fashion, other new agricultural and manufacturing products generate new service jobs.

Second, service industries grow with the shift from unpaid production to *paid production of services*. Sick people were once tended to at home. Today workers from many health occupations provide medical services in hospitals, clinics, outpatient centers, and medical offices. Families once provided most food preparation, child care, gardening, cleaning, and repair services needed by their households. Today many families purchase these services at least occasionally. Because these services can still be produced in the home, during times of hardship some families once again produce these services for themselves, and the sellers of such services may be especially vulnerable to the business cycle. For example, household budgets for many families are balanced by reducing the number of meals eaten away from home.

Third, many service industries arise in response to *higher disposable income* of consumers. For most households, leisure and entertainment services, such as tourism, professional sports, and the theater, are discretionary items purchased if budgets permit. Retailers rely for much of their profit upon consumers' discretionary holiday spending for clothing, gifts, and special foods. Because these expenditures can also be curtailed when consumers are worried about the economy, retailers of these items are especially vulnerable to recessions.

In the next section, we turn to the heart of service work—the interactions between those who provide services and those who receive them.

SERVICE INTERACTION

The essence of service work lies in the interaction between the service provider and the service recipient, who may be called the client or the customer. The workers may interact in person with their customers, they may only hear their voices or read their correspondence, or they may interact only indirectly with a customer. A successful interaction requires that the employer, the customer, and the worker all meet their standards for a successful interaction.

Interaction Standards

Employer Standards *Employers* have several criteria for judging the success of a service interaction. First, how many customers were served? Larger numbers indicate higher productivity and efficiency. Second, how well were they served? Did they spend as much money as the employer hoped? Did the customers feel satisfied with this interaction? Are they likely to be repeat customers?

Customer Standards The *customer* also has criteria by which to judge the success of the interaction. For the most part, customers want to be treated quickly, courteously, and efficiently, but they also want to be treated as individuals. These may be contradictory criteria. The service provider may be able to interact more quickly with more people by being relatively impersonal. Speedy interactions meet the customer's desire for quick service, but they do not allow for individualized treatment of the customer. Handling special requests, explaining the product, or in other ways personalizing the encounter may lengthen the interaction, thus reducing the worker's efficiency. In some service industries (for example, among information operators) the length of interactions is monitored, and an operator who talks too long with customers may be admonished. Operators learn to keep most interactions as brief as possible so that they can spend more time on the occasional complicated request without raising their average time-per-call to an unacceptable level.

Some customers employ additional criteria to judge the success of the interaction. Because service workers serve the public, some people wish to treat them as servants, demeaning them if possible and expecting subservience in return. Employers of maids, for example, may expect their maids to scrub floors on their hands and knees, which the maids may interpret as demeaning. Corporate cleaning services, however, may explicitly instruct their workers to don knee pads so that they may scrub floors (Ehrenreich, 2003). Service workers are prohibited from responding in an angry or reproachful tone to the customer; they must react meekly and according to company procedures, even to a rude or belligerent customer.

Worker Standards   *Service workers* also have criteria for a successful interaction, but these criteria vary depending on the nature of the interaction. Most interactions involve customers who are voluntarily present as shoppers, callers, or diners. Service workers in these situations expect to be treated with at least minimal respect, and they expect the customer to play the accustomed "customer role." Thus, while a certain amount of joking or teasing might be tolerated from a diner, a waitperson would not expect insulting language or physical assault from a customer. Part of the expected customer role in the restaurant also involves paying a tip.

In other situations the "customer" is present involuntarily, and then the service workers may excuse the behavior of the customer. An ambulance driver or an emergency room trauma team might make allowances for unusual behavior among patients who are in a great deal of pain. A police officer would not expect a suspect to be pleased about being arrested, but the officer expects respectful treatment in ordinary interactions with citizens.

Because each of these three parties has different criteria for the success of an interaction, the result may be a struggle for the control of the interaction. As the opening quotation from Daniel Bell suggests, there is a "game among people" that occurs in service work. In the following sections, we examine how employers and workers seek to *control the interaction.*

The Role of Employers

Most service workers are employees. Management tries to control their employees' interactions in three ways: (1) establishing standards for employee conduct; (2) training employees to understand how, from management's perspective, the ideal interaction proceeds, and what to do if the interaction deviates from this norm; and (3) establishing controls to ensure conformity to the standards and practices of the company. Even service workers who are self-employed will often adhere to similar standards because customers come to expect a comparable level of service.

Setting Standards   Setting standards may be formal or informal. Many service occupations require workers to wear a uniform that readily identifies their occupation. Often such uniforms reflect the subordinate status of the worker. For instance, waitresses and barmaids may be required to wear costumes reminiscent of the clothing of maids or clothing that is sexually revealing. Companies deploy slogans such as "the smile comes before the sale" to remind employees that friendliness is part of the ideal interaction. To reinforce this friendliness, management may mandate that employees wear name tags bearing only their first names, thus forcing customers to address the worker familiarly, a practice associated with friendliness.

Companies may also require certain rituals, such as introducing oneself to the customer by name, greeting customers as they enter the store, or asking certain questions in a required manner. One objective of these rituals may be fairness, or treating all customers equally. Service workers may be required to ask all customers the same questions or present them the same information, even if customers are uninterested. For example, the admitting personnel in a hospital may be required to ask questions to determine the patient's financial eligibility. Pharmacists must counsel customers concerning potential side effects of drugs.

Alternatively, the service workers' required "script" may be merely a means to increase the amount of money the customer spends. Telemarketers may receive actual written scripts to read over the telephone. Waitpersons may be required to recite the evening's specials even to a diner who has already announced an order. An insurance sales agent may use a set script to make "cold calls" to prospects who have expressed no interest in life insurance. In such cases, one of the employer's standards for a successful interaction is following the script, even if both the worker and the customer would prefer to omit it.

Training Employers' expectations for performance are set forth in more or less explicit training sessions. Box 10.1 presents an account of the training sessions in two such settings: McDonald's Hamburger University and the training program of Combined Insurance (Leidner, 1993). Aside from the explicit details that apply to the particular firm, such training usually seeks to achieve two goals. The first goal is the *routinization* of the interaction. In fast-food restaurants the exact procedures for preparing each type of food are minutely detailed (Reiter, 1991). Procedures also spell out the ways in which the service worker is to interact with the customer. According to Hamburger University instructors, McDonald's franchisees should buy ten to twelve videotapes a year for training their employees (Leidner, 1993:65). The standardization in training helps ensure not only that the sandwich tastes the same, but also that the customer receives consistent service.

Routinization is the development of a standardized procedure for doing work. The work of telephone operators was routinized soon after telephones became widespread (Norwood, 1990), partly to guarantee comparable standards of service. Routinization may arise to ensure health or safety. Airline mechanics and pilots have checklists to complete before every flight to verify that the airplane is in full working order. The pilot and copilot separately complete a checklist before each takeoff. Hospitals prescribe certain routines of care. Nurses in a surgical ward follow a prescribed routine before sending a patient to surgery; this routine ensures that the necessary preliminary tests have been done and that the correct surgery is performed on the correct patient.

The more recent routinization of sales and personal service, however, is rarely to achieve goals of safety or adequacy but rather to rationalize service delivery to make it as efficient as possible. Rationalization reduces the interaction to its minimum elements and speeds up each transaction, but the resulting interaction may be unsatisfying to the customer because of its impersonality. So a second goal of the training is to teach service workers to program and manage their emotional responses to the customer. "**Face-work**" refers to the management of the facial expression and, by extension, of the worker's emotional reactions (Goffman, 1955). The process of managing one's emotional responses in interactions with customers is called "**emotion work**."

Emotion work plays an important role in the customer-worker interaction, but whether it has positive or negative effects on the workers depends in part on how much autonomy the worker has in the situation (Wharton, 1993). A flight attendant's primary duty is to ensure the safety of passengers. As Box 10.2 indicates, however, an important part of the job is the face-work required to keep passengers feeling satisfied, even if traditional amenities have been eliminated. Part of the flight attendants' emotion work is handling passengers who might themselves be experiencing anxiety, anger, or other emotions (Menon and Duke, 2004). Bill collectors, on the other hand, are encouraged to dominate the interaction, even at the risk of appearing nasty (Hochschild, 1983). The bill collector can freely express negative feelings to the customer, because the bill collector's objective is to induce enough fear, shame, or guilt that the customer will pay an overdue bill. Many companies hire debt collection agencies to collect overdue debts so that their own workers do not have to spoil relations with their customers with unpleasant interactions.

Because feminine gender roles have often been associated with nurturing and soothing others, emotion work, and service work more generally, have sometimes been labeled "women's work" (Fishman, 1978). Some customers may hassle service workers by seeking to redefine the service role as a sexualized encounter (Giuffre, 1994; Williams, Giuffre, and

---

BOX 10.1   Training for Fast Food and Insurance

Sociologist Robin Leidner did fieldwork in two large service firms, McDonald's and Combined Insurance. She recounts some details of the training for both groups of workers. In both cases, the worker is instructed to routinize the interaction to achieve the company's goals.

"Window class" ... lasted a whole morning. My training group included two other newly hired window workers and a grill worker who was being cross-trained to work window. The training began, as usual, with a videotape. It emphasized the importance of the window crew's work, telling us that to guests (McDonald's word for customers), "You ARE McDonald's." Interactive work is only part of the window crew's job, though. In addition to learning about dealing with people, we had many details to learn about dealing with things. The videotape provided instructions on what the various-sized cups and bags were used for, how to stock the counter area, how to work the soda and shake machines, and how to load a bag and set up a tray properly.

Interactions with customers, we were taught, are governed by the Six Steps of Window Service: (1) greet the customer, (2) take the order, (3) assemble the order, (4) present the order, (5) receive payment, and (6) thank the customer and ask for repeat business.... The videotape provided sample sentences for greeting the customers and asking for repeat business, but it encouraged the window crew to vary these phrases. According to a trainer at Hamburger University, management permits this discretion not to make the window crew's work less constraining but to minimize the customers' sense of depersonalization:

"We don't want to create the atmosphere of an assembly line," Jack says. They want the crew people to provide a varied, personable greeting—"the thing that's standard is the smile." They prefer the greetings to be varied so that, for instance, the third person in line won't get the exact same greeting that he's just heard the two people in front of him receive.

Training for door-to-door insurance salespeople was at least as standardized.

In addition to scripting the life insurance agent's words, Combined Insurance tried to standardize how the agents held themselves, how they delivered their lines, how they gestured, and how they used the physical setting. Consider, for example, my field notes from a class lecture on getting through the door:

After ringing the doorbell and opening the screen door, wait for them to answer with your side to the front of the door. Do a half-turn when they open it. It's almost as though they catch you by surprise—nonconfrontational. Be casual.

Lean back a little when they open the door. Give them space.

Attitude: "Be as loose as a goose," Mark says. Be able to respond to what they say.

To get inside, use The Combined Shuffle. It has three steps:

1. Say, "Hi, I'm John Doe with Combined Insurance Company. May I come in?" Handshake is optional. Break eye contact when you say, "May I come in?"

2. Wipe your feet. This makes you seem considerate and also gives the impression that you don't doubt that you'll be coming in.

3. START WALKING. Don't wait for them to say yes. Walk right in. BUT—be very sensitive to someone who doesn't seem to want you to come in. In general, act like a friend; assume you'll come in.

Mark demonstrates one effective technique: he keeps shaking hands while he's walking forward, which makes it hard to stop him.

SOURCE: Robin Leidner, 1993, *Fast Food, Fast Talk: Service Work and the Routinization of Everyday Life.* Berkeley: University of California Press, pp. 68, 111.

Dellinger, 1999). These situations require the service worker to negotiate the boundaries of appropriate interaction with the customer, a situation that intensifies the difficulties of emotion work (Guerrier and Adib, 2000). (Sexual harassment was discussed in Chapter 4.)

Social Control   Managers need to assure themselves that their service workers are upholding the company's standards for interaction, so they may monitor employee behavior. For example, the store manager may observe the greeter in a discount store welcoming customers to a store, or the front-end

---

BOX 10.2 Emotion Work: Flight Attendants

Flight attendants typically work in teams of two and must work on fairly intimate terms with all others on the crew. In fact, workers commonly say the work simply cannot be done well unless they work well together. The reason for this is that the job is partly an "emotional tone" road show, and the proper tone is kept up in large part by friendly conversation, banter, and joking, as ice cubes, trays, and plastic cups are passed from aisle to aisle to the galley, down to the kitchen, and up again. Indeed, starting with the bus ride to the plane, by bantering back and forth the flight attendant does important relational work: she checks on people's moods, relaxes tension, and warms up ties so that each pair of individuals becomes a team. She also banters to keep herself in the right frame of mind. As one worker put it, "Oh, we banter a lot. It keeps you going. You last longer."

It is not that collective talk determines the mood of the workers. Rather, the reverse is true: the needed mood determines the nature of the workers' talk. To keep the collective mood stripped of any painful feelings, serious talk of death, divorce, politics, and religion is usually avoided. On the other hand, when there is time for it, mutual morale raising is common. As one said: "When one flight attendant is depressed, thinking, 'I'm ugly, what am I doing as a

flight attendant?' other flight attendants, even without quite knowing what they are doing, try to cheer her up. They straighten her collar for her, to get her up and smiling again. I've done it too, and needed it done."

Once established, team solidarity can have two effects. It can improve morale and thus improve service. But it can also become the basis for sharing grudges against the passengers or the company. Perhaps it is the second possibility that trainers meant to avoid when in Recurrent Training they offered examples of "bad" social emotion management. One teacher cautioned her students: "When you're angry with a passenger, don't head for the galley to blow off steam with another flight attendant." In the galley, the second flight attendant, instead of calming the angry worker down, may further rile her up; she may become an accomplice to the aggrieved worker. Then, as the instructor put it, "There'll be *two* of you hot to trot."

The message was, when you're angry, go to a teammate who will calm you down. Support for anger or a sense of grievance—regardless of what inspires it—is bad for service and bad for the company.

SOURCE: Arlie Russell Hochschild, 1983, *The Managed Heart: Commercialization of Human Feeling.* Berkeley: University of California Press, pp. 115–116.

---

manager of a grocery store may check to see that cashiers greet every customer and inquire whether the customers found everything that they needed. Managers may electronically eavesdrop on telephone operators' conversations. Airline reservations managers may check electronic logs of the length of calls received by their reservation clerks.

Managers also check directly with the customers to see if service has been adequate. Airlines, restaurants, and hotels distribute surveys to their customers, inquiring about the adequacy of service or particular areas that require further attention. A customer who places an order for merchandise by telephone or online may be contacted later by a supervisor checking on satisfaction with the interaction.

Managers even set up systems whereby customers may report, favorably or unfavorably, on the behavior of specific employees. In effect, customers are recruited to take on part of the monitoring task normally fulfilled by managers. Because customers also have a desired outcome for the interaction, they may welcome the opportunity to become involved in evaluating the service worker. Casino customers, for example, may try to influence the appearance and behavior of cocktail waitresses (Bayard de Volo, 2003) by demanding that the waitresses wear revealing uniforms. Customer evaluation may take the relatively benign form of inviting customers to nominate a particularly helpful worker for "employee of the month." Sometimes there is an incentive for the customer to report employee infractions; for example, a grocery store customer may receive a free quart of milk if the cashier fails to provide a receipt. The general public is sometimes invited to send in anonymous reports on employee behavior. Trucks, buses, delivery vans, and taxis often bear a message such as "Am I driving safely? Call 1-800-RATONME." By this tactic, managers encourage even passersby to help the company enforce its driving standards.

From management's perspective, who should control the interaction between worker and customer? The answer to this question is equivocal. On the one hand, the customer may not always be right, but the customer should always feel well cared for. On the other hand, the service worker needs to control the interaction sufficiently to achieve smooth and efficient functioning of the enterprise. Thus, a prevalent management strategy is to provide the worker with the training to control the interactions while giving the customer adequate opportunities to complain.

The Worker's Perspective

Workers respond in various ways to the pressures of interacting with the public day after day. A worker's response may vary even from day to day, depending upon the characteristics of the customers, the pace of the workday, and how well the rest of the worker's life is going. But several responses are commonplace.

Manipulating the Interaction   One response of workers is to seek the control of the interaction as completely as possible, with the objective of keeping to a work schedule, making the sale, or perhaps increasing the size of a bill and therefore the size of a tip. Keeping to a schedule is especially important when there are many people to be served, and there is no financial advantage to lingering longer with some customers than with others. Workers who are unsuccessful at manipulating the interaction may find themselves caught between company policy and the demands of clients (Troyer, Mueller, and Osinsky, 2000).

The ability to exercise some control over the timing of interactions is often crucial for members of the helping professions. Many patients would like to spend longer with their physician, asking questions and telling the physician more about their health than the physician feels is necessary. Workers learn to manipulate the interaction so that they maintain greater control. A doctor may arrange to be paged after spending a few minutes with a patient or may arrange the timing of a laboratory test so that it will effectively end the interview. Doctors, counselors, and others who often hear the confidences of customers also face the "therapist's dilemma": after hearing the problems of others all day, they need to express their own emotional reactions. Because the implicit terms of the interaction do not permit them to share their emotional state with clients—this is one aspect of face-work—they must make other arrangements in the round of their workday to regroup their resources before helping another client. Social workers, police officers, and health professionals may find that the time necessary for paperwork, documenting the details of each interaction, provides a respite from the emotion work that is otherwise expected of them.

Manipulating the interaction may also help to increase the size of a sale. Automobile salespeople are taught to channel the conversation with potential car buyers to increase the chance of making a sale. Entertainers may sacrifice their artistic preferences to play "square" music requested by the audience and so increase the likelihood of receiving a tip or a longer engagement. Tip-increasing tactics used by waitpersons include suggesting a round of drinks before the dinner is ordered, inviting the diners to order appetizers, salads, or à la carte items, or suggesting dessert (Butler and Snizek, 1976).

Manipulation makes the interactions less personally satisfying for both the workers and the customers. Many service workers prize their regular customers because they have the opportunity to build a personal relationship with them. A genuine relationship then replaces the struggle to control the relationship.

"Losing It"   Occasionally a customer will be so offensive that the worker will "lose it," or fail to maintain the face-work expected in the interaction. Managers may see this as worker error, but workers often justify it by pointing out that the customer's behavior has violated the role expectation for a customer. One waitress described the following interaction when a customer tried to embarrass her:

> I stood up to her regardless of whether they were customers, regardless of whether I lost my job. Nobody's going to degrade me like that because I'm a waitress. And then she started getting loud. And boisterous . . . I said, "Look," and then I put my book down. I slammed my book down. I put my pen on the table. . . . I said, "Look. If you can do a better job than me, you write your damn order down yourself and I'll bring it back to the kitchen." (Paules, 1991:153)

Workers "lose it" because of what they perceive as indifference or hostility on the part of the customer. Some customers expect subservience (Paules, 1991). Others respond to the routinized interaction by regarding the worker as little more than a machine. For example, waitpersons feel depersonalized being called "Girl!" or "Boy!" The terms "boy" and "girl" are demeaning because they imply that the worker is not yet an adult. These terms were also once used by the white community to reinforce the lower status of African Americans in service positions. Calling the worker by the job title "Waitress!" "Cabbie!" is also depersonalizing. One waitress reported the rejoinder "Don't call me Waitress. I don't call you Customer or Eater" (Howe, 1977:104). "We are people too, and we have feelings" is a common response among service workers.

Burnout Chronic stress in a service job, often as a result of interactions that are too frequent, too repetitive, or too upsetting is known as **burnout**. Service workers of all types can burn out, with the result that many of them leave their jobs and seek a different type of work. Burnout seems most likely to occur as a response to an inability to control the number or content of interactions. After the September 11, 2001, terrorist attacks, many New York City police officers and firefighters eventually reported burnout as a result of the stressful, lengthy efforts to rescue and then to recover the bodies of victims. These efforts continued for many months in conditions that were always depressing and often hazardous.

Burnout means more than just having a bad day at work. Bad days can happen to anyone because of issues at home, an unpleasant commute, or other temporary difficulties. Burnout, however, is a response to a long-term set of difficult job situations that might include overwork or an inability to perform the job successfully because of too many customers or other conditions. Burnout becomes more likely when workers have little control over continuously stressful conditions. Workers may try to avoid burnout by reducing their commitment to the job, perhaps by working fewer hours, seeing fewer customers, or refusing extra duties (Ebaugh, 1988). Another response is a change of scene. Workers may seek refuge in another part of the company, in management positions, or in a related field that they perceive as less stressful. For example, some school teachers respond to the stress of the classroom by seeking positions as administrators in the school system. Administrators still have to deal with the public, but they may have more control over their time and their schedule of interactions.

Many service jobs report high levels of turnover, partly because the work conditions are stressful to workers. The personal costs of turnover are always high for the worker. For the employer, with jobs requiring relatively low levels of training, turnover may be a manageable cost of doing business. Turnover among highly trained and experienced workers, however, entails substantial organization costs.

Finding ways to avoid burnout is an especially important objective in organizations that need to retain highly trained service workers. The flight attendants described in Box 10.2 learned to calm one another down and provide social support to one another when there was a difficult passenger. Service workers can be trained to come to one another's assistance when work becomes overwhelming or too difficult. Organizational devices, such as employee lounges, work groups, and flextime, can give employees alternatives to job stress such as group support or longer periods of time away from the job. Informal groups help workers overcome stressful situations (Lewis, 2005).

The nature of service work affects the everyday experience of many workers, and the number of workers who will be affected in the future is likely to increase. Historically, the emergence of a service society occurred over a long period of time and was influenced by many developmental factors already discussed in earlier chapters. In the next section, we turn to the issue of how so many workers have come to be employed in services.

THE RISE OF THE SERVICE SOCIETY

In 1900, 41 percent of the American labor force was employed in agriculture, with 28.2 percent in manufacturing and 30.8 percent in services. Through the twentieth century, there was a long-term shift of employment from agriculture to manufacturing and finally to services. In 1950 over half of the labor force (52%) was employed in services, with only 12.2 percent of the labor force in agriculture and 35.9 percent in manufacturing (Sullivan, 1989). By 2006 service industries employed almost 78 percent of the labor force (BLS, 2006b). Services are projected to grow both in terms of the numbers of service workers and in terms of the proportion of the labor force that delivers services.

Sectoral Transformation

The shift from agricultural through manufacturing to service work is called the **sectoral transformation of the labor force**. The shift proceeds in two ways. First, positions in agriculture, mining, and manufacturing often go unfilled when workers retire, and other jobs have been eliminated through reorganization and plant closings. Second, new jobs have disproportionately been service jobs. Teenagers and young adults are especially likely to enter the service sector. Many new service jobs are knowledge-based and found in fields such as telecommunications or transportation, but others are found in personal services such as restaurants.

Tertiarization

In the developing countries the process of gradual sectoral transformation is often short-circuited by what is called **tertiarization**, or the development of a service economy without a manufacturing base. The term comes from the tripartite division of the economy into primary (agricultural and extractive), secondary (manufacturing), and tertiary (service) sectors. In the United States and many other advanced countries, services did not come to dominate the labor force until after the rise, maturation, and decline of manufacturing. In the developing countries, by contrast, the labor force has frequently shifted from primarily agricultural employment to an urban, service-oriented economy without the intermediate step of manufacturing. In such countries, manufacturing may be limited to the first-step processing of raw materials (for example, rice mills or sugar processing plants), with most manufactured products imported from abroad.

Developing countries face a difficult task if they seek to employ a large fraction of their labor force in manufacturing. To duplicate the heavy industry of industrially advanced countries—such as steel mills and automobile plants—requires huge capital investments and successful entry into the heavily competitive world market for those products. Protecting their new plants from foreign competition by imposing restrictive tariffs may backfire: their country may lose favorable trading terms for its raw products. Another way to industrialize is to engage in joint ventures with large multinational corporations. This strategy may bring only limited investment, often for intermediate-stage assembly plants that export both their products and their profits to the international corporation's headquarters. Often these are "run-away plants" so that their opening in a developing country follows a plant closing in the United States or elsewhere. (The global economy is discussed in greater detail in Chapter 16.)

Such industrialization efforts are often insufficient to absorb much of the labor force. Tertiarization—the rapid growth of the third, or tertiary (service) sector—is the consequence. Table 10.1 shows the sectoral transformation of employment throughout the world. In nearly every region there

TABLE 10.1 Sectoral Shares in Employment, 1995–2005, with the Percentage of the Employment that is Female, by World Regions

| | Employment in sector as share of total employment | | |
Year	1995	2005	Female %
Agriculture			
World	44.4	40.1	39.9
Developed Economies	5.1	3.7	33.6
Central, Eastern Europe	27.9	22.7	45.2
East Asia	54.4	49.5	47.2
Southeast Asia, Pacific	55.3	43.3	39.4
South Asia	64.1	61.2	33.1
Latin America, Caribbean	23.4	17.1	18.9
Middle East, North Africa	30.8	26.3	25.1
Sub-Saharan Africa	70.1	63.6	44.4
Manufacturing			
World	21.1	21.0	31.2
Developed Economies	28.7	24.8	23.1
Central, Eastern Europe	27.5	27.4	32.6
East Asia	25.9	26.1	40.1
Southeast Asia, Pacific	15.4	20.7	36.3
South Asia	13.4	14.1	26.3
Latin America, Caribbean	20.2	20.3	24.5
Middle East, North Africa	20.3	25.0	17.9
Sub-Saharan Africa	8.2	8.9	26.2
Services			
World	34.5	38.9	45.0
Developed Economies	66.1	71.4	52.7
Central, Eastern Europe	44.6	49.9	52.6
East Asia	19.7	24.4	44.0
Southeast Asia, Pacific	29.3	36.0	47.5
South Asia	22.5	24.6	24.2
Latin America, Caribbean	56.4	62.5	50.0
Middle East, North Africa	48.9	48.7	26.7
Sub-Saharan Africa	21.7	27.5	45.9

SOURCE: Adapted from ILO, Global Employment Trends Brief, January 2006, Table 5.

is a continuing shift from agriculture, but there is rarely a corresponding increase in manufacturing. Instead, service work increases. In Latin America and the Caribbean, for example, services employed 56.4 percent of the workers in 1995 but 62.5 percent of the workers in 2005 (ILO, 2006).

BOX 10.3   Improving Life for the Micro-Vendors of Phnom Penh

Many women in Phnom Penh, Cambodia, scrape a living as micro-vendors, or very small vendors, in open-air markets. Most of them specialize in particular food items and must gather the food, sometimes purchasing it from wholesalers in the middle of the night. They also must pay rent and fees for using small stalls in the market. Fees are charged for water, cleaning, and even for using the latrine. Because most of the women are also mothers, they worry about the care of their children. Sochita is a micro-vendor, and she describes her life.

> I am married, have no education and 7 children: the oldest is 17, the youngest is two. My husband used to be a cyclo driver; now he's a cement porter.... My daughter (Bopha) and I leave for the pond to collect edible flowers and grass at 3:00 a.m. each morning. At 5:00 a.m. we bunch them for the market and then go home and get two more children to help sell. We get to the market around 7:00 and sell for about three hours. Then we return home and I cook, clean, and feed my family. After a short rest, we return to the pond to try to collect more for the children to sell in the afternoon. My economic situation is getting worse because I cannot earn enough money to make a living.... I pay 200 riel for taxes, 200 riel for security, 200 riel for latrines, and 200 riel for

space even though I have no particular space to sell from and just put my basket down wherever I can find room. If I refuse to pay, the collectors make me move. If I don't move, the collectors throw away my produce (Banwell, 2001:33).

This hard life is being changed by the development of a credit cooperative. The Asia Foundation's Women's Economic and Legal Rights team interviewed a number of vendors like Sochita and identified, with their help, a great need for credit at reasonable rates. A group-based savings and credit program was launched with five to twelve members "self-insuring" one another's loans. The interest rate is five percent; one percent of the interest is set aside as an emergency fund for use by members of the groups. The women vote on how the money is loaned. The women are also encouraged to learn how to use banks and how to run small businesses. They have received negotiation training to help them advocate for fair treatment from the market committees and tax collectors, and for market improvements, including a childcare program. The women's debt levels are falling, and their incomes are slowly rising.

SOURCE: Suzanne Stout Banwell. 2001. *Vendors' Voices*. Phnom Penh, Cambodia: The Asia Foundation.

Tertiarization rarely results from a deliberate employment policy. Rather, employment in a low-paid service job is often the last resort of the worker. In the cities of developing countries, many workers labor in the most marginal of service enterprises. Urban service workers are often recent arrivals from the countryside who migrate because of the lack of agricultural land, the push of growing population, or the hope for urban opportunities. Some workers find employment as household servants for better-off urban dwellers (Chaney and Castro, 1989).

Another common aspect of tertiarization is self-employment as petty entrepreneurs. Many large cities of developing countries have cadres of self-employed workers who swarm over cars stopped at red lights, offering services such as wiping the windshield, selling flowers or newspapers,

or even selling single cigarettes from a package. Street vendors serve as indigenous fast-food outlets, offering cooked foods that are not generally available in the home. Box 10.3 describes an effort to assist micro-vendors in Cambodia by giving them access to credit, child care, and other assistance.

The lack of jobs in many large cities in developing countries has encouraged the emigration of people to other countries to find work. Women migrants often seek work as domestic workers or nannies (Constable, 2003). Many of these women have left behind their own families to look after an employer's children, but the wages they send home may represent an important source of income to their families and their home countries (Huang, Yeoh, and Rahman, 2005).

Tertiarization in the developing world does not mean the growth of the knowledge-based and business services; instead, the term implies the expansion of poorly paid service work so tenuous that it is sometimes referred to as casual or informal labor. Marginal service employment is found in the United States as well, where homeless people may scavenge scrap metal or recycle aluminum cans to eke out a living (Snow and Anderson, 1993; Rossi, 1989). In Chapter 14 we will discuss peripheral employment in greater detail. The type of service performed is a key indicator in the welfare of the service workers. In the section below, we describe the different types of service industries.

TYPES OF SERVICE INDUSTRIES

Sociologists identify six types of service industries classified by the type of service produced and by the customer to whom it is delivered (Browning and Singelmann, 1975).

Professional Services

Professional services provide knowledge-based services for individuals or firms who are clients. Examples of professional services are medical services, legal services, and counseling services. Not all workers in these industries are themselves professionals. Nonprofessional workers in legal services and architectural services, for example, are employed as aides, helpers, and assistants to the professional staff. In 2006 about 24.7 million workers were employed in professional, technical, education, and health services (BLS, 2006b).

Professional workers are often able to provide services to only one client at a time. Because it is difficult to gauge in advance how much time each client will need, both professionals and their clients often express frustration with the quality of their interactions. One physician describes this frustration:

> You can spend an entire career trying to help the health and mental health of one patient, but the realities of practice are that you have hundreds of patients and I would make an

appointment with someone for twenty minutes for a problem that should have taken five minutes and an hour later I would be sitting there listening to her complaints. (Ebaugh, 1988:55)

Health care is the largest single industry in the United States. Box 10.4 provides more information about this very large service field. We discuss professionals in more detail in Chapter 11.

Business Services

Business services assist individuals, firms, and organizations in carrying out their economic functions. Examples of business services are temporary help agencies, advertising, financial services, accounting, real estate, and insurance. Over eight million people are employed in the finance, real estate, and insurance industries (BLS, 2006b). Other examples of business services are computer and data-processing services and building support services.

Workplace automation is causing rapid change in many of these industries. The insurance industry, which handles massive amounts of data, is one example of an industry in which automation has changed jobs:

> Premium billing and collection, for example, is often performed through the "turnaround" billing system. After a machine-readable notice and payment are returned by the policyholder, the computer stores the payment data for the accounting department, calculates the agent's commission on each premium, and credits the policyholder's account. These computerized billing systems handle about 5,000 remittances an hour, and the capacity of new equipment is over 40,000 remittances an hour.... In general, clerical occupations which involve a variety of tasks, individual decision-making, and face-to-face communication have been least affected by office automation. Those which require routine and repetitive work have been proven most vulnerable. (Cornfield, 1987:118–119)

BOX 10.4 Health Care Services in the United States

The health care industry is the largest in the country, providing over 13 million jobs for wage and salary workers and another 411 thousand jobs for the self-employed. Workers in this industry provide care and service to people of all ages, from the newborns to the elderly, with care being provided around the clock for the most ill. Hospitals make up only 2 percent of all health care establishments, but they account for forty percent of the health care workers.

As Table A indicates, there are many different occupations represented within health care services, from very highly skilled workers, such as physicians and psychologists, to aides and cleaning workers with short periods of training.

By 2014 around 3.6 million new jobs are expected for this industry, more than in any other industry. Of the twenty fastest-growing occupations, eight are expected to be in health care. As the table below indicates, percentage increases greater than 50 percent are expected for four of these occupations. Because of the rapid growth of this industry, jobs are likely to be stable, and the salaries of the most highly skilled workers are expected to rise.

TABLE A Employment of wage and salary workers in health services by occupation, 2004 and projected change, 2004–14. (Employment in thousands)

Occupation	Employment, 2004		Percent change, 2004–14
	Number	Percent	
Total, all occupations	13,062	100.0	27.3
Management, business, and financial occupations	574	4.4	28.3
Top executives	101	0.8	33.3
Medical and health services managers	175	1.3	26.1
Professional and related occupations	5,657	43.3	27.8
Psychologists	33	0.3	28.1
Counselors	152	1.2	31.8
Social workers	169	1.3	29.3
Health educators	17	0.1	27.0
Social and human service assistants	99	0.8	38.6
Chiropractors	21	0.2	47.8
Dentists	95	0.7	18.5
Dietitians and nutritionists	32	0.2	20.1
Optometrists	18	0.1	29.6
Pharmacists	63	0.5	17.3
Physicians and surgeons	417	3.2	28.7
Physician assistants	53	0.4	54.8

TABLE A Continued

Occupation	Employment, 2004		Percent change, 2004–14
	Number	Percent	
Podiatrists	7	0.1	22.2
Registered nurses	1,988	15.2	30.5
Therapists	358	2.7	32.8
Clinical laboratory technologists and technicians	257	2.0	22.7
Dental hygienists	153	1.2	43.7
Diagnostic related technologists and technicians	269	2.1	26.4
Emergency medical technicians and paramedics	122	0.9	27.8
Health diagnosing and treating practitioner support technicians	226	1.7	18.0
Licensed practical and licensed vocational nurses	586	4.5	14.2
Medical records and health information technicians	134	1.0	30.0
Service occupations	4,152	31.8	33.2
Home health aides	458	3.5	66.4
Nursing aides, orderlies, and attendants	1,230	9.4	22.2
Physical therapist assistants and aides	95	0.7	41.0
Dental assistants	257	2.0	43.6
Medical assistants	361	2.8	53.7
Medical transcriptionists	81	0.6	22.1
Food preparation and serving related occupations	462	3.5	12.6
Building cleaning workers	365	2.8	20.6
Personal and home care aides	312	2.4	60.5
Office and administrative support occupations	2,379	18.2	16.2
Billing and posting clerks and machine operators	179	1.4	10.9
Receptionists and information clerks	353	2.7	31.3
Medical secretaries	347	2.7	17.3

Note: May not add to totals due to omission of occupations with small employment

SOURCE: U.S. Bureau of Labor Statistics, 2006a, consulted at http://www.bls.gov/oco/cg/cgs035.htm

Producer Services

Producer services are necessary to help other industries create their products or services. Utilities are a good example of producer services. Providing electricity, gas, water, sanitation, and communication are important services needed by every business and household. Some producer services, such as municipal garbage collection or water treatment, are provided by the public sector and paid for with user fees and taxes. Telecommunications employs nearly one million workers, with Internet service providers, search portals, and data processing employing another 380 thousand (BLS, 2006b).

Jobs in utilities are often well compensated and may require considerable skill with new technology and equipment. The services are so necessary that even brief disruptions in service will inconvenience thousands of people. The development of complex utilities plants can lead to pressure for workers because even small errors on the job can have major consequences.

> Even changing light bulbs has its dangers in these highly engineered, complex systems [nuclear power plants]. In 1978 a worker changing a light bulb in a control panel at the Rancho Seco I reactor in Clay Station, California, dropped the bulb. It created a short circuit in some sensors and controls. . . . The loss of some sensors meant the operators could not determine the condition of the plant, and there was a rapid cooling of the core. (Perrow, 1984:44)

Distributive Services

Distributive services bring about the geographic dispersion of goods. Goods produced in New Jersey reach a consumer in Oregon because of transportation, warehousing, and retail marketing, each of which is an example of a distributive service. Sales workers are discussed in greater detail in Chapter 13. The distributive-service industries include 4.4 million workers in transportation, 5.8 million workers in wholesale trade, and 15.2 million workers in retail trade (BLS, 2006b). Among

the transportation workers, about 1.4 million workers are in trucking.

> You sit in a truck, your only companionship is your own thoughts. Your truck radio, if you can play it loud enough to hear—you've got the roar of the engine, you've got a transmission with sixteen gears, you're very much occupied. You're fighting to maintain your speed every moment you're in the truck. . . .
>
> You have to get all psyched up and keep your alertness all the time. There's a lot of stomach trouble in this business, tension. Fellas that can't eat anything. Alka-Seltzer and everything. There's a lot of hemorrhoid problems. And there's a lot of left shoulder bursitis, because of the window being open. And there's a loss of hearing because of the roar of the engine. . . . There has been different people I've worked with that I've seen come apart, couldn't handle it anymore. (Terkel, 1974:283–284)

Besides truck drivers, transportation services include airlines, passenger and freight railways, and taxi driving. Box 10.5 describes the work of an urban bus operator.

Social Services

Social services are often financed by the government and benefit not only individuals but the society as a whole. An important subset of these services is protective services: the military, police, prison workers, and other law enforcement officers, and also the firefighters, rescue squads, and emergency response services.

Government employees in general are service workers in the sense that they produce services for the entire society. There are 20.8 million government workers, most of whom are engaged in providing social services. Others provide producer services such as utilities and sanitation. There are 13.3 million local government workers, 4.8 million state workers, and 2.7 million federal employees (BLS, 2006b). Important government services include national

BOX 10.5   Bus Driver in Los Angeles: Lupita Perpez

I'm a "bus operator." They don't like calling us bus drivers; they like to call us "bus operators." I have no idea what the difference is. "You're not bus drivers, you're bus operators." Okay, no problem.

I'm thirty-eight years old, and I've been doing this for three years. I used to work at an elementary school as a teacher's assistant. And I liked that okay, but the money wasn't there, you know? And I was looking for a part-time job for the summer and a neighbor of mine told me, "Hey, the Transit Authority is hiring for part-time. What have you got to lose?" So that's how it all started. And it turned out to be a career! [Laughs]

It's not a bad job at all. It really isn't. I mean like anything, you have your good days, you have your bad days. But it's a good opportunity for me and my family. The benefits. The insurance. I'm a single mom, so I have to think of this stuff, you know? This is just a good opportunity. I'm going to stick it out for the long haul and take advantage of this good retirement. When I hit sixty, it's gonna be beautiful. . . .

The route I prefer—everybody thinks I'm nuts—is the 81. The 81 is the Figueroa, and it goes from Eagle Rock Plaza all the way to 117th and Imperial—the middle of Watts. You go through some interesting parts on the 81. It's one of the busiest lines in the country. I like the ones that are very heavy and busy. They're always on the go. Always. And before you

know it, your day is done. I hate the routes that drag— that you hardly pick up anybody, and you got to go really slow—and it's so boring! I don't like those. Not at all.

I really like the people on the 81, too. Because let's face it, the 81, basically we pick up the poor people. The 81 is basically what they consider the low-class passengers. But for being low class, they pay. And they don't give me any hassles. And they always say, "Hi, how are you?" or "Good afternoon," or "Good morning.". . . .

Of course, the biggest thing is learning about how to deal with so many different types of people. You deal with different nationalities. Different kinds of everything. You gotta know the rules. In this part of town, they treat me good. Maybe because they see me as the same nationality. But, like, when I go to South Central, umm, it's okay for them to treat me like garbage. That's the way it is. And then like when I go into San Marino or Beverly Hills, I get treated different, and I have to treat them different, because hey, people act different. And every day is something new. Fortunately, for me, I enjoy that aspect. Most of the time, I love these people. They make the job for me.

SOURCE: John Bowe, Marisa Bowe, and Sabin Streeter, eds. *Gig: Americans Talk about Their Jobs.* New York: Crown Publishers, 2000. pp. 151–153. Reprinted by permission of Random House, Inc.

security and international affairs, justice, public order and safety, postal service, and human resource programs. In addition, nearly 1.8 million people are on active duty as members of the armed forces.

Besides government-provided social services, about three million private-sector workers are employed in social services, including child day care services and residential care. A private nursing home attendant reports her round of daily work activities:

My baby here has cerebral thrombosis. She is ninety-three years old. I get in this morning about eight-thirty. I shake her, make sure that she was okay. I took her tray, wipe her face, and give her cereal and a cup of orange juice and an egg. She's unable to chew hard foods. You have to give her liquids through a syringe. . . .

The first thing in the morning, after breakfast, I sponge her and I give her a back rub. And I keep her clean. She's supposed to be turned every two hours. If we don't turn her every two hours, she will have sores. Even though she's asleep, she's got to be turned.

I give her lunch. The trays come up at twelve thirty. I feed her just the same as what I feed her in the morning. In the evening I go to the kitchen and pick up her tray at four o'clock and I do the same thing again. About five thirty I leave here and go home. (Terkel, 1974:651)

Police work is a major form of public employment in the social services, and private security personnel are a fast-growing area in the private sector. The police receive substantial training in their work. This training has increased in recent decades, as have the efforts of police departments to provide continuing education for their employees. Police officers have substantial autonomy and discretion in deciding how to do their work in the field. Their autonomy results both from their work being spatially dispersed, and therefore difficult to supervise, and also from the fact that they have the training necessary to make quick judgments in a crisis. Private security personnel vary widely in terms of their training and duties. Both police and private security forces encounter danger, many rules prescribing appropriate behavior and proscribing inappropriate behavior, and often harsh discipline for infractions of rules. Because these characteristics are at odds with traditional roles for women, women officers and guards may encounter special problems (Appier, 1998; Britton, 1997). A Toronto alarm seller describes her experience:

> I've had weird cases, where I will go to the home and the male, he'll say, "Oh, I can't believe we've got a female doing this job," which really bothers me.... They don't think females understand the technical side of things. I do.... (Erickson, Albanese, and Drakulic, 2000:315)

As a consequence of the mixed professional and nonprofessional nature of police work, employees have sought to increase their power and prestige through both professional associations and collective bargaining. They have succeeded increasingly in both endeavors, and the status and rewards of police work have risen accordingly.

Personal Services

Personal services are produced principally for a family or for individuals and include restaurants, hotels and motels, entertainment, tourism, and repair services. A subset of personal services are the fewer than one million private household workers—employees of individual households who clean, cook, garden, and care for dependent family members. Over 13.7 million workers are employed in leisure and hospitality, including entertainment and performing arts, accommodations such as hotels

and motels, and food services and drinking places. Laundries, beauty shops, and barber shops are other examples of personal services (BLS, 2006b).

The defining characteristic of personal service work is that the worker delivers a service directly to the final consumer. This service may be operating a movie projector, carrying luggage to a hotel room, or sewing alterations for a suit. Service workers are often expected to address their customers as "Sir" or "Madam" or by their title and last name, but customers usually address the service workers in more familiar terms or by their first names.

> A barber, he has to talk about everything, baseball, football, basketball, anything that comes along. Religion and politics most barbers stay away from....
>
> Usually I do not disagree with a customer. If there is something that he wants me to agree with him, I just avoid the question. (Laughs.) This is about a candidate, and the man he's speaking for is the man you're not for and he asks you, "What do you think?" I usually have a catch on that. I don't let him know what I am, what party I'm with. The way he talks, I can figure out what party he's from, so I kind of stay neutral. That's the best way, stay neutral. Don't let him know what party you're from cause you might mention the party that he's against. And that's gonna hurt business....
>
> When I leave the shop, I consider myself not a barber any more. I never think about it. When a man asks me what I do for a living, I usually try to avoid that question. I figure that it's none of his business. There are people who think a barber is just a barber, a nobody. If I had a son, I'd want him to be more than just a barber. (Terkel, 1974:315–317)

Some service jobs carry a stigma (Saunders, 1981). Like the barber, workers in these occupations find little to brag about concerning their jobs. When asked what they do for a living, their response may be guarded.

Janitorial work is an example of an occupation that is stereotyped as low-status, dirty work. Janitors, however, often play important roles, protecting their buildings from illegal entry, fires, and other hazards. In addition, they are free from direct supervision for most or all of the day. These job characteristics give many janitors a basis for taking pride in their work:

> I make a pretty good buck. I figure if I do my work and do it honestly, I should be entitled to whatever I make. For high-rise buildings, head man makes a thousand dollars a month and his apartment. You never heard of that stuff before. I've turned down high rises by the dozens. I can make more money on the side on walkup buildings. . . .
>
> Today I can walk in the boiler room with clean trousers and go home with clean trousers. You check the glass, you're all set. That's the first thing you do. I check my fires and bring my garbage down right away. I take one of those big barrels on my back, and I bring it up the flight of stairs and back down. I do this on three buildings and two have chutes. . . .
>
> I enjoy my work. You meet people; you're out with the public. I have no boss standing over me. People call me Mr. Hoellen. Very respectable. If I'm a good friend, they say Eric. I'm proud of my job. I've made it what it is today. Up in the morning, get the work done, back home. Open the fires and close 'em. (Terkel, 1974:119–125)

Such positive aspects of janitorial work are counterbalanced by less desirable duties, such as handling garbage and cleaning rest rooms. Some janitors are also expected to be on call for emergencies seven days a week.

COMPENSATION IN SERVICES

Working conditions of service providers vary a great deal. The *high-paying service industries* include transportation, communication, public utilities, wholesale trade, finance, insurance, and real estate, professional and related services, and public administration. The *low-paying service industries* include retail trade, repair

TABLE 10.2 Average Hourly Earnings by Industry Sector, United States, 2006[a]

All goods-producing workers	$18.02
All service workers	$16.38
High Paying Services	
Utilities	$27.00
Information	$23.00
Professional and business services	$18.84
Wholesale trade	$18.73
Financial activities	$18.57
Education and health services	$17.27
Transportation and warehousing	$17.17
Low Paying Services	
Retail trade	$12.61
Leisure and hospitality	$ 9.49

SOURCE: U.S. Bureau of Labor Statistics, 2006b, Table B-3.
[a]Private sector, seasonally adjusted hourly earnings for July, 2006; data collected from payrolls.

services, personal services, and entertainment and recreation services (BLS, 2006b).

Table 10.2 shows the wages and earnings by broad industry group for goods-producing and for the two tiers of service-producing industries. Not every job in the high-paying service-producing industries will pay a high salary, just as not every job in the low-paying service-producing industries will necessarily pay low wages, but the average wages in the table show a pattern of differences. Table 10.2 indicates goods-producing jobs paid more than services, and the low-paying service sector paid even less.

A variety of circumstances explains the differences in the compensation for the different types of service workers. As we have already mentioned, *productivity* tends to be related to higher pay. Telecommunications is an example of an industry in which electronic technology has made workers more productive, and this is a higher-paying service industry.

The nature of the *recipient* is also an issue in how much payment the service provider can demand. In higher-paid fields the worker is often described as working with a *client*; in the lower-paying service fields, the worker is more often described as working with a *customer* (Rothman, 1988). Corporations or businesses are often preferred as clients over individuals because they are better able than individuals to pay high fees. Businesses may also place large orders at one time, making sales to businesses both efficient and lucrative.

The *nature of payment* is a third issue. A waitperson cannot be rude to a customer without jeopardizing the tip, which represents a large proportion of total earnings. A sales worker on commission is under even greater pressure not to ignore any potential customer. A salaried service worker, by contrast, may have more leeway to be rude to customers.

THE FUTURE OF SERVICE WORK

Although service provision may change in some respects—for example, there may be greater application of electronic technology in some service—services will continue to provide employment to the great majority of workers in the advanced economies, and probably in the developing economies as well.

Moreover, most of us encounter service workers regularly in our everyday lives. Especially in large,

urban areas most people will have many transitory encounters with service workers providing many different types of service. Whether these encounters are satisfying or unpleasant will constitute an important part of everyday existence. The concern about the productivity of the service sector has led to greater interest in routinizing and controlling interactions, leading in some large enterprises (such as urban mass transit) to "people processing," an application of assembly-line techniques to *customers* instead of to products. These encounters may leave the consumers feeling that they have never received any human attention. These encounters may also prevent service workers from experiencing the satisfactions of assisting others.

In the next four chapters, we will discuss specific nonmanufacturing occupations in which people work. Chapter 11 will discuss the professions and related occupations. Most of these workers are employed in the high-paying service industries, and their jobs rely on the application of knowledge to serve clients. Chapter 12 discusses managers. Chapter 13 discusses administrative support and sales workers. In Chapter 14 we consider some of the occupations typical of low-paying service industries, including private household workers and personal service workers.

SUMMARY

Workers worldwide, including North America, are increasingly employed in the service industries. Service work differs from goods production in that the service production is time-bound. Services cannot be stockpiled, and the production of the service often occurs nearly simultaneously with the consumption of the service. For services that can be delivered by telecommunications, however, outsourcing offers one way to provide 24–7 customer service.

Service work is often characterized by low productivity, and one response to the low productivity has been routinizing the interactions between the customer and the service worker. Although this routinization has advantages, it also makes the encounters more impersonal and may lead to dis-

satisfaction by the customer and to burnout among the workers.

The process of tertiarization in some developing societies represents a growth of the low-paying service industries without a manufacturing base. The resulting service jobs are often poorly paid and insecure, such as household domestic workers and micro-vendors. In industrially advanced countries, the service industries are more heterogeneous, serving both individuals and organizations with a variety of professional, business, producer, distributive, social, and personal services. Low-paying services have also grown in the advanced industrial countries, but they are also more likely to have high-paying services in knowledge-based fields.

KEY CONCEPTS

services

time-boundedness

outsourcing

routinization

face-work

emotion work

burnout

sectoral transformation
of the labor force

tertiarization

professional services

business services

producer services

distributive services

social services

personal services

QUESTIONS FOR THOUGHT

1. To what extent is productivity the critical variable in the sectoral transformation of the labor force? Is your answer true only for advanced economies, or does it apply to developing countries as well?

2. What factors bring about the different rates of pay in the high-paying service industries and the low-paying service industries?

3. Think of a business or store that you patronize and that you would characterize as having good service. What are the characteristics that lead you to describe the service as "good"? Are these characteristics widespread in the service field?

4. Think of an interaction that you have had recently with a service worker or with a cus-

tomer (if you work in services). Who controlled the interaction, and how? Did you play your role as the other person expected? Was emotion work required?

5. The founders of Federal Express offered a new service to the marketplace. Imagine that you are an entrepreneur. What new service could you provide to the economy? How would you classify this service—as a professional, business, producer, distributive, social, or personal service?

6. Suppose that you manage a service industry. What steps could you take to ensure that customers receive satisfactory service? What steps could you take to prevent burnout among the employees?

MULTIMEDIA RESOURCES

Print

Stephen Coscia. 1998. *Tele-Stress: Relief for Call Center Stress Syndrome.* 2nd edition. New York: CMP Books. This book, written as guidance for call center workers, describes the stressful situations that characterize customer service.

Gary Alan Fine. 1996. *Kitchens: The Culture of Restaurant Work.* Berkeley: University of California Press. This is an analysis of various restaurant occupations

together with a consideration of the role of restaurants in the larger society.

Arlie Russell Hochschild. 1983. *The Managed Heart: Commercialization of Human Feeling.* Berkeley: University of California Press. This study of airline attendants and bill collectors popularized the concept "emotion work" among service workers. She argues that an occupational hazard for such workers is becoming estranged from their own emotions.

Alison Owings. 2002. *Hey Waitress! The USA from the Other Side of the Tray*. Berkeley: University of California Press. Through the words of waitresses, this book explores this service job in terms of the control of the waitress-diner interaction.

Internet

The Newspaper Guild–CWA. *http://www.newsguild.org/* Journalism represents an important service in democracies. This is the website of the Newspaper Guild, a resource for media workers affiliated with the Communication Workers of America and a source of information about current developments within journalism.

U.S. Small Business Administration *http://www.sba.gov/starting_business/employees/finding.html* This is a site for small business owners who need to hire workers, including temporary workers. The larger site of which this is a page was developed by the Small Business Administration to provide would-be entrepreneurs with basic information about starting a business.

RECOMMENDED FILM

Nalini by Day, Nancy by Night (2005). This documentary by Sonali Gulati explores globalization as exemplified by call centers outsourced from the U.S. to India. This film also shows how the call centers are affecting India culturally and economically. 27 minutes. Women Make Movies www.wmm.com

PART IV

Occupations and Professions

Workers who perform the same tasks belong to the same occupation. As you learned earlier, occupation refers to what a worker does. In Part IV we examine occupational groups. The chapters in this part emphasize how occupational groups vary in their worker characteristics, job characteristics, and collective power.

Worker characteristics include how many workers perform specific tasks, the education or skills they possess, and their ascriptive characteristics (for example, gender or ethnicity). *Job characteristics* include how much judgment and autonomy are associated with each job, how frequently the job is a rung on a long-term career ladder, how secure the job is, and how well it is compensated. *Collective power* refers to the control that members of an occupation exercise over the selection and training of new members, the definition of high-quality work, and compensation. Occupations vary greatly in how much control their members exercise over employers, clients, their working conditions, members of other occupations, and one another.

Just as the industrial composition of the developed countries has shifted from agriculture to manufacturing and then to services, so the occupational composition of the labor force has changed. Table A shows that farming, fishing, and forestry occupations, which accounted for less than 1 percent of employment in 2004, are projected to decline further as a proportion of the labor force. The proportion of professional and related occupations, already the largest group in 2004, is projected to experience the largest growth between 2004 and 2014. In the following chapters, we will examine the dynamics that cause these shifts.

Some occupations are widely dispersed among industries, and in the following chapters we will discuss such occupations. Chapter 11 examines professionals, a group of occupations that are distinctive in worker characteristics, job

T A B L E A **Employment by major occupational group, 2004 and projected 2014**
[Numbers in thousands]

Occupation group title	Employment				Change	
	Number		Percent distribution		Number	Percent
	2004	2014	2004	2014		
Total, all occupations	145,612	164,540	100	100	18,928	13
Management, business, and financial occupations	14,987	17,142	10.3	10.4	2,155	14.4
Professional and related occupations	28,544	34,590	19.6	21	6,046	21.2
Service occupations	27,673	32,930	19	20	5,257	19
Sales and related occupations	15,330	16,806	10.5	10.2	1,476	9.6
Office and administrative support occupations	23,907	25,287	16.4	15.4	1,380	5.8
Farming, fishing, and forestry occupations	1,026	1,013	0.7	0.6	−13	−1.3
Construction and extraction occupations	7,738	8,669	5.3	5.3	931	12
Installation, maintenance, and repair occupations	5,747	6,404	3.9	3.9	657	11.4
Production occupations	10,562	10,483	7.3	6.4	−79	−0.7
Transportation and material moving occupations	10,098	11,214	6.9	6.8	1,116	11.1

NOTE: Detail may not equal total or 100 percent due to rounding.
SOURCE: Bureau of Labor Statistics, 2005. http://www.bls.gov/emp/emptab1.htm

characteristics, and collective power. Professionals are well educated, tend to work in jobs that require judgment and autonomy, and exercise control over the specialized knowledge of their profession, the other members of their occupational group, and other occupational groups.

Chapter 12 examines managers, a category that includes a large number of occupational specialties found in every industrial setting, because every industry needs at least some specialized workers who direct and oversee the production of goods and services. The restructuring of many corporations has focused attention on how many such workers are needed.

Chapter 13 discusses administrative support workers, such as clerical workers and customer service workers, who are employed in many industries to manage information and to help market the product or service. Chapter 14 examines workers in marginal jobs. Some of these workers are located in declining occupational specialties such as farm work and household work. Others are located in low-paid jobs, often in small firms, or they are temporary workers. We will discuss ways in which workers may find themselves in peripheral parts of the occupational structure.

The following four chapters do not include occupational groups whose occupations are closely tied to specific industries, such as the manufacturing and service industries. Such occupational groups were already considered in Chapters 8 through 10. The chapters in Part IV do discuss many occupational specialties within the broad occupational categories, however. In Appendix Table 1 you will find a listing of these specialties with the proportion of the workers who are women or members of minority groups.

11

Professions and Professionals

[The professional complex] has already become the most important
single component in the structure of modern societies. It has
displaced first the "state,"...and, more recently, the "capitalistic"
organization of the economy. The massive emergence of the professional complex...is the crucial structural development in
twentieth-century society.

(PARSONS, 1968:545)

I propose that we name the mid-twentieth century. The Age of
Disabling Professions, an age when people had "problems," experts
had "solutions" and scientists measured imponderables such as "abilities" and "needs."...It will be remembered as the age of schooling,
when people for one-third of their lives had their learning needs
prescribed and were trained how to accumulate further needs, and for
the other two-thirds became clients of prestigious pushers who
managed their habits.

(ILLICH, 1977:11–13)

As these quotations indicate, social scientists agree that the professions are
important, but they offer different reasons for the professions' importance.
In this chapter we will explore the characteristics of professions and how
professions are changing. Almost everyone recognizes a few occupations as
being professions. Other occupations have some characteristics of a profession,
and still others seek collectively to acquire more of the characteristics of a
profession. Occupations can be ordered along a continuum from more professional to less professional, and in this chapter we will identify some of the social
forces that push occupations toward one end of the continuum. In particular,
we will examine the semiprofessions and the paraprofessions as groups that
aspire to professional status.

HOW SOCIOLOGISTS RECOGNIZE PROFESSIONS

People often use the word *profession* when sociologists would use the terms *occupation* or *career*. (You can refresh your memory of these terms by consulting the glossary.) Likewise, the term *professional* is used in many different ways. For example, someone may say that a job "was really professional" to indicate that it was well done or refer to "professional athletes" to distinguish them from amateurs. Sociologists, however, assign a specific meaning to *profession*. A **profession** is a high-status, knowledge-based occupation characterized by (l) abstract, specialized knowledge, (2) autonomy, (3) authority over clients and subordinate occupational groups, and (4) a certain degree of altruism. We will refer to these four characteristics as the hallmarks of a profession. A professional is a person who is qualified and legally entitled to pursue a profession (Hughes, 1965:2).

Thus, we can see how the popular usage parallels the sociological definition. A profession is a certain type of occupation. A professional is an expert, and we usually associate expertise with a job well done. Finally, a professional is somewhat like the professional athlete, who is distinguished from the amateur both by higher status and by higher pay. To understand this chapter, it is important to remember that we will always use the terms *profession* and *professional* in the sociological sense of these characteristics. University professors, for example, are professionals who devote their work careers to becoming experts in a particular area, such as sociology. Although others may read sociology books in their spare time, the sociology professor is paid to study society and to teach students about it.

The characteristics we have just presented do not perfectly describe many occupational specialties. Most experts agree that law, medicine, and the ministry possess the characteristics of a profession. Other occupations, including military officer, scientist (Ben-David, 1984), and university professor (Freidson, 1986:15), are widely accepted as fitting the definition of a profession. The members of other occupational specialties may self-consciously seek to elevate the status of their occupation by adopting the characteristics of professionals. **Professionalization** is the process by which an occupational specialty seeks to emulate a profession by demonstrating the four hallmarks of a profession. For now, we will discuss the four hallmarks. We discuss the trend toward professionalization later in the chapter.

Identifying professions by a set of hallmarks is also called the structural-functional approach, the traits approach, or the characteristics approach. In this chapter we will refer to it as the hallmarks approach. Not all sociologists agree that this approach is the best way to understand the professions. Instead, they contend, the professions are merely the *powerful* occupations that are currently winning in the constant struggle among occupations to control preferred types of work. Therefore, the problem is not *which* occupations are recognized as professions but, rather, the *process* by which they gained their recognition. Over time, people may come to change their perceptions about which occupations are powerful (Haber, 1991). In a preindustrial society, for example, the most powerful occupations might have been hunter and shaman, whose members controlled, or were believed to control, the food supply. Hallmark theorists would reply that hunter and shaman were the occupations with the greatest claim to the four hallmarks. We will first present the four hallmarks and then evaluate them in a later section using different approaches.

Abstract, Specialized Knowledge

Our definition states that professions are knowledge-based occupations. But every occupation has its lore—a body of knowledge that its members master. What distinguishes the professions is the *type* of knowledge they master. All societies have both common knowledge (generally known by many) and esoteric knowledge (known by only a few). The traffic laws are common knowledge in most contemporary societies, whereas specific treatments for cancer are more esoteric knowledge, usually known principally by physicians. Not all esoteric knowledge is equally valuable, and people and societies vary in the value accorded different knowledge. For example, astrology is still considered an important body of knowledge in some countries, but in the advanced industrial countries astrology is commonly viewed as merely entertaining. In general, the esoteric knowledge commanded by professionals is considered important, even a matter of life or death, for the well-being of individuals or groups.

In preindustrial societies with a meager accumulation of knowledge, the "professionals" of a tribe or village were sages or shamans, who knew the important lore concerning health practices, weather, the spirit world, genealogies, and the history of the group. It was possible for a single person in every village to know all of these things, and that person was consulted by the entire community. Although others might have aspired to such knowledge, they were often denied it. This specialization met a simple survival need because most members of the tribe were needed to provide food. But the sage also gained an advantage in keeping the information mysterious and inaccessible. Mystification added much to the prestige of the village sage. In addition, the sage had greater discretion to choose his or her successor if everyone believed that the knowledge was important, hard to learn, and entrusted to only a few worthy followers. The selection of apprentices and the teaching of esoteric knowledge were accompanied by rituals, which underscored the importance of the sage and of the knowledge. Later on, monasteries and other formal religious institutions preserved knowledge, maintaining the link between knowledge and religion.

In an industrial society the number and variety of such experts, or professionals, has greatly expanded, and their knowledge has been secularized. As you saw in Chapters 7 and 8, increased productivity in agriculture and manufacturing has shifted the opportunity structure so that more workers can enter the knowledge-based occupations. In addition, the base of human knowledge has exploded and continues to grow exponentially.

Not all of this knowledge will be considered central or important, but most of it can be considered esoteric in the sense that only a few specialists will be aware of it. The very volume of knowledge means that those who transmit knowledge must become increasingly specialized and that advanced industrial societies will require more numerous and more varied knowledge-based workers. Knowledge-based fields have fueled much of the growth of the service sector, especially that part called *professional services* (for example, educational, legal, and medical services). Some authors define the postindustrial society as one that is dominated by professional experts (Bell, 1976).

Knowledge Base The knowledge base of a profession consists of three parts. The first part is *theoretical knowledge*, which may seem rather far removed from the day-to-day activities of the professional. This knowledge is often acquired in college. For example, physicians must have basic theoretical knowledge in biochemistry, mathematics, and physics, which they learn before medical school. Although such bodies of knowledge keep expanding, many physicians keep up with them only in limited ways. Physical and biological scientists, on the other hand, actively maintain and extend this abstract knowledge.

The second part of the knowledge base is *detailed, practical information* that can be applied in serving a client. Physicians, for example, know specific information about diseases and treatments. This part of the knowledge base also expands quickly, but the professional must stay abreast of these developments to provide the best service to clients. One advantage of specialization—for example, in cardiology or pediatric cardiology—is that the professional must keep pace with new information only in a relatively specialized area. Bar associations, medical associations, and other professional organizations frequently require their members to update their practical information in annual continuing education.

The third part of the knowledge base, *technique*, is the application of the knowledge base (Ellul, 1964). Knowing that something must be done is not enough; one must also know how to do it. A physician may know that a certain condition will respond to intravenous medication, but he or she must also be able to inject the medication into the vein. Techniques are learned in an applied or clinical portion of a professional training program. The later years of medical school include clinical rounds during which students learn techniques and skills in many fields of medicine. Technique may also be learned or perfected during an apprenticeship to a more experienced professional. Internships and residencies are opportunities for new physicians to learn techniques. The internship or residency after basic medical training is devoted to learning techniques specific to an area of specialization—for example, orthopedic surgical procedures.

Both the professional associations and the professional schools seek to expand and refine the profession's knowledge base (Hoffman, 1989). The associations may lobby for public funding of research to be conducted by the faculty of the professional schools. The results of the research—new knowledge and techniques—reach the members through professional journals. Professional associations or professional schools publish these journals, available to members through subscriptions and specialized libraries. Formal continuing education is provided through conferences, video and audio cassettes, compact disks, Web-based systems, and electronic mail.

Professionals regard some specialties as "hot" or "exciting" if many discoveries are being made in them. Their intellectual excitement may draw disproportionate numbers of the new professionals to these fields. The very proliferation of knowledge encourages further division of professional knowledge into subspecialties to enable workers to keep up with the volume of knowledge. Specialization encourages a further division of labor among professionals just as it does in other occupational groups.

Professional Culture   Every profession has its characteristic jargon, behaviors, and lifestyles—all elements of the **professional culture**. (More technically, it should be thought of as a subculture within the national culture.) Professional schools convey not only knowledge but also the norms, values, and lifestyles of the profession. The professional school faculty and practicing professionals become role models who demonstrate how to dress and how to interact with clients and peers. Some of this preparation is explicit, such as the requirement that students complete courses in professional ethics. Other information is shared informally. Students learn from the faculty to accord prestige to the visible writers and researchers in the profession, and they learn which specialties have high prestige and which do not. Many law students are initially surprised to learn that criminal law has relatively low status compared with corporate law (Heinz and Laumann, 1982), and some students change their goals accordingly. Learning the professional culture helps neophytes blend in with more experienced professionals.

Autonomy

Autonomy means that professionals rely on their own judgment in selecting the relevant knowledge or the appropriate technique for dealing with the problem at hand. Professionals justify their autonomy by their mastery of the knowledge base. Lay people often accept this autonomy because they assume that professional training is necessary to make decisions. Professional standards limit autonomy to some degree. For example, a physician can use experimental drugs or treatments only within certain well-defined limits but is free to choose among accepted therapies (Freidson, 1970).

Autonomy in decision making generates issues of accountability. The client is sometimes unable to judge whether the professional is providing the correct remedy, and the client must trust either the standards of the profession itself or make judgments based on the personal bearing of the professional—for example, the bedside manner of the physician. When knowledge is important, highly specialized, and inaccessible, its misuse is an important concern to professional and layperson alike. If a physician or a lawyer makes a mistake, the results for the client may be death or severe punishment. Yet clients cannot fully protect themselves against such mistakes, because they do not have the knowledge to judge the medical or legal services they have received. Even a client who seeks the opinion of a second professional will have little more basis for judging the value of the second opinion.

The professions respond to the accountability issue by claiming that they police their own membership in the interests of protecting the public. We discuss this issue further in the section on altruism. For professionals who practice in an organizational setting, such as doctors in a hospital, bureaucratic rules attempt to establish accountability to the organization and, presumably, to the clients. These bureaucratic rules may also infringe on the autonomy of the professional, a topic we will discuss later in this chapter.

Authority

Authority, the third hallmark of a profession, means that professionals can expect compliance with their orders from clients and subordinate occupational groups. Authority is derived from mastery of the body of specialized knowledge.

Authority and Clients   Authority is usually understood to be part of the relationship between a professional and a client. In the doctor-patient relationship, for example, the doctor is usually assumed to have the authority to advise the patient on the proper treatment of illnesses and to prescribe medication. The professional relationship also justifies behaviors that would not otherwise be allowed. For example, the doctor is permitted to view and manipulate the unclothed body. The patient is expected not only to comply but also to trust the professional, even revealing personal secrets (Hughes, 1965:3). A patient with a sexually transmitted disease, for example, will be asked questions about intimate relationships. The expectation that patients will reveal secrets reaches its height in the specialties of psychiatry and clinical psychology.

Professionals seek to expand the scope of their authority over clients. By redefining many aspects of life as health-related or by redefining problems (such as alcoholism) as "medical" problems, physicians claim greater authority over patients and over some other occupational groups. As physicians have become more concerned with maintaining health and preventing disease, they have increasingly offered their patients advice on diet, exercise, smoking, child rearing, family relationships, and many other topics. The doctor-patient relationship implies the doctor's authority to expect compliance with the advice.

Lawyers have defined many transactions of everyday life to contain legal pitfalls or the potential for lawsuits, and they encourage legal audits (analogous to the routine physical health examination) to prevent lawsuits or liability. Other professions seek to increase their authority by claiming dominion over additional areas of life, asserting that people need additional expertise. Members of the clergy propound moral principles that affect many aspects of life, offer counseling to members of their congregations, and explicitly advise exhibiting or avoiding some behaviors. Various professions now compete with expert advice upon such issues as child rearing, care of elderly family members, investing money, and so on.

Just because professionals assert authority does not mean that their clients will accept their authority. Clients may ignore advice on issues they consider to be beyond the professional's domain. Patients may take a prescribed medication but refuse to stop smoking. Professionals consider "compliance," or getting clients to follow professional advice, a major problem.

Authority over Other Occupations   Most professions interact with various other occupational groups, some of which are subordinate to them. In some cases the subordinate worker is an employee of the professional, such as a dental hygienist employed by a dentist. In other cases, however, the subordinate nature of some occupations is established by a larger organization. In a hospital, for example, physicians have some authority over registered nurses, therapists, dietitians, pharmacists, and other health occupations. The physician leaves orders or prescriptions that these workers fulfill. Registered nurses, in turn, have autonomy in certain nursing duties. Frequently, for instance, only nurses are permitted to give particular medications. They also have some authority over other occupational groups, including licensed practical nurses, vocational nurses, nurse's aides, and orderlies.

The dominant profession often delegates what is considered its dirty work to subordinate occupations. Physicians, for example, delegate routine patient care, such as feeding, bathing, and bedpan handling, to the nursing staff. Attorneys delegate routine legal work, such as the completion of standard forms or letters, to paralegals or to legal secretaries. The rationale is that delegation frees the professionals for the more highly skilled work for which they alone were trained. The delegation entails certain risks for the professional. The professional directs the work, at least in theory, and takes responsibility for it (Cassell, 1991). If a paralegal makes a mistake, it is ultimately the lawyer who is responsible to the client. Another risk, however, is that if a specific profession limits the work it will do—or charges a high price for that work—other professions or occupations will take up the work that was abandoned, charge lower prices, and eventually become competitors for the most desired work (Abbott, 1988).

Where these relationships of authority and delegation are accepted, there may be relatively little conflict among the groups. One study comparing two hospitals found less friction among the various occupations than might have been expected, although there was friction *within* some groups (such as nurses) based on their differing conceptions of what the nurse's relationship to the doctor should be (Guy, 1985). Nurses were more likely to disagree among themselves over the proper way to respond to a doctor whose orders were vague than they were to disagree openly with the physicians.

When two or more professions are expanding their spheres of authority into new areas, there are often struggles for domination. If there is no clearly dominant professional group, a struggle may ensue for the authority to claim expertise. In the United States medical doctors historically sought to establish their profession and prevent the recognition of other health practitioners. The medical doctors fought in state legislatures to prevent the licensure and acceptance of osteopaths, chiropractors, herbalists, and lay midwives (Gevitz, 1982). Lawyers have struggled to maintain their professional authority against what they see as infringements of credit counselors (in bankruptcy proceedings), real estate agents (in property transactions), and nonprofit counseling services (in uncontested divorces). *Professional associations,* which are organizations consisting of the members of a specific profession, help maintain the authority of the profession by seeking laws that establish licensing and prohibit practicing the profession without a license. Legal sanctions reinforce the profession's claims to authority and prevent other occupations from competing in their area of expertise.

When professionals declare that more areas of human life need their expertise, new "turf" is created that is not squarely within the province of the existing professions. Nutritionists, dietitians, physicians, and university scientists all offer advice on nutrition. Although charges of "charlatan" and "quack" are sometimes heard among the disputants, there is as yet no clear public acceptance of whose authority should prevail. For many clients the advice given by a nutritionist does not yet carry the weight of advice given by a physician. Other clients, however, may doubt the physician's competence in nutrition. In emerging areas the struggle among occupations for authority can be observed (Zhou, 1993; Lawrence, 2004). Box 11.1 describes the struggle for recognition by nurse midwives.

Altruism

Altruism means concern for others. No one doubts that professionals seek an income from their practice, but the hallmark of altruism implies that they officially see themselves as having additional objectives. Most professions have codes of ethics that express the ideal relationship among the professional, the client, and the community. Altruism

BOX 11.1   Midwives and the Struggle for Professional Recognition

In most of the world and in most time periods, women in birth have been attended by midwives. This process began to change as physicians—who were overwhelmingly male—asserted authority over the birth process. A medical specialty, obstetrics, was devoted to the treatment of pregnancy and childbirth, and expectant mothers were admitted to hospitals to give birth. Throughout the industrialized countries, the redefinition of pregnancy and childbirth as a medical "problem" meant that midwives were displaced by physicians (Patterson, 2004; Jordan, 1992). In some cases, midwives were prosecuted for practicing medicine without a license.

In the 1960s, midwifery began to reappear, in part because pregnant women wanted less medical intervention in birth. Some women wanted to give birth at home instead of in a hospital, and they argued that since birth is a normal process, and most births are medically normal, they should be allowed to choose an attendant other than a physician. All of the hallmarks of a profession came into play in the resulting jurisdictional dispute about who should attend births—did the midwives have sufficient knowledge? Should they work under the supervision of a physician? Was home birth an appropriate risk for the midwife to undertake?

Today there are two main types of midwives (Davis-Floyd, 1998). Direct-entry midwives are educated in a program that prepares them to assist childbirth, but they do not have to be nurses. Certified Professional Midwives have met the standards of the North American Registry of Midwives. Licensed midwives are licensed by the state, and about half of the U.S. states currently provide such licensing. Nurse midwives, by contrast, are advanced-practice nurses, typically with both a registered nursing license and a master's degree. They may work in hospitals or they may deliver at home, but in the latter case they typically have a backup in the form of an obstetrician and a hospital for unexpected complications.

One nurse-midwife recounts the tensions that accompanied her interactions with some obstetricians.

"How far along is your noisy patient in there?" asked Dr. Rider. He frowned and jerked his head toward Room 2 where Latoya sat cross-legged on her bed. We could hear her from the nursing station. She hummed a gospel tune between her contractions and moaned "Ow-ow-ow!" during the peaks....

"Jesus, what's with you midwives that you never medicate your patients? Is it some kind of ritual female bonding? Something about suffering?"

Oh, boy. "Look, she knows drugs are available. She's figured out her own method of handling labor, and it's working. If she's bothering your patient, I'll close her door." ...

There were just three of us midwives, Sandi and Lindy, and I—and about fifty obstetricians. Unspoken issues hovered beneath the surface of nearly every interaction we had with one another.

Doctors looked stunned, incredulous when pregnant nurses bypassed their services and chose instead a home birth with a midwife. Dr. Rider confronted Cherie, pregnant with her first child, and demanded, "How can you consider having your baby at home when you work here and see all that goes on?" Without missing a beat, Cherie answered, "That's the point. It's because I see what happens in hospitals that I'm having my baby at home."

SOURCE: Peggy Vincent, 2002. *Baby Catcher: Chronicles of a Modern Midwife.* New York: Scribner, pp. 185–186.

---

implies that the professional will incur some self-sacrifice to help the client. Altruism also involves the profession's duty to use its knowledge for the public good. On the one hand, because the knowledge is important and is monopolized, the profession has the duty and responsibility to preserve, enhance, and transmit it and to use it in the public interest. On the other hand, the confidential knowledge gained about an individual client must *not* become public; to violate a client's confidentiality is a breach of ethics. The courts uphold some professional-client confidences as privileged; that is, they do not have to be revealed in court.

The first aspect of the profession's altruism is its self-policing. The expression "only professionals can judge professionals" refers to the inability of those without specialized knowledge to evaluate competence. The self-regulation may begin before admission to professional school, when there may be efforts to determine whether the prospective student is "of good character." Further character investigations may take place before formal admission to the profession. Hospital review boards, professional grievance procedures, peer review panels, ethics panels, and disciplinary committees of professional associations police the working professionals. One aim of these procedures is to reassure members of the public that the profession is truly acting in their interest. Many people, however, remain cynical about the effectiveness of such procedures, believing that professionals would rather cover up the incompetence of a fellow worker than admit that one of their number had made a serious error. In some states, there is a state-level board that disciplines members of licensed professionals, and these boards may have lay appointees whose purpose is to represent the general public.

The second aspect of the profession's altruism is its advocacy of community service, called *pro bono publico* work among lawyers. This is professional work volunteered or performed for a lower fee than is usually charged. Local professional associations often arrange for volunteer services to be provided by their members. A local legal society may provide a legal aid clinic, or a local medical society may offer free health screenings or vaccinations to low-income people. Professional associations may also sponsor hot lines, tape-recorded educational messages, and informational materials with professional advice. The association encourages individual professionals to donate their services to the poor or to nonprofit organizations.

Professionals engage in public service in other ways. The professional culture inculcates a sense of *noblesse oblige,* an expectation that elites will be visible contributors to community life. Part of the professional lifestyle includes providing leadership to civic and voluntary endeavors. Thus, physicians often spearhead drives to raise money for new facilities at the local hospital, and attorneys often promote local good-government movements. Clergy members provide significant leadership in public affairs. An example is the leadership of the Reverend Martin Luther King, Jr., in the civil rights movement. Although such activities do not require professional training, the prestige of the professional is believed to add to the success of the efforts.

Such community activities may not be merely altruistic, for they also develop important contacts and referrals. Many professionals become important members of the local business community or candidates for elective office. The prestige they receive from their community service is also a symbolic reward in addition to the monetary rewards of professional work (Barber, 1965:19). Such considerations lead some observers to identify these activities as more self-serving than altruistic.

EVALUATING THE FOUR HALLMARKS

The four hallmarks of a profession are a useful starting point for understanding the professions, but they are incomplete. Taken together, they compose an idealized model that imperfectly describes reality. The type of analysis on which they are based, often called structural-functional analysis, has been criticized by sociologists for ignoring much of the reality of professionals' lives. Many critics use the conflict approach to provide a different perspective on the profession. In this section we will review some of their criticisms.

From the perspective of the conflict approach, the four hallmarks are incomplete explanations of how the professions retain a position of power. The knowledge base, which is monopolized by the professional, is one key to maintaining professional power. Their associations, their political lobbying, and their use of state laws to enforce their privileges serve to maintain their power while denying others a share in it. Maintaining professional power also implies maintaining autonomy and authority. From the power perspective the notion of altruism is mostly a smokescreen to prevent the public from investigating the internal workings of the profession.

How Powerful Are the Professions?

The professions are powerful occupational groups. One criticism of the structural-functional approach has been its neglect of professional power (Geison, 1983:46). Professions are attractive occupations because their members, unlike production workers, clerical workers, and many workers in other occupational specialties, wield real power over their own work lives. They decide what they will do and how, direct other workers in their tasks, and act to prevent other occupations from doing the same sort of work (Halliday and Karpik, 1997).

Monopolizing Knowledge

The professional knowledge base and the esoteric lore of the village shaman both represent a monopoly over knowledge (Larson, 1977). Just as the village sage monopolized the possession and transmission of esoteric knowledge, only the professionals command the knowledge of their field. Because professionals earn their living by applying their professional knowledge in service to paying clients, the professions devote attention to identifying and maintaining their jurisdictions over specific knowledge (Abbott, 1986). Contemporary professionals monopolize knowledge by erecting three basic barriers to laypeople: professional schools, licensure, and mystification.

Professional Schools Admission to the professional schools is intensely competitive. Just as the village sage had to find and train a successor, so the profession must select and train its new members. In advanced industrial societies this training is nearly always accomplished in formally organized professional schools, such as medical schools, law schools, and divinity schools, which are commonly part of universities. There are usually many more applicants than available spaces. Leaders of professional schools reason that by limiting the number of students, the schools will produce highly competent graduates. Aside from maintaining quality controls over the students, the professions also fear producing a glut of graduates who will compete

against one another for clients. The restriction of access helps maintain a higher price for professional services and higher incomes for the professionals. Professional schools are typically located in universities, which function as important allies of the professions (Abbott, 2005).

Professional Associations Professional associations play an important role in the professional culture (Halliday, 1987; Powell, 1999). In addition to representing the interests of the profession to outsiders, they also help form consensus within the profession about norms for practice and the social organization of the work (Halliday, Powell, and Granfors, 1993; Timmermans, 2005).

The professional association is intensely interested in the organization, curriculum, and graduates of the professional schools. The American Bar Association (ABA), for example, accredits law schools. In some states, law graduates are eligible to become lawyers only if they received a law degree from an ABA-accredited school. By accrediting, the professional association seeks to ensure that the school teaches a core body of knowledge and skills and that only qualified faculty conduct the training. Professional associations are often reluctant to accredit new schools or to permit schools to enlarge their student bodies, usually citing concerns over maintaining the quality of graduates. The professional association may be compared to a gatekeeper, deciding whom to admit. Other occupations also have gatekeepers; for example, the unions that conduct the apprenticeship programs in the skilled trades have power over their occupational group because they control access.

Licensing Even if lay people were to acquire professional knowledge, they would not be able to put it at the service of clients. Nearly all professions require licensing by the state, and only those who qualify with educational credentials and by passing an examination can receive a license. There may be additional requirements for periodically updating the license or for completing further training. Most professions are protected by laws that make it illegal to practice without a state license (Zhou, 1993). Thus, the state becomes another important ally of the professions (Abbott, 2005; Krause, 2001). Other professionals such as scientists, although unlicensed, are essentially unemployable without formal schooling credentials, because employers will not hire them. Either through licensing or through convincing employers to hire only trained graduates, the profession can act collectively to restrict access to its knowledge base.

To be certified in a professional specialty within the profession requires further examinations and licensing restrictions. Specialty certification is most widespread in medicine, though some states certify legal specialties. Specialists maintain their own professional organizations (such as the American College of Obstetricians and Gynecologists) as well as the general professional organization (such as the American Medical Association).

Mystification The mystification of professional knowledge comes about because the learning is so specialized and often couched in such difficult jargon that even after being exposed to the information, a lay person might not understand it. Despite their efforts to communicate, the professionals may be so absorbed in their distinctive lingo that their advice is incomprehensible. Lay people then perceive the knowledge as esoteric. Just as the shaman enjoyed increased prestige from possessing esoteric knowledge, a client today may leave a professional's office murmuring, "That person must be smart, because I didn't understand a word of it."

Mystification reinforces the profession's claim to knowledge and discourages efforts by lay people to take upon themselves the tasks that professionals wish to monopolize. In theory these factors that restrict the supply of new professionals should protect the jobs of the existing professionals and help them extract higher fees for their services. Even in the professions, however, employment is not guaranteed. As a rule the unemployment rate for professional workers has been low. Despite the restrictions on supply, however, in some fields there has been an oversupply of professionals.

Power within the Professions

The structural-functional view of the professions is principally concerned with distinguishing the professions from other occupations. It neglects inequality within the professions, except in its observation that some specialties enjoy greater prestige than others. The conflict approach pays more attention to the differential distribution of power within the professions. The most powerful professionals are likely to be those with the greatest control over the sources of knowledge (such as the deans of the professional schools), those who possess the most important and specialized knowledge (Abbott, 1988; Halpern, 1992), those with the most prestigious clients (Heinz and Laumann, 1982), those who have the most challenging and complex intellectual tasks (Cullen, 1985), and those with the greatest organizational or personal resources to maximize their autonomy and authority.

Changes in the Professions

The structural-functional approach best describes the so-called free professions—those in which individuals open their own practice and charge their clients fees for services rendered. This view of professional practice has been criticized because it leads to a romanticized vision of the professional-client relationship and ignores the fact that today there are relatively few independent professionals. Many professionals today are also employees. The movement of physicians into health maintenance organizations (HMOs) exemplifies the new prominence of professional firms and group practices.

An important business advantage of group rather than independent practice is that overhead expenses such as rent, furniture, equipment, support staff, insurance, and accounting can be pooled. Fees for services are also pooled, and the professionals are usually compensated with salaries, although partners may also receive a share of profits. Only licensed professionals can be shareholders (Freidson, 1986:126–129). Some advocates argue that group situations offer better service because the client may be served by whomever of the professionals is available ("on call"). Not all clients prefer such arrangements, for many still favor the personal attention of one professional who is familiar with their case. Moreover, at least some professionals have ethical questions about the advisability of rotating clients among professionals (Starr, 1982:198–232).

The group practice retains control by the profession, in the sense that someone with the requisite professional knowledge makes decisions. The members are not necessarily equal, however. There may be a managing partner or director, and some professionals will be full members or partners with financial equity in the firm, whereas others will be provisional, salaried employees who may not ever achieve partnership. Other arrangements, however, including the so-called "McDoctor Clinics," represent the reorganization of medical practice as franchises or conglomerate organizations. "*McDoctor Clinics*" are small clinics that offer rapid medical attention to minor medical situations, just as fast-food outlets offer quick meals to hungry customers (see Box 11.2).

A related change has been the increased employment of professionals in other organizations, such as universities, research organizations, major corporations, and government agencies. Most big corporations, for example, employ house counsel, lawyers who work as staff members, even though outside law firms may also be engaged for specific tasks. This arrangement subjects the professional to bureaucratic or organizational authority that may supersede their professional authority. Authority is eroded further when the only "client" is also the employer. The house lawyers must provide the best legal advice possible at the risk of losing their employment, but then the top managers of a corporation are free to ignore their advice. A law firm might suggest to a client who consistently refused its advice that it was time to seek new lawyers; the house counsel cannot easily make such a suggestion to the employing corporation.

An additional major change has been the entry of third parties, especially the government, into the professional-client relationship. This role is most evident when a third party pays for the professional services, as when medical care is paid for by an insurance company or by the government through Medicaid or Medicare. Although the professional is still engaged by the client, the economic relationship has changed because the link between receiving and paying for services has been severed. The professional and the client may both be affected by the third party's judgment about whether the services are needed or appropriate and whether the professional is charging too much for them.

The reorganization of professional practice creates new hierarchies within the professions, with some professionals exercising managerial control and commanding higher salaries than other professionals who are salaried employees (Bok, 1993). These newer forms of professional practice may, however, improve the availability and affordability of professional services.

Are the Professions Meritocracies?

The hallmark view implies that the professions are **meritocracies**, that is, groups in which rewards are based on achievement. In the profession the achievement is presumably mastering the core of knowledge, contributing to new knowledge or techniques, or practicing the profession well. Because these tasks are difficult, high monetary and symbolic rewards are used to attract the most capable individuals into the profession. The achievements are presumably independent of such ascribed characteristics as race, gender, and social class; and the selection criteria of the professional schools presumably identify actual and potential accomplishment.

In fact most professions have been dominated by white men, often with origins in the upper middle class. As Figure 11.1 indicates, in most professional specialties the proportion of women and minority group members remains small. This lack of balance has raised concerns that the professions are not really recruiting the "best and brightest," but instead recruiting the children of professionals to take the coveted seats in professional schools.

Since the 1960s and the changes generated by the civil rights movement, minority group members and women have sought better representation in the professions. The exclusion of women from the professions was not only customary but also had

BOX 11.2 The Rise of Corporate Medicine

One important influence on the professions is the organizational climate in which they work. In this selection, sociologists George Ritzer and David Walczak describe the ways in which bureaucratic initiatives are changing the practice of medicine.

[A] significant development is the rapid growth of Health Maintenance Organizations (HMOs). While these organizations take a variety of forms, all of them are business organizations in which the objective is to work within budgetary constraints, or in some cases to turn a profit. As business organizations, HMOs are under tremendous pressure to rationalize operations in order to minimize costs and maximize income. Physicians, and their norms of autonomy, authority and altruism, are affected by these pressures whether they control these HMOs or are employed in them.

Also significant is the "conglomeration" of medicine. Whether it be through expanding product lines or product-line diversification, these large medi-giants aim to integrate healthcare facilities, medical services, and insurance functions on a regional or national basis. Large corporations are not only coming to own for-profit hospitals, McDoctors, and HMOs, but they are also controlling health insurance firms, buying out hospital supply companies, and generally moving in the direction of controlling as many components of the "medical-industrial complex" as possible (e.g., nursing homes, laboratories, emergency-room services, home care, etc.). To compete, large, non-profit organizations like Blue Cross-Blue Shield are expanding in a similar manner. In either case, ever larger conglomerates are increasingly impelled to rationalize their operations. This means that they are going to seek ways of becoming more efficient, predictable, and calculable, of replacing more and more workers with nonhuman technologies, and of exerting increasing control over employees (including physicians).

Those physicians who resist moving into various large-scale capitalistic medical organizations are going to have to be even more entrepreneurial than they have been in the past. Physicians have, of course, always combined professionalization and entrepreneurialism, but in order to compete in today's more market-oriented medical system, physicians are going to be forced to emphasize business practices more with the result that professionalization may suffer. Further, physicians in private practice are coming to emphasize entrepreneurialism more and professionalism less by opening an array of free standing or noninstitutional health care centers including pathology labs, health promotion centers, obesity and substance abuse clinics, chains of local emergency rooms, proprietary hospitals, and nursing homes.

While good empirical data do not yet exist on the impact of entrepreneurialism on patient care decisions, there is suggestive evidence that "is adequate to confirm the common sense conclusion that investments and economic arrangements that reward physicians financially for making certain patient care decisions (e.g., ordering lab tests) will bias physicians in favor of making such decisions." In any case, it is clearly harder for primarily entrepreneurial physicians to continue to claim to adhere to substantive rationality, especially the value of altruism. Furthermore, the demands of the capitalistic marketplace will exert greater control over an entrepreneurial physician.

One result of this control by the market is the expansion of advertising by the various agencies involved in the medical business, including even physicians in private practice. While there are still strong negative sentiments among some physicians toward advertising, there is evidence to suggest that advertising is gaining acceptance in the medical profession. Here is one of the clearest examples of the erosion of the distinction between professional and capitalist, between substantive and formal rationality. Again, with physicians advertising their services on television and in the newspapers, it will be harder for them to continue to put forth an altruistic image. Patients are more likely to question the authority of physicians selling their services side-by-side with used car salespeople. [Notes and references omitted.]

SOURCE: Excerpt from George A. Ritzer and David Walczak, 1988, "Rationalization and the Deprofessionalization of Physicians." *Social Forces* 67, 1 (September):10–11. Copyright © 1988 by the University of North Carolina Press. Used with permission of the publisher. www.uncpress.unc.edu

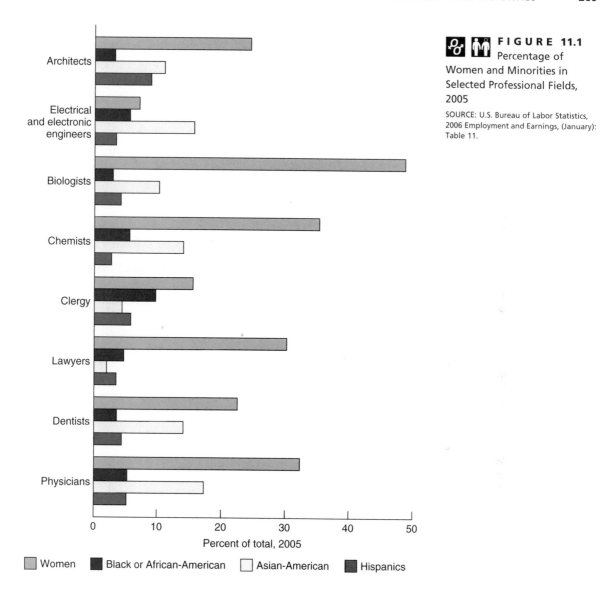

FIGURE 11.1
Percentage of Women and Minorities in Selected Professional Fields, 2005

SOURCE: U.S. Bureau of Labor Statistics, 2006 Employment and Earnings, (January): Table 11.

Women   Black or African-American   Asian-American   Hispanics

legal sanction. At one time the courts upheld the rights of the state legislatures and of the bar associations to exclude women from professions. In 2005 women accounted for about 30 percent of all lawyers and 32 percent of physicians, though the median weekly earnings of women lawyers were 77 percent those of male lawyers, and the median weekly earnings of women physicians were 61 per-

cent those of male physicians (based on full-time workers only; BLS, 2006c). Women are now better represented in the professional schools, where they constitute half or more of many medical, law, and dental classes.

The number of minority professionals has increased in recent years, but the *proportion* of professionals who are minorities has changed slowly. One reason for the underrepresentation of racial and ethnic minorities in the professions is the cost of professional school. African Americans, Hispanics, and other minorities are often reluctant or unable to incur heavy debts to finance professional schooling.

Once in the professions, moreover, women and minorities tend to have lower incomes and to occupy the less prestigious specialties. Many female and minority-group professionals do not specialize but maintain general practices. One reason for this may be a preference to serve in a minority community whose members have fewer resources for professional services (Weinfeld, 1999).

Women who specialize within medicine are often steered to such "appropriate" specialties as pediatrics, gynecology, family practice, and psychiatry (Lorber, 1984). Specialties such as dermatology or ophthalmology, which involve relatively little emergency or weekend work, are thought to be more compatible with marriage and a family. Within law, female and minority professionals are overrepresented in the specialties of family law and real estate law (Epstein, 1970:160–161). Women and minorities are also more likely to work with individual clients, although work for corporations is more lucrative and prestigious. Women and minorities entering professions may find themselves isolated and even harassed by colleagues (Booze, 2000; Davey and Davidson, 2000).

A 1986 U.S. Supreme Court decision made it illegal for the partners in a law firm to discriminate against a qualified woman lawyer by preventing her from becoming a partner in the firm. (A new lawyer is usually an associate of a firm for six or more years before being considered for a partnership.) Even so, women and minority professionals sometimes experience discrimination. Direct verbal aggression occurs when women physicians are referred to as "hen doctors" or when male law school professors use hypothetical examples that depict women as stupid (Benokraitis and Feagin, 1986). Women and minorities are somewhat more likely to have their authority challenged, both by clients and by coworkers from subordinate occupations (MacCorquodale and Jensen, 1993).

Within the profession they may be less likely to receive referrals of potential clients from other established professionals (Epstein, 1982:65). Perhaps for these reasons many female and minority professionals maintain neither traditional solo practices nor even group practices, but seek employment in government agencies and other workplaces. Box 11.3 discusses the issues that arise for professionals in a multicultural information technology workplace in Great Britain.

CHANGING DEGREES OF PROFESSIONALIZATION

Regardless of the sociological approach taken, virtually everyone agrees that professionals play a key role in the occupational structure. Because of their monopoly of knowledge, the desirability of their knowledge, and the barriers to entry into their professions, many professionals have been able to earn high incomes. In addition, professionals have been able to retain substantial control over their daily work lives and to delegate some unpleasant parts of their jobs to others (Stevens et al., 2000). Many workers would like to have these generally pleasant work conditions, and many occupations have sought to develop similar work conditions by emulating the professions. As noted earlier, this process is called professionalization.

Professionalization

Professionalization can be understood as the effort by an *occupational group* to raise its collective standing by taking on the characteristics of a profession. Professionalization needs to be distinguished from **professionalism**, which is the competence and effectiveness of *workers* in their job performance. Professionalization often occurs in occupations with considerable heterogeneity in the training and background of their members (Barber, 1965:23). It often takes the form of seeking to adopt the four hallmarks of the professions to a

B O X 11.3 How Racism Persists in Professional Workplaces: A Case of Information Technology in Britain

Journalist Roisin Woolnough interviewed nonwhite workers in Britain's information technology (IT) industry, and found that they experienced racial prejudice from clients who expected to work with white consultants. In addition, minority workers were sometimes excluded from important business-related networking.

The IT industry is perceived as a particularly open industry where race is not an issue. Look around most IT departments and they are invariably made up of a diverse workforce. But take a closer look, and it often becomes apparent that white IT professionals are the ones who are making it to executive positions of authority.

Rene Carayol is a man who's made it. Formerly the managing director of IPC Electric, the 42-year-old is now chief executive of an e-business consultancy. Yet he says he is still shocked by the amount of racial prejudice that he comes across and how it affects his career. "When I go to new offices with my team, receptionists instantly turn to a member of my team to see who is the boss," he says.

"At IPC Electric, many of the suppliers didn't expect to be working for a black IT director and they had to get used to working with me. It took some time."

Carayol thinks it took him longer to get where he has in the IT industry than it should have done, because of the colour of his skin. "I found that I stayed a junior for what appeared to be a little too long. It is a persistent complaint that I hear from ethnic minority people. It's taking minorities longer to reach top positions than it should."

Sarabjit Ubhey, head of operational control at Bupa, thinks institutional racism has to be at the root of it. "People expect racism not to be an issue in IT because it's a new industry," she says. "But the so-called glass ceiling seems to be there as there are so few senior non-white IT people. The numbers don't match up and that shows there is an issue.

"The fundamental problem is awareness. For example, my team has a very diverse make-up. There are a couple of Muslims and people from other parts of the world. As a manager, if I am organising an event I have to make sure I don't arrange things that will exclude some of them. For example, if I invite people to the pub I'm sending out the wrong message because I could be excluding the two who don't drink. So, I make sure we go out to eat instead."

A lot of networking goes on at work social events, and if someone is not party to that interaction, it could mean they are missing out on opportunities.

Paul Riddell, the Commission for Racial Equality's (CRE) head of legal strategy and private sector in London and the South, says one of the most common complaints from people who approach the CRE is that they feel excluded from these networks and that this has hindered their career progression.

Riddell agrees it is the responsibility of managers to make sure people are not being excluded. "They shouldn't use the informal stuff as a forum for career advancement," he says. "Instead, they need to do performance assessments. The ideal is career appraisals every six months with a yearly in-depth one."

Some companies have support groups for ethnic minority groups. While these can be a good forum, they can also have the negative effect of compounding a "them and us" situation.

Some companies have also appointed diversity managers. But Carayol thinks companies should not need to employ a diversity manager: "Diversity should be on every manager's list."

The more role models that non-white people have, Carayol says, the easier it becomes for them to progress up the career ladder. The good news is that he thinks it has already begun.

"Get a couple of people through and it opens the floodgates. The race issue is in the same place as women were 10 years ago. It is taking longer than I expected and it is more painful, but it has started to change."

SOURCE: Roisin Woolnough, November 4, 2000, "Racism Reinforces the Glass Ceiling," *The Guardian* (London), p. 301. Used by permission.

greater or lesser extent, and it is usually initiated and maintained by the leaders of the occupation.

Steps in Professionalizing   The first step in professionalizing is forming an organization or strengthening an existing one. The professional association seeks to convince the general public, the state legislature, and perhaps other professions that its claims to professional status are legitimate. Some occupational groups are divided over the best way to achieve greater status, and they may have several competing organizations, not all of which adopt professionalization as a strategy. In the efforts to improve the status of elementary and secondary teachers, for example, some organizations are oriented toward a union model, and others explicitly follow a professionalization strategy.

A second step is standardizing the body of knowledge that the members of the occupation should have. This step requires developing courses of training, and most professionalizing groups seek to locate the training in universities. One step toward professionalizing nursing was developing university schools of nursing to supplement the traditional training provided in hospital nursing schools. Although the university-trained nurses would also receive clinical training in hospitals and take the examination to become registered nurses, they would have a bachelor's degree in nursing. A professionalizing occupation usually develops its own research program, journals, and continuing education. Registered nurses constitute an occupation that has done all of these things.

Developing the body of knowledge is not sufficient (Torstendahl and Burrage, 1990). The professionalizing group must convince the public that the knowledge is important and that only those graduates who complete the certification process should be assumed to possess the knowledge. The association must develop a certification process or specify a credential required to be in good standing. This credential might be a degree from an accredited school, passage of an examination, or a state-issued license (Freidson, 1986:63–88). Ideally, the professionalizing association would like to convince

a legislative body to mandate that workers in the occupation be licensed or certified.

A field that is currently seeking some type of licensure is marriage counseling; in many states people who open a practice in marriage counseling need not have any training nor any certification. Certification laws would protect those already in the field from further competition. Certification would protect members of the public from unqualified persons whose marriage counseling might be of poor quality, even emotionally damaging. In this particular example certification would probably encourage heightened competition for "turf" among social workers, clinical psychologists, clinical sociologists, clergy members, and others who offer marriage counseling.

Most occupational groups must negotiate their status with the public and with legal authorities for many years. In the health fields it is important to gain the approval of health insurers to receive reimbursement. Even more formidable opposition, however, is likely to come from the established professions, especially if the "newcomer" occupation potentially threatens the control of a well-established profession. Established professional associations often oppose efforts by new professions to achieve a measure of autonomy or authority (Smith, 1993). For example, dentists have challenged the right of dental hygienists to open their own offices. Medical societies have prevented the granting of hospital privileges to chiropractors and have questioned whether the work of nurse-midwives constitutes practicing medicine without a license (see Box 11.1).

Some new professional specialties have been allowed a toehold of recognition only because they have agreed from the beginning that their members will be supervised by a licensed member of an existing profession. For example, nurse practitioners usually work under the direct supervision of a physician, although they have advanced training and are able to take on some tasks once reserved for physicians. We will discuss such workers further in the section on paraprofessions.

Professionalizing occupations often adopt codes of ethics and internal disciplinary procedures, but employers may not defer to these procedures when an issue arises. These codes and procedures often seem to be part of the apparatus of convincing others that misuse of the knowledge base could endanger the public. Professionalizing associations may encourage their members to be publicly visible and to donate their time and talent, both because that is something that many professionals do and because it again brings their occupational group before the public eye.

Most professionalizing groups achieve autonomy and authority slowly, if at all (Forsyth and Danisiewicz, 1985). Sometimes they succeed because of a lack of personnel in the established professions. For example, osteopathic doctors (also called osteopaths) have achieved standing in many small, rural communities because they provide medical services where medical doctors are unavailable (Gevitz, 1982). In rural communities pharmacists may have higher standing than do their counterparts in large cities because people rely on them for advice and not just for filling the prescriptions of a medical doctor.

Obstacles to Professionalizing   Perhaps the most significant obstacle to professionalization is the opposition of existing professions. The established professions not only monopolize knowledge but are also gatekeepers for other functions. Physicians, for example, control access to prescription drugs, to the functioning of hospitals, and to key government health agencies. These gatekeeper functions may frustrate the efforts of other would-be health professions to professionalize.

A second obstacle is the fact that most would-be professionals do not practice independently but work for large organizations. Teachers, for example, are employed by school boards. The claims of such workers to professional authority clash with the bureaucratic authority exercised by administrators. The employing organization is especially likely to resist claims of professional expertise that disrupt its operating procedures. For example, although teachers contend that their expertise should influence curriculum revision and textbook adoption, these functions are often performed by state and local educational bodies that give the teachers' opinions little weight.

A third obstacle to professionalization is public skepticism. At a time when many occupational groups argue that the public needs more experts to provide advice on more subjects, some consumers have come to resent the professions. Claims to expert knowledge, autonomy, and authority are likely to be taken with a grain of salt by both the public and elected officials. This skepticism is one element of a countermovement called "deprofessionalization."

Deprofessionalization

Deprofessionalization, sometimes called postprofessionalism, is the process of weakening or eliminating the professional characteristics of an occupational group (Kritzer, 1999). It is rare for an occupational group to deprofessionalize itself, although efforts to unionize such professionals as veterinarians or dentists are sometimes interpreted in that light. Rather, deprofessionalization usually originates with the general public, the government, or organizational employers (Ritzer and Walczak, 1988).

Demystifying and Empowering   One form of deprofessionalization is demystifying the esoteric knowledge of the profession so that it is more widely shared among the general public. This process could also be described as *empowering the consumer,* because consumer knowledge redresses the imbalance in the professional-client relationship. Publicized cases of malpractice may induce consumers to learn more about professional knowledge so that they will make better choices in seeking professional advice. Sometimes professionalizing occupations competing for the same clients will seek to inform the public, bolstering their own public-service claims.

Non-competing occupations may also be responsible for demystifying knowledge. For example, journalists and broadcasters now disseminate medical and health information through specialized websites, health newsletters, general magazines, newspapers, and radio and television programs. Even without consulting physicians or other health professionals, consumers can become more informed about improving their health through lifestyle changes and preventive medicine. Marketing developments and changes in government regulation may also help demystify knowledge. Home medical screening kits, the reclassification of some prescription drugs for over-the-counter sales, and new courses in home health care and disease prevention have also empowered clients to take more responsibility for their own health. Do-it-yourself kits for wills, divorces, and bankruptcies may serve the same function in deprofessionalizing some aspects of the law.

The response of the professional associations to disseminating information has been ambivalent. Although most professional associations encourage clients to learn more, they also warn against the possibility of misusing information or being misled by the do-it-yourself approach. Physicians and lawyers alike warn about the dangers of self-diagnosis. Doctors worry that home births, unauthorized cancer or AIDS treatments, and similar experiments outside the sphere of professional medicine may increase the risks to patients. Attorneys worry that uninformed clients actually worsen their legal situations because they do not understand the potential legal ramifications of such actions as writing their own wills or representing themselves in court.

One response by the professional associations has been to undertake their own programs to demystify their knowledge and to empower the client. Medical associations often sponsor hot lines with information and referrals, and many hospitals offer courses on medical self-help. Physicians have begun some health-related websites. Legal associations offer "ask-a-lawyer" services featuring brief, often free, consultations; and they provide publications that describe the circumstances under which a consumer needs to consult a lawyer. But the professional associations remain vigilant to see that other occupational groups are not taking advantage of consumer empowerment in ways that might constitute practicing law or medicine without a license.

Regulation A second form of deprofessionalization is regulation, which attacks both the autonomy and the authority of a professional. Regulation does not refer here to the self-regulation of the professional association but rather to rules imposed by government agencies, insurance companies, or other third parties.

Government agencies indirectly regulate physicians through such measures as approving drugs, regulating hospitals, and placing some restrictions on medical schools that receive tax dollars. Government regulations also mandate additional self-regulation by physicians on such issues as the length and appropriateness of hospital stays. Medicare and HMOs reduce fees by refusing to pay charges they consider excessive. Many HMOs further restrict their members to physicians who have agreed in advance to accept lower fees. The courts, too, have been active in changing the professions. The U.S. Supreme Court ruled in 1978 against professional associations' bans on advertising and price competition among their members.

Private insurance companies may require second opinions before surgery or prior approval of hospital stays. Many insurance carriers will not reimburse patients for experimental treatments, and they may dispute what doctors view as appropriate care. Many proposals for health-care reform advocate more extensive use of such regulation.

Managerial Control A third form of deprofessionalization comes from managerial control, which constitutes a potential threat to the autonomy and the authority of the professional. Managerial control may compromise the right of the professional to be judged by others in the professional community. In bureaucratic organizations the professional employee has a supervisor who is likely not to be a professional and not to have access to the same body of knowledge. In making decisions, the supervisor is likely to consider many organizational factors and not just the recommendations based on professional knowledge. Far from prescribing what a client should do, the professional in the bureaucratic organization is reduced to making suggestions or offering recommendations.

In addition, the employing organization will have its own set of rules and procedures that may occasionally conflict with the ethics or sound practice of the profession. In Chapter 5 you learned about the problems of role conflict between work and home. Many professionals may find themselves in a conflict between two work roles if they are expected to behave in one way as a professional and in a different way as an employee. A potential case of role conflict faces a company physician. The company wishes to minimize time lost to sickness, to avoid worker compensation claims, and certainly to avoid liability for permanent damage to employees' health. The physician may be pressured to determine that symptoms are not job-related but instead are caused by non-job factors, such as lifestyle. A physician may feel pressure, for example, to find that lung cancer was caused by smoking, not by the worker's involvement in the production of a carcinogenic chemical.

To resolve role conflict, many large organizations assign some managerial functions to experienced professional scientists (Shenhav, 1988). For the individuals such assignments may constitute another aspect of deprofessionalization because they gradually draw them away from the professional community, with its norms and values, and into a different occupational group with different norms and values. Professionals in large organizations may find their loyalties shift from their profession to the company and, perhaps, to trade associations with which their company is affiliated. In Chapter 12 we will discuss why the concerns of managers are likely to be different from those of the staff professionals.

A new concern for professionals is the growing ability of large firms to outsource professional work to other countries. Early examples of such outsourcing included accounting and computer programming. Today, medical images may be transmitted electronically for interpretation by physicians abroad, and some paralegal work is also routinely outsourced. The face-to-face interaction between the professional and the client is unlikely to be outsourced, but some of the work necessary for this interaction may be done abroad. The principal reason for outsourcing is to cut costs, but a secondary reason is to access key skills that may be available in a foreign population (Kobayashi-Hillary, 2005:147). To the extent that such tactics reduce the autonomy of the professionals, they are a potential form of deprofessionalization.

Deprofessionalization is an important counterprocess to professionalization. Whether the professions of the future will have more or less prestige and influence than they have today will depend in large part on the relative progress of the two movements.

THE SEMIPROFESSIONS AND THE PARAPROFESSIONS

The Semiprofessions

The term **semiprofession** refers to an occupational group that has achieved some of the characteristics of a profession or possesses each hallmark of a profession in a limited way. Many people accord workers in these occupations professional status, and studies conducted by government statistical agencies classify them as professional workers. Sociologists, however, do not always classify them as true professions, principally because of limits on their autonomy. Most of these occupations are professionalizing, and some of them may gain recognition as professions.

Among the major semiprofessions are elementary and secondary schoolteachers, librarians, social workers, and registered nurses. These occupations are sometimes called the female semiprofessions because they are dominated by women. Figure 11.2 shows both the sex and minority composition of these semiprofessions. The few men in these occupations may earn higher salaries and receive rapid promotions, unlike the women who dominate the occupation (Williams, 1992). Because women generally exercise less social power than men, however, the female semiprofessions face obstacles in gaining collective power.

Members of minority groups have the greatest relative representation among social workers, dietitians, and nursery school and kindergarten teachers. Their levels of representation are lower in other areas. Only 5.8 percent of librarians were black, for example, and only 4.3 percent of registered nurses were Hispanic.

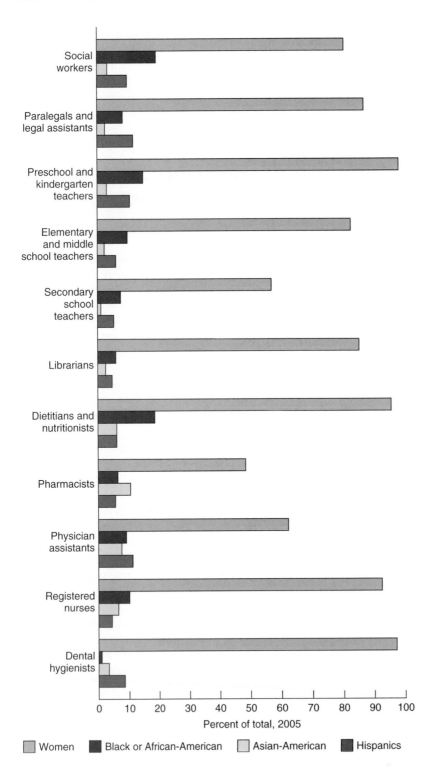

FIGURE 11.2 Percentage of Women and Minorities in Semiprofessions and Paraprofessions, 2005

SOURCE: U.S. Bureau of Labor Statistics, 2006 Employment and Earnings, (January): Table 11.

Other occupations that could qualify as semiprofessions are opticians, human resources officers, and systems analysts. Some sociologists would add other occupations to the list of semiprofessions. Rather than comparing various lists, however, it is probably more useful to talk about distinctions between semiprofessions and other occupational types.

The semiprofessions have a body of knowledge, but they usually do not monopolize it, nor do they erect effective barriers to entry. Elementary teachers have a college degree and usually have completed specialized work in education; in addition, they have a body of knowledge on child development, teaching methods, and many other fields. But the basic information they teach in class is known to most adults. Most adults can count, add, subtract, and read. Although some adults accord authority to the elementary teacher who has mastered the specialized knowledge of teaching, many more do not, presumably on the erroneous assumption that "anyone could do that." Most parents would probably give greater deference to the advice of pediatricians than to the advice of elementary teachers concerning a child's performance. Teachers do have to be licensed by the state, however, so there is some barrier to entry.

Librarians pose a somewhat different case (Winter, 1988). Although librarians with a degree in library or information science are trained in a variety of techniques that others will not know, many libraries (especially smaller ones) will hire people without librarian credentials. In this case the credential and the specialized knowledge it represents are not considered necessary. There are few barriers to entry for many librarian jobs (Abbott, 1988).

A still different problem occurs when the semiprofession's body of knowledge closely overlaps that of an established profession. For example, both registered nurses and pharmacists share many aspects of professional knowledge with physicians. But because physicians are an established, powerful profession, they can limit nurses' and pharmacists' abilities to use the information. Both nurses and pharmacists may know what medication would normally be effective for a specific condition, but only the physician can diagnose the condition and prescribe the

drug. In some states, however, there are limited experiments in allowing nurse practitioners and some others who are not physicians to prescribe drugs.

The semiprofessions enjoy only partial autonomy because most of their members work in bureaucratic settings. For example, a teacher has guidelines about what must be taught, what books to use, and how students may be disciplined. A social worker has agency policies and rules to follow concerning how outreach is to be done, how new cases are written up, and what reports must be filed. A registered nurse has hospital policies and physician orders to follow. Even so, within these constraints the semiprofessional has discretion in dealing with individual clients. Teachers diagnose the learning problems of specific students and work to overcome them. Physical and occupational therapists work under a physician's direction but with autonomy to apply their specialized knowledge of rehabilitation techniques to a patient's needs.

Some semiprofessions have adopted the trappings of the professions, including those associated with altruism. Many semiprofessions have codes of ethics and disciplinary rules enforced by the members of the professional associations. Unlike the established professions, however, their associations are less likely to be politically powerful. The semiprofessions' associations are often circumspect for fear of offending major employers or the more powerful professions. They must also continue to educate the public and negotiate with state agencies to improve their position regarding licensure and recognized authority. To make the process even more difficult, in some semiprofessions several associations compete for the allegiance of the members.

In summary, semiprofessions have eagerly sought professionalization. Many semiprofessions are accorded status and legitimacy. Some have considerable authority over clients, and most have some form of licensing or credential. But they still lack the autonomy and organized power of the professions. From the power perspective the lack of collective power is the single most important difference between the semiprofessions and the professions.

The Paraprofessions

A different group of near-professions are the **para-professions**, which might be considered occupational helpers of the major professions. Members work closely with professionals, usually doing tasks delegated and supervised by them. The dominant profession dictates the conditions of training and certification, if any, and provides the guidelines mandating the point at which a member of the dominant profession must be consulted.

Paramedics and paralegals are examples of paraprofessionals. Paramedics provide emergency medical care until a patient can be transported to a hospital or other site for treatment by a physician. In many large American cities the paramedic is linked by radio to physicians who can give precise instructions for emergency care. Paralegals work under the direct supervision of an attorney. They may do some legal research, prepare and file forms, interview clients on routine matters, and perform other necessary tasks, but always under the direction of the supervising attorney. Paralegals are considered cost effective because they can perform routine duties and free the attorneys for more specialized tasks.

The paraprofessions, unlike the semiprofessions, have a more limited body of knowledge, although much of their knowledge base overlaps that of the dominant profession. Paramedics, for example, know how to treat accident victims for shock, and they are able to use a number of medical techniques on an emergency basis. Paraprofessionals have substantial responsibility but little autonomy because of the close supervisory relationship with a professional. They may have some authority over the client, but it is sharply circumscribed and is derived from the authority of the supervising professional. Finally, they have limited ethical codes and lobbying efforts and are not generally self-disciplining, because the supervising professional is presumably responsible for the quality of their work.

Workers in many occupations share some characteristics of paraprofessionals. These workers include physician's assistants, teaching assistants and teacher's aides, library aides, caseworkers without formal social work credentials, and various technical assistants who work with engineers, architects, or accountants. Many occupational coding systems do not list them as professional workers; instead, they may be listed as technical workers, administrative support workers, or service workers. What these occupations have in common is their close association with one of the professions or semiprofessions (Creighton and Hodson, 1995).

As with the semiprofessions, many of the paraprofessions are predominantly female. In the United States over 97 percent of dental hygienists and 86 percent of legal assistants, for example, are women. The representation of minority group members in these fields is still relatively small, however. Among dental hygienists, only 1 percent are black and 8.5 percent are Hispanic, and among legal assistants, 8.2 percent are black and 11.5 percent Hispanic. In Canada dental hygienists are also predominantly female and seeking professionalization (Brownstone, 2000).

THE FUTURE OF THE PROFESSIONS

The two opening passages of this chapter summarize two views of the future. The quotation from Talcott Parsons suggests that the future will be the age of professions, with professionals exercising great social control over work and knowledge. The quotation from Ivan Illich suggests that deprofessionalization will accelerate so that we will come to wonder why we accorded so much authority to the professions. There is support in this chapter for both of these views.

The professions constitute the backbone of the **"new middle class"** that is based on knowledge rather than on the possession of capital (Mills, 1951; Gouldner, 1979; Brint, 1984). Among sociologists there is considerable disagreement about the importance of this new class (Bruce-Briggs, 1979; Kellner and Heuberger, 1992), but some aspects of the importance of professionals are well established. Professionals in private practice constitute an important segment of small-business owners, and they tend to be respected members of local business communities. Many professionals are also in demand to play important roles in large organizations, either because their expertise is central (for example, a physician who is a hospital administrator) or because they perform important functions on the staff (for example, a lawyer who is house counsel to a corporation). Both individually and collectively, professionals have gathered considerable political and economic power (Polsky, 1991). Through their professional associations they have become adept at maintaining their power through promoting favorable legislation and educating the public about their importance. Their self-organization and substantial control over their work environment are often envied by other workers with little such control.

But the professions are changing. The relentless creation of new knowledge has many consequences. One consequence is *intellectual obsolescence* among professionals who cannot keep up with the new knowledge (Leventman, 1981). A second consequence is the *routinization of professional judgment.* New computer applications in the professions provide a good example. Interactive computer terminals can be used to take medical histories, and computers can search databases for illnesses consistent with a cluster of symptoms. Databases of legal opinions can identify, within seconds, opinions that once would have required a lawyer many hours of library research. Although these developments may make more information accessible to professionals, they may also deskill the individual professional by putting more control in the hands of managers, technicians, and bureaucrats and leading to a lower value being placed on professional knowledge and judgment (Derber, 1982:196–198; Ritzer and Walczak, 1988).

A third consequence of the new knowledge is *pressure for specialization.* Specialization, in turn, has created the need for integration. The knowledge needed to solve a particular problem may require that the efforts of several specialists be integrated by a generalist. This integration has often taken place in large professional firms, where the professional values may remain paramount, but the individual professional has much less influence. Another alter-

native is to train more professionals who are generalists rather than specialists (Winslow, 1993). A key unanswered question for the future is who will exercise the social control over esoteric knowledge (Rueschemeyer, 1983).

Many professionals no longer work in independent practices nor even in large-scale professional practices; instead, they are employees of a large organization or government agency. They may often find that much of the self-regulating nature of their profession is lost, because for them the bureaucratic rules of the employing organization often take precedence (Leventman, 1981). Physicians who manage HMOs or hospitals may find themselves in conflict with physicians who remain in practice; professions are not so cohesive under such conditions as they once were. Globalization, information technology, and other developments challenge the traditional practices of the professions, with some professional jobs already being outsourced (Greenwood and Lachman, 1996).

Other pressures may eventually diminish the authority and autonomy of professions. New bodies of knowledge may challenge the old claims of the professions, as alternative medicine has done for physicians. Acupuncture, herbs, and various forms of therapy now compete with traditional Western medicine and its practitioners. Semiprofessions and other occupations seeking professionalization add to the pressures against the established professions. Deprofessionalization affects both the established professions and the occupations seeking to professionalize. There is a renewed demand to demystify professional knowledge so that clients can be more intelligent consumers. Court decisions and third-party regulation have affected the autonomy of the professional to set fees. Professions with adjacent bodies of knowledge often do battle with one another to claim the unique right to advise clients in their areas of expertise.

Nevertheless, the professions retain considerable strength. The fact that their knowledge bases continue to grow leads to more specialized services. Most have been successful in convincing consumers that professional services are needed and worth the price. Legislatures still defer to professional authority.

Individual professionals are accorded high prestige within the community, and most continue to command incomes well above the median income for all workers. Despite their problems, professionals remain among the most powerful of workers (Halliday and Karpik, 1997).

SUMMARY

Professions are high-status, knowledge-based occupations. According to the hallmark or structural-functional approach, they are characterized by a monopoly over advanced, specialized knowledge; by autonomy; by authority over clients and often over subordinate occupations; and by a degree of altruism. The professions attempt to be self-regulating by using associations to maintain their legal perquisites, erecting barriers to entry, and policing the behavior of their members. The associations monitor professional schools, promulgate codes of ethics, and educate the public about the importance of the profession. The professional association also seeks to maintain the authority of its members against the competing claims of other occupations whose members may have mastered similar bodies of knowledge.

The conflict approach emphasizes that the four hallmarks of the professions are more the result of professional power than its cause. In contrast to the structural-functional approach, the conflict approach stresses the struggle for power among members of a profession, among professions, and between the professions and other occupational groups.

Professionalization is a process by which occupational groups seek to improve their collective status by more closely resembling a profession. Semi-professions are lower-status, knowledge-based occupations seeking to emulate the professional model. Most semiprofessionals are employed by organizations, and both bureaucratization and unionization may undercut their drive for professional recognition. Some of the semiprofessions may also be regarded as lower status because they are primarily female in composition and have traditionally been relatively low paid. Paraprofessionals are more closely tied to a particular profession, which oversees their training, day-to-day performance, and evaluation. Although the paraprofessions may have a professional organization and credentials, they are really helper occupations in the service of the dominant profession.

Deprofessionalization poses an important challenge to the existing professions and to the professionalizing occupations. Individual professionals may lose autonomy when they are employed in large, professional firms. Autonomy is also lost when they are employed in bureaucratic organizations or in government agencies. Professions as a group face challenges from the general public, many of whom are no longer content to be passive clients. Professions are also challenged by new government regulations, the demands of third-party payers, and the unionization of their own members.

Professionals are significant models for other workers because they have achieved so much autonomy and authority. Being a member of a profession confers considerable prestige and usually higher income. Many of these individuals are considered members of the so-called "new middle class," which achieves its position based on knowledge and achievement and not on the acquisition or inheritance of capital. The future of the professions may take either of two paths. The process of deprofessionalization challenges the position of the professions and of other occupations that seek to emulate them. As knowledge expands and the economy is increasingly transformed by both knowledge and the service industries, however, the collective prestige of the professions is likely to grow.

KEY CONCEPTS

profession	meritocracy	semiprofession
professionalization	professionalism	paraprofession
professional culture	deprofessionalization	new middle class

QUESTIONS FOR THOUGHT

1. According to the definitions used in this chapter, is a professional football player a professional worker? Why or why not? Would your answer to this question differ if you adopt the conflict approach instead of the hallmark (structural-functional) approach?

2. The professions are based on knowledge. How does the spread of information technology affect the professions? What is the effect of the Internet on the monopoly of professional knowledge?

3. Choose an occupation you have considered for your own career. Describe how this occupation might try to develop abstract knowledge,

authority, autonomy, and altruism in its efforts to professionalize or to maintain professional status.

4. Suppose an occupation did not require a state-issued license for its practice. Could it still be considered a profession? Can you think of any examples? What are the alternatives to professionalization for an occupation that wishes to elevate its status?

5. Many semiprofessions are predominantly female. How might the gender of the workers affect the ability of an occupation to professionalize?

MULTIMEDIA RESOURCES

Print

Andrew Abbott. 1988. *The System of Professions: An Essay on the Division of Expert Labor.* Chicago: University of Chicago Press. A systematic discussion of the role of professionals, their work, and the process of professionalization.

Paul Starr. 1982. *The Social Transformation of American Medicine.* New York: Basic Books. A Pulitzer Prize-winning account of the struggles of the medical profession to gain recognition in the United States.

Internet

American Bar Association. *http://www.abanet.org* The American Bar Association is one of the most powerful professional associations. Although many

parts of this site are password protected for ABA members only, visitors to the site can read articles in the Association's *Journal* and find information about legal practice and how to locate an attorney. There is a section with student resources that includes information on attending law school.

American Medical Association. *http://www.americanmedicalassociation.org* This site provides information on finding a doctor, recent medical news, and other information for both members and non-members. There is also a section with information for medical students.

Aboriginal Nurses Association of Canada. *http://www.anac.on.ca* This professional association seeks to encourage nursing careers among aboriginal peoples and is also concerned with improving health for aboriginal peoples.

RECOMMENDED FILM

Patch Adams. 1998. Robin Williams plays a medical student who confounds professors at his medical school with his belief that laughter is the best medicine. He eventually founds the Gesundheit Institute to further his cause. Rated PG–13.

12

Managers

First-line managers and supervisors—those who supervise direct
production and service workers—have always had a difficult job:
exhorting workers to live up to standards and demands thrust upon them
by higher levels. The new corporate era, which has brought new work
systems, only increases the pressures on them.
(KANTER, 1983:56)

Workers have willingly offered themselves complete with their personal
and family lives to the organization. Like children, they have allowed the
organization to determine what is best for them and how they should best be
used. They expected no power and that is precisely what they got. Such childlike
passivity is fading fast. The divine right of kings or managers no longer exists.
(KELLEY, 1985:93)

Ask most workers who has power in their workplace, and they will answer
"the boss" or "the management." **Executives, managers**, and **administrators** form the occupational group that organizes and coordinates work and makes
decisions about production, finances, and the hiring, firing, and deployment of other
workers. These responsibilities do not imply, however, that managers act unilaterally.
As the quotations above emphasize, even the boss faces limitations. In this chapter we
will examine the types of jobs managers hold, the changes they are experiencing in
their roles, and the ways in which managerial performance can be evaluated.

TYPES OF MANAGEMENT ROLES

Many kinds of workers can be classified as executives,
managers, or administrators, from the cabinet-level
head of a government agency to the self-employed
proprietor of the corner candy store. In this chapter
we will use the generic term *manager* to refer to all
three groups. Managers may have responsibility for
millions of dollars in a budget, or they may manage the
evening shift in a round-the-clock restaurant. In labor

relations, management is distinguished from labor by its responsibility for the maintenance and growth of the employing organization. Managers are defined principally in relation to their employing organization and not in relation to their occupational group. Unlike a professional, for example, who shares common training and credentials with other members of the profession, managers do not necessarily share a common base of knowledge, nor is entry into management restricted by need for licensure or certification. The most common credential associated with managers is the MBA (Master of Business Administration) degree, but it is by no means essential for many management jobs.

What is essential to management jobs is providing leadership and coordination, one of the oldest specializations in work roles. (See Chapter 1 on the emergence of leader roles in pre-industrial societies.) Leadership roles have especially been associated with political and government work, and the terms *executive* and *administrator* are likely to be used for such workers. In the private sector, and especially in corporations, the terms *executive* and *manager* are more common. All such work involves integrating, synthesizing, and coordinating information, but different terms are used depending on the type of organization and on how highly placed the worker is within the bureaucracy of the workplace.

Executives

The term *executive* refers to workers at the very top of a workplace bureaucracy. The president of the United States is called the chief executive. Within the government, the executive branch refers to all the agencies that carry out government programs. Even very small tribes typically have a chief who serves the executive function. Agrarian societies developed hereditary monarchies to provide leadership. Later, executive roles in the modern political states were filled by government officials. The top officers of most corporations and agencies are also called executives.

Managers

Although management as a field of study for specific occupational preparation is relatively new, for centuries managers ran farms and factories in the absence of owners. Even a large household might have had a steward as manager. These prototype managers had foremen, supervisors, or servants reporting to them. The functions of these managers included purchasing and accounting; directing which products were produced and how; hiring, monitoring, and disciplining workers; and providing for the maintenance, repair, and replacement of equipment. These functions are similar to those of the managers of today's bureaucratic enterprises.

But today's management is different for two basic reasons. First, today's enterprises are far larger and more geographically dispersed than the early farms and factories were. Second, today the ownership is also more dispersed (Hansmann, 1996). The old-style managers reported directly to the owners, but the modern corporation is owned by many shareholders who elect a board of directors. The board in turn appoints corporate executives. The rest of the management team, who are appointed by the executives and their subordinates, ultimately receive their authority from the board of directors.

Administrators

Government administrators historically assisted rulers in developing and enforcing policies, and today they serve the same functions, but in much more elaborate and complex government agencies. In Europe the medieval church developed separate offices to perform specific functions, and each office had a cadre of administrators. The local bishop served as the executive, in the sense of having final authority, and other members of the clergy administered parishes, charitable institutions, or church offices. Today the top managers of nonprofit firms are often called administrators. For example, at your college or university there are probably specialized officers who are recognized as administrators. Hospitals, nursing homes, and similar institutions are often headed by an administrator.

Staff and Line Managers

Executives, managers, and administrators can also be distinguished by whether they have line or staff responsibility. A **line manager** supervises workers directly involved in the production process. Line managers typically have substantial numbers of "direct reports"—lower-level supervisors and workers who report directly to them. A **staff manager**, by contrast, provides or supervises a service that is necessary to the organization but is not the organization's major product or service. Examples of staff specializations that may have managers include accounting, personnel, purchasing, maintenance, and legal services. When restructuring, many companies have sought to outsource such staff functions. For example, instead of maintaining an accounting staff, the company may find that a contract with an accounting firm provides high-quality accounting services at a lower cost.

EXECUTIVES, MANAGERS, AND ADMINISTRATORS AT WORK

Management attracts workers because its duties are varied and carry some measure of autonomy and authority. Managers also tend to be well compensated, although compensation depends upon experience and the employing organization. In 2005 the median weekly earnings of all full-time workers were $651. For male managers the comparable earnings were $1,230, and for women they were $902 (BLS, 2006b). For top executives, the compensation is much higher.

Top corporate executives are among the best compensated workers in the United States, with salaries, bonuses, fringe benefits, and perquisites that easily make them multimillionaires. An important issue is whether investors and tax officials even know how much the top executives earn. One study, for example, estimated that the CEOs of the largest companies had a median pension plan value of $15 million, a figure that had not been routinely disclosed (Bebchuk, 2006).

Demand for Managers

Within a firm, the demand for managers is a function of technology and organization. Technology may alter a manager's **span of control**, the number of workers a manager supervises, and this in turn will affect the number of managers required. For example, a manager who supervises workers through telecommunications links to computer workstations can supervise a greater number of workers, and workers dispersed over a geographically larger area, than can the manager who must supervise in person. This is a case in which technology might increase the span of control.

The bureaucratic complexity of the work organization also affects the number of managers who are needed. The nature of the bureaucratic hierarchy determines how many levels of supervision are required, referred to as the **steepness of hierarchy**. A firm that has many establishments needs a set of managers at each location. If the divisions or product lines of a company are autonomous, the company needs to hire a set of managers for each one.

Changes in organizations also affect the number and type of managers needed. Reorganizations, especially acquisitions and mergers, are often followed by the layoff of some managers in the acquired firms. In a merger, where both firms already have a full set of managers, there is usually lengthy negotiation over how many managers will be retained. Executives, or **top management**, may negotiate in their hiring contracts for compensation if a merger, acquisition, or reorganization leads to the loss of their jobs. These severance pay provisions are sometimes so lavish that they are called "golden parachutes." Lower-level managers and ordinary production workers receive much smaller severance benefits, if any. Within the labor force as a whole, the demand for managers is a function of the number of firms as well as of technology and organization. Even very small firms must have managers. The term *manager* appears in the job title of many workers—office managers, sales managers, apartment complex managers, restaurant managers, and so on—even though these jobs may seem to have relatively little in common with managing a large corporation.

Managers' unemployment rates have traditionally been very low. Managers, who are responsible for personnel decisions, traditionally tended to lay off themselves last. In 2006, when overall unemployment was 4.1 percent, the rate for managers was 1.5 percent (BLS, 2006b). As an occupational category, managers are unlikely to suffer from involuntary part-time employment or very low pay (Clogg and Sullivan, 1983).

More recently, however, *organizational changes* have reduced the long-term job security of managers. Mergers and acquisitions often result in laying off "redundant" managers (Newman, 1993). Foreign competition, new technology, deregulation, and other financial issues have led some organizations to reduce their managerial staff. During the 1990s **downsizing** was the term used to justify flattening the company hierarchy by reorganizing in ways that laid off some middle managers. These layoffs followed decisions by many executives that their companies were too fat and needed to become "lean and mean." Interestingly, the willingness to lay off more workers is becoming a sign of managerial ability. One observer comments that "job shedding has become fashionable—the mark of a good manager" (Uchitelle, 1994). Middle managers continue to be targets for job shedding.

The Self-Employed Worker

The self-employed proprietor is often classified as a manager, even if there are no employees and the proprietor is managing accounts, inventory, and the daily business rather than people. Less than 8 percent of the U.S. labor force is self-employed, but the proportion of self-employed is much higher among certain groups, such as immigrants.

The self-employed tend to be concentrated in a few industries. About one-third of the self-employed are in professional and business services or in education and health services. In these industries the self-employed worker may be both a professional, in the sense we used the term in Chapter 11, and the manager of a self-owned accounting service, law firm, or clinic. The self-employed are also commonly found in retail trade, in which the manager-proprietor operates a small retail store. The most common retail stores run by the self-employed are grocery stores and small clothing stores. A third common industry for the self-employed is personal services, in which the proprietors may own bars or restaurants, dry-cleaning establishments, or small motels.

Some ethnic communities have developed specializations as small-business owners (Butler, 1991; Woodard, 1997). Examples include Koreans who specialize as greengrocers (vegetable sellers) in New York City, or Indians who specialize as small motel owners (Kim, Hurh, and Fernandez, 1989; Yoon, 1991; Dong, 1992). The ethnic community often provides the capital needed by these small-business owners (Light, 1972). The term *ethnic enclave* is used to describe a cluster of ethnic business in which the owners provide substantial employment to other members of their ethnic community. Cuban Americans in Miami represent an example of an ethnic enclave (Portes and Jensen, 1989).

The failure rate of small businesses is very high. Between 2000 and 2001, entrepreneurs began 585 thousand new businesses, but another 553 thousand small businesses closed (Census, 2006). Although many small businesses fail, some entrepreneurs become extremely successful and begin what are destined to become large companies.

One way the small business owner can share some of the risks of ownership is by becoming the *franchisee* of establishments affiliated with large chains of restaurants, convenience stores, motels, and other businesses. Franchises are also available in personal services, such as weight-loss groups or house-cleaning services. Although some franchise establishments are owned by the company, in over 80 percent of the cases the franchisee buys the business or a share in the business by investing a large sum of money. The franchisee agrees to run the business according to the conditions set down by the national organization. The franchisee receives the know-how and advertising of the national organization. This is a great advantage in building a clientele because customers will recognize the nationally advertised product. The national organization controls much of the business operation, including what training is necessary, what products must be bought, and how they may be sold. The national company may require not only that all materials be bought from them, but they may also receive a portion of the profits without risking much capital. The franchisee provides the capital; hires and supervises the workers; arranges purchasing, accounting, and other services; and oversees the daily operations of the business. Although many franchisees hire additional workers, they also commonly work long hours, often alongside other members of their families.

Supply of Managers

Management was historically the preserve of white Protestant men. One reason given for this exclusivity is that managers must work in a climate of uncertainty, and homogeneity among themselves was believed to reduce one source of miscommunication and uncertainty (Kanter, 1977). This view cannot explain, however, why the preferred group is white, Protestant, and male. Even white Protestant men sometimes miscommunicate and may hold different values, however, so the workplace culture may be more important than recruitment in maintaining homogeneity. The fundamental reason is simply that a group with power tends to maintain its power.

Partly because of legislative pressure and partly because of popular sentiment, women in recent years have made notable advancement into management positions. About 37 percent of all managers are women. Figure 12.1 shows the percentage of women and minorities in some managerial specialties. Firms that use formal recruitment procedures for managers, as opposed to using more informal networks, tend to have higher proportions of female managers (Reskin and McBrier, 2000). Some firms have a two-track system of promotion to management, with the "mommy track" (or *ippanshoku* in Japan) having less complicated jobs, lower mobility requirements, and a ceiling on managerial promotions (Strober and Chan, 1999).

Management remains a field with relatively little minority representation. In 2005 only 6 percent of all executive, managerial, and administrative workers were black, and only 6.3 percent were Hispanic. About 4.2 percent were Asian. Black managers had their greatest representation among education administrators (13.4%) and social and community service managers (11.7%). By contrast, only 2.7 percent of the nation's 805,000 construction managers were black. Hispanics had their greatest representation (13%) among property, real estate, and community association managers (BLS, 2006b).

One study has indicated that fluency in English is the most important explanatory factor affecting the chance that Hispanics will be employed as managers, whereas for non-Hispanics years of schooling is the most important variable (Mundra, Moellmer, and Lopez-Aqueres, 2003). Box 12.1 reports the experience of some black women executives who must negotiate acceptance from white male colleagues.

Social class also affects recruitment to management. Virtually all executives have a college education; controlling for education, people with upper-class origins are more likely to be found within the top ranks of corporate management. Top managers are more likely to have a bachelor's degree from a top-ranked college and a master's degree in business administration (MBA) or a law degree from a leading institution. Holding an advanced degree and having an upper-class background are useful for joining formal and informal networks inside and outside the corporation; these networks are useful for career moves within the corporation (Useem and Karabel, 1986).

Middle managers are more likely to have come from middle-class or even working-class origins and to have graduated from small state colleges or denominational colleges. There are also industries, such as auto repair services and retail sales, in which the majority of managers do not have college degrees (Verma and Boyer, 2000). Their position within management may represent social mobility from their parents' positions. Both they and their spouses often feel proud of their success, and this keeps them committed to the corporation even though the conditions of work may be grueling (Newman, 1988). Middle managers often experience a "ceiling effect," reaching a level within the firm above which they cannot pass.

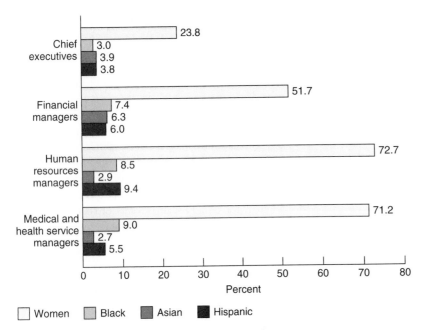

FIGURE 12.1
Percentage of Women and Minorities in Selected Managerial Specialties, 2005
SOURCE: BLS, 2006b.

Business proprietors and other self-employed workers often have backgrounds similar to those of the middle managers, although they are not so likely to have college degrees. Many of the self-employed, however, have specialized training and skills; examples include independent craft workers and professionals.

The Managerial Career

In large firms the managerial career is marked by a number of transitions. In her interviews with managers at a firm she calls "Global Products, Inc.," Diane Margolis (1979) identified critical stages in the managerial career. During the *initiation*, the period immediately after hiring, the manager comes to adopt a self-identity as a manager and, equally important, an identification with the firm. The initiation is intended to be a training period, and the firm uses every expedient to train the new manager as quickly as possible. The new manager works long hours and often long weeks; to refuse these assignments, without very good cause, makes superiors wonder aloud if the manager is really committed and "has what it takes." (This period of overwork also characterizes some professional workers, including medical interns, new associates in law firms, and beginning university faculty.)

Frequent transfer is a technique firms use to train managers and to initiate them into the company. Ostensibly, the transfer provides adequate staffing for the firm and teaches the manager new skills. As one executive put it:

> We have somebody in the Philippines we're going to replace. This situation has triggered five moves. The man in the Philippines is being let go because he's not doing the job. A man in Hong Kong will be moved to the Philippines. A man in Singapore will go to Hong Kong. One in Thailand will be transferred to Singapore, and a man somewhere else in Asia, we haven't decided who yet, will replace the man in Thailand (Margolis, 1979:48–49).

But the frequent moves also serve other functions. They make it difficult or impossible for the manager's family to become involved in the

---

B O X   12.1   Black Women Executives Negotiate a Place

Patricia Parker is Associate Professor of Communication Studies at the University of North Carolina at Chapel Hill. In her study, she conducted in-depth interviews of fifteen African American women who were senior executives in private companies or government agencies. Her research centered on the challenges these women faced in their workplaces and in the communication strategies they used in dealing with their challenges.

A common challenge to the women executives was having their ideas ignored or inappropriately claimed by others during meetings to discuss workplace issues. One of the women said,

> You will sit in a meeting or be on a conference call, and the White males will be talking loud and aggressively, and they like to hear each other talk. You can say something, and get no attention. They will say something, restate [what you said] (not as clearly as you did, by the way) and all of a sudden it's a good idea. (Parker, 2002:257).

Because it is necessary for executives to work together with others on the leadership team, this type of exclusion makes the job harder for African American women, and it also deprives the workplace of the value of their ideas.

Based on her interviews, Professor Parker identified three ways in which the women executives dealt with this problem: avoidance, indirect strategies, and direct strategies. An example of an avoidance strategy was being quiet in a meeting and waiting to be called upon. One example of an indirect approach was humor:

> Sometimes, if possible, I'll be humorous, and, you know, insert myself and say, listen you guys just are not listening. Listen, this is your mother speaking, this is a woman's voice, you know, just kind of making light of it. (Parker, 2002:258).

The more direct strategy involved a certain amount of confrontation. One executive described her use of this strategy in this way:

And so [for our next meeting] we went to the training center.... [for a special executive meeting], and the guy that was leading it, I made sure that I sat next to him. We were at a rectangular-shaped table; he was at the head of the table, and I sat to his right. I wanted to be right close to [the person in charge of the meeting]. So we're having the meeting, and the guys are just ... they just love to hear themselves talk, you know. You can just tell. It didn't matter too much what they were saying as long as they each talked. And one would make a point, and somebody else would have to reiterate it in their own words, and a lot of that going on. Not a lot of facilitation ... you know. Well, in any event, I sort of took my time, the only female ... at the table, and we got to some area and I thought, in my mind, I said, "I know this is a good idea. I mean, I'm gonna throw this one out because this is good." Threw it out there, and true to form ... [the person] who was leading the meeting, when I said what I had to say, he kind of said, "Oh. That's a thought." And then moved on, and I just touched his arm, and I said, "Does that mean that we're not going to consider what I just said?" He said, "Oh. No." And you know what ... this is the truth, he said, "No. You know, that really did have a lot of merit, and I can see how that'd work." And all of those guys jumped in, and they all said it was good, and they all ... for some reason they just didn't want to acknowledge it ... And after that, they voted for me Employee of the Month. I was just their best girl-friend, you know ... and so I won them over in that split ... you know, it was probably two minutes ... and that was it (Parker, 2002: 258–259).

Although their strategies differed, each executive found her way to communicate effectively with other executives and to have her ideas taken seriously.

SOURCE: Patricia S. Parker, 2002. "Negotiating Identity in Raced and Gendered Workplace Interactions: The Use of Strategic Communication by African American Senior Executives within Dominant Culture Organizations." *Communication Quarterly* 50, (Summer–Fall): 251–268. Used by permission.

---

community or even to have many friends outside the company. When moves come as often as every six months, the family relies on itself or on other

corporate families for companionship (Newman, 1988). Limiting ties with other groups increases the manager's reliance on the corporation as a source of income, identity, and prestige. More recent research suggests that the mergers and reorganizations have also become a frequent cause for moving managers (Stovel and Savage, 2006).

Most successful managers have *mentors* during their early years in the company (Kanter, 1977). The typical **mentor** is two levels higher than the manager and serves as a role model, protector, adviser, and cheerleader. Many management jobs are ambiguous, and companies often act in a climate of uncertainty. In such a climate the mentor can provide a road map for understanding. Mentors may warn their young protégés of interdepartmental "dirty tricks" and teach them how to protect themselves while giving every appearance of being good team players. Eventually, as a part of developing the protégé's separate identity, the protégé and mentor part ways, sometimes with hard feelings on both sides. But as the protégé becomes better established, a rapprochement and lasting friendship sometimes develop between the former protégé and the former mentor (Harriman, 1985).

On the other hand, management careers also encounter barriers. Table 12.1 indicates the percentage of male and female managers who reported experiencing various career barriers. Typically, managers endure a period of *pruning* during which some managers are promoted and others are encouraged to leave (Margolis, 1979). Many firms do not actually fire managers, but they manage to convey their sense that the manager "doesn't fit in" or "would be happier elsewhere." Many large corporations now have their own outplacement offices for helping to relocate displaced managers. Some corporations also have a group of smaller client firms to whom they refer surplus managers. These smaller firms, often unable to afford the training that large firms provide, are eager to cooperate because they can receive managers who will be better trained than if they had been promoted from within the smaller company. Short of being pruned, managers may also find themselves *rechanneled* into the by-waters of the corporations, often into staff positions that are not rungs on a career ladder.

The manager who survives and is promoted walks a fine line between too obviously advancing his or her own career and being perceived only as

TABLE 12.1 Percentage of Respondents Mentioning Principal Barriers in Their Corporate Career, by Sex

	Men	Women
Inflexible working patterns	6	14
Family commitments[a]	44	43
Lack of adequate childcare	3	8
Lack of career guidance	45	50
Lack of training provision	9	5
Prejudice of colleagues	11	23
Lack of personal motivation/confidence	19	21
Senior management seen as a "club"	32	54
Social pressures	1	10
Sexual discrimination/harassment	1	17
Insufficient education	11	13

SOURCE: Judy Wajcman, 1998, *Managing Like a Man: Women and Men in Corporate Management*. Cambridge, England: Polity Press, p. 88. Used by permission.

[a]Here the data refer only to respondents with children.

a team player. The manager who is too self-interested is suspected of being insufficiently loyal; the one who is too little interested in promotion is insufficiently ambitious. Box 12.2 presents a situation in which a subordinate had to convince his superior that an appropriate decision was the superior's idea. Promotion to higher levels within the firm is often acknowledged to be "political," in the sense that it is based on something other than technical qualifications. The corporate ideology, however, is that the firm is a meritocracy and that those who hold the highest positions have achieved them by their accomplishments. This means that for the many managers who never make it to the top, the ideology encourages the belief that they themselves are to blame for their failure (Newman, 1988).

Many questions have been raised in recent years about the ethics of managers in American corporations. Revelations of managers' bribery of overseas officials, their tolerance for unsafe products and production facilities, their efforts to

---

BOX 12.2   Whose Idea Was It?

Sociologist Calvin Morrill studied conflicts and their resolution among executives. Successful subordinates learn not to outshine their superiors, and yet they still need to achieve recognition to advance. In this excerpt, a younger executive explains how he led his boss to a good decision.

Asymmetries between superiors and subordinates also occur in written communication. First, more memos (almost ten times as many) travel upward than downward. Second, the qualitative content of executive communication differs among superiors and subordinates. Upward communication tends to be informative *and* covertly rhetorical. Conversely, superiors are overtly rhetorical in their downward communication and less informational. An example of the differences in the contents of upward and downward communication was related by this senior vice president of branch management in dealing with his superior, an executive vice president division manager:

> We had a branch that I wanted to close. The place was in an older part of [a city] and wasn't doing very well. We had a newer branch nearby that could serve the needs of the older branch's custo-

mers. It would be too expensive to make the older branch state-of-the-art. I could have written a memo that argued for closing the old branch, but I didn't. I had my staff make up a report using several sets of data—cost accounting, depreciation of the building, customer profile—you name it, it was in there; lots of pretty graphics from our new graphics software, nicely bound cover, the works. I sent that up to the EVP. It said what I wanted it to say—Close the old branch down—without saying it in so many words. [Smiling as he tells this part of the story.] Led the EVP right to the decision to close the sucker. It's a different story when he wants to get something done. One time he wanted to close a branch that had been a flagship at one time for a small bank we acquired a few years ago. He writes a one-page memo telling me to start disentangling ourselves from the branch after the new year. No figures, no nothing, just do it. Obviously, if I know what's good for me, I don't write a memo like that to him. It's simple here, really, at Old Financial; different rank, different way to get things done.

SOURCE: Calvin Morrill. 1995. *The Executive Way: Conflict Management in Organizations.* Chicago: University of Chicago Press, pp. 102–103.

---

evade regulation and statutes, and their conspiring to fix prices have led many to ask whether the lengthy process of initiation into management does not destroy the moral sensibility of managers (Jackall, 1988). More recent accounting frauds further undermined public confidence. The spectacular collapse of corporations such as Enron, Global Crossing, Tyco International, and World-Com, which caused the loss of billions of dollars to shareholders and the loss of employees' jobs and pensions, have led to substantial rethinking about the role of ethics in corporate life. The subsequent jailing of prominent corporate executives has made national headlines. New government regulations, especially the Sarbanes-Oxley Act, have been developed to try to eliminate false or misleading misrepresentations of the corporation's

financial condition and to require corporate boards to justify executives' compensation and perks (Lander, 2004).

CONTINUITIES AND DISCONTINUITIES IN MANAGEMENT ROLES

The jobs done by village chiefs, medieval bishops, plantation overseers, and ships' captains are still recognizable as management tasks. There are also, however, important differences between the types of work historically done by managers and administrators and the types of work they do today. These differences stem from changes in scale, environment, specialization, and technology.

Changes in Scale

Compared with the governments and companies of the past, today's agencies and corporations are larger and more bureaucratized. They are likely to have several layers of managers, each coordinating and monitoring the layers below. This leads to the distinctions among *top management* (the executives), **middle management**, and *first-line supervisors* (foremen and the like). The **first-line supervisors** are not considered part of management by some organizations. The larger the organization, the more likely it is to develop a more complex bureaucratic structure. In large, bureaucratized workplaces, the average manager must think about both the **horizontal** and **vertical dimensions of bureaucracy**.

The horizontal dimension refers to the development of functional units that perform discrete tasks. For example, a factory may have separate departments for purchasing, personnel, production, maintenance, and shipping. It is the manager's job to ensure not only that his or her own unit is doing its job but also that its job is coordinated with those of other units. If purchasing has failed to buy needed raw materials or if completed products cannot be shipped, the production manager may still receive part of the blame. In many corporations considerable friction exists among the functional units over budgets, resources, and perceived importance within the company.

In a conglomerate (a company made up of companies), the functional units may be separate corporations. If these units are autonomous or nearly so, coordination among them may be a key function of top management. One accusation leveled at American corporations is that their quality control suffers because of uneven attention given to the functional units. As more and more chief executive officers are recruited from among lawyers and financial experts, their attention shifts to those areas and away from the basic products of the company (Hayes and Abernathy, 1980).

The vertical dimension refers to the bureaucratic hierarchy of the workplace. Before workplaces became large and functionally specialized, different functions were often performed by small, separate firms, and the coordination was done by impersonal market forces. As workplaces have become larger, the coordination has been done by managers and has been accomplished by adding layers to the hierarchy (Chandler, 1980). The typical manager reports to a superior who in turn reports to other superiors. The number of layers in a hierarchy is referred to as the steepness of an organization, and steeper organizations experience greater communication problems because of the necessity for information to be accurately transmitted up and down the corporate chain of command. "Flatness," or relatively few vertical layers, is now considered a desirable attribute for corporations that wish to be agile in responding to changing conditions.

Changes in Environment

Managers, and especially executives, must be concerned not only with what is happening within their organization but also with the relationships between their organization and other organizations. The organization will not survive unless it is successful in its exchanges with others in its environment. These exchanges can be cooperative, as are those between suppliers and purchasers. They can also be competitive, particularly those between organizations producing the same goods or services. Relationships with regulatory agencies, consumer organizations, lenders, and possible merger partners may range from quite friendly to fiercely adversarial. Top levels of management must pay serious attention to each of these relationships. Managers also try to monitor what changes are being made within other organizations, especially if similar innovations might benefit their own company.

Monitoring and responding to changes in the environment are among the most important

---

B O X 12.3 Staying Competitive

By 1966, as a result of a string of technical innovations, Corning [Glass Works] held the dominant position in the U.S. market for television picture tubes, and the television business provided nearly three-fourths of Corning's earnings. Yet by 1975, in less than a decade, the television business was almost all gone. Corning no longer made black-and-white tubes, and its color tube business was losing money. What happened? The Japanese. In the late 1960s they had started entering the U.S. market for televisions, first for black-and-white and then for color sets. Gradually they began running American manufacturers like Corning's customers out of the market. By the mid-1970s, imported televisions accounted for over 60 percent of the market for black-and-white sets and over 40 percent of the market for color sets. Corning's base of customers for its tubes had shrunk from twenty-eight manufacturers to only five.

With the decline of its television business and setbacks in other areas, Corning set out in the late 1960s and early 1970s to do what it had always done: it launched a wide-ranging search for new products, hop-

ing to accomplish the next big breakthrough. During this period, Corning pursued development of such diverse products as safety windshields, glass-based lasers, ceramic roofing shingles, ceramic heat exchangers for turbine engines, optical fibers, medical instruments, optical waveguides for telecommunication, glass razor blades, integrated circuits, and even computer terminals. Many of these development efforts, such as the safety windshields, ceramic components for glass turbines, and glass lasers, built upon Corning's historical core competency in glass-ceramics technology. Others, such as efforts to establish a position in medical equipment, integrated circuits, and biotechnology, did not.

There was a marked difference in Corning's success with these different types of ventures. When Corning pursued ventures that were consistent with its core competencies, it was usually successful. When it ranged far afield and abandoned its expertise, it wasn't.

SOURCE: Joseph H. Boyett with Himmie T. Boyett, 1995, *Beyond Workplace 2000: Essential Strategies for the New American Corporation.* New York: Dutton, pp. 44–45.

---

tasks of contemporary managers. These tasks are especially pressing in industries in which the environment is changing rapidly through technology, organizational realignments, expansion of the world market, or other developments (Madu, 1996). Box 12.3 discusses how Corning Glass Works successfully changed in response to international competition for one of their products.

Changes in Specialization

In small firms a manager may handle all decision making and day-to-day administration. Many management tasks require controlling and organizing the flow of information about accounts, products, timetables, productivity, and so on. These tasks may be barely distinguishable from clerical work. Management also requires a great deal of work with other people. In the small firm, the manager handles all personnel decisions in addition to interacting with customers, suppliers,

the owners, and others. Managers in small firms know the entire business, and their job is varied and requires a range of skills.

In large firms, however, no single manager serves as the jack-of-all-trades. As you saw in Chapter 11, the increasing base of knowledge forces professionals to specialize. Increased size and complexity of the organization cause managerial functions to become more specialized. Whether in the professions or in management, specialization always implies acquiring a particular type of knowledge. The two major ways to acquire this knowledge are through training and experience. Some managers are trained to assume one particular management function. For example, personnel training prepares a person to recruit, test, train, assign, and evaluate workers. The public relations manager needs knowledge of press relations, advertising, community service activities, and so on. The training to perform these functions goes on in specialized programs, often within university business schools. The graduates could perform these functions, with minor modifications, in other firms. The phrase *professional manager* refers to someone who is occupationally prepared to take on such management functions within any workplace, regardless of what good or service is produced there.

Alternatively, managerial specialization comes about through *experience* in one or more specialized departments within the organization. Experience implies specific, detailed knowledge of how things are done and how people work within a particular organization. Managers' experience depends on the functional departments they have managed during their careers. For example, the circulation manager of a large metropolitan newspaper develops expertise in setting up distribution networks, but not in gathering news. An account manager in an advertising firm becomes very familiar with the products and plans of a specific clientele, but might know very little about other clients. An administrator in a state highway department performs some of the same functions as an administrator in the state human services agency, but the specific decisions made in the two agencies are quite different. Managers' importance to their firms depends partly on their experience within the specific organization.

These two types of specialized knowledge— training and experience—presuppose very different types of preparation for management. The former approach assumes that the professional manager is one who has been trained, perhaps in an MBA program, to perform tasks needed in every organization. Such a manager should be able to move laterally from one organization to another, applying the expertise to the new situation. The latter approach assumes that the most important managerial knowledge comes from experience in a specific firm: knowledge of its production practices, its clientele, and its workers.

These two types of knowledge are not incompatible; someone trained as an MBA with specialized knowledge of one management function may also gain valuable experience in a particular organization and have both types of knowledge. But more typically, the professional manager is seen as one who moves laterally from one firm to another,

joining the company's labor force through distinct entry ports. The manager with local experience is more often seen as one who stays with a single company for long periods, often being promoted within the company's internal labor market. In recruiting new managers, the existing management must choose between the options of lateral recruitment and internal promotion, or "outside" versus "inside" (Menon and Pfeffer, 2003).

These two types of knowledge are related to the traditional distinction between staff and line managers. The line managers exercise bureaucratic authority over production of goods or services. The staff managers provide ancillary or auxiliary services, often of a skilled or technical nature, to the line managers. So, for example, line managers know about the technology and equipment used in production. Staff managers may be responsible for accounting and other financial services, legal services, personnel management, safety, marketing, and public relations. Staff managers may be blocked in seeking promotions because they lack knowledge about the central production processes, and so they seek advancement by moving laterally to similar (but perhaps better paid) positions in other firms. Line managers are more likely to be promoted from within; moreover, their specialized knowledge might not be transferable to other firms.

If line managers can be promoted, a first-line supervisor could theoretically rise to chief executive officer because the first-line supervisor is most likely to have actual experience both with the production process and with managing workers. At one time, first-line supervisors were promoted within the ranks, at least in large manufacturing firms. Today, a sharp distinction is frequently made between first-line supervisors, who are often at the top of a job ladder, and middle managers. Middle managers are recruited directly from the outside, often because they have credentials, such as college degrees, that foremen do not have. The first-line supervisor is at the top of one career ladder. The middle management positions, because they are filled from the outside, are entry ports and the first rung on another career ladder within management.

Changes in Technology

Managers manage people, production, and information. Technology affects how managers perform all three functions. It typically does so by increasing the amount and type of information available to managers. For years, a manager's *knowledge of workers* was a function of direct observation and perhaps rudimentary productivity records. New technology makes it possible for the manager to electronically eavesdrop on the speed and accuracy of clerical workers, technicians, and machine operators. Managers potentially possess far more information about workers now than they ever did (Zuboff, 1988). On the one hand, it is argued that the new technology expands the span of control of the managers, which would mean that fewer managers would be needed. On the other hand, the increased volume of information implies the need for more managerial interpretation, and so the argument could be made that more managers are needed. One reason for the succeeding rounds of employment and layoffs in the management suites is indecision about the appropriate staffing ratios of managers to other workers.

Technology affects a manager's *knowledge of production* in many ways. New machines, procedures, and methods of organizing work require the manager to learn new techniques. Technology may lead to changes in the size and characteristics of the company's workforce. What form that transition takes will depend on managerial decisions to purchase and deploy the new equipment. Managers sometimes decide that a different number of workers, with a different skill mix, will be needed (McElroy and Hazzard, 1994). For example, the adoption of electronic data-processing equipment may lead to fewer clerical workers and more equipment operators. Jobs may require fewer skills, or deskilling (as discussed in Chapter 7), or the jobs may be upgraded to require new or different skills.

The staffing ratio (the ratio of workers to capital) is a management decision, and it has a major effect on the number of managers. Upgrading may involve giving new supervisory skills to some workers so that the overall number of managers can be reduced. Managers may also decide that each production process can now use fewer workers. In the short run, this will result in a smaller workforce, although in the long run new jobs may be created if the product market expands or the company becomes more competitive (Osterman, 1986). Alternatively, managers may decide to maintain the same staffing ratio but to reorganize the work into two shifts or in other ways redeploy the existing workforce.

Finally, technology affects a manager's *information base*, usually by dramatically expanding the available information. Managers may receive so much information that they simply cannot use it and suffer from "information overload." If managers receive conflicting types of information, they must decide which data are reliable. For example, a manager may demote or discipline all slow or inaccurate workers to boost average speed and accuracy. Another manager who relies on "eyes and ears" may judge a worker to be valuable to a company, even though electronic monitoring may reveal the worker to be only average or even below average in speed or accuracy.

Many service firms now use customer responses as a source of information for managers.

> A customer relations manager in a hotel chain reported that when top management received the results of customer surveys they wrote to middle managers, informing them of problems and complaints; middle managers had to respond within 30 days, notifying upper-level management of the ways in which they had rectified any faulty conditions. If no response was made by 30 days, corporate management instigated formal proceedings against the manager (Fuller and Smith, 1996:83).

How information will be used within various organizations is at least partly a function of the prevalent models of management. In the next section we discuss different distinctions among management models.

TRACKING MANAGEMENT PERFORMANCE

Because managers are found in every type of establishment and industry, the actual content of their jobs ranges widely. As we have just discussed, four factors that affect management jobs are scale, environment, specialization, and technology. Each of these factors, however, is little more than an external constraint on how managers play their roles. In practice, even a manager with a very specialized job will find that the job contains many different tasks. The manager must simultaneously seek good relationships with subordinates, peers, and superiors, without neglecting the specific tasks of directing, coordinating, and planning. The manager must pay attention to day-to-day operations without forgetting about the company's long-range goals and objectives. High-quality job performance depends on juggling many tasks. Depending on the company, the season, and the business climate, the aspects of the job that must be emphasized will change from day to day.

Sociologists have developed several ways to investigate the job performance of managers, and we will describe two of them. One is *behavioral*, and it is based on the observation of individual managers; thus, it is a micro-level approach. The second emphasizes *organizational cultures*, and it generalizes from studies of large numbers of managers to develop more abstract management styles. It is a macro-level approach that is concerned more with the social structures that make managers behave similarly than with variations among individuals.

The Behavioral Approach

The behavioral approach to managerial performance emphasizes how managers choose to perform their roles from among the many activities available to them. Mintzberg (1973) observed five chief executives over five-day periods. As a result of this work, he reports that managers play ten major roles: figurehead, leader, liaison, monitor, disseminator, spokesperson, entrepreneur, disturbance handler, resource allocator, and negotiator. Each of these roles requires specific activities. Box 12.4 provides a checklist for specific activities that managers perform. So many activities are part of the managerial role that there can be substantial variation in how two managers perform the same job. The checklist in Box 12.4 can be used by trained observers to record the actual behavior of managers at time intervals throughout the day. Another way to gather such data would be to ask managers to recall their activities or to log them during the day. Using trained observers is probably better, however, because it avoids the problems of recall or of subtle distortion of one's performance.

Several studies have observed the actual time managers allocate to these tasks. Mintzberg (1973) found that top-level managers spend relatively little time on the activities traditionally thought to be central to the managerial role, such as long-range planning. Instead, their work consists mostly of superficial and reactive encounters. Later studies have confirmed that top managers spend much of their time interacting with others through short conversations and jokes that seem unrelated to work but actually help to build workplace communication (Kotter, 1982).

One study of successful and unsuccessful managers used the checklist in Box 12.4 to identify what successful managers really do on the job. In this study the successful managers were identified through their relatively rapid promotion records. They were more likely than others to engage in "networking" and other forms of informal contact with their subordinates and superiors. Although this networking often did not involve exchanging any substantive information, it served to maintain good horizontal and vertical relationships. Then when there was important information to be exchanged, the successful manager was in a good position to receive or transmit it. Successful managers were also more oriented toward conflict management and spent more time on decision making, planning, and coordinating (Luthans, Rosenkrantz, and Hennessey, 1985). Studies of

B O X 12.4 A Checklist for Managers

1. Planning–coordinating
   a. setting goals and objectives
   b. defining tasks needed to accomplish goals
   c. scheduling employees, timetables
   d. assigning tasks and providing routine instructions
   e. coordinating activities of different substitutes to keep work running smoothly
   f. organizing the work
2. Staffing
   a. developing job descriptions for position openings
   b. reviewing applications
   c. interviewing applicants
   d. hiring
   e. contacting applicants to inform them whether they have been hired
   f. "filling in" when needed
3. Training-developing
   a. orienting employees, arranging for training seminars, and the like
   b. clarifying roles, duties, job descriptions
   c. coaching, acting as a mentor, "walking" subordinates through tasks
   d. helping subordinates with personal develop-ment plans
4. Decision making–problem solving
   a. defining problems
   b. choosing between two or more alternatives or strategies
   c. handling day-to-day operational crises as they arise
   d. weighing trade-offs, making cost-benefit analyses
   e. deciding what to do
   f. developing new procedures to increase efficiency
5. Processing paperwork
   a. processing mail
   b. reading reports, emptying the "in box"
   c. writing reports, memos, letters
   d. routine financial reporting and bookkeeping
   e. general desk work
6. Exchanging routine information
   a. answering routine procedural questions
   b. receiving and disseminating requested information
   c. conveying the results of meetings
   d. giving or receiving routine information over the phone
   e. attending staff meetings of an informational nature

7. Monitoring–controlling performance
   a. inspecting work
   b. walking around and checking things out, touring
   c. monitoring performance data (computer printouts, production, financial reports)
   d. preventive maintenance
8. Motivating–reinforcing
   a. allocating formal organizational rewards
   b. asking for input, participation
   c. conveying appreciation, compliments
   d. giving credit when due
   e. listening to suggestions
   f. giving feedback on positive performance
   g. increasing job challenges
   h. delegating responsibility and authority
   i. letting subordinates determine how to do their own work
   j. sticking up for the group to superiors and others, backing a subordinate
9. Disciplining–punishing
   a. enforcing rules and policies
   b. nonverbal glaring
   c. demotion, firing, layoff
   d. any formal organizational reprimand or notice
   e. criticizing a subordinate for negative performance
10. Interacting with others
    a. public relations
    b. contacting customers
    c. contact with suppliers, vendors
    d. external meetings
    e. community service activities
11. Managing conflict
    a. managing interpersonal conflicts between subordinates or others
    b. appealing to higher authority to resolve a dispute
    c. appealing to third-party negotiators
    d. trying to get cooperation or consensus between conflicting parties
    e. attempting to resolve conflicts between a subordinate and oneself
12. Socializing–politicking
    a. chitchat about family or personal matters
    b. informal joking around
    c. discussing rumors, hearsay, grapevine
    d. complaining, griping, putting others down
    e. politicking, gamesmanship

SOURCE: Fred Luthans and D. L. Lockwood, 1984, "Toward an Observation System for Measuring Leadership in Natural Settings." In *Leaders and Managers: International Perspectives on Managerial Behavior and Leadership*, edited by James G. Hunt, D. Hosking, C. Schriesheim, and R. Stewart. New York: Pergamon, p. 122. Reprinted by permission of the publisher.

BOX 12.5 How American Managers Fail

Many reasons have been offered for the decline in competitiveness of American industries. In this selection, two professors of business administration at the Harvard Business School argue that American managers should claim some responsibility for the decline.

In the past, American managers earned worldwide respect for their carefully planned yet highly aggressive action across three different time frames:

Short term—using existing assets as efficiently as possible.

Medium term—replacing labor and other scarce resources with capital equipment.

Long term—developing new products and processes that open new markets or restructure old ones.

The first of these time frames demanded toughness, determination, and close attention to detail; the second, capital and the willingness to take sizable financial risks; the third, imagination and a certain amount of technological daring.

Our managers still earn generally high marks for their skill in improving short-term efficiency, but their counterparts in Europe and Japan have started to question America's entrepreneurial imagination and willingness to make risky long-term competitive investments. As one such observer remarked to us: "The U.S. companies in my industry act like banks. All they are interested in is return on investment and getting their money back. Sometimes they act as though they are more interested in buying other companies than they are in selling products to customers."

We refuse to believe that this managerial failure is the result of a sudden psychological shift among American managers toward a "super-safe, no risk" mind set. No profound sea change in the character of thousands of individuals could have occurred in so organized a fashion or have produced so consistent a pattern of behavior. Instead we believe that during the past two decades American managers have increasingly relied on principles which prize analytical detachment and methodological elegance over insight, based on experience, into the subtleties and complexities of strategic decisions. As a result maximum short-term financial returns have become the overriding criteria for many companies.

For purposes of discussion, we may divide this new management orthodoxy into three general categories: financial control, corporate portfolio management, and market-driven behavior.

Financial Control

As more companies decentralize their organizational structures, they tend to fix on profit centers as the primary unit of managerial responsibility. This development necessitates, in turn, greater dependence on short-term financial measurements like return on investment (ROI) for evaluating the performance of individual managers and management groups. Increasing the structural distance between those entrusted with exploiting actual competitive opportunities and those who must judge the quality of their work virtually guarantees reliance on objectively quantifiable short-term criteria.

Although innovation, the lifeblood of any vital enterprise, is best encouraged by an environment that does not unduly penalize failure, the predictable result of relying too heavily on short-term financial measures—a sort of managerial remote control—is an

successful managers also emphasize that beyond the horizontal and vertical relationships within the firm, managers must also maintain good relationships with outsiders who affect the environment of the firm.

The *simulation* is a type of study in which a group of managers is asked to solve a hypothetical problem. The managers are observed during this exercise to better understand how they reach a decision, which factors they consider most important, and how they interact with one another. This technique can also be used to encourage team development, which has become more important with the advent of such practices as matrix environment in which no one feels he or she can afford a failure or even a momentary dip in the bottom line.

Corporate Portfolio Management

This preoccupation with control draws support from modern theories of financial portfolio management. Originally developed to help balance the overall risk and return of stock and bond portfolios, these principles have been applied increasingly to the creation and management of corporate portfolios—that is, a cluster of companies and product lines assembled through various modes of diversification under a single corporate umbrella. When applied by a remote group of dispassionate experts primarily concerned with finance and control and lacking hands-on experience, the analytic formulas of portfolio theory push managers even further toward an extreme of caution in allocating resources. "Especially in large organizations," reports one manager, "we are observing an increase in management behavior which I would regard as excessively cautious, even passive; certainly over-analytical; and, in general, characterized by a studied unwillingness to assume responsibility and even reasonable risk."

Market-Driven Behavior

In the past 20 years, American companies have perhaps learned too well a lesson they had long been inclined to ignore: businesses should be customer oriented rather than product oriented. Henry Ford's famous dictum that the public could have any color automobile it wished as long as the color was black has since given way to its philosophical opposite: "We have got to stop marketing makeable products and learn to make marketable products."

At last, however, the dangers of too much reliance on this philosophy are becoming apparent. As two Canadian researchers have put it: "Inventors, scientists, engineers, and academics, in the normal pursuit of scientific knowledge, gave the world in recent times the laser, xerography, instant photography, and the transistor. In contrast, worshippers of the marketing concept have bestowed upon mankind such products as new-fangled potato chips, feminine hygiene deodorant, and the pet rock...."

The argument that no new product ought to be introduced without managers undertaking a market analysis is common sense. But the argument that consumer analyses and formal market surveys should dominate other considerations when allocating resources to product development is untenable. It may be useful to remember that the initial market estimate for computers in 1945 projected total worldwide sales of only ten units. Similarly, even the most carefully researched analysis of consumer preferences for gas-guzzling cars in an era of gasoline abundance offers little useful guidance to today's automobile manufacturers in making wise product investment decisions. Customers may know what their needs are, but they often define those needs in terms of existing products, processes, markets, and prices.

Deferring to a market-driven strategy without paying attention to its limitations is, quite possibly, opting for customer satisfaction and lower risk in the short run at the expense of superior products in the future. Satisfied customers are critically important, of course, but not if the strategy for creating them is responsible as well for unnecessary product proliferation, inflated costs, unfocused diversification, and a lagging commitment to new technology and new capital equipment.

SOURCE: Robert H. Hayes and William J. Abernathy, 1980, "Managing Our Way to Economic Decline." *Harvard Business Review* 58, 4 (July/August):68, 70–71. Copyright 1980 by the President and Fellows of Harvard College. All rights reserved. Used with permission of the *Harvard Business Review.*

---

supervision (see Chapter 7). It can also be used to diagnose problems. For example, one simulation revealed that the managers were too deskbound, wrote too many memos, and did not communicate well along the horizontal dimension because they preferred vertical communications (Kaplan, Lombardo, and Mazique, 1985).

American managers, when compared with their counterparts in other industrialized countries, have been accused of failing to make good decisions or failing to emphasize the most important roles in their job. Box 12.5 presents one criticism of the behavior of American corporate managers.

The Organizational Culture Approach

The organizational culture approach to managerial behavior identifies unique features of organizations that affect the behavior of managers as a group. The organization, like a society, is said to have a distinctive culture (Trice and Beyer, 1992). The major components of the organizational culture that have been identified are climate, supervisory leadership, peer leadership, and group process. These components affect production costs and financial outcomes, and they also affect individual workers and their behaviors, including absenteeism and job satisfaction (Steele and Hubbard, 1985). These analyses of management behavior are closely tied to the sociological study of organizations. We will review three classifications of organizational culture: scientific management, human relations, and the so-called "new school." These ideas were presented in Chapter 7 as Theories X, Y, and Z. Here we examine the effect of these approaches on managerial behavior.

Scientific management (Theory X) has already been discussed as one stage in the development of labor control. It also has important implications for how managers behave. The scientific management school views workers as motivated principally by economic rewards, and it views management's task as distributing rewards and punishments. This version of management is paternalistic, in that managers make virtually every decision about workers' performance and the views of workers are rarely sought. This approach is also very hierarchical, and lower-level managers often find themselves treated paternalistically by those above them in the hierarchy. In this culture the successful manager pays close attention to vertical relationships, watches productivity carefully, and emphasizes rewards and punishments. The successful manager looks for ways to simplify tasks and to make the workers as interchangeable as possible—processes that minimize the costs of turnover. High turnover is attributed to worker behavior, not to poor management.

The *human relations approach* (Theory Y) views workers as motivated both by extrinsic rewards, such as money, and by intrinsic rewards, such as job satisfaction. In this approach, management's task is to make work more pleasant while still maintaining productivity. This might mean redesigning the jobs to make them more interesting, but it is more likely to mean that the company provides amenities such as an attractive lunchroom, lighted parking areas, and so on. Managers are expected to pay close attention to their relationships with their subordinates. High turnover is considered a warning signal.

Theory Z emphasizes a much closer, almost fraternal relationship between managers and workers (Kelley, 1985; Ouchi, 1981). This school views workers as highly trained and skilled partners in production; it is often recommended for managers who supervise knowledge workers. Some examples of appropriate work sites include research and development departments, high-technology firms, organizations that hire many professionals (hospitals or universities), or organizations composed chiefly of professionals (law firms or engineering firms). This organizational culture recognizes that managers have responsibilities to allocate resources and otherwise maintain the organization but that efficient production requires teamwork by both managers and workers. Because the workers have specialized knowledge, their cooperation and performance are essential to the firm's success. In this model, high turnover is a sign of management failure.

The argument has been made that this model is appropriate for all work sites, even those in which most workers are not highly trained, because all workers have developed some expertise and may have insights into improving the production process. Workers can respond in different way to workplace change, and how they respond can be as important as how managers behave (Vallas, 2006). Because Theory Z emphasizes worker participation in some types of decision making, it is associated with such organizational practices as climate surveys and quality circles. Climate surveys assess workers' perceptions of the workplace. Quality circles are informal discussion groups composed of workers from various levels of the organization who meet periodically to discuss improving the work process and product.

Studies of organizational climate are very important for understanding the managerial occupations. A good climate may boost productivity and morale, just as a poor climate may lead to deterioration in the quality of products and in relationships among the workers. No single manager, however, can change the organizational climate to any significant degree. Instead, new managers learn to work within an organizational climate, and their success usually requires being able to reproduce that climate. Studying organizational climate illuminates the roles managers play in maintaining the climate and the ways in which they organize to influence change in the climate.

THE FUTURE OF EXECUTIVES, MANAGERS, AND ADMINISTRATORS

The Bureau of Labor Statistics predicts continued growth in the managers from 9.1 million in 2004 to 10.1 million in 2014, a growth rate of 11.3 percent (Hecker, 2005). Every organization will continue to need workers who provide leadership and coordination. But as organizations change, the number of managers will change, and the tasks they perform will also change. As this chapter indicates, alterations in the characteristic forms of technology will also change the day-to-day functioning of the manager. Many of the information tasks formerly done by managers are today delegated to administrative support workers, who also serve an important function in coordinating information. We will consider administrative support workers in greater detail in Chapter 13.

A different issue will be how much the public trusts top managers and what is the perceived accountability of managers. The very high salaries of top managers are not always correlated with top performance, but corporate boards and the general public are not fully able to evaluate performance. Within the firm, a large spread between the best-paid and the worst-paid people—usually produced by the actions of managers—is associated with lower performance, lower productivity, and decreased employee engagement (Pfeffer and Sutton, 2006). Valuing short-term results and profits at the expense of human costs are management values with negative consequences for firms and for other workers.

SUMMARY

Managerial occupations vary a great deal among firms and even within firms. Self-employed proprietors are considered managers, but so are the top executives of multinational corporations. What the jobs have in common is the necessity to make decisions, control and allocate resources, and provide for the maintenance and growth of organizations. Concern about management behavior and organizational climates is directed toward understanding how managers can improve on juggling the many roles of their job. Serious issues of values have been raised about top executives, including excessive pay, lack of accountability, deliberate falsification of corporate information, and an overemphasis on short-term results. Although women and members of minority groups are beginning to enter the management suite, management jobs have tended to be strongly limited by race, gender, and social class.

KEY CONCEPTS

executive	staff manager	downsizing	horizontal dimension of bureaucracy
manager	span of control	mentor	
administrator	steepness of hierarchy	middle management	vertical dimension of bureaucracy
line manager	top management	first-line supervisor	

QUESTIONS FOR THOUGHT

1. Why is technology likely to change the span of control and the steepness of hierarchies?

2. Layoffs of managerial personnel have tended to target middle managers and staff managers. Why are these positions being eliminated?

3. How do managers' responsibilities differ when considering the vertical dimension of bureaucracies as opposed to the horizontal dimension of bureaucracies?

4. A researcher discovers that minority and women managers are more likely to be in staff positions than in line positions. How would you explain this finding?

5. Is there such a thing as a "dead-end job" in management? What kind of job might it be?

6. Why should the public and elected officials be concerned about the honesty of accounting in large corporations?

MULTIMEDIA RESOURCES

Print

Marilyn J. Davidson and Ronald J. Burke (editors). 2004. *Women in Management Worldwide: Facts, Figures, and Analysis.* Aldershot, Hants, England and Burlington. VT: Ashgate. This book assembles statistics to support an analysis of women's progress in management throughout the world. Each chapter is devoted to a different country.

Anne E. Kingsolver (editor). 1998. *More Than Class: Studying Power in U.S. Workplaces.* Albany: State University of New York Press. Each chapter of this edited volume offers an ethnographic glimpse of a workplace and how power is distributed within it.

Michael Woodard. 1997. *Black Entrepreneurs in America: Stories of Struggle and Success.* New Brunswick, NJ: Rutgers University Press. Interview studies of black business owners and the struggles they faced to become entrepreneurs.

Internet

Academy of Management. *http://www.aomonline.org* The website of the professional group that studies management and publishes some of the leading journals reporting research on business.

Forbes. *http://www.forbes.com* This website for Forbes magazine provides listings such as the Global 2000 largest corporations and a list of the top 100 private companies.

U.S. Small Business Administration. *http://www.sba.gov* This website provides a variety of information for entrepreneurs interested in starting or growing a small business.

RECOMMENDED FILM

In Good Company (2004). When Dan Foreman is demoted from head of advertising sales because his company is acquired in a corporate takeover, he acquires a new boss who is half his age and a business school star. PG-13 (sexual content, drug references). 1 hour 50 minutes.

13

Administrative Support and Sales

Secretaries, clerks and bureaucrats were once grateful for having been
spared the dehumanization of the factory.... They had higher status
than blue-collar [workers].... But today ... such positions offer little in
the way of prestige ... imparting to the clerical worker the same
impersonality that blue-collar workers experience in the factory.
(U.S. DEPT. OF H.E.W., *WORK IN AMERICA*, 1973:38)

Selling is when it matters how you and your customers get to know
each other, how you help her find what's best, how you understand
what she needs. Here you no sell. Lady comes in, asks for something,
you go find it, she buys or no buys, that's it, you never see her no
more. That's not selling. That's showing. But most of the time,
you don't do that even. You're putting labels on corsets, or
emptying boxes.
—MARLENE, A DEPARTMENT STORE SALES WORKER (HOWE, 1977:87)

Administrative support is a term to describe jobs that are principally clerical
and sales occupations and are sometimes called lower-level white-collar
jobs. Figure 13.1 shows the rapid growth of these occupations. By 2005 the 19.5
million workers in the "office and administrative support" occupations accounted
for nearly 14 percent of workers, and the 16.4 million sales workers accounted for
nearly 12 percent of workers (BLS, 2006a).

Numbers alone make administrative support jobs important to study. In
addition, these occupations have in common a rapid transformation driven by
technology and the reorganization of offices and stores. The outcome of this
transformation is by no means certain. Demand for workers could increase, or

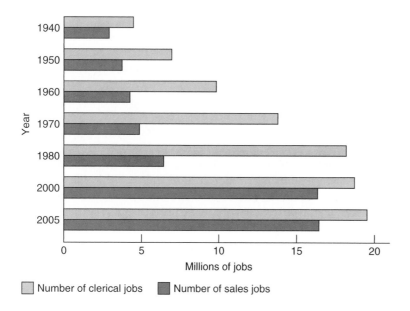

FIGURE 13.1 Number of Clerical and Sales Jobs, 1940–2005

SOURCES: Census, 1952, *Statistical Abstract of the United States, 1952*. 73d edition. Washington, D.C.: U.S. Government Printing Office, table 221, p. 186. Also, Census, 1981, *Statistical Abstract of the United States, 1981*. 102d edition. Washington, D.C.: U.S. Government Printing Office, table 673, p. 401; and BLS, 2001, BLS, 2006a. *http://www.bls.gov/pdf/cpsaatll.pdf.*

new machinery could displace them. Newly created jobs might be more interesting and have more complex tasks, giving each worker greater autonomy. Or the transformation might result in ever more specialized, boring, repetitive jobs that reproduce factory-like conditions in offices and stores. The outcome depends on choices made by different firms. The new technology also blurs the distinction between sales and clerical jobs. In this chapter we will examine the origins of administrative support work, the demand for and the supply of these workers, and the ways that technology is changing these jobs.

HISTORY OF CLERICAL WORK

Clerks have been needed since written records were developed. When most Europeans were illiterate peasants, work requiring literacy was performed by clerics—that is, those in priestly orders. The terms *clerk* and *clerical* originated at that time. Bishops, abbots, and other high church officials required secretaries for such tasks as reading and writing correspondence and keeping business accounts and inventories. Thus began a close association between clerical work and the professions—the clergy, in this case. We preserve this older use of the term *clerk* today when we speak of

law clerks, who are usually new attorneys performing tasks under the close supervision of a judge. Unlike other clerical workers who work with professionals, many law clerks will eventually assume full-fledged professional positions. In general, however, clerical workers cannot expect to become professionals without acquiring further training. Remember that we discussed the training of professionals and entry into the professions in Chapter 11. The work of clerks is subordinated to that of professionals, although professionals require an administrative support staff in addition to any semiprofessionals or paraprofessionals whom they employ.

Clerks were also needed in secular pursuits. Monarchs and their representatives needed clerks to transcribe rulings and laws, send messages, and maintain records. Feudal lords required at least rudimentary records on land use, agricultural products, and taxes. When mercantile capitalism arose, merchants and shippers needed clerks to keep accounts, maintain correspondence, and generally assist the owner. This was the origin of the close association between clerical work and managers still retained today in job titles such as confidential secretary.

Clerks replaced neither professionals nor managers in making decisions. Nevertheless, when reading and writing were uncommon skills, the clerical worker was well regarded and had relatively high prestige. A good clerk might expect preferment and possibly promotion, because the clerkship served as a training ground and apprenticeship for management. The clerk knew intimately how the business was run and what its financial situation was, so the clerk often became the confidant of the business owner and sometimes even a business partner.

Work requiring reading and writing was perceived as clean; hence, the clerical worker could wear a white collar and dress in a style approaching that of the gentleman. Before the U.S. Civil War, the clerical occupations were almost entirely male. Among other reasons, men were more likely to have received the needed education and to be viewed as appropriate apprentices. Although some women completed high school or even college at this time, they were much less likely to be working outside the home. As we shall see, the male domination of the clerical occupations was completely reversed, and that reversal was accompanied by a number of other changes.

Demand for Clerical Workers

The first change was a large increase in the number of clerical jobs. Between 1870 and 1930 the clerical workforce grew from 76 thousand to 3.8 million (Census, 1943). The last half of the nineteenth century saw the growth of large, bureaucratic organizations, the forerunners of today's giant corporations. Government agencies, too, proliferated. As

we saw in Chapter 7, the growth of these organizations resulted in a heightened need for more efficient organization. Just as there was a division of labor among workers, there came to be a division of labor within companies. Departments and other organizational groups specialized to perform certain tasks (Chandler, 1980). For example, a typical company developed divisions for sales, orders, production, billing, and shipping. Both the larger scale and the new organizational units increased the requirements for coordination and record keeping (Mills, 1951). There were also changes in record keeping. After 1910 the widespread use of cost accounting led to a greatly increased volume of financial records (Strom, 1986). These changes required additional workers to prepare, send, receive, and file the records.

With greater numbers of clerical workers came increased specialization. During the mid-nineteenth century there were four basic clerical occupations (Davies, 1982). Before telephones the *messenger* (or office boy) was needed to run errands; he also did housekeeping chores around the office and often studied bookkeeping or another office skill for internal promotion. Before the invention of carbon paper or photocopying, *copyists* wrote and rewrote documents. Some copyists could aspire to become clerks, but many never advanced. Most businesses kept only rudimentary accounts, but later there was a need in larger firms for *bookkeepers* to track sales, debts, and credit extended. Because bookkeepers' training was so specialized, some business owners preferred not to promote them to supervisory or management positions. Finally, the *clerk* often had minor management authority and had acquired all or most of the other skills. In small firms one person did the tasks of all four specialties.

Greater occupational specialization arose from two sources. One source was the *reorganization of work*. Some occupations, such as personnel worker, file clerk, or record clerk, arose from the consolidation of similar tasks previously performed in each division of the organization. A second source of specialization was the development of *office equipment*. New occupations were developed and named according to the technology that was used. Typists operated typewriters, and there were billing-machine operators and switchboard operators; today there are word processors. Appendix Table 1 shows the variety of such occupations that still exist.

In smaller firms the clerical worker (often called a *secretary*) was not so specialized and often performed all sorts of tasks, including transcribing dictation, keeping records, filing papers, and running personal errands for the boss. A small office had neither enough workers nor enough office machines to justify a higher degree of specialization. For the all-purpose secretary, clerical work provided variety and autonomy. Specialized clerical workers in large firms, however, were more likely to find their work monotonous and closely supervised.

The demand for clerical workers remains strong. One indication of strong demand is their relatively low unemployment rate, which in October 2006 was 4.6 percent, identical to the overall average rate (BLS, 2006b).

Supply of Clerical Workers

The increased demand for clerical workers during the latter half of the nineteenth century led employers to identify new sources of workers. During the Civil War, the U.S. Treasury Office became one of the first employers to experiment with female clerks. The young males who were the usual clerical workers were in uniform, and women were sought as their replacements. Box 13.1 discusses this experiment, which the government evaluated and generally considered to be successful. Other employers repeated the experiment and a major change was under way—the change in clerical occupations from mostly male to mostly female.

The Feminization of Clerical Work Today's clerical labor force has been largely *feminized*. In 2005, 97 percent of all secretaries and administrative assistants were female, and 75 percent of all office and administrative support workers were female. Over 22 percent of all women workers were in the office and administrative support occupations (BLS, 2006a). How did this transformation occur? During the last quarter of the nineteenth

century, the demand for clerical workers increased sharply. Employers wanted both to find a good supply of skilled workers and to minimize their wage costs. Native-born white women became the predominant clerical labor supply for four basic reasons: (l) they had the necessary training; (2) they accepted lower pay than men; (3) clerical work was attractive to them because of its relatively high status; and (4) competing sources of labor lacked the necessary language skills and education.

First, because of educational reforms, women had the *necessary training* to perform clerical tasks. Universal public education provided both girls and boys with the necessary skills for clerical work. By 1890 the number of women receiving high school diplomas had outstripped the number of men receiving diplomas (Davies, 1982:57). When clerical work required skills beyond literacy, "commercial courses," such as shorthand and bookkeeping, were added to the high school curriculum. By 1924 women composed two-thirds of the enrollment in high school business courses (Strom, 1986:85). Many proprietary trade schools offered secretarial training to high school graduates, and by about 1895 a third of their students were women (Davies, 1982:73).

The provision of high school and post-high school business education for young women allowed employers to reduce costs by providing minimal training for new workers. Before being hired, a new secretary was expected to have already mastered the basic office skills, and that mastery could be verified with tests of typing, spelling, and dictation. This result was important because employers expected women workers to have high turnover, which increases training costs because workers must be trained even if they stay on the job for only a few months.

Second, women workers traditionally received *lower pay* than men. Employers expected women to accept less pay because the women would work only briefly before getting married. Employers also justified lower pay because typewriters were thought to make the job easier by reducing its physical demands. When bookkeeping machines were introduced, they were usually operated by women, who were paid less than the more skilled male bookkeepers who knew

B O X 13.1   The Civil War Experiment with Female Clerks

Today most clerical workers are female, but once they were mostly male. This selection discusses an experiment made by the U.S. government in hiring women for clerical work. This passage hints at some of the reasons for the "feminization" of clerical work: there was a willing supply of well-trained women workers available, and they were willing to work for lower pay than the male clerks received.

During the Civil War the U.S. Treasurer General, Francis Elias Spinner, confronted a severe labor shortage caused by the large numbers of men in Union uniforms. He decided over considerable opposition to hire some female clerks, who worked at relatively mechanical tasks such as sorting and packaging bonds and currency. This "experiment" was continued after the war, and in 1869 Spinner declared "upon his word" that it had been a complete success: "Some of the females doing more and better work for $900 per annum than many male clerks who were paid double that amount." Such wage figures indicate one of the

reasons Spinner thought so highly of his experiment: female labor was cheaper than male. A contemporary claimed that most of these early female clerks got their positions through political patronage, with the result that some of them were not well trained for their jobs and had to take writing lessons after they were employed. But a study of federal government clerks from 1862 to 1890 has found that, by and large, the women did have sufficient education for clerical work, most of them having remained in school at least until the age of sixteen. They came overwhelmingly from white, native-born, middle-class families and were the daughters of men with jobs that ranged from clerks to judges; almost none of them were the daughters of craftsmen, much less unskilled laborers. Whether they were single, widowed, or, less frequently, married, these women sought clerical jobs out of economic necessity.

SOURCE: Margery W. Davies. 1982. *Woman's Place is at the Typewriter: Office Work and Office Workers, 1870–1930.* Philadelphia: Temple University Press, p. 51. © 1982 by Temple University Press. Reprinted with permission of Temple University Press.

how to do all of the bookkeeping tasks by hand (Strom, 1986). Thus, even though the numbers of clerical workers grew, substituting women for men often reduced total labor costs.

The rate of substitution varied, however, depending on how intensively the firm used clerical workers. In industries with large numbers of clerical workers, large savings resulted from hiring women. In industries with only a few clerical workers, men tended to keep the clerical jobs because the differential in the wage bill was relatively small (Bridges, 1982; Cohn, 1985a, 1985b; Lowe, 1987). Some clerical specialties continue to be predominantly male; these are often in workplaces in which the clerical specialty is closely related to production and in which production workers may be promoted into clerical jobs. Examples of such jobs are meter readers; clerks who do traffic, shipping, and receiving; and transportation dispatchers.

Third, office work had relatively *high status* that attracted women from both the middle class and

the working class. Although men of all classes and working-class women had other job opportunities, few appropriate jobs were open to middle-class women who wanted to work. Offices were seen as "clean," suitable workplaces for women workers. Middle-class women were not interested in the many jobs available as factory hands and servants, but they perceived office work as consistent with their position in the community. For working-class women, office work provided a channel of upward mobility from the farm or the factory. In addition, although women were paid less than men, a woman could still earn more in clerical work than she could in almost any other line of work. In 1883, when the average weekly wage for women in manufacturing was about $5, copyists earned $6.78 and cashiers $7.43 (Davies, 1982:64). The difference in weekly wages was even more pronounced because women in offices often worked shorter hours than the women in manufacturing. Ironically, however, as women came to dominate the clerical fields, these occupations often lost status, at least in part because they became identified as "women's work."

Fourth, language skills and education requirements generally disqualified the competing sources of cheap labor. At the time that the clerical labor force was feminizing, immigrants from Europe were a major source of new workers. The immigrants, however, were less likely than the native-born to be literate and much less likely to have an excellent command of English. Good written and spoken English remained an important qualification for nearly every clerical job. Moreover, prejudice against the immigrants was strong. Even immigrants who spoke English, such as the Irish in Boston, were far less likely than native-born women to be hired for clerical work (Handlin, 1973).

Another possible pool of competing labor, African Americans migrating from the South, were generally not considered by employers for clerical work. One rationalization for this discrimination was the belief that the schooling of African-American women had been so inferior that they could not perform clerical tasks adequately. It is equally likely that both the employers and their white employees resented sharing the relatively high status and clean white-collar office work with African Americans. After World War II, however, minority women became an important source of clerical labor.

Minority Women as Clerical Workers African-American women were first hired as clerical workers in businesses owned by other African Americans. When these women were hired in other firms, they tended to be in "back-office" jobs that required few face-to-face interactions with the public. As an example, one relatively early employer of African-American women was the Montgomery Ward mail-order house (Strom, 1986:85). These workers filled mail orders and never interacted with the customers. The Depression in the 1930s reinforced the disadvantaged position of African-American women who wanted clerical jobs, because competition for the jobs with white women was intense.

The continued growth of clerical employment after World War II drew minority women into the field, but they entered the least skilled and most poorly paid jobs. White women are still more likely to be found among secretaries and receptionists, and minority women are more likely to be found among typists, file clerks, and mail clerks. Nevertheless, the number of minority women has been rising rapidly, even in the predominantly white specialties: between 1970 and 1980 the number of black female secretaries increased by 148 percent and the number of Hispanic female secretaries by 131 percent (Murphree, 1987:100). By 2005 nearly 13 percent of all office and administrative support workers were black, and over 11 percent were Hispanic (BLS, 2006a).

Opposing Trends in Clerical Work Continued occupational specialization within the clerical fields tended to follow one of two trends, each of which affected the supply of workers. One trend was *professionalization*, which we explored in Chapter 11. One objective of professionalization is to "reserve" jobs for members of a profession and to protect them from competition. Members generally seek to upgrade entry requirements and to define an area of knowledge that only specialists master. By restricting the supply of workers only to those with the appropriate skills, professionalizing fields are able to command higher wages. Among clerical workers, professionalization predominated in areas dominated by men. Such areas as accounting, personnel administration, and actuarial science became men's fields. After women started entering bookkeeping in large numbers, male accountants professionalized the work of bookkeepers, which protected accountants' positions and limited access by women.

Professionalization also had its disadvantages. The newly professionalized specialists were often passed over for promotion to management jobs because their knowledge was narrow. Furthermore, the specialized training required for professionalization was provided by colleges and universities that did not necessarily restrict the entry of women into the field. Women have now gained access to these

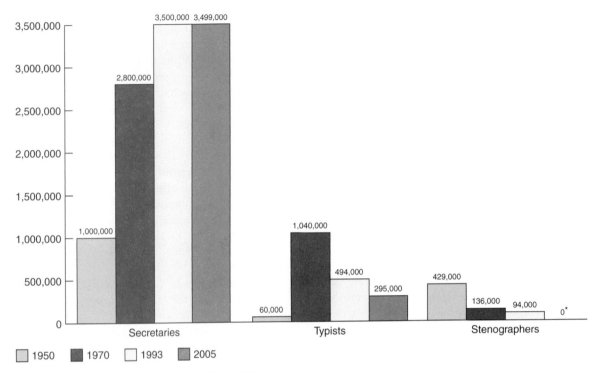

FIGURE 13.2   Rise and Fall of Clerical Occupation, 1950–2005

SOURCE: Allan H. Hunt and Timothy L. Hunt, 1987, "Recent Trends in Clerical Employment: The Impact of Technological Change." In *Computer Chips and Paper Clips*, edited by Heidi I. Hartmann. Washington, D.C.: National Academy Press, pp. 228–229. Also, BLS, 1994a. *Employment and Earnings* 41.1 (January): 206–207, table 22; BLS, 2006a, *http://www.bls.gov/pdf/cpsaatll.pdf*.

fields through university training. By 2005 over 62 percent of accountants and auditors were women, and women accounted for over 73 percent of human resources managers and 72 percent of insurance underwriters (BLS, 2006a).

The second trend was the development of *clerical operatives* who specialized in specific pieces of office equipment. Their jobs were defined by the machinery they operated rather than by their knowledge. This knowledge was available in vocational education courses or in commercial business schools. Many office workers learned to use only one or two pieces of equipment. New technology could render them and their machinery obsolete. Because they were often perceived as relatively

short-term workers, however, their narrow specialization was a distinct advantage from management's perspective. If they left for marriage or family reasons, their replacements could be quickly trained.

Figure 13.2 shows the continued rise and fall of various clerical occupations between 1950 and 2005. Secretaries peaked in 1993, but their numbers remain high. Typists rose and then declined. As word processing became more prevalent, they declined once more. Stenographers declined steadily after 1950, and by 2005 the occupation was no longer separately reported by the Bureau of Labor Statistics.

Changing gender composition in the occupation of computer operator represents an interesting mix of both trends. Computer work was originally seen after World War II as a rather limited clerical specialty, and the first computer operators were women. Within a few years, however, men were being hired in these jobs. In 1960 the occupation was 65 percent female, but by 1970 it was only 29 percent female. The expansion of computer science, an engineering-based and hence stereotypically male area of endeavor, may have stimulated the influx of men. The trend again reversed, however; in 1994, 62 percent of computer operators were women, but by 2005, 55 percent were women (BLS, 2006a).

What is common to both the professionalization and specialization trends is that the costs of training are borne by the workers themselves as part of their job preparation. The employer provides relatively little training except when new technology is being introduced.

TRANSFORMING ADMINISTRATIVE SUPPORT

The office is sometimes portrayed as playing the same role for an organization that the brain plays in a living organism. The office receives information flowing in from all parts of the organization (or organism). It processes that information and sends back responses, instructions, and commands through established channels of communication, which serves as a nervous system (U.S. Congress, Office of Technology Assessment, 1985:7). In every industry an office, or a central location for information management, is needed. New technology and work reorganization continue to change office work.

Office Technology

For the last 125 years, technology has had an important impact on the nature of clerical work. Figure 13.3 indicates the relative timing of the introduction of new office technology. The typewriter and the telephone, both introduced between 1870 and 1880, revolutionized even the small office, where the all-purpose secretary had to master them. In larger offices the new machines became the basis for displacing some workers, changing the job content for others, and developing new specialties. The telephone made the office messenger obsolete, changed the receptionist's job, and created jobs for switchboard operators. The typewriter made the copyist obsolete, changed the job of secretaries, and created a whole new occupational specialty, the typist. As you learned in Chapter 7, these three outcomes—displacement, transformation, and the emergence of new occupational specialties—are typical responses to new technology.

The typewriter initially sparked controversy, and its adoption was relatively slow (Davies, 1982). Managers were skeptical that typing would ever be faster than handwriting. There were also concerns that typed correspondence, instead of the handwritten letter, was inappropriate business etiquette. Similarly, some managers expressed parallel concerns about the propriety of computer-generated correspondence, but today a printed letter appears more businesslike than a handwritten one. Eventually, the typewriter was widely adopted, typing courses were instituted in many schools, and the same keyboard is now used in desktop and laptop computers.

Typist training programs, such as those offered by YWCAs beginning in 1881, were specifically oriented toward teaching young women a useful job skill. Offering such courses was controversial, for several reasons. Giving women job skills was seen as encouraging them not to marry and have families; it was also seen as a way to deprive men of jobs needed to support their families. Concerns were expressed that working at a typewriter might endanger the health and constitution of young women and, by implication, their childbearing. An equally stereotypical counter-response emerged: women had more manual dexterity and endurance for repetitive tasks, and so they were better "suited" to be typists. Typing, it was argued, reduced the stamina required for clerical work and made clerical jobs more suitable for women.

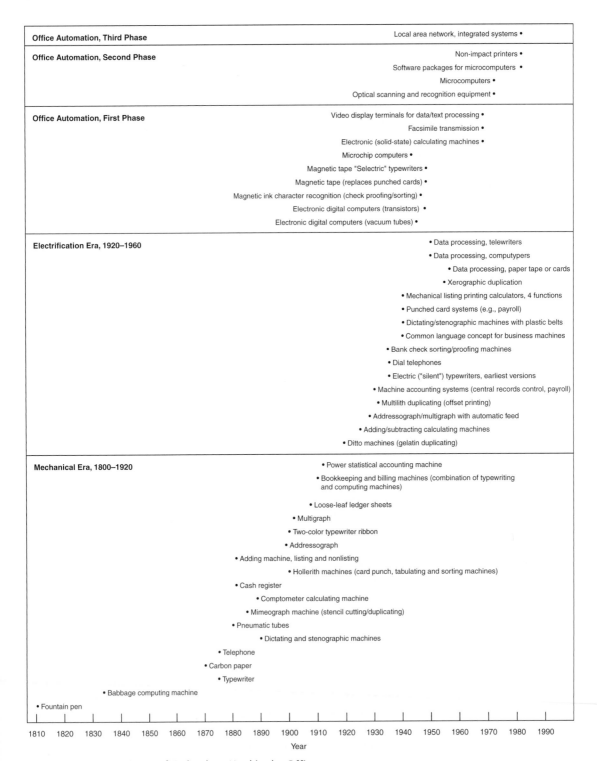

FIGURE 13.3 History of Technology Used in the Office

SOURCE: U.S. Congress, Office of Technology Assessment, 1985, *Automation of America's Offices*, 1985–2000. OTA-CIT-287, December, Washington, D.C.: U.S. Government Printing Office, p. 9.

Much like the copyist, who was unlikely to experience upward mobility in the office, the typist was expected to do only one thing and might not learn any other office skills. Many offices sought to reduce the number of skilled office workers, such as stenographers, while increasing the number of typists. Typists, in turn, were vulnerable to further technological changes. Electric typewriters were faster and made better carbon copies than manual machines. Duplicating machines, and then later photocopying equipment, eliminated the retyping of multiple copies. The development of self-correcting or programmable typewriters reduced the need for typing accuracy and spelling skills. Audio recording equipment made shorthand obsolete. Figure 13.3 shows the years these new technologies were first introduced.

Recent Innovations Word processors, micro-computers, and electronic workstations are more recent office technologies. Although these innovations do not require the worker to be "computer literate" in any sophisticated sense, they require skills that older workers may not have mastered and might be reluctant to learn. Innovations can penalize older workers if the skills they worked for years to master have been rendered obsolete, putting them on a par with the newest young workers hired. Some studies indicate that women experience more negative effects of technology introduction than men, perhaps because they are less often invited to participate in the decision-making process (Zaucher et al., 2000).

Although the demand for workers with computer skills is rising, the increased demand may not translate into a better competitive position for the existing workers. The skill levels in some jobs may be reduced. For example, bank clerks once received up to six months' training in back-office functions before becoming tellers. The back-office functions included processing canceled checks, checking records, and performing other functions not requiring personal interaction with customers. With the computerization of many of these functions, the needed training can now be provided in four months (Rajan, 1985). Automated teller machines (ATMs) further reduced the number of tellers, although the remaining tellers perform more non-routine activities.

More innovations in office technology lie ahead. One trend is the linking of office machines into a continuous-process system. (Recall that this term was introduced in Chapter 9.) The "capture" of data at the point of origin eliminates the need for repeated data entry. For example, an order can be electronically tracked from receipt to delivery, producing automatically along the way the mailing labels, invoices, and inventory adjustments required by the transaction. Electronic files and records have largely replaced paper records in many offices. Voice recognition systems can produce correspondence that is virtually untouched by human hands, while electronic mail systems have already made some correspondence obsolete. It is difficult to completely forecast the effect of these innovations on administrative support workers without considering the key role of **work reorganization** in the adoption of new technology.

Work Reorganization

In a majority of institutions, pre-computer clerical jobs were extremely routine. Far from deskilling clerical workers, new technology can be viewed as their liberator. However, this is not to imply that technology's impact is predetermined—far from it. Much depends upon foresight and planning on the part of management (Rajan, 1985:413).

Many studies have shown that clerical workers feel powerless and experience considerable stress (Reeves and Darville, 1986). The adoption of new technology and accompanying restructuring are often stressful for both clerical workers and management (Fox and Sugiman, 1999). Whether technology can change this situation is a function not only of what devices are developed but also of how they are used. Technology can degrade and deskill workers when it forces narrow specialization with accompanying loss of variety, autonomy, and job mobility. By one estimate, 90 percent of clerical workers using a computer have little autonomy or discretion in their work (Crompton and Jones, 1984).

Work reorganization affects three major dimensions of the job: job content, supervision, and relations among workers.

Job Content For a quarter of a century following World War II, secretarial work was commonly organized by assigning one secretary, who was usually a woman, to a particular "boss" or perhaps a group of "bosses," who were nearly always men. Secretaries had many jobs to perform in a day, but they could schedule them as they wished, subject to the boss's priorities. A trusted secretary made decisions about screening calls and mail and bringing urgent matters to the attention of the boss. Thus, the job had both variety and some degree of autonomy. Upward mobility was possible because successful managers usually took their secretary with them to a more responsible position (Kanter, 1977). Secretaries gained prestige from the position of their boss, and the nature of the supervisory relationship was a very personal one, with the boss and the secretary often protecting each other from the demands of superiors or office rivalries among competing departments (Glenn and Feldberg, 1977).

The secretary remained dependent on the goodwill of the boss. Many male bosses viewed the secretary role as an extension of the wifely role and expected personal services (Pringle, 1988). Although secretaries made many decisions during the course of a day, they worked in a gray area that made them liable to accusations both of having exceeded the limits of their authority and of not having taken sufficient initiative. The secretary's role under such conditions required considerable interpersonal skills.

When clerical work was **rationalized**, the constituent parts of secretarial work were identified as discrete tasks, and workers specialized in those tasks. Thus, a receptionist greeted all visitors. Switchboard operators answered all phone calls and took messages. Typing was centralized in a typing pool, where all work to be typed was delivered. Any typist might be assigned the work of any manager. This system resembled an assembly line. Today in many offices a word-processing pool has replaced the typing pool.

The purpose of rationalization is to make the office more efficient, but that is not always the result. In theory, specialization makes clerical work more productive; in practice, this does not always happen. Receptionists and switchboard operators often lack sufficient knowledge to screen out nuisance calls or alert managers to urgent ones. Because of their lack of full information, they sometimes annoy or frustrate potentially important customers. Typists, because they work for many managers, never learn to read all their different handwriting idiosyncrasies, and they never gain the overall sense of projects or priorities that the individual secretary had.

Supervision The rationalization of office work also changed the supervision of clerical workers. Some clerical workers in banking and retail trade, because they deal with money, have always been closely supervised, and the new technology has merely changed the means of control. Electronic monitoring of workstations has replaced record checking by a supervisor. But in other cases the nature of supervision has changed more dramatically. Whereas many clerical workers once experienced one-to-one supervision by the boss, today clerical workers are more likely to be supervised as a group. First-line supervision is often provided by another clerical worker who has been promoted to supervisor. The supervisor is often forbidden to socialize with the former co-workers but is also unlikely to receive any further promotions (Glenn and Feldberg, 1977).

The old-style manager knew whether the work was being satisfactorily done and whether secretaries made good use of their time. In the rationalized office supervisors rely on more abstract, measurable criteria of productivity, many of which are generated by the new technology. For example, typists can be rated in terms of keystrokes per minute. Telephone company supervisors can "listen in" or have machinery automatically measure how long operators spend in responding to each call. Timing devices attached to a workstation can monitor how long a clerical worker is absent to collect supplies, change positions, or take a break. Some workplaces block certain websites during business hours and forbid games on company computers.

Relations among Workers   When office technology reproduces factory-like conditions, the relationships among clerical workers also change. Under the old system, secretaries had sufficient autonomy to schedule breaks and lunches together and to cover for one another when the office was particularly hectic or when one of them needed to run an errand (Strom, 1992). It was also true that the secretaries of higher-ranking managers often did not associate with the secretaries of lower-ranking managers and that the rivalries among managers at the same level might be carried over to their secretaries. But for many clerical workers, one attraction of the job was being part of a work group.

By contrast, supervision using the new office technology often makes clerical work lonelier by preventing employees from socializing. Some offices stagger breaks and lunch periods so that workers cannot eat together or even drink coffee together. Some firms also stagger shifts so that there can be more or less continuous typing, photocopying, and word processing available, with the result that clerical workers have fewer chances to meet and know other workers. Workers cannot add variety to their jobs by covering for a busy coworker or trading jobs with another worker.

In some firms the clerical workers are forbidden even to talk to one another during working hours. Some managers justify these practices by claiming that socializing on the job (or even during breaks) reduces the productivity. But such practices also reduce the flexibility of the clerical workers to respond to busy periods, and they eliminate the opportunity for workers to learn from one another. Further, the depersonalization of their jobs may contribute to lower work commitment and less cooperation with management (Costello, 1984:119).

Alternatives to Rationalization   When efficiency becomes the overriding issue in adopting new technology, many of the more pleasant aspects of clerical work are eliminated, and the work is monitored, monotonous, and meaningless. How technology is adopted is a management decision, however, not an outcome predetermined by the equipment. When clerical work has been rationalized, the clerical worker spends the time saved by doing still more of the same specialized tasks. Nevertheless, it is possible to adapt the technology so that the quality of work life is improved for clerical workers and their work has more variety and autonomy.

Many technological changes have the potential to make clerical jobs more interesting and less tedious. For example, the development of photocopying equipment freed typists from retyping or handling carbon paper. Using liquid correction fluid and later self-correcting typewriters freed typists from tedious erasing and retyping. Because of word processing, typists no longer type the same form letters over and over, or retype all of the drafts of a report or manuscript. Many secretaries report that these innovations make it possible for them to use the time saved to perform tasks that need to be done but for which there is never time. Telecommuting clerical jobs may allow workers more flexibility in combining work with other responsibilities (Steward, 2000).

In general, less routinized work is associated with greater job control by workers (Simpson, 1985:421). As alternatives to routinization, **job enlargement** and **job enrichment** can change the content of clerical work when technology is introduced. Job enlargement makes a job "bigger" by adding on previously separate tasks. In banking, a teller may now handle many different types of transactions once handled by separate departments or specialized tellers. A routinized solution, by contrast, would route all transactions of one type to just one teller, who would repeat the tasks over and over. With job enlargement, the bank teller learns to perform a wider variety of transactions. These two strategies take advantage of the time saved by the technology to make the job content more interesting and challenging.

Job enrichment, a strategy that has been used in the insurance industry, is similar to job enlargement except that it gives the worker a fuller cycle of sequential tasks. For example, a single worker in an insurance firm may handle a customer's account from initial inquiry through underwriting and claims processing (Baran, 1986). In the routinized solution, all inquiries would be referred to one worker, who would channel them to different workers who would write and process policies, adjust claims, and so on. Job enrichment, like job enlargement, requires workers to acquire a wider range of skills and knowledge to perform their jobs. Job enlargement and job enrichment are alternatives to routinization that have applications to many occupations and not just to clerical work.

The Persistence of the Secretary The preceding discussion may seem to imply that the all-purpose secretary is extinct, but such is not the case. Smaller firms employ clerical workers who do all clerical tasks around the office, from greeting visitors to typing (Simpson, 1985). In these settings new technologies have been most welcome because these workers have many uses for the time they have saved. Box 13.2 provides an account of some of the satisfying aspects of a secretary's job.

The secretary's level of job satisfaction, however, is again a matter of how the time and skills of the secretary are used. Legal secretaries are highly skilled clerical workers who prepare legal documents for an attorney's signature. With technological innovation, the documents have become permanently available in the word processor, and the blanks can be quickly filled in. In many law firms this technology has freed the legal secretary to do more skilled tasks, some of which may be considered paraprofessional tasks of the sort we discussed in Chapter 11. In other law firms, however, the secretary's job description now requires relatively fewer skills. The personal secretary is still found working for the highest-ranking lawyers, but the secretary's job description has often changed dramatically. The specialized knowledge of the legal secretaries is no longer needed, but their jobs persist because having a personal secretary is a sign of prestige among the top lawyers in a firm. The secretaries' functions have come increasingly to revolve around welcoming clients, serving them

coffee, and running errands for the attorney (Murphree, 1981).

THE FUTURE OF CLERICAL WORKERS

Technology can change not only the *content* of jobs but also the *number* of clerical jobs. One issue that is inevitably raised is whether new technologies will displace workers or will lead to the creation of jobs. This issue is of particular interest because, as mentioned earlier, clerical occupations are a large segment of the labor force.

The pessimistic view is that the new office technology will increasingly displace workers so that the job of record keeping and coordination currently done by many clerical workers will be done mostly by machines. In general, the occupations that work with obsolete equipment, such as typists and switchboard operators, are more likely to decline than the occupations that require interpersonal interactions, such as hotel clerks (Silvestri, 1993).

A more optimistic view is that the new office technology will eventually lead to the creation of clerical jobs, although these jobs may differ in content from the older jobs. Better coordination, resulting from both human and machine contributions, may improve efficiency, leading to lower prices and increased demand for the goods or services produced. In this case, the increased demand may require the hiring of new workers (Osterman, 1986).

There are not enough studies so far to let us choose between the pessimistic and optimistic views. One study of U.S. data between 1972 and 1978 suggests that both displacement and new jobs will result from new technologies. In the short run some clerical workers are displaced when computers are deployed. Shortly thereafter, employment expands because coordination needs increase after the technology is adopted. Jobs are created to meet these needs (Osterman, 1986). Automation will affect clerical specialties differently. According to the Bureau of Labor Statistics, two of the fastest-growing

BOX 13.2 Living What You Do Every Day

Yolanda O'Bannon is executive assistant to a biochemist. In this essay she explains how she finds satisfaction in the tasks associated with this job, even though she is aware that some people do not accord high status to her job.

Last year, I left a job I hated as a programmer for a job I love as an executive assistant, which is just a fancy word for secretary. I still feel a little embarrassed when people ask me about my new job. Not because of what I do, but because of what some people, including myself, have thought of secretaries.

I had always thought that secretaries were nice and maybe competent, but not smart or strong or original. I have a master's degree in English literature, have interviewed the Dalai Lama, and co-founded a nonprofit organization. People who know me wondered why I would go for what seemed to be such a dull and low status job. Even my new boss asked if I would be bored.

Why would I want to be a secretary? Because it fits me like a glove. I get to do what I love best all day, which is organize things. I like the challenge of holding the focus on the top priorities in my boss's wildly busy schedule. I can function with a high degree of chaos.

Untangling finances feels like playing detective to me. I find filing restful.

The only hard part is dealing with my own and other people's stereotypes, and learning to focus on internal rewards rather than humble appearances. I admit that I feel vaguely embarrassed bringing the faculty lunch or serving coffee to my boss's visitors. But deep down I don't believe that serving food is humiliating. Really, I think of it as a practice in humility. My husband is Tibetan. In Tibetan communities, you serve each other tea as a form of respect. When I'm serving coffee at work, I imagine that I'm serving a monk.

Whenever I get down or defensive about being a secretary, I think of those sharp, fast-talking assistants on *The West Wing*, and how they speak in paragraphs and remember everything, and I feel pretty cool. Sometimes I just look around at my fellow secretaries—savvy and articulate women who are masters at multi-tasking. I know I'm in good company.

SOURCE: Yolanda O'Bannon, "Living What You Do Every Day" © 2006 Yolanda O'Bannon. Reprinted by arrangement with This I Believe, Inc. (www.thisibelieve.org)

occupations between 1998 and 2008 will be office clerks and computer support specialists. Whether the future contains more elements of the optimistic scenario or of the pessimistic scenario, changes in clerical work will continue to be guided by the three factors that have so far transformed it: the technology, the work organization, and, to a lesser degree, the size of the firm. These three factors, in turn, affect the degree of specialization, variety, and autonomy on the job.

An additional complicating factor is the ability of North American firms to outsource some types of office work overseas to English-speaking populations in countries such as India, Singapore, and Ireland. The development of the Internet and distributed computer networks has made it possible for routine office work to be done in locations that

are geographically remote from headquarters. Routine paperwork in industries such as automobile insurance, credit cards, and travel reservations are often outsourced. Outsourcing permits clerical processing to proceed 24/7, which is important in some industries, but it also allows for the expansion of the workforce to take place in countries whose pay scale is generally lower.

HISTORY OF SALES WORK

Goods and services were originally sold by their producers, who were often farmers or craft workers. Learning to price, display, and sell the wares was part of the training for most occupations. Even merchants were not principally sales workers, and their apprentices often functioned mainly as clerical workers, although attending customers was one of their duties. Just as sales work was not necessarily distinguished from management, farming, and craft work in the past, so sales work today is not always clearly distinguishable from clerical work. This is especially true when the sales worker has responsibility for shipping, maintaining inventory, and keeping track of orders. Cashiers are an example of workers whose jobs have aspects of both clerical work and sales work.

Sales work became a separate occupation when some workers began to sell what others had produced and traded. The earliest sales workers were probably itinerant peddlers. In rural areas the peddler was a source of entertainment and news as well as goods. The peddler might carry imported or otherwise hard-to-find products such as spices or salt. Later, peddlers carried small manufactured goods. Specialized sales jobs accompanied the expansion of mercantile capitalism and manufacturing, extensive export-import trade, and large-scale distribution systems. Large-scale trade created tiers of merchandising, with both wholesale and retail levels. The traveling salesman solicited orders for wholesalers or manufacturers, much as the peddler retailed goods to individual consumers. The new forms of merchandising, in turn, created specialized jobs for brokers, agents, and salespeople. Mail-order houses and telemarketing have created new sales jobs.

DEMAND FOR SALES WORKERS

The U.S. Bureau of Labor Statistics predicts that retail sales jobs will be the occupation adding the largest number of jobs by 2014, with a projected increase of 736 thousand jobs, which represents an increase of 17.3 percent over 2004 (Hecker, 2005:77). Sales work has grown because of the proliferation of trade and the elaboration of distribution systems. Worldwide improvements in transportation, communication, and credit arrangements widened markets. Specialized sales workers became

more important in reaching and developing new market areas.

Sales work is a heterogeneous occupational category. The U.S. Bureau of Labor Statistics classifies many sales occupations by the product sold, such as insurance, parts, securities, and so on. In addition to the variety introduced by different products, sales work is organized in many different ways. Sales workers may work for either wholesale or retail concerns; they may be paid on commission, on salary, by the hour, or in some combination of ways; they may work either as regular employees or independent contractors. Many factors contribute to this diversity, including the marketing history of the product, the type of firm, and the knowledge expected of the sales worker. These factors affect earnings and benefits, the full-time or part-time nature of the work, job security, and mobility.

Product Marketing

Several innovations in marketing affect the organization of sales work. The development of the department store was accompanied by the specialization of sales workers into *departments*, such as men's clothing, home appliances, jewelry, and cosmetics. Department stores first appeared in the 1860s and 1870s. Some began as single-line retail outlets, while others were part of wholesale operations. By 1900 the largest department stores employed as many workers as the largest manufacturing establishments (Carter and Carter, 1985:587). Employment growth was rapid and specialized (Benson, 1986). By the 1980s department stores accounted for 80 percent of the general merchandise market (Hartmann, Kraut, and Tilly, 1986:48).

Many store departments adopted the sales practices of the specialty stores they partially replaced. For example, furniture stores usually paid their sales workers on commission, while clothing and fabric stores paid their workers an hourly wage. In the department store, departments that sold "big-ticket items," such as appliances and furniture, followed the custom of paying the sales worker a commission on the products sold. In the clothing departments, it was customary to pay an hourly wage.

This practice was justified by the argument that customers required more individual attention when they were buying expensive items, and so the sales worker had to work harder for fewer sales.

These differential payment practices came to be reflected in gender segregation within sales work. In department stores men worked in such departments as appliances, furniture, and automotive sales, where they were paid on commission. Women worked in women's and children's clothing, housewares, cosmetics, stationery, and similar departments, and they were usually paid hourly wages. Men's clothing represented an intermediate case. For the customers' comfort and privacy, men worked in the men's clothing department, and selling clothing was usually compensated on an hourly basis. But selling the more expensive menswear, such as suits, sometimes brought the sales worker a commission. In 2005 male sales workers earned $690 weekly, compared with $420 for women (BLS, 2006a).

A second innovation was the development of the *self-service store*. The modern supermarket is an achievement of this innovation; customers roam the aisles looking for the merchandise they want, and they rarely request sales personnel to help them. In fact, even if the customer needs help in locating an item, no sales personnel may be available. The self-service concept has been adopted widely throughout the retail trade industry. The old style "five and dime store," which had a salesperson at every counter, has been replaced by discount department stores that feature self-service and checkout counters. The new self-service stores have also proven a major economic challenge to the older, established department stores, in part because fewer sales workers are needed, and the workers are often paid lower wages than the department stores pay.

Self-service systems work best if the consumers receive sufficient information to select products for themselves. The more information provided directly to the customer, the less important it is for the customer to find a sales worker. Fixed prices, price tags, location maps and signs, and alphabetical arrangements of goods or departments are ways to make the customer self-sufficient inside the store. Depending on the goods they sell, different stores will seek new ways to inform their customers. Food stores often post information on the price per unit (such as the price per ounce) and on nutritional composition. Ready-made clothing carries tags with information on fabric care and cleaning. Many retail stores use advertising both to lure the customer to the store and to extol the virtues of various products. Consumer-oriented magazines and newspaper articles also help take the place of the knowledgeable salesperson.

A third innovation with an impact on retail sales workers is the creation of large *chains* of stores, located in separate cities or parts of the city but all carrying the same products and following similar personnel policies. Chains offer considerable advantages that locally owned stores have trouble duplicating. Because of their size the chains can buy in large quantities at discounted prices, advertise widely, and develop uniform policies for workers and transactions in each store. Customers are attracted by national brand names such as Wal-Mart. Chain and discount stores account for 89 percent of the department store trade, and have now largely displaced the traditional department store.

A fourth change is the *longer operating hours* kept by many retail establishments. Routine retail shopping has usually been a task for women, but working women often find it impossible to shop during traditional business hours. In response retail stores have extended their closing times and added weekend and holiday hours. Many convenience stores and some grocery stores are open twenty-four hours a day. In shopping centers and malls, tenant stores may be required to stay open during all the hours the mall is open, perhaps as many as twelve hours a day. These expanded hours require retail establishments to hire more workers and arrange work shifts. One solution is hiring part-time workers, especially for peak days or hours. Another solution is to hire temporary workers for especially busy seasons, such as Christmas. More than a third of department store employees work only three months of the year, usually around Christmas

(Hartmann, Kraut, and Tilly, 1986). The temporary and part-time workers are usually poorly paid, which is sometimes a source of friction between full-time and part-time workers.

A fifth change is the resurgence of *long-distance merchandising*. Beginning in the nineteenth century, firms sold goods through the mail. Catalogs were a convenience for people living in rural areas or small towns. As more Americans settled in large metropolitan areas, many observers assumed that the mail-order business would become obsolete. Instead, consumers today have many options for shopping over the Internet or by phone and mail order. These options attract busy workers who have little time to shop. Catalog companies are easier to access because of toll-free telephone numbers and fax numbers, with payments made by credit card. The Home Shopping Network and similar sales networks on cable television, and the eventual development of interactive video, promise additional types of retailing and new sales jobs.

Type of Firm

Sales work is organized differently in different firms. In general, the larger the employing organization, the more secure the sales job. Large corporate employers usually have a sales department or division; if there are several product lines, there may be a sales division for each product. Workers who sell to other companies tend to have especially stable jobs. Sales workers in large organizations have relatively secure full-time positions with salaries and fringe benefits, and they may also earn commissions. For example, sales of securities usually carry a commission. In many companies, upward mobility is possible from sales work to sales management and even into general management.

At the other extreme are the relatively insecure commission jobs with locally owned firms, where the customers are likely to be individuals. In sales positions with real estate agencies and car dealerships, for example, earnings depend on one's social networks (to find customers) and skill. The significance of networks is emphasized by this insurance salesperson:

This may sound silly, but my manager considered it very important for me to join community organizations and boards: Little League, PTA, school boards, hospital boards. You get your face in the local paper. You know: "Ralph Williams and what he's doing now." It tells people you're with Pilgrim Mutual. You're well-established, you're a family man, you're part of middle class America ... you can be trusted. This is the guy you go to for insurance. So networking yourself into all these small community organizations is very important. (Oakes, 1990:22–23)

In effect, these workers are independent contractors who are offered a desk in the firm but who have few of the other characteristics of steady employment. These jobs often cannot be categorized as "full time" or "part time," because some workers will work up to sixty hours a week, while others work fewer hours.

There are other independent contractors in sales work, many of whom think of themselves as self-employed workers. In many cases their sales work is a second job and is called *direct sales*, meaning that they have no store or other permanent place of business. Direct sales jobs include distributing household products and cosmetics, often through a merchandising technique called a *house party* held in customers' homes. Door-to-door sales workers still exist, although with the decline in full-time homemakers and the fear of crime in urban neighborhoods, the door-to-door technique is becoming less common. *Telemarketing*, or the use of the telephone to make sales calls, is becoming the technique of choice for many workers in direct sales. On the one hand, direct sales work is financially risky if the worker is expected to buy all the inventory in advance. The work carries uncertain earnings and no benefits; the pay for telemarketing, for example, is often based entirely on commission. On the other hand, direct sales companies argue that their workers have the advantage of selling known products that are nationally advertised and that they provide sales training and incentives. Box 13.3 describes one prominent direct sales firm, Mary Kay Cosmetics.

BOX 13.3 Direct Sales and Mary Kay Cosmetics

Entrepreneurs provide a new product or service or innovative marketing for old products, and, in the process, they provide jobs for themselves and others. Some entrepreneurs become executives of their own companies, and a few become extremely wealthy. Mary Kay Ash, the founder of Mary Kay Cosmetics, was a successful entrepreneur listed among the one hundred wealthiest Texans before her death in 2001. Despite her wealth, Ash encouraged people to use her first name. "I'm called Mary Kay. If people call me Mrs. Ash, I think they either don't know me or they're unhappy with me" (Ash, 1981:168).

Mary Kay Cosmetics manufactures cosmetics and sells them through direct sales. In 2005 Mary Kay was the best-selling cosmetic line in facial care and color cosmetics with revenues of $2.2 billion. Mary Kay products are not sold in stores, but through "beauty shows" to which hostesses invite five to six friends. Mary Kay's 1.6 million sales workers, called beauty consultants, earn commissions on their sales of cosmetics. Their sales directors also earn commissions from the cosmetic sales of other beauty consultants whom they have recruited. In effect, the organization is made up of many entrepreneurs, and the success of each consultant's own business adds to the success of the company. Such arrangements are sometimes called *pyramids*.

Some sales directors have earned over one million dollars in commissions, and many are enthusiastic and loyal members of the organization. Most consultants, however, do not earn very much money; for many of them, Mary Kay Cosmetics is a second job or a part-time job (Coughlin, 1989). The organization provides consultants with training, products, and sales incentives, such as diamond bumblebee pins, furs, and pink Cadillacs. Especially successful sales workers are termed "royalty" at annual seminars, which have been described as "a combination of the Academy Awards, the Miss America Pageant, and a Broadway opening" (Ash, 1981:191).

Mary Kay Ash's own career included working in direct sales for Stanley Home Products and World Gifts. As a divorced mother with three children, she had to hold three Stanley house parties a day to make ends meet. Mary Kay's mother had worked long hours as a restaurant manager to support Mary Kay's invalid father and the children, and Mary Kay credits her mother's example with giving her the encouragement she needed. But she found it irritating that she would be asked to train men who would then become her superiors, and she wrote of this time, "It seemed to me that a woman's brains were worth only fifty cents on the dollar in a male-run corporation" (Ash, 1981:24).

Mary Kay began her cosmetics business in 1963 after buying a formula for skin care products from the daughter of a hostess for a Stanley Home Products party. By the second year, her company's sales had reached $800,000. During her work in direct sales for other organizations, she developed an upbeat, enthusiastic philosophy about direct sales work and its benefits for women workers. Stating her belief that a woman's priorities should be "God, family, and work," in that order, she noted that direct sales work is flexible in scheduling. Mothers can be home when their children arrive from school or are sick. Husbands are also considered a part of the Mary Kay team; Mary Kay has been quoted as saying, "A woman who has her husband with her is a woman and a half. A woman who doesn't have her husband with her is half a woman" (Ash, 1981:70).

Mary Kay Cosmetics tries to maintain a personal relationship with consultants, who receive birthday greetings, newsletters, and frequent praise from their sales directors and from the national organization (Ash, 1984). This philosophy differs sharply from that of many corporate employers, but it is very attractive to many potential workers. The charismatic personality of entrepreneurs like Mary Kay has been a key to the success of direct sales organizations (Biggart, 1989).

Not every direct sales worker is happy. Turnover is high, and sales workers become discouraged. Beauty consultants are not employees in the usual sense of the word. They receive no compensation other than their commissions, and they receive no fringe benefits. Consultants must pay cash to keep an inventory on hand. Because cosmetics are available in every drugstore and supermarket, Mary Kay's beauty consultants must convince their customers that Mary Kay's products are worth the extra expense and trouble of direct sales. Despite the drawbacks, however, entrepreneurship offers them the opportunity to be their own boss and to succeed through hard work.

Intermediate in working conditions are the many local retail stores. Most of them hire many part-time workers who have few benefits, little job security, and no prospects for promotion. Except for the smallest firms, however, most retail stores also have full-time employees who run the store when the proprietor is absent. These full-time jobs may carry fringe benefits; they provide a measure of autonomy and variety, and they may lead to supervisory positions or to positions such as store manager, shift manager, or department manager. They also provide useful experience for the worker who wishes eventually to open a small business. Although these jobs may carry the title of manager, sales remains an important part of their job content.

Knowledge Base

Except in large companies, most sales workers experience little upward career mobility. The pay in many of the jobs is minimum wage, and many jobs are part time. Turnover is high, especially in retail trade, so employers invest little in training. With high turnover, hiring is nearly continuous, and in recent years the sales force has become younger and often better educated. In general, those sales workers with greater knowledge of the business and the product are the ones most likely to find upward mobility with the employer.

Sales jobs requiring detailed or specific knowledge are more likely to be secure and well compensated, often with a combination of salary and commission. Some sales workers are highly trained and are expected to know and understand the products they sell. They need to understand the manufacturing process, the features of the product, and how the product compares with those of the competitors. In highly competitive markets, the sales workers provide management with important information concerning the inroads that competitors are making in the market and the new features that the customers want. Because of their knowledge of production and of the market, these workers may be promoted, either within the sales department or to management positions.

Most sales jobs, however, require little or no training, and the sales worker may not be expected to know much about the products. Indeed, many managers believe that good sales skills are not learned and therefore cannot be taught; thus, there is little value even in having experienced workers train new workers. Sales skills are viewed as deriving from one's personality; workers either have it or they don't. Sales workers are expected to have generic knowledge about maintaining displays and inventories, keeping the store tidy, and opening and closing cash registers. Because sales work is a common part-time job for students, many young workers have already learned these generic skills and are employable in a variety of sales jobs. Because these skills are considered to be widespread, they do not command high wages or job security.

One competitive technique used in some lines of retail trade is hiring well-trained and knowledgeable sales personnel to assist the consumer (Carter and Carter, 1985). For example, more expensive clothing stores often feature salespersons who are well versed in fashion, alterations, and other relevant areas. These shops may even provide such special services as wardrobe consultation, closet coordination, and home visits. "Personal shopping" is another service that can be offered to the busy consumer. The personal shopper has a list of the customer's needs and knows the customer's preferred styles, colors, and price range. The personal shopper provides the customer with a selection of possible purchases, which are often brought to the customer's home for approval. These additional services may lead to higher wages for the sales worker, even in competitive product markets with slim profit margins. Box 13.4 discusses the development of customized services in the case of supermarkets.

SUPPLY OF SALES WORKERS

The low-paid sales jobs that require little knowledge and skill are disproportionately filled by women, minority workers, and young people. By 1900 women were already one-fourth of the retail sales force (Carter and Carter, 1985). Today, about half of all sales workers are women. Women are more likely

---

BOX 13.4 Grocery Sales

Grocery stores employ 133,000 workers, most of whom do retail sales work for at least part of their jobs. Grocery store workers are expected to do many other things, however, including stocking shelves, carrying groceries to the customers' cars, and cleaning the stores. These jobs are often strenuous and may help explain why grocery stores have relatively high levels of occupationally related injuries (Company and Personick, 1992).

Grocery stores are extremely competitive and have small profit margins. One innovation in grocery sales has been the expansion of the stores into one-stop shops that offer lines of sundries, toys, video rentals, flowers, and specialty foods (Strople, 2006). Many larger stores feature prepared foods as a direct response to fast-food restaurants. One interesting feature of the prepared foods is the new autonomy it has allowed some of the workers.

Q: Have you got a grill there?

Deli clerk: Yes, we've got everything, a grill, stove, oven, deep fryer. We make doughnuts, fried chicken. We have hot lunches. We can make whatever we want.

Q: Do you have recipes?

Deli clerk: No, whoever works the morning, just makes what she likes, what will sell. You try to make something simple, brats and sauerkraut, fried chicken, meat loaf.... You get two vegetables and a potato, or roll. It's about $2.00. We sell out every day. When you're cooking, you decide what you want, and then take it off the shelves. You can make anything, because we have all the equipment, and we have a whole grocery store.... We also make all our own salads. We make pasta salad, antipasto, potato salad, ambrosia, stuffed shells.

There is a great deal of worker autonomy at the store level. While the division office issues a marketing plan, the store employees have a lot of flexibility within that plan... Given a bundle-of-tasks perspective, in-store production seems to make more sense for many types of foods. Because service work is intermittent, adding production tasks to the bundle allows workers to spend their down time on production. Not only is their labor used more effectively, but the complexity of the jobs also increases, and therefore, presumably, job satisfaction increases. Similarly, the move into catering allows the firm to diversify in a way that takes advantage of both the production skills of employees (who must be able to cook for in-store production) and the down time in their schedule (because the meals can be made between waiting on customers during the morning and still be ready for an event that afternoon or evening). Employees of specialty shops can also use this time to prepare goods for self-serve, such as pre-made sandwiches or fresh pizzas. In-store production not only allows better time management by employees, it also allows a better fit to the variation in local demand, since employees know what sells in their market... Finally, in-store preparation allows firms to take advantage of mistakes in other parts of the store. For example, the deli can run a special on meat loaf if the meat department grinds too much meat or make a Polynesian stir-fry if the produce department over orders pineapples. Thus, while interdependencies in the store are limited (the different departments are not forced to interact very often), synergies exist among the departments that can increase productivity or buffer the store from uncertainties.

SOURCE: John P. Walsh. 1993. *Supermarkets Transformed: Understanding Organizational and Technological Innovations.* Copyright © 1993 John P. Walsh. Reprinted by permission of Rutgers University Press.

---

than men to be counter clerks and to work in apparel sales. Women make up 76 percent of all cashiers. Men still dominate jobs selling higher-priced items, such as motor vehicles, furniture, appliances, and parts (BLS, 2006a). Most insurance sales workers and securities sales workers are men.

The trend to part-time jobs has certainly affected the labor supply. About 28 percent of all sales workers are part-time, but 38 percent of women sales workers are part-time. The development of telecommuting for mail order and Internet sales opens up more flexibility in hours and the possibility of working from home for some sales workers (Webster, 2000).

Over 11 percent of sales workers are black, and 11 percent are Hispanic. Both groups are overrepresented among cashiers, of whom 15.5 percent are black and over 16 percent are Hispanic (BLS,

2006a). In other sales occupations, however, minority group members are underrepresented. This situation is long-standing, and some employers justify it by arguing that sales workers from a minority group might alienate white customers. Because white customers have accepted minority group members in clerical and technical specialties, however, it seems likely that they would also accept sales workers. Fears about "acceptability," because some employers assume that racism is a constant, help perpetuate racist stereotypes in the workplace.

Because of the high turnover in sales jobs, finding sales work is relatively easy for youth, but they are unlikely to seek a sales career. As more part-time sales jobs are created, there are more opportunities in sales for students in school, but part-time work is less attractive to many other workers. Hourly wages are often only marginally above the minimum wage (Hartmann, Kraut, and Tilly, 1986:50). It seems unlikely that these jobs will become more attractive as careers because of the changes being wrought by new technologies.

THE FUTURE OF SALES WORKERS

As with clerical workers, two important issues affecting sales workers are how technology is changing the content of their work and how it will affect the number of available jobs. New technology is simplifying sales work. Many retail establishments already use "intelligent" cash registers that simultaneously check stocks, control inventory, check customers' credit records, and record the transactions. Sales workers do not even need to know the price of items if they use scanners that read product codes or icons that picture the product (Sullivan, 1996). In stores that already employed low-skilled workers, even the routine sales skills will become less important.

Technology has changed the traditional work of the traveling sales worker. Telecommunications, computer networks, and modern transportation have diminished the amount of time that many sales workers must spend on the road while also providing better means for keeping track of sales contacts, orders, and so on. One effect of these changes may be to reduce the size of the sales force, even when a national or international market is being served.

"Customer service" represents a new job category that combines elements of both sales and clerical work, and which can be exported to call centers located in a distant state or located overseas. Sales jobs come to resemble clerical jobs because the sales worker also performs many of the tasks of information management and processing. The job of cashier, one of the fastest growing, is a good example of this trend. The cashier's job is to complete a sales transaction, but today the cash register may also perform a number of other information-processing tasks at the same time. The cashier may not even be aware of the many uses of the data being recorded by the register.

SUMMARY

Clerical and sales jobs have grown rapidly in the recent past and can be expected to grow in the near future, but the content of these jobs is being changed by developments in work organization and technology. Among clerical workers, job content has been changed by the imitation of assembly-line production techniques. The introduction of new technology may worsen this trend, or it may reverse it through the processes of job enlargement and job enrichment. Among sales workers, work reorganization has led to a proliferation of part-time jobs, and the new technology has made possible greater use of relatively unskilled and inexperienced workers. On the one hand, the business of "selling" the consumer has often been taken over by advertising, self-service retailing, and other innovations that have reduced the need for skilled sales workers. On the other hand, where products require sales workers with specialized or technical knowledge, the workers are likely to retain greater control and influence.

KEY CONCEPTS

feminization of clerical work

work reorganization

rationalization

job enlargement

job enrichment

QUESTIONS FOR THOUGHT

1. A number of analysts have tried to explain the feminization of clerical jobs. Can you think of other jobs that might be "feminized" or "masculinized" in the future?

2. Some experts think the number of clerical jobs will increase, while others expect a decrease. With which position do you agree, and why?

3. Under what conditions does technology improve clerical and sales jobs?

4. Think of a recent technological innovation. How could this innovation improve or worsen the work life of a clerical or sales worker?

5. Under what conditions is a clerical or sales job a dead-end job? Under what conditions is it part of a job ladder?

MULTIMEDIA RESOURCES

Print

Evelyn Nakano Glenn and Roslyn L. Feldberg. 1977. "Degraded and Deskilled: The Proletarianization of Clerical Work." *Social Problems* 25, 1 (October) 52–64. A classic description by two sociologists of how the factory model of clerical work comes to affect the workers.

C. Wright Mills. 1951. *White Collar.* New York: Oxford University Press. A classic work in sociology by a leading American analyst. Examines the transformation of the occupational structure and the nature of white-collar work.

Internet

Monster.com. *http://www.monster.com* A large website with information for job seekers. Visitors may look for jobs by occupational title.

RECOMMENDED FILM

The Devil Wears Prada (2006). Meryl Streep is a difficult boss for her young assistants; this film also provides an interesting view of the fashion industry. 1 hour, 50 minutes. Rated PG-13.

14

Marginal Jobs

Many employers carefully select a core group of employees, invest in them, and take elaborate measures to reduce their turnover and maintain their attachment to the firm. Many of these same employers, however, also maintain a peripheral group of employees from whom they would prefer to remain relatively detached, even at the cost of high turnover, and to whom they make few commitments. The strength of these employers' desire to avoid attachment to this peripheral group is such that a broad spectrum of institutional arrangements has emerged to facilitate this relationship.

(MANGUM, MAYALL, AND NELSON, 1985:599)

Now the delegation of dirty work to someone is common among humans. Many cleanliness taboos, and perhaps, even many moral scruples, depend for their practice upon success in delegating the tabooed activity to someone else.

(HUGHES, 1958:2)

Most of us carry a mental image of a bad job. In this chapter we use the term **marginal jobs** to refer to jobs considered undesirable because they are boring, low-paid, intermittent, dead-end, and they lack autonomy. The term *marginal* describes only the job; in many circumstances, the workers in such jobs are hardworking and capable, but unable to find better jobs.

As the quotations above suggest, marginal jobs originate in different ways. In this chapter we will discuss types of marginality, marginal occupational positions, the marginal jobs that arise from employer decisions, and some characteristics of workers found in marginal jobs. We will also discuss trends in the numbers and types of marginal jobs and the hardships faced by workers who have no job at all, not even a marginal job.

WHAT IS A MARGINAL JOB?

A marginal job departs significantly from the employment norms of the society in which it is located. Because jobs are governed by many norms, there are also many ways in which jobs can depart from these norms. Four important norms commonly apply to jobs: (1) the job content should be legal; (2) the job should be institutionally regular; (3) the job should be relatively stable; and (4) the job should provide adequate pay with sufficient hours of work every week to make a living. We will discuss marginal jobs in terms of their departure from one or more of these norms. It is important to note the subjective evaluation involved in identifying marginal work; for example, two people might disagree whether a wage level is sufficient to provide an adequate level of living.

Illegal or Morally Suspect Occupations

The longer you strip, the harder it is to retain a positive view of men. Long-term dancers—at least the ones I met—were all bitter. . . . Be a waitress. —A former stripper (Bowe, Bowe, and Streeter, 2000:368–369)

Some jobs are marginal because the content of the work is socially defined as improper, immoral, or illegal. Drug pushers, pimps, and prostitutes can be considered members of **deviant occupations** (Miller, 1978). Deviant work includes criminal behavior, such as the production and distribution of illegal goods or services. Criminologists sometimes speak of criminal "occupations" and criminal "careers."

Which occupations are considered deviant varies depending on which groups have the power to define deviance, and the labels may change with time and place. During Prohibition, legislators defined as deviant the people who sold alcoholic beverages, because the law prohibited the production and sale of alcohol. Today, although regulated by federal and state law, producing and selling whiskey, wine, and beer are legal. Liquor stores alone employ 130,700 workers (BLS, 2006b). Nevertheless, people might view these jobs as deviant if they regard alcohol as a source of social ills.

There are other occupations, such as stripping that, although legal, violate some people's norms of modesty. Some occupations, such as prostitution and playing professional poker, are legal in some geographic areas but not in others (Hayano, 1982). Box 14.1 discusses a bordello in Guatemala (where prostitution is legal) that leaves its owners with little profit.

Workers may be recruited into deviant occupations because other, more legitimate economic opportunities are blocked. Some observers argue that minority youths are susceptible to recruitment into deviant occupations because other jobs have migrated to the suburbs, far from the inner-city homes of many minority families (Wilson, 1987; Andersson, Holzer, and Lane, 2005). Deviant jobs may also be more lucrative than regular jobs. Even if they are not officially in the labor force, those in deviant occupations may be considered workers.

Unregulated Work

They pay us once a week by check. You can't open a bank account without a Social Security number so it's a little difficult. You can cash the check in the company bank. . . . There are also a lot of thefts—people break into our houses and steal our money, because they know we can't keep our money in banks. —An undocumented immigrant working in a poultry processing plant (Bowe, Bowe, and Streeter, 2000:190)

A second norm, that the job be covered by government work regulations, recognizes the government interest both in protecting workers and also in collecting taxes. Especially in industrialized countries, state regulation is supposed to ensure that "regular jobs" meet minimum standards for pay, working conditions, and safety. The employer is also expected to withhold employees' income and Social Security taxes and to pay taxes on its own payrolls (such as unemployment compensation insurance). Thus, a second dimension of marginality characterizes jobs that lie outside this system of regulation.

In the **shadow economy**, jobs are not necessarily illegal, but they are institutionally irregular because the income is not reported, there is no

BOX 14.1   **Prostitution and the Informal Economy in Guatemala**

This selection from an anthropological study of Guate-mala highlights one of the many strategies that the residents of San Pedro use to make a little extra money. San Pedro's economy is characterized by many small enterprises and an informal economy. Prostitution is legal in Guatemala, but the business failed anyway, and the anthropologists try to explain this failure.

The local house of prostitution, known as *La casa de las mujeres* or *la casa*, is a business owned and operated by a woman and her three daughters. None of these women are prostitutes. They employ four women from out of town for that work ... the women have no control over the prostitutes, who only work when they want to. If the employees leave for the coast for a few days, the owners can-not open and thus lose money. They complain quietly, but have neither the leverage nor the powers of persuasion normally attributed to a male owner, madam, or pimp. This lack of authority is not simply a matter of physical strength, nor of sexual exploitation, but is as well a result of the absence of female experience as employers. Women have tradi-tionally worked with relatives alone. As minimally capitalized corporate enterprises made up of the labor of family members, by their very nature these businesses make employment of outsiders difficult. Where *muchachas* or apprentices are hired, authority may be expressed almost as an option of status or class. With a group of independent, strong-willed prostitutes, however, this is an improbable response for *Sampedranas* [women residents of San Pedro] who do not have the experience of men as employ-ers. Rather than being bosses and strong authority

figures, *la casa* owners are as kind and considerate to the prostitutes as they would be to any house-guests. They are not coercive. To the contrary, Doña Tila tries to take good care of them by feeding them lunch and dinner, with sizeable portions of meat as often as she can. She pities the prostitutes for their way of life and in small ways attempts to ease their situation by doing things like caring for the son of one of them while his mother works. In another sense, their being women is fiscally limiting as they have little access to capital to invest in the business. They have only debt.

La casa is both like other female family busi-nesses and different from them. It does contain many of the common elements of female business systems, such as minimal capitalization, donated labor of educated and uneducated women, and reciprocal responsibilities. On the other hand, it was begun a year ago as a desperate investment by a group of inexperienced women. It has no history in their family nor do any of the members express loy-alty or dedication to it. It has been a difficult and unsatisfying business experience that has barely sustained them financially. Interestingly, the pending disappearance of the eldest daughter from *la casa* is typical of the trend away from female family businesses and toward modern employment among *Sampedranas,* as is the closing of the business that will doubtless accompany her resignation.

SOURCE: Tracy Bachrach Ehlers, 2000. *Silent Looms: Women and Produc-tion in a Guatemalan Town* (Austin: University of Texas Press), pp. 91–93. Used by permission.

official monitoring of health, safety, or working conditions, and other institutional regulations of work are missing. In developing countries, terms such as the *informal* or *submerged economy* are more commonly used (Portes and Sassen-Koob, 1987). What is common to all the definitions is that the shadow economy is neither accountable to state authorities nor subject to economic measurement. Examples include maids or gardeners who work for cash or whose employers refuse to withhold taxes.

The shadow economy is illegal in the sense that it evades normal regulation; the goods and services produced may not be illegal. In the case of the undocumented worker quoted above, working in the chicken processing factory is certainly legal, but the worker has no Social Security number, and so there is an irregularity involving the payroll tax, if not other things as well.

The shadow economy may or may not use money as a medium of exchange. Workers may receive their wages in cash or in goods, services, discounts, or even barter. However, even if the workers are paid in money, one goal of the shadow economy is to evade the monetary costs of regulation. The head of one French firm remarked, "If I were to declare the work of a translator, it would cost me 5,000 francs, and she would get only 3,500 francs net. I'd rather give nothing to the taxman and pay her about 4,200 francs" (DeGrazia, 1980:550). No one can be certain how large the shadow economy is, but estimates run into the billions of dollars.

Contingent Work

> The worst part about temping is probably when people ask you what you do. It's, like, "Oh, you're doing temp work still, Lillian? Can't you get a job?" That's a big one. My parents are, like, "Lillian, let's get a job." (Henson, 1996:145)

The third norm, stability, implies that a job should represent a relatively long-term relationship between the employer and the employee, assuming that both of them perform their contractual responsibilities. A corresponding dimension of marginality is **job instability** or **contingent employment**, terms that imply that the job inherently carries no job security.

Unstable jobs are not the same as probationary jobs. Many employers initially hire workers for a probationary period that may vary in length from a few weeks to several years. Typically, the probationary period is relatively short for unskilled or semiskilled jobs but often quite long for professional workers. Among university professors, for example, the usual probationary period is six or seven years. During the probationary period, the professor receives no job security. But a professor who is given a tenured position has a secure job unless the school faces severe financial problems, eliminates a program, or finds the professor guilty of serious misconduct. By contrast, the unstable job never offers job security.

SOURCE: Distributed by United Feature Syndicate, Inc.

Employers have many devices for getting work done without making a long-term commitment to the worker, including the use of consultants, short-term contracts, and temporary workers (Sullivan, 2004). In a business that is about to fail, all of the jobs in the firm may become unstable jobs. A job may be unstable because the work is seasonal or because intense foreign competition makes it difficult for the employer to guarantee continued work. Even in successful businesses, an employer may for strategic reasons refuse to guarantee continued work, reasoning that workers who are fearful for their jobs will be more docile and compliant. Some employers are willing to provide stable employment for those workers whom they consider central to the workplace, but they are unwilling to provide similar stability to other workers.

In 2005 there were 5.7 million workers holding contingent jobs, about 4 percent of total employment. A majority of them (55%) said in a survey that they would have preferred a permanent job, and only 33 percent said that they preferred their current arrangement. Full-time contingent workers earned on average about 81 percent of what full-time workers earned in permanent positions (BLS, 2005).

Underemployment

[I had a] definite decrease in weekly hours. Used to work a lot of hours, 50 to 75 hours per week. Regular work week is 40 hours. For the past year my hours were cut dramatically. Work never picked up again. —Julie Keys, filing for bankruptcy (Sullivan, Warren, and Westbrook, 2000:95)

A final norm is that work be adequate in terms of the hours of work available and the compensation offered. **Underemployment**, or inadequate work, describes jobs that may be permanent but that provide fewer hours of work than the worker would like (a condition called *involuntary part-time employment*) or that provide very low wages. Involuntary part-time work may occur in several ways (Reynolds, 2003). Workers may have their hours of work reduced, as Julie Keys did. A company may reduce every worker from five work shifts per week to only four, with pay reduced proportionately. Such partial layoffs may be imposed because of bad weather, shortages of needed materials, and other conditions beyond the worker's control. When the weather turns rainy, construction workers may be sent home until conditions are better for working. Involuntary part-time work also occurs if a worker who wants a full-time job can find only a part-time one. Very-low-wage workers are called the working poor. We will use the term *inadequate work* in this chapter to refer to these types of underemployment. The word also describes the situation of a worker who has greater qualifications or skills than the job requires (Sullivan, 1978; Clogg, 1979; Handel, 2003).

Many jobs are characterized by more than one dimension of marginality. In the following sections, we examine entire occupational groups in which the jobs are marginal. Later we discuss the expected growth of marginal employment and the possible division of employment into a stable sector and a growing marginalized sector.

HOW DO JOBS BECOME MARGINAL?

Some jobs are marginal because of the content of the work. Undesirable or "dirty" work is likely to be delegated by more powerful occupational groups to less powerful groups, and the resulting jobs are also likely to be unstable, poorly paid, and in other ways marginal. Other jobs become marginal because new technology changes the job content, making the workers' skills obsolete. While it is not possible to predict with certainty just which skills will become obsolete, we will examine some of the major effects of technology on marginality. Finally, some employers create marginal jobs, regardless of the job content (Appelbaum, Bernhardt, and Murnane, 2006). Thus, given two identical jobs, the job at one firm might be secure and stable while the same job at the other firm might be marginalized.

Marginal Occupational Groups

In every advanced capitalist society, somebody has to haul sheetrock, clean toilets, can fruits and vegetables, harvest crops, cut up and package animals and fish to be trucked to supermarkets. Turks sweep Switzerland's streets. Jamaicans pour concrete in London. In central Texas, illegal immigrant Mexicans carry turkey semen from barn to barn to inseminate female birds. Puerto Rican women working in the tuna canning plants of Mayagüez smell so bad that the local taxi drivers refuse to carry them in their cars, and North Carolina women working in crab houses throughout the eastern part of the state acquire a rash they simply call "crab rash" for lack of a medical term (Griffith, 1993:4).

In terms of the dimensions of marginality we have discussed, some farm work, private household work, and some manufacturing and service jobs can be defined as marginal. As Figure 14.1 shows, one characteristic of these occupational groups has been that they have generally declined as a proportion of

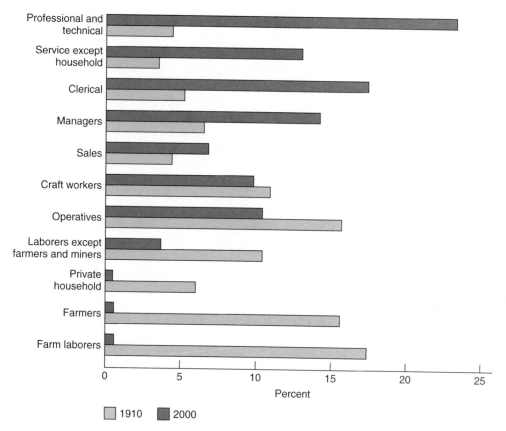

F I G U R E 14.1 Proportional Employment in Occupational Categories, 1910 and 2000

SOURCE: U.S. Department of Labor, Monthly Labor Review, April 2006, accessed at http://www.bls.gov/opub/ted/2006/apr/wk1/art04.htm

the labor force. As they have declined numerically, they have also been increasingly marginalized for three related reasons: (1) their jobs are generally considered unattractive, (2) they work in relatively small, non-bureaucratized workplaces, and (3) they have little occupational power.

Unattractive Work Nearly every job contains at least some dull or unpleasant tasks. People in more powerful occupations, however, are often able to delegate unpleasant tasks to other occupations. Physicians delegate personal care tasks to registered nurses; nurses in turn may delegate the tasks of making beds, giving patients baths, and emptying bedpans to nursing assistants or orderlies. A similar

process occurs in many industries, with some jobs consisting mostly of dirty, repetitive, or boring work. Because others often perceive these jobs as unskilled and poorly paid, these occupations share relatively low occupational prestige. It is important to note, however, that these generalizations are being made about the occupation in general. It is not necessarily the case that every individual within the group will have few skills and receive low pay.

Small, Non-Bureaucratized Workplaces Members of these occupational groups are among the least likely to work for large bureaucratized firms. Their employers are likely to be farm owners, small-business owners, or homeowners—people who may not even think of themselves as employers because they employ only one or a few workers. The employer and the worker might have expectations that the work will continue, but there is no commitment to provide permanent jobs or full-time work. To make ends meet, each worker may have to work part-time or part of the year for several employers, ensuring that every employer has even less incentive to improve job conditions. Even if a large bureaucracy is the employer, such as agribusiness corporations that employ farm workers, marginal work may be treated differently from other work. Many farm workers, for example, are excluded from unemployment coverage and minimum wage laws.

Low Occupational Power   Marginal occupations tend to have little occupational power. The workers are rarely in a good position to bargain for better wages. Workers who have small, unbureaucratized employers find collective organization difficult (see Chapter 6). The workers' knowledge is not sufficiently specialized to claim the sort of monopoly that professionals or craft workers have. Nor are their skills believed to be in great demand. Further, the workers rarely own the capital necessary for doing these jobs, such as the farm land and equipment, the restaurant equipment, or the cleaning equipment and appliances. As a result, the workers are in a poor position to improve their jobs.

Because of poor working conditions, employers often claim an inability to find enough people to fill marginal jobs. Large growers petition Congress to admit immigrant farm workers. Owners of restaurants and hotels complain of high turnover among their employees. Homeowners complain that "there's no good help available." Their comments suggest that there is actually high demand by employers for marginal jobs, but only if wages are kept low and job security is avoided (Waldinger and Lichter, 2003).

Farm Laborers   *Farm laborers* are employees, permanent or casual, of farm owners. Their principal duties include the planting and harvesting of crops and the other tasks needed to bring the crop to harvest, such as weeding, fertilizing, irrigating, and crop dusting. Farm laborers also care for livestock

and help prepare animal products (such as eggs or milk) for market. As you saw in Chapter 8, agriculture is declining in terms of the numbers of workers employed. This decline is attributed to the increased use of machinery for planting, fertilizing, and harvesting. But farm workers are still needed to harvest the soft fruits and vegetables that would be damaged by equipment, and they are needed to run the equipment.

Farm work is considered low-skilled work, but it would be more accurate to say that farm work does not require formal schooling credentials but instead is learned on the job (Smith, 1987). The work is often dangerous because of exposure to pesticides, the use of heavy equipment, and other potential hazards. Farm work involves heavy manual labor and exposure to all kinds of harsh weather.

Farm workers can be defined as marginal in terms of job regularity, stability, and adequacy of employment. The organization of agriculture has affected the *regularity* of farm work. There are several types of farm laborers, but each of them is irregular in some aspect of the job. The **migrant farm worker** follows the crops as they mature and works for a succession of employers. Different migration patterns follow the seasonal harvests through continental climatic zones, up the East Coast, the West Coast, or the Midwest. The **sharecropper**, a decreasingly common type of farm laborer, lives on the land owned by another and is paid partly with a portion of the harvest. Between harvests, sharecroppers often borrow money, seed, and fertilizer from their landlords, and pay them back with interest after the harvest (Jaynes, 1986). Because of high interest rates and occasional crop failures, many sharecroppers were once so far in debt to their landlords that they were virtually captive labor. The most "regular" farm labor jobs are those of the **hired hands**, including cowboys. These are often relatively permanent positions. Hired hands receive wages, but part of their pay is sometimes room and board.

More recently, farm workers have been employed by large-scale agribusiness firms. Three types of corporations now dominate agricultural production. The first is the family corporation engaged in farming. This is the "family farm," but usually on a very large scale, often with thousands of acres under cultivation. The second type is the publicly owned corporation primarily engaged in food processing, such as Del Monte and the Ralston-Purina Company. The third is the publicly owned conglomerate whose major operations are in non-food industries. An example of such a company was Gulf and Western Industries, which before its acquisition by Viacom owned sugar plantations, clothing manufacturers, and entertainment firms. Despite having large, corporate employers, however, farm laborers still do not have completely "regularized" jobs because they are explicitly excluded from many types of labor legislation. Most laws regulating pay and working conditions exempt farm workers. Thus, a farm worker employed by a subsidiary of a large corporation might not receive the fringe benefits available to its other workers.

Farm jobs are often *unstable*. Some instability arises because farm work is seasonal; in addition, if the crop fails or is poor, there will be little or no work at all. Some instability results from the economics of agriculture, which encourage each farm owner to hire workers only for the peak periods. To make ends meet, each worker may have to work part time for several employers, and the worker is often legally defined as an independent contractor rather than an employee. Independent contractors are ineligible for fringe benefits and responsible for their own taxes. Migrant workers often make a customary annual circuit of employers. The employment relationship, however, is asymmetrical. The workers may be bound to the employer through contracts and sometimes through loans made to provide for the workers during the off-season. But the employer has no reciprocal requirement to rehire the workers. Most farm jobs offer no opportunities for career advancement, except perhaps for promotion to crew leader.

Many farm jobs would also be considered *inadequate employment*. Full-time farm workers in 2003 earned total compensation of $27,500, just 54 percent of the average total compensation (U.S. Census, 2006). The compensation is often based on piecework rates, with a sum of money paid for each bushel harvested or each acre planted. In practice farm laborers may literally work from sunrise to sunset without earning enough money to support themselves. The farm worker may lose wages if a piece of equipment breaks down or if a thunderstorm begins. If the farm worker is paid an hourly wage, an additional problem is finding and keeping full-time work. Many farm workers are not even covered by Social Security, let alone pension plans, so that when they become too old to work, they must rely on public assistance. Migrant workers may not even qualify for public assistance because of ambiguities about their legal residence (Nordheimer, 1988).

Given the prevailing work conditions, recruiting workers to farm laborer jobs is difficult. Many farm workers are members of minority groups; 40 percent of them are Hispanic and 5 percent are black. Teenagers are recruited for work during school vacations. Large growers have persistently sought to recruit unskilled immigrant workers as temporary or permanent farm laborers (Majka and Majka, 1982). The late civil rights leader, Cesar Chavez, helped organize the United Farm Workers, and the union has been able to negotiate improved wages, working conditions, and some pension coverage for its members (Nordheimer, 1988). Only 3 percent of all farm, forestry, and fishing workers are represented by a union (BLS, 2006b).

Marginal Service Workers As we noted in Chapter 10, the service industries and the service occupations are growing rapidly. Many service jobs are not marginal. Professional services, including medical, legal, social, and educational services, are rarely marginal (see Chapter 11). Similarly, business services, such as finance, insurance, and real estate, are rarely marginal, nor are such protective services as the police and fire departments.

Three groups of service workers are more likely to be marginal: (1) personal service workers, (2) food service workers, and (3) cleaning service workers. Personal services are provided directly to individuals and include porters, dry cleaners, shoe shiners, beauticians and barbers, and repairers. Food service workers include cooks, dishwashers, waiters and waitresses, and counterpersons. Cleaning services, another major category, need to be distinguished from private household workers who clean homes. The principal difference is that cleaning-service workers are paid by cleaning companies and have no direct employment relationship with the people whose dirt and mess they clean (Hood, 1988).

Service work may not be *regular* in terms of its legal protection. In 2005 there were 1.4 million workers who were paid hourly rates below the federal minimum wage, most of them in services such as restaurants, personal services, and other service industries (BLS, 2006b). In other ways, however, greater regulation is beginning to characterize these jobs. For example, cleaning companies that hire maids and send the maids to work in private households provide more regular employment than do private household employers, because the wages and working conditions are established by the cleaning company and not through individual negotiation with the homeowner. Illegal aliens have often been employed in positions such as hotel porters, dishwashers, and janitors. With the passage of the Immigration Reform and Control Act of 1986, it became illegal for employers to hire aliens without work authorization, although this provision is irregularly enforced.

Service jobs are often *unstable*, in the sense that the work contracts are rather informal and employers easily breach them. Many service jobs are seasonal, especially in the resort industry and in some restaurants. Other service jobs are intermittent, with the employer calling on workers when they are needed. When business is slack, the workers are furloughed. Finally, service jobs are often *inadequate* in wages and hours. In 2005 the median weekly earnings of full-time service workers were $423, compared with $675 for the entire labor force (BLS, 2006b).

Service work differs from the other marginal occupational groups in that the number of jobs is expected to grow. Demand for services such as food preparation increases as home meal preparation diminishes. Precisely because these jobs were once in home production, however, the knowledge and skills needed are often widespread. Workers who are technologically displaced from manufacturing or

clerical work are available to take the service jobs. Thus, even though demand will remain strong, wages and conditions of employment in many of the services may not improve.

Some service workers are highly skilled. Some cooks may have learned their trade on the job, but others are graduates of culinary arts institutes or chefs in four-star restaurants. Barbers and beauticians are skilled and must hold a state-issued license. Unlike professionals, however, service workers have difficulty limiting access to their occupations. The higher wages that skilled service workers might otherwise command are often undermined if the service can be mass-produced. For example, a fine restaurant can seat relatively few diners in an evening, compared with a fast-food outlet. A barber can cut hair for relatively few customers in a day, but a chain-operated discount haircut store can employ more barbers and serve dozens of customers every hour. Because of such open competition, even skilled service workers may not be highly compensated.

There is a growing unionization movement among some service workers, including hotel workers, food preparation workers, and cleaning-service workers. As mentioned in Chapter 6, organized labor has targeted the service industries as a major site for organizing.

Private Household Workers Private household workers are distinguished from other service workers because they are never employed by corporate employers. Less than one-half of one percent of the U.S. labor force now falls into this category. Private household workers include launderers, cooks, maids and housekeepers, gardeners, babysitters and nannies, chauffeurs, butlers, and other household servants. The occupational group is dominated by women; only 3 percent of private household workers are men.

Private household workers are marginal in terms of job regularity, stability, and adequacy. Their jobs are *irregular* because, like farm workers, they are excluded from many forms of labor legislation. The 1938 Fair Labor Standards Act extended minimum-wage coverage to nearly 87 percent of all non-supervisory

---

B O X 14.2   Household Workers

The [employer's] use of domestics' first names, calling domestics of all ages "girls," the encouragement of performances of subservience, demands of spatial deference, perceiving domestics as childlike, giving domestics used household articles—all of these conventions of domestic servitude have in common the quality of affirming the employee's inferiority. And this purpose is further served by other conventions: the domestic is asked to be inferior in her material conditions, in her intelligence, her appearance, and sometimes even her character....

The low regard for this sphere of labor—whether paid or unpaid—has been well documented. The female employer, regardless of the degree to which she may have chosen to buy her way out of it, knows that she is seen as responsible for all household maintenance and that this is devalued work. She perceives the person she hires to do such work as doing *her* work in a way the male employer does not. The domestic is something more than an employee; she is an extension of, a surrogate for, the woman of the house. And she operates in what is increasingly the least prestigious realm of women's activities. This view of the domestic on the part of the employer—as an extension of the more menial part of herself rather than as an autonomous employee—may help to explain why the women tend to see domestic service as a more informal arrangement than other occupations....

The employer's low regard for this "women's work" can combine with her own sexism, racism, and class prejudice to further degrade the work and the

groups already subordinate in the "three structures of power" in the United States (women, people of color, and the lower classes). For some employers, like Alberta Putnam, it is incongruous to hire a man to do such work: "I would feel uncomfortable with a man in that position. I wouldn't feel right giving him orders like that. I even feel funny asking my husband to clean the dishes." For some, like Holly Woodward's husband, it is incongruous to hire a middle-class person: "Then there was Patricia, a fascinating British girl. Her father was an actor, and she wasn't sure what she wanted to do. My husband was against hiring her. He told me, 'You don't want help like that around.'" And it may be assumed that for some employers—particularly in the South, Southwest, and Far West, where the servant population has been almost exclusively black, Mexican-American, Native American, and Asian-American—it is incongruous to hire a white. One can begin to see why the lower-class woman of color, just *because* of this society's sexism, racism, and class prejudice, might be psychologically the most desirable "type" for a position of servitude and why being associated with this archetypical "women's work" further degrades her— even, or perhaps especially, in the eyes of her female employer. The employer benefits from the degradation because it underscores the power and advantage (easily interpreted as the rightness) of being white and middle class.

SOURCE: Judith Rollins. 1985. *Between Women: Domestics and Their Employers.* Philadelphia: Temple University Press, pp. 183–184, 194. © 1985 by Temple University Press. Reprinted with permission of Temple University Press.

---

employees as of 1985, but private household workers continue to be exempt (Levitan and Shapiro, 1987). When many household workers were covered by minimum-wage legislation for the first time in 1974, employers often responded by reducing their hours of work (Rollins, 1985:68). Live-in maids may be expected to work for ten or twelve hours a day and to accept part of their pay in room and board. Another form of **payment in kind** that flourishes is the giving of gifts, such as old clothes, cast-off furniture, and leftover food. Private household jobs rarely have job descriptions. Workers complain that their employers increase the amount of work expected from week to

week. Even lunch hours and coffee breaks may be frowned on (Rollins, 1985).

Even if some benefits are mandated, employers often fail to comply. For example, employers of private household workers are supposed to pay Social Security taxes, but many do not. One private employer said, "No, I've never taken out for Social Security or made arrangements about sick pay or vacation. In this kind of job, there is no formalized arrangement. I can't imagine what the [vacation arrangement] would be, but certainly time without pay" (Rollins, 1985:78). The workers themselves may resist deductions from their already meager wages.

Private household work is *unstable*. There is little or no job security. If the employer is out of town or on vacation, the worker is effectively laid off. Workers who want full-time work must often arrange to work in several different households. There are no career opportunities. Employers can discipline or fire their household workers at will, and the workers have no grievance procedures or other protection beyond the goodwill of the employer (and perhaps the difficulty of "finding good help"). The work is rarely dangerous, but it may be heavy work, and it is often isolated. Private household work is often *inadequate* in terms of the hours worked and the pay received.

Private household employment is similar to unpaid housework in that it is generally devalued. The work is described as being unskilled, and it has typically been relegated to the least powerful members of the society: women in general and women of lower-class or minority origins in particular. Private household workers were traditionally recruited from among immigrant groups or from racial minorities (Romero, 1992). Box 14.2 discusses the association of domestic work with gender, race, and class inequalities.

EMPLOYERS WHO MARGINALIZE JOBS

INTERVIEWER: Do you feel like you get a message from higher up in terms of what you should be doing with part-time and full-time? STORE MANAGER: Oh, without a doubt. They want to decrease the full-time ratio and increase the part-time ratio. . . . We would want at least a three-to-one ratio, here part-time to one full-time. (Tilly, 1996:110)

Many workers in the three marginal occupational groups just discussed will have a marginal job. In other occupational groups, *some* workers will have marginal jobs. For example, many factory workers are unionized and paid well. They have

monetary and other benefits and at least some promotion opportunities. But others work in sweatshops with unstable, poorly paid jobs and without grievance procedures or other protections.

We have already seen that the members of the same occupation may be employed in many different industries and in many different firms. Because industries and firms may have independent effects on which jobs are marginal and which are not, there are occupations in which a large proportion of workers hold very good jobs, but in which another large proportion could be classified as holding marginal jobs. In terms of job stability and permanence, there are probably no occupations that are "safe" from marginality. Rather, some form of economic vulnerability is possible in every occupation as a function of the industry, firm, or employment contract.

Employers create marginality by dichotomizing workers, jobs, firms, and even industries into categories that are "more important" or "less important." A single employer can divide jobs into **core and peripheral jobs**, carefully rewarding the core job holders to minimize their turnover. The peripheral workers, by contrast, may be laid off, intermittently laid off, or even temporarily hired from an outside agency. Firms can be classified along similar lines. The core firm employs a permanent workforce but may also subcontract to peripheral **client firms** when there is too much work or if it wants to hire temporary workers. Even industries can be divided into those that are most central to the economy and those that are more peripheral.

This argument suggests a symmetry such that the preferred workers are able to secure core jobs within core firms and within core industries. The preferred workers are usually able-bodied, well educated white men between the ages of twenty-five and sixty-four. The remaining workers are marginalized—not because they are necessarily poor workers, but because they have the highest probability of filling marginal jobs. The symmetry between core and peripheral economic structure and the division of labor force into "preferred" and "unpreferred" workers is sometimes called the **dual labor market**.

In this section we consider three sources of marginality: industry, firm, and temporary employment contracts. In later sections we will consider in more detail the workers who become marginalized, and we will also return to the dual labor market hypothesis.

By Industry

Some industries are associated with high proportions of marginal jobs, and the common characteristics of these industries can help us understand why the jobs are often marginal. In general, modern industrialized economies can be thought of as having certain *core,* or essential, *industries* (Averitt, 1968). These industries are responsible for a disproportionate share of the wealth produced, and, given the international division of labor, they also tend to be the industries in which the home country is most specialized. *Peripheral industries,* by contrast, are considered less important. They might have unstable product markets, perhaps because they produce a seasonal product or because the market is subject to sudden, intense price fluctuations. Three additional characteristics of peripheral industries seem to be associated with marginal jobs: competitiveness, low productivity, and a management orientation toward low-wage, low-skill production practices.

Marginal jobs tend to be more common in highly competitive industries. Within a single country, very competitive industries often have many small firms in economic competition with one another. In industries dominated by only a few firms—industries called *oligopolies*—individual firms often face much less competition for domestic markets. Even oligopolies, however, may now face stiff foreign competition so that the number of firms within an industry in any given country is no longer a sufficient index of competitive pressures. Competitive conditions can lead to more marginal jobs in two ways.

First, a firm may respond to competition by creating low-wage, unstable jobs and laying off its workers as a result of even minor economic setbacks. This strategy gives management some additional flexibility in meeting competitive conditions; it is easier to move the company in a new direction, modify the product lines, or simply conserve capital. But it is costly in terms of employee morale and loyalty. When managers follow this strategy, competition results in increased numbers of marginal jobs.

In the second case, the firm may decide simply to export its jobs to another region or even another country, often a developing country, where workers will accept lower wages. In this case, the effect of competition is to decrease total employment in the first location by eliminating some jobs, although a number of lower-paid jobs will be created elsewhere. It is difficult to predict which strategy will be used more frequently, and so it is difficult to assess whether the effect of competition is more marginal jobs or a decreased total number of jobs. A combination of both effects may occur.

Some foreign competition is real. Multinational corporations may, however, use the *fear* of foreign competition to justify lower wages and poorer working conditions within the industrialized countries. The multinational corporation can produce its products and create jobs in a number of countries. Its profit margin may not depend on the locale of production. Its managers may threaten workers with the possibility of closing its North American plants and moving them to the Dominican Republic or Malaysia (Adler, 2000). In such cases, workers or their union representatives may agree to lower wages and poorer working conditions to keep their jobs. By participating in this process, workers may seem to acquiesce to their own marginality, but they may see a marginal job as preferable to unemployment.

The *less productive* industries are also more likely to create marginal jobs. Lower productivity may result because the work itself is labor intensive and makes little use of technology. Many of the service industries, for example, are less productive than manufacturing industries because no technological process is available to improve the delivery of the service. In other cases, productivity-enhancing technology exists, but firms have lacked the ability or willingness to invest in this technology.

There are also industries with a tradition of creating low-income, unstable jobs, usually as a result of *management practices* (Appelbaum, Bernhardt, and Murnane, 2006). The textile industry, which was once thought to be a low-wage industry only because of intense competition, actually employs quite productive workers who are, however, paid less than they would be paid in other industries (Galle et al., 1985). They often cannot seek higher wages because the textile mill is the only major employer in the town. It appears that the practice of paying textile workers low wages, and of seeking lower-wage labor markets for textile mills, is, in part, a result of management tradition. Textile mills have traditionally sought to hire low-wage workers—first native-born women, then immigrant women, then minority women, and now workers from developing countries. The practice of always seeking out lower-wage workers keeps prices (and wages) low.

By Firm

Firms were once ready to keep extra, trained workers on their payrolls so that the workers would be available to progress through the **internal labor market**. This practice ensured a ready supply of skilled labor when work was plentiful. More recently, firms appear oriented to shrinking rather than to maintaining employment levels (Osterman, 1999). When markets are slack, employers cut their wage bills rather than support a large number of employees. When there is an abundance of work, employers want to get the work done without permanently expanding the size of their workforce. This means that a large firm may identify some of its workers as being more important than others. They are considered the core workers in the firm, and they often have job security, good pay and benefits, and a career line. Less important workers, perhaps those with lower skill levels, may have unstable jobs, be subject to layoff and recall, or in other ways be deemed expendable in times of slack demand.

Firms sometimes seek institutional mechanisms, such as subcontracting the work to client firms (Averitt, 1968), for dealing with extra work. In 1992, for example, Aetna Life & Casualty Company decided to "reengineer" by cutting 2,600 jobs and

then contracting out the work (Zachary and Ortega, 1993). **Client firms** are smaller, more specialized firms that contract to perform work as needed when demand is high at the larger, patron firms. For example, a local mailing service may be called in to address and mail an unusually large number of notices for a bank, although the bank would normally use its own clerical staff to prepare smaller mailings. Client firms tend to have weak market positions because they are dependent both on the level of demand and on larger firms. As a result, the client firm often creates marginal jobs and resorts to even more stringent layoff policies than those that characterize the patron firms.

Small firms may be more likely to rely on part-time or seasonal workers and to offer lower wages and fewer benefits. Client firms and small firms more generally are often legally exempted from providing some fringe benefits. Small firms are also less likely to be bureaucratized, to have established personnel policies that provide trained supervisors, and to have grievance procedures. This can have the advantage of a friendly, less bureaucratized work environment. Alternatively, it can mean that a worker is relatively powerless and must continually appease a capricious employer. For all these reasons, small size is one important characteristic of firms that is associated with marginal jobs.

By Employment Contract

Another way the large firm may deal with peak periods of work is to contract with temporary workers. Temporary workers may also be hired to fill in for sick or vacationing permanent core workers. In one survey, the leading reason employers gave for hiring temporary workers was "to alleviate an overload of work" (70% of employers); the second most frequent response was "to cover for workers on leave" (52% of employers) and "to cover for vacationing workers" (51% of employers) (McCarthy, 1988). The percentages overlap because employers could give more than one reason in this survey.

Temporary workers work sporadically, often moving from one location to another. Most school systems, for example, have a group of substitute teachers who can be called on as needed, usually to fill in for a full-time teacher who is ill. Unions and some professional groups also maintain rosters of members available for temporary work. Temporary workers who perform manual labor are often called *day laborers,* and most cities have an area where homeless people congregate to be hired for manual day labor. This hiring is often called casual hiring because the job will last for only one day.

In a more bureaucratized form, temporary hiring agencies serve as intermediaries between the workers and the employers. Once popular chiefly in clerical work, now there are temporary workers in many types of jobs, including nursing and other health professions, computer programming, retail sales, and manual labor (Henson, 1996). Box 14.3 recounts the experience of Robert Parker, now a professor at the University of Nevada at Las Vegas, whose doctoral dissertation research consisted of participant observation as a temporary manual worker.

"Temps" differ from others in the same occupation not primarily in their skills or knowledge but in their job contract. Their contract is with the temporary agency rather than with a specific employer. The pay is not high, and a portion of the pay goes to the agency. Although some temporary agencies provide training, most provide few benefits beyond Social Security coverage, some worker's compensation coverage, and, after completion of a minimum number of work hours, unemployment compensation (Henson, 1996).

Temporary workers may work a full work week, but they may also work only one or two days a week, depending on demand. In 2002 there were about 3.5 million workers employed by personnel supply services (BLS, 2006b). Many temps are available only for part of the year. School teachers and students, for example, may hold temporary jobs during school vacations. Many retail stores offer temporary employment during the holiday rush season.

Temporary workers can be in a variety of occupations and may work in a variety of industries. We can consider them marginal to the extent that their work is involuntarily intermittent, and their pay and hours of work are substantially below what they might obtain in a more conventional, permanent job. Some "temps," however, choose this type of work because it is compatible with their other home or school respon-

sibilities and because it provides useful job experience and exposure to a variety of workplaces and potential employers. In one survey, the most important reason given for working for a temporary help service was "freedom to schedule my work in a flexible manner." Fewer than 10 percent reported that temporary work was "a stopgap measure until I can obtain a permanent job" (Gannon, 1984).

UNEMPLOYMENT

Unemployment, as you learned in Chapter 2, is officially defined as the search for work by people who have no job. Workers in marginal jobs may experience frequent periods of unemployment between jobs, but many workers will experience unemployment at some point during their working lives. The unemployment rate in the United States has stayed at relatively high levels during much of the period since the end of the Second World War. In the mid-2000s the overall unemployment rate ranged between 4 and 6 percent, but rates were much higher in some occupations and regions, and among some demographic and age groups (BLS, 2006b). A rate of 5 percent translates to about 7.6 million unemployed individuals. High levels of unemployment have plagued many industrialized countries, including 8.1 percent in 2004 in the European Union and 6.8 percent in Canada in 2005 (Census, 2006). Unemployment rates remain high in less economically developed countries.

Some unemployment results from unavoidable delays between jobs, such as when a worker moves to a new location. Such **frictional unemployment** is said to account for an unemployment rate of about 3 percent. Frictional unemployment could be reduced by programs designed to match job openings more quickly with qualified candidates. During the Second World War the unemployment rate fell to 1.6 percent, perhaps a more realistic minimum level of frictional unemployment. The greatest share of unemployment, however, stems either from a chronic gap between the number of jobs available and the number of people seeking work or from short-term downturns in the

---

B O X 14.3 Working as a Temporary $

On the morning of the assignment, I phoned the temporary company's office at approximately 7:20 A.M. to check on work availability. The firm's job dispatcher said there would be no assignments available for me that day. About twenty minutes later, however, she called back with an offer for a light industrial assignment. This quick turnaround of circumstances illustrates both the uncertainty surrounding the availability of work and the short notice commonly given to and expected of temporary employees. Taking note of the employer's address and other details, I accepted the temporary firm's offer and headed for the warehouse.

On this assignment I had been hired by a major U.S. retailer that needed so-called special-project workers. The retail company's regional management was attempting to reorganize a local warehouse more efficiently. In addition to myself, more than a dozen other temporary workers were assigned for this special project. I later learned that I had been called in as a last-minute replacement for a temporary employee who, according to another worker's account, decided that yet another mid-August day in a warehouse was intolerable. His reluctance to return was easy enough to comprehend; by the time I arrived at 8:30 A.M., the temperature had surpassed ninety degrees.

The job site on this occasion was easy to find, but the pervasive uncertainty that characterizes temporary employment was evident in my inability to locate the appropriate personnel upon arriving. Furthermore, the temporary company that had arranged the assignment did not know to whom I should report. For roughly ten minutes I toured the inside of the warehouse before finding the supervisor. After being introduced to the

first of several managers from whom I would take instructions that day, I began my day-long task—rotating appliances from one part of the warehouse to another. At first I moved water softeners, then tool kits, and later drill presses, but it was all the same kind of manual work.

Despite the mundane activity, this assignment was particularly valuable to me in terms of understanding light industrial temporary employment from the workers' perspective. It allowed me to interact informally with as many as twelve other temporary workers at one time. In addition, it eventually yielded several in-depth interviews. One of the interviews that stemmed from this assignment was with Ron, a recently divorced, black Vietnam veteran. Compared with most temporary employees I met, Ron displayed a significant level of assertiveness at the workplace. For example, he remarked to one of our supervisors that "we don't need any of your help" (in organizing the warehouse). Also on this assignment, I worked alongside, and interviewed, Tim, a white thirty-two-year-old who had recently quit working as a drummer for a local band. Another light industrial temp in my immediate work group was Tom, a white forty-year-old who had recently been operating a small engraving shop in the central city. The four of us had lunch at my house that day and became better acquainted.

Back on the job that afternoon, rotating stock grew increasingly disagreeable, difficult, and dangerous as the temperature eventually climbed to 103 degrees. Much of the work we had earlier accomplished solely with our backs was increasingly relegated to forklifts and dollies. In mid-afternoon, as a

---

economy. The chronic gap between the number of jobs the economy provides and the number of people seeking work is called **structural unemployment** (BLS, 1998). This gap between available work and people needing work widens in times of economic downturn. This latter component of unemployment is called **cyclical unemployment**.

Published unemployment rates understate the true number of people who would work if they were able to find employment. Many people give up

actively seeking employment because of discouragement and an absence of opportunities. These potential workers, sometimes called **discouraged workers**, *desire* work and are available to start work immediately, but are not *actively* looking for work because they are in school, ill, disabled, keeping house, or are convinced they cannot get a job. Adding these discouraged workers to those actively seeking employment would increase the unemployment rate by about three-tenths of a percentage point (BLS, 2006b).

collective expression of our discomfort, we took a half-hour break, a period three times longer than officially allowed. When we returned to work, one manager began to convey his disapproval about the time our idleness had consumed. Before he could finish, however, he was cut off by Ron, who relayed his own displeasure about this particular supervisor's "rigid" managerial approach. Given the heat and arduous labor, none of the temps expressed feeling particularly guilty about slowing the work pace or taking extra time off. Even the head supervisor was sympathetic. And despite our transgression, the retailer's management encouraged all of us to return the following day.

After the late afternoon break, we were instructed to perform tasks typically forbidden by temporary help firms. Specifically, the work involved disassembling two sections of scaffolding that rose more than thirty feet. All the storage components were made of long, awkward, heavy pieces of metal and required at least two workers to handle them safely. Accomplishing this task genuinely aroused concerns within me about safety, especially because of the lateness of the day and the fact that the work required all the stamina and alertness that could be mustered. Nonetheless, we successfully completed the last-minute chore without incident, signed out, and left for home by 5:00 P.M.

Two general points are noteworthy about this temporary assignment. First, the work itself was dead-end, offering no opportunities to learn new skills or hone existing ones. With the exception of operating a forklift, the work required no formal skills and no ongoing supervisory instruction or on-the-job training. Several co-workers that day expressed the belief that temporary employees are often assigned dead-end work because permanent employees cannot handle its stifling character.

Second, there were several modes of worker resistance that temporary employees exercised to protest the numbing assignments. As intimated earlier, temporary workers have several methods at their disposal to resist the exploitative working conditions they encounter. The most fundamental form is to simply resign from temporary work arrangements or routinely decline assignments when temporary firms make an offer. This is not a practical strategy, however, since most temporary employees are working out of economic necessity. Another form of worker resistance I witnessed was the deliberate restriction of output. At the warehouse assignment workers cajoled one another not to give more than $4.00 an hour worth of effort (the wage they were being paid that day). Taking frequent breaks is yet another method. On the warehouse assignment described here, workers left early for lunch, extended allotted breaks beyond specified time limits, and took numerous other informal breaks. When resistance surfaced, temporary workers often discussed the issue and justified their behavior in terms of the lack of employer commitment and the generally dubious working conditions they confronted.

SOURCE: Robert E. Parker, *Flesh Peddlers and Warm Bodies*. Copyright © 1994 Robert E. Parker. Reprinted by permission of Rutgers University Press.

Occupational Differences in Unemployment

Unemployment is distributed unequally across occupational groups, with the least-skilled workers facing the highest rates of unemployment. In November 2006, when the overall unemployment rate stood at 4.5 percent, the rate for professionals was 1.7 percent; for sales and office workers, 4.2 percent; for service workers, 5.9 percent; and for construction laborers, who face high levels of seasonal unemployment, 6 percent. The highest rate was 13.4 percent, for farm, fishing, and forestry workers, who also experience seasonal unemployment (BLS, 2006b).

Differences between Social Groups Unemployment is also unequally distributed across social groups, with minority workers typically experiencing twice the rate for majority whites. Youth unemployment poses a special problem in industrial societies. The unemployment rate among those sixteen to nineteen years old is two to three times as high as the overall rate. Unemployment among young people has had devastating consequences: "The incidence of unemployment among youth and the difficulties which youth face in entering the world of work continue at levels which are unacceptable. The consequences, in terms of antisocial behavior, alienation, lost output, and reduced social mobility, impose both short- and long-term consequences that cannot be ignored" (National Commission for Employment Policy, 1979). One-third of African-American teenagers were unemployed in 2006 (BLS, 2006b).

Layoffs

Unemployment can occur for many reasons. These include quitting a job, being fired, becoming disabled, being laid off, and moving to be with a spouse or loved one. **Layoffs** are a particularly important cause of unemployment because they affect so many people at one time and in one place. Older workers who are laid off may have an especially difficult time finding new employment. Layoffs have become more common as bankruptcies have increased among small companies and as large companies have downsized and become more mobile both within the United States and overseas (Sullivan, Warren, and Westbrook, 2000). Plants are routinely closed in high wage areas and reopened in areas where companies can pay lower wages (Bluestone and Harrison, 2000).

Layoffs vary in their duration. Short layoffs are more typical for skilled workers employed in large firms that operate in key industries; longer layoffs, including "indefinite layoffs," are more typical for workers in less-skilled occupations employed in smaller, more peripheral firms (Grunberg et al., 2000). The term *indefinite layoff* is frequently a euphemism for permanent termination, and even if a recall is possible at some future date, few of the laid-off workers can afford to wait indefinitely.

Social Consequences of Layoffs Layoffs can have devastating consequences for workers. The loss of income and fringe benefits, such as health insurance, often results in a rapid deterioration of lifestyle, including potential loss of home, delayed educational plans, reduced medical care, and reductions in food and recreation budgets. Speaking of his choice of police work as a career, a young African-American worker reports: "I like the work and I like the job security. When I see others laid off and jobless, I feel I made the right choice" (Martin, 1980:70). Workers often feel helpless and worthless in the face of lengthy unemployment. In combination these factors can have severe negative health consequences. Based on a study of how unemployment affects health, the American economists Barry Bluestone and Bennett Harrison (1982:65) projected that a 1 percent increase in the unemployment rate over a six-year period results in 37,000 total deaths (including 920 suicides and 650 homicides), 4,000 state mental hospital admissions, and 3,300 state prison admissions.

Layoffs can have devastating consequences for workers:

> Very little has gone right for Harold Johnson, 38, since he was laid off by GM. . . . He has not held another job for more than a few months since. After briefly working at a seasonal job at a boatyard, Johnson worked at a small factory in rural Michigan that was on strike, but he was laid off when the strike ended. Later, he briefly worked for a small recreational vehicle manufacturer, but the firm quickly went bankrupt. He went to a trade school to become a furnace repairman, but could not find any work in the field because he lacked experience. . . .
>
> In the meantime, his wife has divorced him, and he's had to sell his house to avoid losing it to the bank and he has developed high blood pressure.
>
> "I'm out there looking, but it has sure been discouraging," says Johnson with quiet understatement. (Risen, 1984:I19)

Careers, marriages, families, and communities are all vulnerable to the devastating effects of layoffs. Higher unemployment rates produce increases in family violence, family breakups, and divorce. Community services such as education, parks and recreation services, police protection, and street repair all typically deteriorate because tax revenues decrease when many workers are laid off.

Replacement Jobs?   Some jobs lost due to layoffs reappear in other parts of the country as plants close in the Northeast and Midwest and reopen in the South and Southwest. Some workers regain employment by following these jobs, almost always at significant personal costs of severed ties with their families, friends, and communities. Unfortunately, many laid-off workers do not have the qualifications for the new jobs. As workers move out of the unionized Northeast and Midwest and into the predominantly non-unionized South and Southwest, they may also often find their wages dramatically reduced. Workers may also find social services in the new industrial boomtowns, as well as the overall quality of life, to be lacking.

Other jobs lost through layoffs reappear overseas when American companies move their operations to other countries. The volume of manufacturing activity exported overseas has been staggering. The value of goods and services produced overseas for American multinational corporations is now larger than the gross domestic product of every country in the world except the United States itself (Barnet and Cavanagh, 1994). This export of jobs overseas to lower-wage areas is one of the reasons that unemployment rates in the industrialized nations have risen and stayed at high levels. We discuss the nature of the world economy and its implications for workers in Chapter 16.

Coping with Unemployment

When workers become unemployed, they have a rather limited set of resources to ease their way. Perhaps the most important of these are financial and emotional help from their families (Kenyon, 2000). Some workers also receive **unemployment compensation** from the state. However, only about 78 percent of workers are covered by unemployment insurance, and for those who *are* covered, payments expire after a set period of time (generally twenty-six weeks). Unemployment insurance laws vary from state to state, but those not covered typically include farm workers, personal service workers (such as in-house child-care workers), part-time workers, workers who have been employed for less than a specified number of weeks, and workers who quit their jobs voluntarily or were fired for cause. As a result, at any given time only about one out of four unemployed workers receives unemployment compensation. Some union contracts also provide *supplemental unemployment benefits* that extend the amount or duration of unemployment insurance benefits. If workers are permanently laid off, they may also receive **severance pay**, a lump sum payment to help pay bills until they can find other work. Finally, workers may also be able to receive some form of welfare (public aid). However, this relief provides minimal benefits, is generally unavailable until the worker and his or her family are completely destitute and have sold off nearly all of their possessions, is time limited, and is often contingent on a willingness to accept work at subminimum wages and conditions.

Legislative Controls   Laws that help reduce unemployment are almost nonexistent in the United States. A partial exception is provided by the administrative procedures of state unemployment insurance systems. The procedures mandate that a company's unemployment insurance premiums reflect the number of workers it has laid off in the past. High premiums thus discourage companies from casually laying off workers.

Some states have also enacted more specific legislative measures to discourage layoffs. Several states have enacted plant-closing legislation that limits the ability of companies to close plants without providing substantial advance warning, compensating workers, and sometimes even compensating communities for their investments in the company's success. There has even been precedent for awarding claims to workers for unjust layoffs where such plant-closing legislation does not exist. In California 537 laid-off workers at an Atari home computer manufacturing plant were awarded four weeks' pay and legal fees on the basis of a little-cited provision in the state's labor code requiring advance notice of termination of employment (Walsh, 1986:7). Atari had repeatedly assured workers that their jobs were secure, but it suddenly closed the plant and moved its computer assembly operations to Hong Kong. The workers were told at mid-morning on a Tuesday that they could pick up their final checks that Friday at a local high school and then were made to leave by the back door (Lewin, 1984:G6). California Superior Court Judge Peter Stone ruled that Atari's repeated assurances to workers constituted a contractual obligation to provide advance notice of termination.

WHY ARE SOME WORKERS CONSIDERED MARGINAL?

Certain segments of the labor force are more likely than others to fill marginal jobs. It is safe to say that these are the workers who have fewer options than other workers, but reasons for their limited options may vary. Workers in marginal jobs are more likely than other workers to have characteristics of geographic isolation, low education, disability, and recent job displacement, and to be distinctive in terms of demographic characteristics such as age, race and ethnicity, and gender (Kalleberg, 2006).

Geographic Isolation

Some workers may find that because of their place of residence the only available job opportunities are marginal jobs. For example, workers in rural areas or in inner-city ghettos may find that only unstable, low-paid jobs are available within their immediate vicinity. Better jobs may be available in other parts of the same metropolitan area, but workers who do not already have good jobs may not be able to afford the transportation costs. Even in large cities with good mass transit systems, ghetto residents often find that the good jobs are being created in suburban areas and that outbound transportation to the suburbs is available only at the end of the day when suburban commuters return to their homes.

Geographic isolation also occurs in small towns, where the only available jobs may be unstable or poorly paid. Because employers do not have to compete to attract laborers, there may be little improvement in the jobs, even over a long period. In economically depressed areas, high levels of unemployment may be chronic. In such circumstances, workers or their organizations may turn to the political process to lobby for economic development funds or for other efforts to bring employers to the community.

People with limited geographic mobility may be in a difficult employment situation even if they are not actually isolated. As you learned in Chapter 5, two-career couples are a notable group with limited mobility. Married women, in particular, are likely to find their job prospects limited by their husbands' jobs. Better jobs might be available to these women if they were free to move, but there may be a poor match between their skills and the jobs available in their local labor market.

It is difficult to estimate how many workers have marginal jobs because of their isolation. Many workers do not know that their wages would be higher in another location, and, indeed, they cannot be certain that they would be hired for more attractive jobs even if they moved.

Educational Level

Workers' education and skills are highly correlated with their annual earnings. Earnings increase with level of schooling. In 2003 a worker with a bachelor's degree earned an average of $51,206, compared with $27,915 for high school graduates (Census, 2006). Workers with a professional degree, by contrast, earned $115,212 on average.

Years of schooling is not a sufficient indicator of education, however, because as many as 10 percent of high school graduates are functionally illiterate, and as many as 44 percent of the foreign-born in the United States may be functionally illiterate in English. Productive workers need to be literate even for marginal jobs; farm workers, service workers, and household workers are exposed to toxic pesticides and chemicals daily, and at a minimum, they need to be able to read the labels of the products they use. In one reported instance, a herd of cattle had to be destroyed because an illiterate worker unknowingly added a poisonous compound to their feed (Levitan and Shapiro, 1987:70). National surveys of new workers, young adults aged twenty-one to twenty-five, showed a decline in literacy over time ("Lowdown on Literacy," 1994).

Disabling Conditions

Although many people suffer some work disability, about twelve million American adults report themselves unable to perform an instrumental activity of daily level, such as using the telephone or doing light housework (Census, 1999:Table 235). These workers tend to be overrepresented among the underemployed (Sullivan, 1978; Levitan and Shapiro, 1987). Nearly 12 percent of American workers report that they have some kind of disability (Census, 2006). But those with severe disability have difficulty finding work, especially full-time work. The disabled may not be able to compete well for the most desirable jobs. On the one hand, they may be disqualified from some jobs because of their specific disability; for example, a paralytic might not be able to move heavy objects. On the other hand, there are many jobs that the disabled could handle but that are not offered because the employer has many able-bodied applicants. Some disabled people are forced to accept marginal work because no better job is available.

Job Displacement

Displaced workers are defined as workers who are at least twenty years old who lost their job because their company closed or moved away, or because there was insufficient work for them to do, or because their position or shift was abolished. A Bureau of Labor Statistics survey conducted between 2003 and 2005 found that there were 3.8 million displaced workers who had worked for their employer for three or more years. About 70 percent of them were reemployed at the time of the survey. Just under half of the displaced workers had lost their jobs because of plant or company closings (BLS, 2006d). Manufacturing accounts for a disproportion-ate number of the displaced workers, although service industries also displace workers. About 30 percent of the displaced workers are earning 20 percent less than they made on their last job, although 51 percent are earning as much or more on their new jobs as they had on their lost job.

Age

Age tends to be correlated with skills and experience, and for this reason young and inexperienced workers are most likely to be recruited for marginal jobs. Some observers believe that it is an advantage for teenagers that many farm labor and service jobs require little training, because these jobs are then available to provide job experience to teenagers. Some legislators have proposed the **subminimum wage**, a wage lower than the current minimum wage, as a way to increase the number of jobs available to young people. To be sure, these jobs would be poorly paid. But because the jobs would provide work experience and some on-the-job training, proponents of the subminimum wage hope that young workers would be able to move into more responsible and better-paid jobs.

Child labor laws have limited the types of work and the hours of work for teenagers. In general, any worker aged eighteen or older may do any job for unlimited hours, but sixteen- and seventeen-year-olds may not do jobs declared too hazardous by the U.S. Department of Labor. Such jobs include logging, slaughtering livestock, work with explosives, and mining. Younger teenagers are covered by state labor laws that affect what type of work children may do, but the level of protection varies. In Oregon, children as young as nine may pick berries and beans (Mendoza, 1997).

Elderly workers whose experience might be considered obsolete may be hired for marginal jobs, although older workers may also respond to job loss by retirement. As we mentioned in Chapter 5, the decline in births during the 1970s meant fewer entry-level workers in the 1990s, so the firms and industries that traditionally hired teenagers began to view the rapidly growing elderly population as a potential source of workers. Advertisements for several fast-food chains depict older employees. These ads are part of a campaign to convince elderly and retired people to consider part-time work. Although elderly workers might be good for the firm, it is not clear that the jobs will be attractive to them. Low-wage jobs are less likely to have job-related medical insurance and pension coverage, benefits of particular interest to elderly workers.

Race and Ethnicity

The overrepresentation of minority groups in marginal jobs stems directly and indirectly from discrimination (DeAnda, 1996). Discrimination directly affects minority workers through a cultural division of labor that relegates them to the less desirable jobs. These marginal jobs may even become a sort of occupational ghetto, with employers refusing to hire members of the majority group. Discrimination may also indirectly influence the job opportunities available to members of minority groups. Inferior school preparation may keep many minorities channeled into less attractive jobs. Residential segregation, and especially the concentration of minority workers in inner-city neighborhoods, isolates many minority workers from attractive jobs in the suburban ring.

Immigrants are a special case of the general marginality of minority groups (Lamphere, 1992). Generations of immigrants have taken relatively menial and marginal jobs because they were the only jobs available. This is still true of immigrants who have relatively low levels of education or job skills. (Immigrants with professional skills, on the other hand, often earn on a par with native-born workers within a few years.) Immigrants continue to be recruited as temporary farm workers and household workers. Discrimination, poor ability in the language of the host country, and unfamiliarity with host country regulations may all be reasons why immigrant workers tend to be channeled toward marginal jobs. Employers often assume that the native born will not take a marginal job. For example, some U.S. agricultural growers recruit foreign workers because they assume that no American workers would take the jobs. Such employers may prefer to recruit aliens instead of making the job more attractive to the native born. More attractive jobs may not be open to immigrants because of legal barriers. For example, jobs with the government or within certain core industries may be limited to citizens, or the licenses required for some jobs may specify that the credential be obtained from training schools within the host country.

Gender

Women have traditionally been more likely to be involuntary part-time workers and to be among the working poor (Sullivan and Mutchler, 1985). Constraints unrelated to the job market, such as family responsibilities, may influence women workers to take temporary or part-time jobs that are marginal. Women are disproportionately found, for example, in household service, other personal service jobs, and temporary work. This circumstance ironically reinforces the stereotype that women work only for "pin money," that their work roles are secondary to their family roles, or that they are unreliable. Such stereotypes reinforce gender discrimination if female workers are relegated to lower-paying, marginal jobs within a firm.

Interacting Characteristics

Some of these characteristics interact with each other. For example, young African Americans are more likely to be geographically isolated and somewhat more likely to be school dropouts than are other workers. Having several of these characteristics increases the probability that a worker will be employed in marginal jobs. As we have discussed in Chapter 4, some employers use a type of statistical discrimination in hiring for core jobs. To minimize the time spent in interviewing and hiring candidates, these employers simply hire according to their stereotype of the appropriate worker for the job. Because the gender, race, education, and other easily verifiable characteristics of productive workers are already known, employers prefer to hire workers who have these same characteristics. When they are hiring for their more peripheral jobs (such as cleaning staff), women, minority workers, disabled people, immigrants, and others may be considered appropriate.

Such stereotypes may become self-fulfilling prophecies. Some observers fear that marginal jobs "train" people to be bad workers who do not take initiative, move from job to job, and fail to develop good work habits. For example, if the employer shows little interest in the jobholder, and the job offers no chance for advancement, the employee may likewise see little reason to be punctual or to work very hard. Employers who expect little from their workers often find their expectations fulfilled. Such workers may quit frequently to seek better jobs (Brown, 1982). In fact, some research indicates that low-wage workers are more likely to be successful if they do try to move to a new job (Andersson, Holzer, and Lane, 2005).

Unfortunately, the record of frequent resignations and job changes may backfire if employers assume that the worker lacks job commitment. Thus a worker who begins work in a series of marginal jobs may have difficulty finding a more permanent, core job. A vicious cycle develops in which good habits are not rewarded and bad habits are reinforced. As bad as this situation is for the individual worker, it also perpetuates stereotypical thinking about whole groups of workers. This stereotypical thinking eventually leads to the occupational steering of some people toward marginal jobs.

MARGINAL WORKERS AND SOCIAL CLASS

There were times when people who worked for GM, in the factory, might have been good [role] models. But those factory jobs are gone. So what you have to impart to a young guy is that they have to have diverse skills and flexibility. My own experience is that I had

to confront a very painful need to figure out how to exist economically without having to go and apply for what is considered a job. —Hakim, a street vendor of books (Duneier, 1999:40–41)

Jobs and occupations are important indicators of social class. In Chapter 4 we discussed this important influence on workers. Many social scientists are concerned about what the persistence or perhaps increase of marginal jobs will mean for the class structure of advanced industrialized societies. They express this concern in two basic ideas: the development of an underclass, and the shrinking middle class.

The **underclass** is a social group that cannot achieve social mobility because of lack of personal resources and opportunities (Wilson, 1987; Auletta, 1982). To be a true social class there must be a mechanism to transmit the status of parents to their children. Children of marginal workers are less likely to finish school than are children of non-marginal workers. Teachers and employers may believe that these children cannot achieve much in the workplace and may steer them away from college preparatory curricula and other opportunities that would prepare them for better jobs. In addition, the low income levels of the families may mean that the teenage children must take available jobs—often marginal ones—and then never advance into good jobs. If their parents' jobs have reinforced bad job habits and negative attitudes, their children may not learn better habits or more positive attitudes. In addition, the children are less likely to have a network of contacts who might help them get a better job (Granovetter, 1995). The occupational inheritance of marginal jobs from generation to generation is thought to be part of the formation of an underclass.

Welfare reform has required many former welfare recipients to seek employment. The employment most often available to them consists of relatively low-paid jobs, often in the service sector (Edin and Lein, 1997; Newman, 2000). The former welfare recipients who go to work often face much higher costs for transportation, child care, and other expenses, and the jobs available to them are not always sufficient to cover the additional expenses. These working poor compete with teen-agers, recent immigrants, and inner city residents for the available low-level jobs.

The hypothesis of a **shrinking middle class** suggests that the traditional middle-class jobs are declining in relative numbers or, at least, are becoming more unstable (Levy, 1987). Some jobs are being reduced to marginal jobs, perhaps because they become part time or temporary. Some jobs are being deskilled because of technological advances. Other jobs are being lost altogether to overseas competition, perhaps because they are being exported to other countries (Leicht and Fitzgerald, 2006). Whereas the underclass argument suggests that class position is inherited, the shrinking middle-class argument suggests that middle-class children may *not* be able to inherit their parents' class position and may slip below their parents' class position. At the same time, mobility into the middle class will become even more difficult (Rosenthal, 1985; McMahon and Tschetter, 1986).

There is as yet little convincing evidence to support or refute these two hypotheses, although some social scientists have argued that public policy tends to emphasize the problems of the poor and the goals of the rich, to the exclusion of the middle class (Skocpol, 2000). As you learned in earlier chapters, technological advances can lead, in some situations, to improved jobs; the creation of a marginal job is not inevitable. Moreover, at least some marginal jobs may still provide an opportunity for young people and immigrants to gain experience before moving into more permanent employment. Social scientists still disagree whether workers in marginal jobs will become a permanent underclass or whether they can "cross over" into better employment. Plainly, the class arguments hinge on the future number and characteristics of marginal jobs.

THE FUTURE OF MARGINAL JOBS

Every society has some work that is defined as "dirty work," and dirty jobs will probably always have low status and may also have low pay. But as we have seen, many marginal occupational categories are declining in numbers. Thus, the important issue in assessing the future of marginal jobs would seem to be marginality that results from the organization of employment. Two types of theories are useful in understanding this type of structural marginality: dual labor market theory and internal labor market theory.

Dual Labor Markets

Sociologists have hypothesized that labor markets are increasingly demarcated between good jobs and bad jobs (Hodson, 1983). In developing countries the distinction is sometimes made between formal and informal sectors or between modern and traditional sectors; in industrialized countries the distinction may be expressed as one between primary and secondary economic sectors or between core and peripheral industries. Productivity and profits are expected to grow in the primary sectors, permitting the substitution of more capital for labor. Thus, the number of good jobs may not grow rapidly, although the existing good jobs might become even better in terms of pay and fringe benefits. Meanwhile, there is concern that the more labor-intensive secondary sectors, which are poorly paid and less secure, will absorb disproportionate numbers of the workers. Such theories attribute job creation and job quality to the overall trajectories of economic growth.

Internal Labor Markets

According to internal labor market theories, the decisions of employers are crucial in determining the relative proportion of good versus bad jobs. Even an employer in a profitable, core industry may have two kinds of jobs to offer. Employers may provide good jobs for the workers with many skills or with rare skills and marginal jobs for unskilled workers. In times of economic recession, the workers holding the more peripheral jobs are laid off. Even if the workers in the marginal jobs have high turnover, the production process will be little affected. According to internal labor market theories, even substantial economic

B O X 14.4 New Products and New Jobs from Serving the Poor

Professor C. K. Prahalad is a professor of business at the University of Michigan. In looking at the large numbers of very poor people throughout the world, he argues that they are resilient entrepreneurs and value-conscious consumers. He believes it is possible to partner with them so that the poor actively participate in developing profitable goods and services. This effort could produce an enormous market and also a solution to poverty. Prahalad refers to poorest in the world as the "Bottom of the Pyramid" (BOP), because a diagram of the world's population by purchasing power would show a pyramid with a very large base of about 4 billion people with the very lowest earnings.

Dealing with the BOP requires changing many common assumptions of western firms and also developing an openness to learn from the poor themselves. As the example below indicates, the new products developed may then also find a market in the advanced countries.

BOP consumers in Latin America are careful in their use of diapers. They use one or two changes per day compared to the five or six changes per day common among the top of the pyramid consumers. Because they can afford only one or two changes, they expect a higher level of absorbency in the diapers and an improved construction of the diaper that will accommodate additional load. This means that the firms have to technically upgrade the quality of their diapers for the BOP consumers compared to the products they currently sell to the rich in those markets. Needless to say, the new product built for the BOP market is higher in quality and provides a better price-performance proposition.... The BOP can be a source of innovations for not only products and processes, but business models as well.

SOURCE: Prahalad, C. K. 2006. *The Fortune at the Bottom of the Pyramid: Eradicating Poverty through Profits.* Upper Saddle River, N.J.: Wharton Publishing. pp. 48–49. Used by permission.

growth might not be translated into better jobs. The critical issue will be the decisions of employers or, possibly, the countervailing power of workers.

The International Factor: Beyond Outsourcing International competition and technology may lead either to the creation of more marginal jobs or to the creation of better jobs. It is not yet certain which of these is the more likely outcome (Levy and Murnane, 2005). Many North American workers view outsourcing as principally a source of marginalization. If globalization were to enhance all of the

world's economies—an outcome many economists believe will eventually come about—then the long-run result could be the development of better jobs.

Most of the world's marginal jobs are found in the developing countries, where most of the world's population lives, often on very low wages. One issue of considerable interest is whether better jobs could be made available to these people, increasing the size of the worldwide middle class and providing markets and economic stimulus for the advanced industrial countries. Box 14.4 presents a look at the world's poor as a potential source for positive economic growth.

SUMMARY

A marginal job is not defined by job content alone but also by organizational and economic context. The job content, job regularity, job stability, and adequacy of employment in some occupations

mark them as marginal jobs. These jobs include farm labor, service work, and household work. Other occupations may be employed either in marginal jobs or in stable, "good" jobs. For these occupations, the industry, the firm, and the nature of the labor contract determine whether the job is marginal. For example, temporary clerical workers may do the same job as others in the same clerical occupation, but their jobs are intermittent because of the nature of their job contract.

Many of the same conditions that lead to marginal unemployment also lead to complete unemployment, which varies depending upon location, industry, and business conditions. Frictional unemployment occurs when workers are changing jobs and is often temporary, but chronic structural unemployment occurs when there is a persistent gap between the number of available jobs and the number of people seeking work.

The workers most often hired for marginal jobs tend to be distinctive in terms of their race, immigrant status, gender, birthplace, schooling, disability status, previous work history, and geographic loca-tion. Employers may use statistical discrimination as a means of channeling workers into the good jobs and the marginal jobs. Many of the same characteristics associated with marginal work are also associated with unemployment.

Social scientists do not yet know whether the relative number of marginal jobs is increasing. Dual labor market theory suggests that economic growth and concentration encourage the creation of marginal jobs. Internal labor market theory emphasizes the labeling of jobs within the firm. International competition and technology will surely affect the eventual outcomes. If the relative number of marginal jobs increases, the middle class is likely to shrink, and the underclass may grow. The interaction of the advanced industrial countries with the world's developing countries will play an important role in the balance of marginal versus good jobs.

KEY CONCEPTS

marginal jobs

deviant occupations

shadow economy

job instability

contingent employment

underemployment

migrant farm worker

sharecropper

hired hand

payment in kind

core and peripheral jobs

client firm

dual labor market

internal labor market

client firm

frictional unemployment

structural unemployment

cyclical unemployment

discouraged workers

layoffs

unemployment compensation

severance pay

subminimum wage

underclass

shrinking middle class

QUESTIONS FOR THOUGHT

1. If my salary is not as high as the salary I want, is my job a marginal job? Why or why not?

2. Why are female and minority workers more likely to find employment in marginal jobs?

3. What are the differences between a secondary sector and an internal labor market? Why are these concepts important for a study of marginal jobs?

4. Suppose that a temporary worker receives employment on every day that he or she wishes to work. Could this person's job still be called a marginal job? Why or why not?

5. How is it that an industry could be peripheral in one country and core in another country? Are there marginal jobs even in core industries?

MULTIMEDIA RESOURCES

Print

Newman, Katherine S. 2006. *Chutes and Ladders: Navigating the Low-wage Labor Market*. Cambridge, MA: Harvard University Press. This ethnographic study follows a number of low-wage workers to see whether they would benefit from the general economic prosperity of the 1990s.

Shipler, David K. 2005. *The Working Poor: Invisible in America*. New York: Vintage. Reprinted. This book identifies the barriers that keep workers poor even if they work full-time. Some of these barriers come from the nature of the employment, but others come from the network of transportation, housing, medical and other services that are available to poor workers.

Venkatesh, Sudhir Alladi. 2006. *Off the Books: The Underground Economy of the Urban Poor*. Cambridge, MA: Harvard University Press. This study of a Southside Chicago neighborhood shows how the poor make ends meet through networks that involve both legal and illegal work.

Internet

United Farm Workers. *http://www.ufw.org/* The website of the United Farm Workers. Founded by Cesar Chavez, this union seeks improved working conditions for farm laborers.

Citizenship and Immigration Canada (CIC). *http://www.cic.gc.ca/english/work/index.html* This website provides information from the Canadian government to foreigners seeking to work temporarily in the country.

RECOMMENDED FILM

Maquilapolis (City of Factories). Tijuana, Mexico, is home to 800 factories that produce components of various types, usually for export to the United States. The filmmakers worked collaboratively with the factory workers, providing them with cameras. For five years the women documented their daily lives, including the uncertainty of their jobs, the difficulties they had in earning a living, and the movement they eventually began. Directed and Produced by Vicky Funari and Sergio De La Torre. Spanish with English subtitles.

PART V

Work in the Twenty-First Century

The three chapters in Part 5 discuss the forces that are shaping the future of work. Chapter 15 examines huge corporations and their continuing movement toward both greater size and greater diversification. Chapter 16 shows how events in any one part of the global economy have consequences for all the other parts. Along with technologically based automation (Chapter 9), corporate concentration and globalization represent some of the most significant challenges to the jobs of workers in industrialized nations. Chapter 17 projects current developments into the future in an attempt to understand the face of work for future generations.

Chapter 15 focuses on the emergence of the large diversified corporation as the dominant organizational form in the modern economy. We discuss both positive and negative consequences of large corporations for workers and for society. Workers are frequently paid better in large firms than in smaller firms and have better benefit and retirement packages. Their work, however, may be more alienating and repetitive. The buying and selling of subsidiary companies by large corporations also creates new vulnerabilities for workers as jobs appear and disappear overnight for reasons that may be quite remote from workers' own efforts and productivity. Mergers, acquisitions, and divestitures also absorb financial and human resources that might be better invested in innovation and in activities that directly increase productivity. What is good for any given conglomerate may be destructive for the economy as a whole. In the concluding section of Chapter 15 we examine the small-firm sector and its potential importance for the future.

Chapter 16 explores the origins of a world economy in the colonial empires developed by Britain and other European states. We examine the replacement, after the Second World War, of direct political control of colonies by a system of economic neo-imperialism with the United States at its head. Under neo-imperialism, former colonies are at least nominally politically independent, but are often still highly economically dependent. Economic development, which is highly uneven across world regions, has greatly increased the complexity of the world economy. In conjunction with the rebuilding of Europe's industrial base after the Second World War, these economic and political developments have created a world economy with greater diversity than ever before. In the concluding section of Chapter 16 we examine some of these diverse systems of industrial organization, including those typical of Japan, China, Sweden, Germany, and the former Soviet Union.

Chapter 17 discusses possible futures for advanced industrial societies. These futures include both the possibility of a large innovative sector with rapid growth and good jobs and the possibility of a large marginal sector with low pay and unstable jobs. We argue that the choice between these futures depends on the concerted actions and political choices of all of us as producers, consumers, and citizens.

15

The World of the Large Corporation

"It's mine. I built it. You bump it down—I'll be in the window with a rifle. You even come too close and I'll pot you like a rabbit."

"It's not me. There's nothing I can do. I'll lose my job if I don't do it. And look—suppose you kill me? They'll just hang you, but long before you're hung there'll be another guy on the tractor, and he'll bump the house down. You are not killing the right guy."

"That's so," the tenant said. "Who gave you the orders? I'll go after him. He's the one to kill."

"You're wrong. He got his orders from the bank. The bank told him, 'Clear those people out or it's your job.'"
"Well, there's a president of the bank. There's a board of directors. I'll fill up the magazine of the rifle and go into the bank."

The driver said, "Fellow was telling me the bank gets orders from the East. The orders were, 'Make the land show a profit or we'll close you up.'"

"But where does it stop? Who can we shoot? I don't aim to starve to death before I kill the man that's starving me."

"I don't know. Maybe there's nobody to shoot. Maybe the thing isn't men at all. Maybe, like you said, the property's doing it. Anyway I told you my orders."

In the world of work you may find that your life is strongly influenced by economic forces outside your control. You may not be able to see who or what these forces are or to personalize them so that you can understand them. As an individual, you may be nearly powerless in the face of these forces. At some point in your career you may well work for a large, seemingly impersonal corporation. Such corporations exercise immense power because of their size. They exercise this power in relation to their employees, their customers, and smaller firms. Their actions determine the availability and quality of employment opportunities for the citizens of the nations in which they operate.

In this chapter we discuss the increasing size of corporations and its implications, focusing on the issue of corporate power. We also examine how large corporations have expanded their influence over time. These ways include mergers, diversification across product lines, overlapping boards of directors, and the use of subcontracting. A clear understanding of the nature and power of large corporations is essential for students as they grapple with issues concerning both their own careers and the economic future of our nation and the world. We conclude the chapter with a consideration of the continuing importance of small firms and their role in the economy of the future. Together the giant corporations and the constantly reemerging small-firm sector set the organizational stage for the economy of the twenty-first century.

CORPORATE POWER

Some workers are self-employed. Some work for small employers. Others work for huge corporations with holdings around the world. As one moves along this continuum, the worker and the owner of the enterprise are increasingly distant. When the worker labors alongside the owner in a small shop, there is often a personal link between them (though it may be a paternalistic link and entail very unequal power). In the large corporation workers may not even know who

the owner or owners are. They may know who their immediate supervisor is or the plant manager's name, but they typically have little knowledge of who actually owns and controls the corporation.

Public Concerns about Corporate Power

Large corporations have immense power in society. Concerns about their role in society were enunciated during the earliest stages of industrial capitalism by Adam Smith (1776), who believed that monopolies were detrimental to the free development of productive forces. These concerns have led the governments of industrialized nations to enact legislation to prevent or regulate monopolies. Most businesses would like to be monopolies if they could, and it has become one of the roles of the government to prevent this. In spite of these concerns and regulatory actions, corporations have continued to grow in size and their sheer size alone often gives them many of the powers of monopolies (Carruthers and Babb, 2000).

The power of large corporations creates the possibility of corporate actions with grave consequences. Public concerns about the abuse of corporate power involve many issues, including (1) the exercise of concentrated economic power, (2) the exercise of concentrated political power, (3) the creation of highly bureaucratized organizations that are inflexible and resistant to change, (4) downsizing of workforces and the intensification of work, (5) the exploitation of consumers, (6) the degradation of the environment, and (7) control of the media (Jacoby, 2004).

Public concern about the **economic power** of large corporations arises from their powerful influence on smaller businesses and on aggregate employment levels. Large firms have the economic power to prevent the entry of smaller firms into their industries and to undermine the viability of small firms already operating there. This power rests on economies of scale and on control of technologies and markets. Such exclusionary power reduces price competition, innovation, economic growth, and job creation. In aggregate it displaces economic resources from other segments of society to large corporations (Connor, 2001).

The *downsizing* of large companies through cutting back employment has become a source of public concern in the 2000s. Downsizing can entail plant closings (discussed in Chapter 8), technological displacement (discussed in Chapter 9), the intentional increase of marginal employment positions (discussed in Chapter 14), and the movement to "lean" managerial structures (discussed in Chapter 12). Downsizing can also be achieved through moving production to lower-wage locations around the world (see Chapter 16). Taken as a whole, these tendencies have produced chronically high unemployment rates and eroded the quality of available jobs. In the 2000s the largest U.S. firms have achieved increasing profits but are employing fewer workers.

Large corporations exercise **political power** first and foremost through lobbying to promote favorable legislation and through campaign contributions to influence elections. To avoid federal and state limits on corporate political contributions, this money often flows indirectly through lobbyists, law firms, and corporate executives, rather than coming directly from the corporation. Exercise of political power by corporations results in subsidies for favored industries and often to the control of regulatory bodies by the very corporations they were intended to regulate (Mizruchi, 2004). At the local and state level, this power translates into favorable tax treatment and state subsidies for needed infrastructure such

as roads and utilities. There is a general public awareness that large corporations are frequently able to shape government policy in their own interests. For instance, oil companies are seen as having inordinate influence on America's energy policies—moving these toward development of new gas and oil fields and away from energy conservation. Large corporations have also sometimes damaged America's image abroad by manipulating local governments away from policies more favorable to their own citizens and workers (Anderson and Cavanagh, 2005).

Fears about the **inflexibility** of large corporations focus on the possibility that they may become inefficient, overly bureaucratic, and overly generous with salaries for top executives. Such organizations may be unable to change rapidly, adapt to new situations, or innovate. In some ways large organizations may be less efficient and less nimble than smaller ones (Kanter, 2001).

Many people also fear the power of large corporations to *exploit workers* and to reduce the quality of working life. Large corporations can undermine the sense of community among workers. As you saw in Chapter 3, people dislike working for large corporations because of alienating and rigid procedures and reduced opportunities to use their abilities. Many employees in large corporations have also experienced an intensification of work and increased expectations for extraordinary effort as a part of corporate downsizing.

Finally, many people are concerned that large corporations *exploit customers* through high prices and the production of shoddy goods and that they *damage the environment*. These outcomes can occur through a disregard for the long-run consequences and side effects of economic activity. For example, a corporation may lower its costs of production by avoiding environmental regulations and allowing the public to absorb these costs in the form of increased pollution, noise, and ugliness. Many people believe that large corporations stress profit to the point of neglecting moral, social, and aesthetic values (Ritzer, 2004). Large corporations have also been argued to have *excessive influence*

TABLE 15.1 The 20 Largest U.S. Corporations (ranked by sales) $

2000 Rank	Company	Sales	Profits	Assets	Employment
1	Exxon Mobil	$339,938,000,000	36,130,000,000	208,335,000,000	83,700
2	Wal-Mart Stores	315,654,000,000	11,231,000,000	138,187,000,000	1,800,000
3	General Motors	192,604,000,000	−10,600,000,000	475,284,000,000	335,000
4	Chevron	189,481,000,000	14,099,000,000	125,833,000,000	59,000
5	Ford Motor	177,210,000,000	2,024,000,000	269,476,000,000	300,000
6	Conoco-Phillips	166,683,000,000	13,529,000,000	106,999,000,000	35,600
7	General Electric	157,153,000,000	16,353,000,000	673,342,000,000	316,000
8	Citigroup	131,045,000,000	24,589,000,000	1,494,037,000,000	299,000
9	American Intl. Insurance	108,905,000,000	10,477,000,000	853,370,000,000	97,000
10	IBM	91,134,000,000	7,934,000,000	105,748,000,000	366,345
11	Hewlett-Packard	86,696,000,000	2,398,000,000	77,317,000,000	150,000
12	Bank of America	83,980,000,000	16,465,000,000	1,291,803,000,000	176,638
13	Berkshire Hathaway	81,663,000,000	8,528,000,000	198,325,000,000	1,500
14	Home Depot	81,511,000,000	5,838,000,000	44,482,000,000	345,000
15	Valero Energy	81,362,000,000	3,590,000,000	32,728,000,000	22,068
16	McKesson Pharmaceuticals	80,515,000,000	−157,000,000	18,775,000,000	26,400
17	J.P. Morgan	79,902,000,000	8,483,000,000	1,198,942,000,000	168,847
18	Verizon Communications	75,112,000,000	7,397,000,000	168,130,000,000	217,000
19	Cardinal Health	74,915,000,000	1,051,000,000	22,059,000,000	55,000
20	Altria Tobacco	69,148,000,000	10,435,000,000	107,949,000,000	199,000

SOURCE: www.fortune.com and www.primark.com Also, *Moody's Industrial Manual*, 2000, New York: Mergent.

on the media and on public discourse through concentrated ownership of the print and broadcast media and through their ability to influence the terms of debate about the issues of our times. See particularly the works of Noam Chomsky (2001) and Richard Sennett (1998) for elaboration of these and related themes on corporate social and cultural power. The twenty largest U.S. corporations, ranked by sales, are listed in Table 15.1, along with their profits, assets, and employment. Note that the largest private employer in the United States is Wal-Mart with more employees than its five nearest competitors combined—IBM, Home Depot, General Motors, Ford Motors, and General Electric.

Types of Corporate Market Power

The power of the large corporation—including its influence on governmental policy and public debate in the media—is ultimately based on its economic power. Social scientists conceptualize economic power in terms of large firms having either a **monopoly** or **oligopoly** position in relation to their environments. In addition, large size, in and of itself, confers many elements of power.

Monopoly Power A monopoly firm holds a dominant position in the market for the goods it produces. Where a few firms dominate an industry, these firms can also be said to hold monopoly power. For instance, if the top four firms in an industry control more than 70 percent of industry sales, we would consider these firms to have monopoly power (Borjas, 2005). Technically, such a situation is called an oligopoly (dominance by a few firms), but the economic power these few firms exercise is virtually identical to that of a true monopoly. Steel manufacturing in the United States was organized in this way before the introduction of foreign competition in the 1980s. On the basis of market domination, monopoly firms are able to restrict competition and charge higher prices for their goods. The result is that goods are sold at higher prices, output is reduced, jobs are lost, and the real income of consumers declines (North, 2005). In a competitive economy such industries would be flooded with potential producers. What keeps this from happening in monopoly industries?

Barriers to Entry A variety of barriers keep competing firms out, thus protecting the monopoly firm's profits. One important barrier is provided by *economies of scale*, which make it difficult for small firms to compete with large firms. Large batches are generally less expensive to produce per unit than small batches. Monopoly firms may also own important sources of supplies and may make it difficult for competing firms to secure access to these supplies; this is called a **backward linkage**. They may also control wholesale or retail outlets for goods, a **forward linkage**. Companies with forward and backward linkages are said to be **vertically integrated**. Monopoly firms may also have privileged access to credit. A firm can also achieve monopoly power through advertising. If the firm is able to differentiate its products from competitors' products through advertising, customers may come to prefer these products and consider them superior to otherwise equivalent goods. Monopoly firms may even engage in **predatory pricing**, in which they intentionally underbid competitors until they drive them out of the market. At that point, no

longer facing any competition, the monopoly firm recoups its losses by charging higher prices.

Industries with monopolistic structures typically have several distinctive characteristics (Averitt, 1968). These industries operate independently of suppliers, wholesalers, and distributors because they own their supply sources and control their own sales networks. They engage in a larger-than-average share of research and development. They set wage standards against which other firms have to compete. And they have a strong influence on the overall health of the economy, because so many subordinate and supporting firms depend on their economic success.

Monopoly control is unlikely in some industries. For example, residential construction precludes monopoly structure because the work is relatively unstandardized, production sites are geographically dispersed, and entry is relatively easy. Monopolies are also difficult to secure and maintain in personal services and in other services that cannot be standardized, such as repair. In the United States antitrust laws have retarded the growth of monopolies in other industries such as interstate trucking.

Perhaps surprisingly, concentration within industries has not increased greatly over time. In 1935 the average four-firm **concentration ratio** in manufacturing industries was 37 percent. In other words, on average, the four largest firms in a typical manufacturing industry controlled 37 percent of industry sales. By the 1970s this figure had risen only to 39 percent and has remained relatively stable since then (Census, 2000). One reason for this relative stability is that antitrust laws have limited the growth of concentration in key industries. As a result large companies have grown by moving into related (and sometimes unrelated) industries rather than by further increasing their share of their primary market. In this way antitrust laws may have actually promoted the growth of large conglomerate corporations sprawling across several industries. Such corporations are the focus in the section on mergers.

Oligopoly Power An industry is considered dominated by *oligopolistic* corporations if its output is controlled by few enough firms that they can engage in collaborative price setting. Such price fixing would be illegal if done openly. However, collaborative pricing more typically involves informal agreements to hold prices at a specified level and is rarely caught or prosecuted. The maximum number of companies that can effectively coordinate such collaborative practices varies from industry to industry but may be as high as a dozen or more in some industries. This practice creates market conditions similar to those in monopoly industries. Cigarettes and processed foods are familiar examples of industries with oligopolistic structures.

Two types of activity are particularly distinctive of oligopoly corporations. First, oligopolies tend to divide the available market into spheres of influence. These spheres may be defined by geographic regions or by different product lines. Second, oligopolies take advantage of reciprocal buying agreements in which suppliers or purchasers are pressured into allowing the oligopoly company privileged access to their markets (Scherer, 2000). For example, a wholesaler may be pressured to sign an agreement to buy high-priced microwave ovens from an oligopoly manufacturer as a prerequisite for obtaining their popular washing machines at the volume and price desired. The oligopoly company is thus able to use its power base in one industry to increase its profits in other industries.

Size as Power   Large firms, regardless of their market share, have a great deal of power. This power is similar in many respects to monopoly or oligopoly power. For example, before Japanese firms began competing in the automobile industry, a third of the labor force in the United States was engaged in manufacture, sale, or distribution of automobiles or automobile parts. This gave the industry tremendous leverage in Washington:

> [If a large firm were to fail,] the effects on workers, shareholders, creditors, suppliers, and distributors of and tangent to the stricken organization would be intolerable. . . .
>
> Moreover, the dominant position of the giant firms demands that public policy-makers avoid at all costs decisions which could con-

ceivably jeopardize the financial integrity of the mammoth. As a practical matter this means that most doubts in divestment proceedings, in antitrust prosecutions, or levies on unreasonable accumulations of surplus, or the vigor with which certain regulatory statutes will be applied . . . will be resolved in favor of the corporation in question. In a real sense the giant enterprises have achieved many of the privileges accorded the publicly regulated utilities without at the same time being burdened even by formal restraints imposed on the latter. (Baratz, 1971:151–153)

Thus size, in and of itself, confers tremendous power on large corporations, both in relation to the government and in relation to their competitors, customers, and employees.

In the 1980s the Chrysler Corporation, after decades of bad management and bad decision making, used its influence to secure low-interest loans from the government and wage concessions from its employees in order to restructure its activities. Smaller firms in less important industries would simply have been allowed to fail. In the late 1990s, subsequent to its publicly financed revival, Chrysler was acquired—at great profit to its shareholders—by the Daimler-Benz Corporation of Germany. More recently, Wal-Mart has come under public scrutiny for pressuring its suppliers to lower their prices to the extent that they are forced to move to China or other low-wage locations to meet Wal-Mart's demands (see *walmartwatch.com*). Although microeconomic theory focuses on market share as the basis of monopoly power, sociologists and institutional economists point out that many aspects of economic power, such as those benefiting Chrysler and Wal-Mart, result directly from size alone.

Corporate Law

Business corporations in the United States, including the very largest ones, have had legal protection as "persons" for over a century (Santa Clara County v. Southern Pacific Railroad Co., 1886). This legal status serves at least two important functions. *First,*

BOX 15.1 **America's Worst Corporations**

Each year in their December issue, the Multinational Monitor *presents a list of the nation's ten worst corporate "price gougers, polluters, union-busters, dictator-coddlers, fraudsters, poisoners, deceivers and general miscreants." Five of these are listed below. See the website in the source line of this box for the remaining five and for yearly updates. (Click "Archived Issues" and go to the latest December issue.)*

Abbott Laboratories: Raised the U.S. price of its anti-AIDS drug Norvir (generic name ritanovir) by 400 percent. The drug was invented on a grant from a government contract.

Coca-Cola: Complacency (if not complicity) in 179 human rights violations of Coca-Cola workers in Columbia, including nine murders and the abduction and torture of union activists and their family members.

Dow Chemicals: Continuing refusal to take responsibility for 8,000 deaths and for injuries to at least 150,000 people with ongoing health problems result-

ing from a 1984 pesticide spill in its Union Carbide factory in Bhopal, India. Its former executives are still fugitives from justice refusing to appear before Indian courts.

Riggs Bank: Has used its reputation based on its 1840 founding and service to such historic figures as Abraham Lincoln (and 19 other Presidents) to avoid prosecution for laundering money for dictators around the world, including General Augusto Pinochet of Chile, Equatorial Guinea dictator, President Teodoro Mbasago, and members of the ruling family of Saudi Arabia.

Wal-Mart: By paying below market wages to its workers and providing health benefits that few can afford, Wal-Mart has forced taxpayers to pick up the tab for its workforce in terms of public housing, free school lunches, state supported health care, and federal tax credits for low-income families.

SOURCE: Multinational Monitor, "The Ten Worst Corporations of 2004," December 2004, *multinationalmonitor.org*

---

the owners of the corporation (stockholders) are protected from bearing full legal or fiscal responsibility for the actions of the corporation. Because the corporation is recognized as a person, debts, bankruptcy, and lawsuits are restricted to the corporate entity itself and do not carry over to its owners. This concept is known as **limited liability**. Critics argue that the lack of full accountability for owners allowed by limited liability encourages irresponsible actions on the part of large corporations. Some of the worst abuses of corporate power are described in Box 15.1.

The Cloak of Privacy A *second* consequence of their legal status as persons—along with limited liability for owners—is that corporations receive certain rights granted to people, including the right to privacy. This right is particularly important for the large modern corporation:

> The nature of oligopolistic competition generates pressure on management to maintain

secrecy. In the oligopoly, product price is a collective decision reached overtly or covertly through "oligopolistic pricing mechanisms." . . . Competition in such industries occurs through non-price mechanisms (e.g., advertising) and rivalry over product quality. . . . Strategies for non-price competition (e.g., choice of product lines or marketing policies) must be secret until they are implemented. The need for secrecy in oligopolistic firms is heightened by their clear visibility among each other and before the public. (Cornfield and Sullivan, 1983:258–259)

Such secrecy, however, is not granted to *public* bodies; indeed, it would be considered a breach of democratic principles. The legal status of corporations as persons, along with the principle of "employment at will," which allows the firing of employees without justification (see Chapter 4), provide a legal framework for corporations that allows them immense power in modern society.

Antitrust Regulations   In an effort to prevent single companies or small groups of companies from using their monopoly power to eliminate competitors and exploit consumers, the government attempts to regulate the existence and behavior of monopolies through the activities of the Federal Trade Commission (FTC, *www.ftc.gov*). This agency performs important functions in reducing unfair practices of large companies against both their competitors and against consumers. Antitrust regulations are a bulwark in maintaining a competitive market system through preempting its tendency to self-destruct through the excessive concentration of economic power. For instance, in 2000 the FTC found Microsoft guilty of anticompetitive practices against its competitors by putting "technological roadblocks" in its Windows operating system that prevented competing programs from operating effectively.

Owners versus Managers

The owners of a large corporation (the stockholders) do not, in general, directly control the daily activities of the corporation. Rather, hired managers control daily activities. Control by hired managers has been alternatively interpreted as encouraging lesser and greater corporate responsibility. The economists Adolf Berle and Gardiner Means coined the term **management-controlled firms** in 1932 to describe corporations in which the stockholder with the largest holding owns *less than 20 percent* of the corporation's stock. Using data from 1929, they found that 44 percent of the two hundred largest United States nonfinancial corporations were management controlled. They were concerned that this management control might make corporations less responsible (in relation to their stockholders). Berle and Means (1932:9) argued that management control "destroys the basis for the old assumption that the quest for profits will spur the owner of industrial property to effective use."

These findings have been intensely debated since Berle and Means put them forward over half a century ago. According to a reanalysis by sociologist Maurice Zeitlin (1974), only 22 percent (rather than 44%) of the corporations studied by Berle and Means were actually management controlled under their own definition. Further, researchers have argued that management-controlled corporations may, in fact, be just as oriented toward maximizing profits as owners ever were (Mizruchi, 2004). If profit rates are not as high as possible in a management-controlled corporation, owners may withdraw their holdings from the company and invest in other companies. This will cause a decline in the price of the corporation's stock and jeopardize the interests of management. Under this thesis, the profit motive is built into the system of ownership and investment—it does not reside in individual owners' desires for accumulation.

Other researchers argue that managers are more interested in maximizing *revenues* than *profits* because maximizing overall revenues enhances their own power, prestige, and earnings. This thesis may explain in part the corporate drive toward increased size. Corporate executives who successfully increase the size of their companies are paid salaries unthinkable to the average wage or salary worker. As noted in Chapter 4, chief executives can be paid up to 25,000 times the salary of an average worker in their company (see Table 4.1).

MERGER MANIA

Corporations have increased in size and concentration in a halting and irregular fashion patterned by periods of rapid activity and relative quiet. Corporate growth occurs not only through expansion of a firm's operations but also through the acquisition of other firms, and these **mergers** tend to occur in distinct cycles. The history of these cycles reveals not only dramatically changing merger rates but also the emergence of different kinds of corporations during six distinct periods.

The First Five Merger Waves

The *first* wave of mergers occurred in the early 1890s in metallurgy, chemicals, and electrical machinery. Technological breakthroughs in these industries made large-scale mass production more efficient than smaller-scale production. Large firms also benefited from their domination of emerging national distribution and marketing systems based on transcontinental railway and telegraph lines. With these advantages large firms succeeded in forcing smaller ones out of business by using practices that would clearly be illegal under today's antitrust legislation. Such practices included buying up railroads and charging competitors higher freight rates in order to put their products at a price disadvantage. Large firms also increased profits by using new mass-production technologies to replace skilled workers with less well paid semiskilled workers. This merger wave resulted in a dramatic increase in the size of firms in the affected industries (Zunz, 1998). Westinghouse, International Harvester, and Standard Oil emerged as industrial giants at this time.

The *second* merger wave occurred between 1897 and 1905 and was again based on economies resulting from large-scale production, marketing, and transportation systems. This wave had a wider industrial base than the first one and created monopolies in industries such as tobacco and food processing that are still in existence today. Companies emerging as giants at this time include American Can, United Fruit, and American Tobacco. The first billion-dollar merger occurred in 1901 when the Carnegie Steel Corporation combined with its leading rivals to form United States Steel (*www.ussteel.com*) (Geisst, 2004).

Vertical Integration   During the first two merger waves large companies acquired smaller companies engaged in similar lines of production. The *third* merger wave occurred in the 1920s and introduced a new corporate form: the *vertically integrated* company. Vertically integrated corporations come into being with the acquisition of firms supplying raw materials or component parts (backward linkages) and firms engaged in further processing or in selling the manufacturer's products (forward linkages). Steel companies were central actors in this wave of vertical integration. They acquired ownership of mines, ore transportation networks, finishing mills, metal fabrication plants (which turn raw steel into pipe, construction girders, and other products), and

even transportation and distribution systems for finished steel products. Bethlehem Steel and Republic Steel emerged at this time.

Diversification   The *fourth* merger wave stretched across two decades, starting in the Great Depression and extending into the early 1950s (Chandler et al., 1997). **Diversification** was the keynote in this merger wave, and the first multi-industry companies emerged at this time. During the Depression large companies were scrambling for opportunities to remain profitable. Many smaller companies were facing bankruptcy and could be purchased for a song. Mergers tended to follow lines of technologically similar products or production processes and resulted in the expansion of companies into related product lines. The movement continued through the boom period of the Second World War when the larger and more diversified companies were able to take advantage of the heightened demand for new military weapons and supplies more effectively than were smaller companies. This merger wave created E.I. du Pont and many other corporate giants of today's economy.

Industrial Conglomerates   The *fifth* merger wave, which occurred in the late 1960s and early 1970s, created a new type of company, the *industrial conglomerate*. These mergers occurred between firms engaged in completely distinct product lines; for example, in 1973 Mobil Oil purchased the Montgomery Ward retail chain, which it subsequently sold as unprofitable. Later Montgomery Ward was acquired by General Electric, which dismantled the company and liquidated its assets through bankruptcy proceedings in 2001. These mergers were motivated by the desire of large firms to move into profitable areas regardless of the field. Vastly different levels of economic power between the acquiring companies and their acquisitions allowed many "hostile takeovers" to take place. In such takeovers the acquiring company subsequently exploits the cash reserves and market position of the newly acquired company for its own profit without necessarily making even minimal investments to maintain the acquired company's productive capacity.

The history of corporate mergers has witnessed a movement from growth through expansion within product lines, to forward and backward linkages, to expansion into related product lines (diversification), to expansion into totally different product lines (conglomeration). Large companies today operate across a variety of related and sometimes unrelated product lines. Diversification and conglomeration are essential concepts for understanding the modern corporation and the modern economy.

The Current Megamerger Frenzy

The fifth merger wave ended with the recession of the mid-1970s brought on by the end of the Vietnam War and rising oil prices. However, after a very brief respite, a *sixth* wave took off in the early 1980s, quickly accelerated to even higher levels of activity, and has continued unabated since then creating a seemingly permanent culture of mergers. The sixth wave is distinguished by what have come to be known as megamergers between some of the largest companies in the economy, such as Conoco Oil and E.I. du Pont.

By the 2000s a high level of merger activity had become an established fact of economic life in the United States and Canada. Top managers and rank-and-file employees alike have become increasingly nervous about the future of their companies as corporations came to be defined as salable bundles of assets (Krantz, 2006). Takeover bids have assumed an increasingly hostile character. In this section we explore the nature, causes, and consequences of this megamerger frenzy.

Why is a wave of megamergers occurring at this time? One reason is that it had become cheaper to buy a company in financial trouble than to build new plants and facilities. Many such troubled companies exist in North America because of increased international competition. The value of the acquired company on the stock market may greatly underestimate the actual value of its assets. Tax laws also encouraged conglomerate mergers. Part of what may make a company an attractive acquisition is its package of tax losses, write-offs, and deprecia-

tions. These are also acquired, along with the plant and equipment, and can be used to offset the taxes of the acquiring company. In some cases it has even been possible to purchase bankrupt companies for less than the value of their tax losses and use the losses to offset income of the acquiring company (Geisst, 2004).

Alternatively, a successful company with a secure market niche but a low growth potential (and therefore a low price) may provide a desirable target for a corporation interested in growth through conglomerate expansion. Such a company can serve as a **cash cow**. The acquiring corporation can use the regular profits and depreciation allowances of the acquired company to fund expansion in other areas rather than setting them aside to maintain or upgrade the productive capacity of the acquired company (Bluestone and Harrison, 2000).

A Permissive Legal Environment   Changes in the interpretation and enforcement of antitrust laws have also encouraged the megamerger frenzy. In the 1980s the Federal Trade Commission stopped targeting industries as a whole for antitrust investigation and instead only went after companies already identified as having engaged in specific illegal activities (Blau and Moncada, 2005). Antitrust cases were left to the Justice Department, where presidential appointees gave low priority to enforcement of this legislation:

> There is no doubt that the administration of President Ronald Reagan brought with it a different attitude toward antitrust legislation.... [The] case against IBM, the world's largest computer company, was dropped. A case against the eight largest oil companies in the United States was abandoned just before the trial was set to begin. And an investigation of the major American automobile companies was closed. In their place, the FTC sued a small group of attorneys who represent indigent criminal defendants in Washington, DC and started investigations of state boards that license taxicabs and optometrists and of labor unions that represent actors. (Davidson, 1985:125)

The fact that mergers between large corporations are less often legally contested today greatly accelerates their occurrence.

Merger Targets   How are targets identified for corporate mergers? One criterion is technological compatibility between two companies. Other criteria include shifting consumer preferences or a desire to control resources critical to the acquiring company's survival. In conglomerate mergers between companies in unrelated industries, the target company is selected by financial advisers who appraise it as a "good buy" (Mizruchi, 2004). Conglomerate firms interested in acquiring subsidiaries will often pay a premium for their purchase, typically buying large blocks of the company's stock at between 40 percent and 50 percent more than their current market price. This provides a lucrative, almost irresistible situation for shareholders in the company being acquired who are interested in quick profits. Even if the firm wishes to avoid such a takeover bid, its only option may be to search out a "white knight"— a more congenial firm with which to merge. A highly colorful language has emerged to describe the takeover process: "'White knights' rush to rescue 'sleeping beauties' from 'black knights'—or end up 'paying ransom' as a last defense" (Hirsch, 1986:814). (Some contemporary merger jargon is decoded in Box 15.2.)

At the beginning of the megamerger frenzy, many of the corporations acquiring other companies were in the oil industry. These corporations had amassed huge stockpiles of profits as a result of rapidly rising energy prices. American companies received windfall profits from the sale of cheaply produced American oil at prices leveraged upward by the Middle Eastern oil cartel. With this money the oil companies expanded into non-energy areas. In 1981 Mobil, Seagram, and DuPont entered into a bidding war for the purchase of Conoco, the fourteenth largest industrial corporation in the United States. By mid-year DuPont had won the battle but had to borrow the $3.9 billion it needed to buy out Conoco shareholders, thus increasing its long-term debt nearly fivefold. The total deal cost $7.2 billion and was the largest corporate acquisition to that point in history (Sikora, 1993:6).

In the early 1980s the airlines also began to move toward a diversified structure as they bought hotel chains, restaurant chains, and car rental companies. By the mid-1980s the merger frenzy seemed to be consuming the industry. In 1986 Southwest Airlines bought Muse Air. People's Express bought Frontier. Texas Air bought People's Express. Piedmont acquired Empire Airlines. Northwest agreed to buy Republic. By the summer of 1986, the sizes of the acquired companies had gotten even larger. Texas Air agreed to buy Eastern for $800 million, and this deal was quickly followed by TWA's offer to purchase Ozark Airlines. By the 1990s virtually all of these recombined airlines had entered bankruptcy proceedings at least once.

The impact of these mergers extends beyond the immediate companies and workers involved. The fifty to one hundred hostile takeover attempts of large firms each year indirectly affect virtually *all* companies, because companies must spend a great deal of time, money, and energy to make themselves unlikely or difficult takeover targets. This pressure moves their investment decisions toward more conservative options (Taggart, Berry, and McDermott, 2001). It also drives up the price of credit, making productivity enhancing investments more expensive.

In 1998 two of the world's largest automakers combined when Chrysler and Daimler-Benz merged. In 2000 the largest merger ever occurred when the two telecommunication giants Time Warner and America Online combined their resources. The combined company was valued at $350,000,000,000, nearly ten times the value of the Chrysler-Daimler conglomerate. Some of the subsidiary companies of the Time Warner-AOL telecommunications empire are listed in Box 15.3.

In this context of heightened megamergers, workers, shareholders, and communities have come to fear that top managers are motivated more by greed and ambition than they are by the security and prosperity of the organizations they head. Top executives control important information that they use to plan their own career strategies but that they may choose to share only selectively with stockholders or employees. They may even engage in whitewashing the facts and in stonewalling (withholding information or comment) while working furiously to feather their own nests. More concretely, they may hide real operating costs as costs

---

BOX 15.2   Merger Jargon

The following terms describing key merger events were identified by the sociologist Paul Hirsch. His research attempts to unravel the symbolic social meanings through which those involved in mergers make sense of the merger events and their aftermath.

afterglow post-merger euphoria of acquirer and/or acquiree, usually soon lost.

ambush swift and premeditated takeover attempt.

bear hug hostile tender offer, usually with considerable muscle behind it.

big-game hunting plotting and executing takeovers of large companies.

black knights unfriendly acquirers drawn to a target by news that the company is already being propositioned by others.

bring to the altar consummate a merger, usually friendly.

chain letter effect apparent growth, through acquisition of other companies rather than high internal performance and productivity.

confetti stock traded by acquirer for that of acquiree, particularly if thought of as having little value (see also "Russian rubles").

cyanide pill antitakeover finance strategy in which the potential target arranges for long-term debt to fall

due immediately and in full if it is acquired (see also "scorched earth").

double Pac-Man strategy target firm makes tender offer for the stock of its would-be acquirer.

dowry outstanding assets that the target may carry into a merger (e.g., low-interest loans, long-term contracts).

friendly offer merger proposal cleared in advance with the target company's board and top management; usually leads to the firm's recommending it favorably to shareholders for approval.

golden parachutes provision in the employment contracts of top executives that assures them a lucrative financial landing if the firm is acquired in a takeover.

greenmail a firm's purchase of its own stock, at a premium, from an investor who it fears will otherwise seek to acquire it or else initiate a proxy fight to oust its present management.

hired guns merger and acquisition specialists, other investment bankers, and lawyers employed by either side in any takeover.

hot pursuit warfare image for strenuous campaign by aggressive, would-be hostile acquirer to obtain shares of the target firm.

---

of mergers and reorganizations, thus making profits appear high (because ongoing costs of production are hidden). The result is a speculative rise in the stock price of the company for which the executives take credit but which is eventually paid for by lost wages or earnings for employees and investors (Krier, 2005).

Increased Conglomeration

Many large firms have become increasingly diversified across a variety of industries. This tendency has intensified dramatically during the megamerger wave of the 1990s and 2000s. Corporations view diversification as providing protection from vulnerabilities in any one industry and as leading to greater

stability for their assets and profits. Business schools have also promoted this view and trained two decades of managers in its tenets. This distinctly American view of corporate strategies has encouraged a shift of attention away from production and toward the financial manipulation of assets.

Growth through increasing conglomerate diversification is largely responsible for the increasing domination of the economy by a few large firms. Industrial concentration has two distinct meanings. In many industries a few key firms hold a dominant market share but operate solely or primarily within that industry. This type of concentration has been relatively stable since the Second World War. The second type of concentration

junk bonds high-risk, high-yield debt certificates traded publicly, so called because they are rated below investment grade, either by Moody's or by Standard & Poor's; junk bonds are often used to help finance hostile takeovers.

mushroom treatment post-merger problems from an acquired executive's standpoint: "first they bury us in manure, then they leave us in the dark awhile, then they let us stew, and finally they can us."

pigeons highly vulnerable targets.

pirates hostile acquirers.

rape forcible, surprise hostile takeover, sometimes accompanied by looting of acquiree's profitability.

rescue party a company that provides any substantial aid to the target firm in a takeover battle (see also "white knight").

safe harbor antitakeover advantage: company owns a subsidiary in a heavily regulated industry (such as broadcasting or interstate trucking). The long time required for government approval to transfer its ownership delays the acquisition process and decreases the chances for successful takeover.

Saturday night special a fast and predatory merger.

scorched earth policy whereby the target company would rather self-destruct than be acquired (e.g., all personnel threaten to quit).

shark repellent protective strategies for preventing or combating a hostile tender offer.

sharks takeover artists.

shoot-out climax of a takeover battle, usually conducted by "hired guns."

sleeping beauties vulnerable targets (see also "pigeons").

summer soldier executive of target company offering only token resistance against a takeover.

takeover the purchase of majority ownership in a corporation; usually resisted by the target company but accomplished nonetheless by paying a premium above the current market price for the firm's shares.

white knight acceptable acquirer sought by a potential acquiree to forestall an unfriendly takeover; the preferred suitor.

wounded list executives of an acquired firm who develop health or career problems from the deal.

SOURCE: Adapted from Paul M. Hirsch, 1986, "From Ambushes to Golden Parachutes: Corporate Takeovers as an Instance of Cultural Framing and Institutional Integration." *American Journal of Sociology* 91:800–37. Used with permission of The University of Chicago Press.

results from a few huge firms, which operate across a variety of industries, controlling a large share of the economy. This type of concentration has increased dramatically. Economic theory has largely ignored the growing importance of the concentration of economic power in diversified conglomerates (Scherer, 2000).

Growth within an industry (and growth through vertical integration) occurs either through *expansion* of the core firm or through the acquisition of enterprises doing related work that subsequently become **divisions** of the larger corporation. Growth through conglomerate mergers occurs through the incorporation of acquired companies as semiautonomous *subsidiaries*. This autonomy may only mean having a separate corporate name, or it may entail substantial latitude in decision making for the subsidiary company.

The Effects of Size and Concentration

What effects have the increasing size and concentration of enterprises had on workers and on organizations? Overall, the effects have been enormous. One need only reflect for a moment on the difference between working in a small corner grocery and a new multiservice shopping center that is part of a national retail chain. These changes can have both positive and negative aspects.

BOX 15.3   The Companies of Time Warner $

As the core of the economy has shifted from the production of goods to the production of services, mega-mergers have followed into these new industries. A partial list of the companies of Time Warner subsequent to its merger with America Online (AOL) is presented below.

America Online

    America Online, Latin America
    America Online, UK
    AOL Web Access
        CompuServe
    MapQuest.com

Home Box Office
MovieTickets.com
New Line Cinema
Time, Inc.

    Time Warner Books

Time Warner Cable

    Road Runner

Turner Broadcasting

    Cable News Network
    The Cartoon Network
    Turner Network Television
    Turner Sports
        Atlanta National League Baseball Club

Warner Brothers Entertainment

    Castle Rock Entertainment
    DC Comics
    Movielink, LLC
    Warner Brothers Interactive Entertainment
    Warner Brothers Television Production
    Warner Brothers Home Video

SOURCE: www.hoovers.com; www.moodys.com

Effects on Organizations   Size has a huge impact on the organization of the workplace for two major reasons. First, specialization (division of labor) and complexity increase as more activities have to be coordinated (Carruthers and Babb, 2000). Second, bureaucratic rules and procedures replace direct control by owners, as you saw in Chapter 7. Increasing specialization, formalization of rules, and the prevalence of impersonal authority may result in reduced employee identification with the enterprise. Reduced identification makes employees reluctant to take on the values and goals of their increasingly impersonal employer. In addition, workers in large corporations often have fewer opportunities to develop their abilities across a wide range of activities than do workers in smaller firms (Hodson and Sullivan, 1985).

On the positive side, large firms typically pay their employees more than smaller firms. This difference allows large firms to compete successfully for the best workers and also helps motivate these workers. Large firms also typically have fairer and more equitable compensation schemes than smaller firms, resulting partially from the prevalence of bureaucratic rules and procedures that help limit favoritism. The unionization of large firms and the role of unions in negotiating compensation contribute to pay equity. Large firms also typically have better affirmative action records. Again, this results from their formalized procedures, which limit blatant forms of discrimination, and also from their greater public visibility. Finally, workers in large firms are typically more highly rewarded for their formal education than are workers in small firms. This difference results from the premium placed on educational credentials in bureaucratic settings.

In recent years, increasing size and concentration have taken on new implications for job security, replacing the image of the large corporation as a source of opportunities for lifelong careers with the image of the large corporation as faithless to its workers. Many large corporations have searched for flexibility to confront rapidly changing product markets by reducing their core labor force, employing workers on short term or temporary contracts and subcontracting many production and support functions. Such strategies also cut the total wage bill, thus increasing short-term profits and making the company appear attractive to investors. While this strategy may achieve some flexibility for the firm, it does so at the cost of decreased job security for employees. Such arrangements are sometimes called **lean production** in recognition of their reduced use of a core labor force and intensification of work for the remaining employees (Cohen and McBride, 2003).

Critics of lean production suggest that this organizational strategy may be shortsighted (Watson, 2003). Making a distinction between short-term and long-term flexibility, they argue that lean production may achieve short-term flexibility but at significant costs in terms of long-term adaptability. Adapting to rapidly changing market conditions requires retaining employees and investing in their human capital development to create the conditions for increased commitment and innovation. Without employee loyalty and commitment, innovation and adaptation become harder, not easier.

Effects on Society   Mergers between economic organizations can have both positive and negative consequences for society. The buying and selling of companies may help move capital from stagnant industries into growing ones. Conglomerate mergers can even intensify competition in an industry that has been dominated by a few giant firms. Large, multi-industry conglomerates may be able to compete on the home turf of monopoly companies in a way not possible for smaller companies.

Profit-making organizations, however, do not undertake mergers with the public interest in mind. Mergers are rarely peaceful, and they can have many negative consequences for society. Conglomerate mergers can hurt the economy because they divert resources from productive investment to the buying and selling of corporate properties. Managers who are focused on these transactions may overlook opportunities to increase productivity more directly. This preoccupation retards innova-

tion. And it results in skyrocketing prices for companies that look like good merger targets, thus adding to inflationary pressures. Most researchers believe that excessive mergers damage the underlying health of the economy (Mizruchi, 2004). Mergers have also been implicated in the pattern of *"jobless growth"* typical of the early twenty-first century—economic activity appears to be increasing and the stock market rises, but few if any new jobs are being created.

Conglomerate mergers can also damage the social fabric of local communities. If a conglomerate buys a local plant and then shuts it down after a few years, the loss of jobs and revenue can devastate a community. (Recall the discussion of layoffs and plant closings in Chapter 8.) Outside ownership of local enterprises also undermines the participation of local business leaders in civic affairs. The welfare of the large, outside-owned corporation and its managers does not depend on the cultivation of local contacts. Local civic participation declines accordingly. Outside-owned firms are also less likely to look to local sources for supplies, labor, and expertise than are locally owned firms.

An additional long-term consequence of mergers is that they damage public confidence in the economy. The movement toward mergers is based on the pursuit of profits through financial manipulation rather than through producing products or services of economic value. In addition, a great many mergers are obvious failures. Employment is often reduced, and stock prices of the acquiring company sometimes plummet. For the company being acquired, there is little indication that negative outcomes are reversed at any point in the future. A noted historian of corporate mergers observes: "Mergers produce wonderful results for . . . the executives involved as well as the investment bankers. Shareholders in the new companies created by merges are less than thrilled by many of the subsequent results, however" (Geisst, 2004:310).

The telecommunications giant WorldCom provides a useful example. Following a period of aggressive acquisitions, WorldCom entered the largest bankruptcy proceeding ever after the disclosure that the company had fabricated $11 billion in

DILBERT reprinted by permission of United Feature Syndicate, Inc.

profits. The litigation resulted in creditors being paid thirty-six cents on the dollar to clear $35 billion in debt. The squandered capital was then unavailable for other potentially more productive uses. The remnants of WorldCom were reorganized as MCI with a workforce of 55,000—down from 85,000 before the bankruptcy (Padgett, 2005).

Effects on Employees Top managers often fare very well in merger situations. Mergers increase the size of the company and thus provide justification for higher pay for chief executives. Managers in the acquired firms are increasingly protected by **golden parachutes** that guarantee compensation to top corporate executives if the firm is taken over (see Box 15.2). The provisions are basically lump sum severance packages and are called golden because they are so generous. For instance, when Gillette was acquired by Proctor and Gamble in 2005, its chief executive, James M. Kilts, was awarded a severance package worth $165 million (Symonds and Berner, 2005).

The potential effects on lower-level employees, however, can be devastating. Both material and psychological losses are likely. Employees may be laid off or transferred against their wishes, often abruptly. Seniority acquired over years or even decades may mean nothing in such situations. Benefit programs are also likely to be disrupted. New work rules and disciplinary procedures are likely to be implemented. Pensions, sick leaves, vacations, holidays, and employee stock purchase plans may

all be subject to change. Pension funds have been a frequent target in mergers. Reorganization of these funds often makes large amounts of money available to the new owning firm at the expense of workers who had counted on these funds for their retirement. Mergers are thus especially dangerous for the well-being of older workers, who may have few options for regrouping, and workers with families, who may be heavily dependent on health insurance and other benefits that are put at risk by mergers. Some of the euphemisms companies use to describe terminations in the wake of corporate downsizing are presented in Box 15.4.

One of the most notorious examples of the loss of benefits, jobs, and retirement savings occurred in the early 2000s with the collapse of energy giant Enron. This company had grown rapidly through the acquisition of other companies in the oil industry. The company's problems resulted from the gross exaggeration of its assets by its chief executives, in collusion with its accounting firm, Arthur Andersen, both of whom were found guilty of fraud (Dahl, 2004). Revelation of the misaccounting caused Enron's inflated stock to collapse, sending the company into bankruptcy. Retirement accounts in the company had been heavily invested in now worthless Enron stock and became virtually of no value. And since the company was basically a holding company for other oil firms, there were no physical assets to reorganize into a new post-bankruptcy enterprise (Vollmer, 2004). Some small

---

BOX 15.4 Euphemisms for Termination $

Michael Moore compiled the following euphemisms for employment termination in the wake of downsizing for his best selling book, Downsize This!

Downsized
Rightsized
Destaffed
Degrown
Dehired
Involuntarily separated
Personnel surplus reduction

Transitioned
Resource reallocation
A save (as in a "savings" to the company)
Displaced
Dislocated
Disemployed
Redundancy elimination
Workforce imbalance correction

SOURCE: Michael Moore, 1997, *Downsize This!* New York: Harper.

---

measure of justice was achieved in 2006 when Jeff Skilling, Enron's former Chief Executive Officer, was sentenced to twenty-four years in a Federal Penitentiary for fraud connected to misrepresenting Enron's assets to inflate its value for his own personal gain.

A merger can also result in reduced trust among all levels of employees (Cook, 2001). Employees are likely to doubt that the company has their best interests at heart and may adjust their own commitment downward in response. Ambiguity in role expectations is also likely to increase after a merger, and the company may seem relatively aimless during the transition period. Morale and motivation typically decline (Taggart et al., 2001). Mergers threaten employees' relations to their jobs and their economic security. Fear spreads.

Even contractual labor relations may be disrupted by mergers. The National Labor Relations Board recognizes only the obligation of a new parent company to bargain with an existing union, not an obligation to recognize existing labor contracts. The "obligation to bargain" gives legal recognition to the union as the workers' legitimate representative. It does nothing, however, to help secure a new contract. The terms of any new contract are completely open for renegotiation, and the increased power of the enterprise as part of a larger organization may substantially alter the terms of the

agreement. Bankruptcy, as well as mergers, also renders labor contracts, including pensions and benefits, open for renegotiation, often under highly limiting and unfavorable circumstances (Sullivan et al., 2000).

A Modest Backlash?

By the 2000s, forty states had enacted some form of antitakeover legislation (Taggart et al., 2001). These laws provide safeguards against the most financially irresponsible and predatory takeovers. In addition, unions and consumer and environmental groups have increasingly publicized illegal and morally questionable practices of large corporations, ranging from the exploitation of child labor in poor nations to the killing of dolphins in tuna nets. Inside large corporations, **whistle-blowing** by employees has also increased. Whistle-blowing refers to employees reporting illegal activities inside the organization to law enforcement or regulatory agencies. This activity sometimes halts the activity and brings about change, but it also typically results in significant emotional, financial, and personal losses for the person blowing the whistle (Rothschild, 2000).

Increased scrutiny and demands for accountability by the government, by employees, and by the public have forced companies to become increasingly interested in corporate ethics. As a result large

---

BOX 15.5 Sanctions against Executives for Financially Irresponsible Mergers

Directors of large corporations are sometimes held accountable for some of the financial consequences of poorly considered mergers. The following discussion is from a book intended as a guide for managers to help them avoid such accountability. The book also reports, somewhat ironically, on the high profile—and financially rewarding—careers of the executives subsequent to the sanctions against them.

The Supreme Court of Delaware found several of Chicago's best and brightest business leaders guilty of breaching their duty to the company's shareholders. Nine directors of the billion-dollar blue-chip Trans Union Corporation were held liable for agreeing to sell the company without careful review of its value and ordered personally to pay the difference between the per share selling price and the "real" market value of the company's shares....

What became of Trans Union? Less than a year after the merger, Bruce Chelberg, Trans Union's president and chief operating officer, became senior vice president, international, for IC Industries, Inc. A few months later, Jack Kruizenga, once head of Trans Union's tank car division, was recruited to run Pullman Standard Inc., a railcar industry giant that Kruizenga proceeded to overhaul. On the recommendation of Van Gorkom's good friend George Shultz, President Ronald Reagan appointed Van Gorkom Under Secretary for Management of the State Department. *The New York Times* reported that Shultz wanted a "trusted confidant" in this State Department senior management position.

SOURCE: Arthur Fleischer, Jr., Geoffrey C. Hazard, Jr., and Miriam Z. Klipper, 1988, *Board Games: The Changing Shape of Corporate Power.* Boston, MA: Little Brown and Company, pp.15, 37.

---

corporations have had to reconsider many of their activities. Many large corporations have hired experts in ethics to help internally monitor their own behavior; books and business school courses on ethics have flourished. The lasting impact of these developments is yet to be evaluated. History suggests that the driving force of corporate activity—profits—is strong but that corporate activity can be channeled and regulated if there is sufficient public will. Box 15.5 reports on a partially successful lawsuit establishing the personal liability of corporate directors for financially irresponsible mergers.

INTERCORPORATE LINKAGES

We have observed that large corporations have tremendous power in relation to customers, competitors, and workers. If a corporation is spread across several industries, its employees may even have difficulty understanding for whom they actually work. Interlocking directorates, bank ownership of major-

ity shares, and subcontracting provide other sorts of linkages between firms that may be even more difficult for employees and communities to perceive and comprehend (Useem, 2004). These links, however, may be very important in determining investment decisions and, thus, in determining the fates of employees and communities (Keister, 2000).

Interlocking Directorates

Many executives sit on the boards of directors of several different corporations, creating *interlocking directorates.* For example, many members of the Rockefeller family, and other wealthy families, are incumbents of the boards of directors of a variety of corporations. Such interlocking directorates can coordinate the policies of supposedly competing large corporations. Thus, they are important in facilitating various oligopolistic practices. The networks defined by such interlocking directorates are often quite intricate. Banks or enterprises competing in one country may be parts of international networks collaborating with one another in markets elsewhere in the world. Employees may find it extremely difficult to understand and support the goals of their company when those goals are in part determined by its role in an international network of corporations and banks.

The Role of Banks

Bank control of large corporations removes ownership and investment decisions yet another level from the actual site of production. The Chase Manhattan Bank alone has predominant control in more than 10 percent of the two hundred largest industrial corporations in the United States. Such power provides investment bankers with the ability to influence the allocation of resources over a significant portion of the economy. Such influence has contributed to the ascendancy of short-term financial considerations in production decisions over all other factors, including human capital development and investment in the long-term adaptability of the firm (Fligstein, 2001). The banking industry is concentrated much as industrial enterprises are. As of 2005, the one hundred largest commercial banks in the United States held approximately 75 percent of all the deposits in the 7,769 commercial banks in the country (Census, 2005).

Outsourcing

A final type of corporate linkage is based on **subcontracting** (also called outsourcing, see Chapter 10 on services). Subcontracting arrangements exist in two forms. In *industrial subcontracting* the subcontractor manufactures parts that will be incorporated into a final product made by the principal company. This type of subcontracting is common in the manufacturing of equipment and machinery, including automobiles. In *commercial subcontracting* the subcontractor manufactures an entire finished product that the principal company markets, typically under its own brand name. Commercial subcontracting occurs in a wider range of industries, including clothing, footwear, toys, plastic articles, and many other consumer goods. Many large firms also subcontract out peripheral support functions, including food service, security, janitorial services, and accounting services.

Lower Wages and Reduced Benefits   The motivation for subcontracting is often to avoid the higher wage and benefit packages secured by workers in the principal firm. Subcontracting sets up relationships of subordination between companies. The job security of workers in the subcontracting firm may be highly vulnerable to decisions made by the principal firm. Workers in subcontracting firms face problems resulting from their employer's weak position in relation to the principal firm. Subcontracting firms often achieve a competitive price for their product or service by cutting back on wages and sometimes by failing to pay Social Security, medical insurance, pensions, unemployment insurance, and other worker benefits. The companies involved in subcontracting arrangements, because of their small size, are frequently exempted from legislation protecting workers and working conditions. Subcontracting may also be a tool for undermining a union in the principal company. The union may find that its membership is being eaten away by employment losses as work is subcontracted outside the company. Unions generally bargain for some say in the subcontracting arrangements entered into by their employing company. Unless the union is very strong, however, it rarely succeeds in securing a contract that excludes subcontracting.

The growth of subcontracting undercuts the sort of permanent employment relations that once typified jobs in industrialized nations. Insecure employment contracts have contributed to the stagnation or decline of real income for workers in these nations. Many of those who lost jobs previously considered permanent have been able to secure employment only at much reduced wage rates (Newman, 1999). This tendency has contributed to the growth of contingent and peripheral employment discussed in Chapter 14.

Even when the subcontracting arrangement is relatively favorable, the subcontracting firm still exists in a very precarious relationship to the market for its product. In the face of an economic downturn, orders from subcontractors can be cut rapidly to preserve available work for the principal firm. Subcontracted production is relatively easy to enter into because of the specialized nature of the production process involved and because of the guaranteed market. But, for these same reasons, it is also extremely vulnerable to market downturns. In the early 2000s many developing countries encouraged the rapid development of semiconductor component subcontracting as a route to industrial development. However, cyclical downturns in the semiconductor industry have caused widespread cutbacks in this industry, and many nations, most notably South Korea, have been left with large, expensive plants standing idle.

Subcontracting relationships are not necessarily exploitive. In the international context such arrangements can have positive consequences for developing nations (Gereffi et al., 2002). Such arrangements may provide an opportunity for the developing nation to manufacture items it was previously importing as well as encouraging raw material production and component production for international markets. Because relationships between international corporations and domestic subcontractors in developing nations generally involve a strong imbalance of power, Gereffi et al. argue that the governments of developing nations should closely regulate such agreements in order to ensure that they are economically beneficial.

THE SMALL-FIRM SECTOR

Our discussion in this chapter has so far focused on the increasing importance of corporate entities that can swallow up individuals in large, monolithic organizations. This focus should not lead us to assume, however, that the small-firm sector is on the verge of extinction. Small companies come into existence every day, and though many of them also quickly pass out of existence, their aggregate effect on employment is quite significant. This is especially true in the service sector, which employs an increasing proportion of the labor force. The subcontracting relationships we have just discussed are also important in the ongoing reproduction of a small-firm sector. One factor that heightens the significance of small firms is that employment is relatively stagnant in large firms, in spite of the increasing proportion of production that takes place there—large firms typically use capital investment to increase productivity, and the employment-generating consequences of this strategy can be marginal or even negative (O'Connor, 1998).

Small firms are generally run by a single individual or family. Profits and retained earnings are typically lower than those in larger firms. Long-term borrowing is difficult and expensive, and the firms are heavily affected by local economic conditions. Production and marketing strategies may be outdated, and the firm generally maintains its competitive position by cost cutting rather than by brand-name recognition or by the other forms of competition available to large firms.

Such small firms, however, employ a significant share of the labor force and are thus an important part of the economy. The competition between small, medium, and large firms contributes significantly to the dynamism of the economy. The small firms sector also provides opportunities for those with entrepreneurial abilities and adds to the economy's potential for innovation and creativity (Keister, 2005).

Satellites, Loyal Opposition, and Free Agents

Three types of small firms exist in the economy: satellites, the "loyal opposition," and free agents (Averitt, 1968). **Satellite firms** engage in subcontracting relationships with larger firms, supplying either components or distribution and marketing services. Satellite firms may either be attached to a particular principal company or may "float," being dependent on sales to a certain industry but not to any particular firm in that industry. Companies supplying glass, upholstery, and electrical components to the automobile industry are good examples of satellite firms. **Loyal opposition firms** provide competition to larger firms in their own industries. Such companies rely heavily on local sources of supply, have technically inferior equipment, have limited access to foreign markets, surrender the power to set prices to the larger firms, and as a group have lower profits than larger firms. American Motors occupied such a position relative to Ford, Chrysler, and General Motors for decades. (In the early 1980s the remnants of American Motors were divided between the French automobile manufacturer Renault and Chrysler.) **Free agents** are a diverse group of firms that spring up in the nooks and crannies between large firms. In manufacturing, free agents often specialize in small-batch production. In retail trade and services they serve markets in which it would be difficult for larger firms to achieve economies of scale. The economic niches occupied by free agents typically have too low a profit rate to attract larger firms. Specialty tool shops, repair services, and residential construction firms are all examples of free agents.

How Important Is the Small Firm Sector? Simply counting the number of firms in the economy would indicate that small firms are of overwhelming importance (Freel, 2003). Eighty-five percent of U.S. businesses have fewer than twenty employees (Census, 2005). However, recall that in earlier sections on the increasing concentration of corporate assets, we arrived at different conclusions. Large firms control the greatest share of productive assets in industrially advanced societies and are responsible for most economic production. How are we to resolve this seeming anomaly? What are the economies of industrially advanced nations actually like? Are they dominated by a few large corporations or by many small firms? By looking at the distribution of employment, we will be able to arrive at a clearer answer to these important questions.

The proportion of manufacturing employees in establishments smaller than one hundred employees declined from 39.2 percent in 1904 to 27.4 percent in 2005. An establishment is a plant or other place of employment (see Chapter 2). But as you have learned, employment growth has increasingly shifted toward services and wholesale and retail trade over this period. In these sectors small firms are more typical, and the decline in their role has been slower. In services, for example, the proportion of employees in establishments with fewer

than one hundred employees decreased from 83.5 percent in 1948 to 62.4 percent in 2005. In the economy as a whole approximately 50 percent of the labor force is employed in establishments with fewer than one hundred employees, and 50 percent is employed in establishments with more than one hundred employees (Census, 2005).

Different Places, Same Company? These figures are for employment in different physical *establishments*. Such figures, however, do not take into account the reality of **multiplant companies**. For example, individual Wal-Mart discount stores may have only a few dozen employees at any one establishment, but Wal-Mart has tens of thousands of employees nationwide. Nor do these figures take into account the type of concentration in which one company owns the production facilities of other subsidiary companies. Thus, in manufacturing only 16.2 percent of employees actually work in *firms* with fewer than one hundred employees. In the economy as a whole, while approximately 50 percent of the labor force is employed in *establishments* with fewer than one hundred employees, only 40 percent work in firms with fewer than one hundred employees (Granovetter, 2000).

How many workers are employed in the largest industrial corporations? The list of the largest U.S. corporations published in *Fortune* magazine provides a common frame of reference. If we include the top 1,500 *Fortune* companies (identified by total revenue)—made up of the top 1,000 industrial companies; the top one hundred firms in diversified services, commercial banks, and diversified financial companies; and the top fifty firms in life insurance, retail, transportation, and public utilities—approximately a third of the labor force is employed in the 1,500 largest industrial, financial, and service firms (Hodson, 1983).

What do these figures tell us about the organizational locus of work in modern society? There are a great many small firms in industrially advanced societies. Assets and sales, however, are concentrated in the very largest corporations. Approximately a third of the labor force is employed in small firms (those with fewer than 100 employees).

Another third is employed in intermediate-sized firms (those with between 100 and 5,000 employees). The final third is employed in large firms (such as those identified in the *Fortune* list). Thus, employment positions are spread relatively evenly across the three types of organizations. Many workers are employed by one of the huge corporate giants. Numerous others work in one of the small firms that fill the niches between these giants. And many others are employed in intermediate-sized firms.

The Creation of New Jobs

In recent years the news media have given a great deal of attention to the importance of small firms in creating jobs. Much of this discussion was sparked by research by economist David Birch (1987). His most widely cited conclusion was that firms with fewer than twenty employees create two-thirds of the new jobs in the U.S. economy. We must interpret this conclusion with caution, however. Many of these jobs disappear just as quickly as they are created. Accordingly, some of these jobs represent a circulation of workers among small firms as new firms are born, struggle, and die rather than a lasting contribution to the supply of available jobs (Bednarzik, 2000).

Small firms create many jobs. Some of these jobs, however, are of short duration; many others are relatively marginal and offer only low wages and meager or non-existent benefits. Seasonal hiring in small retail stores of part-time or temporary workers provides a good example of this latter type of job.

Economic Revitalization

As researchers and policymakers have increased their awareness of the continuing importance of small firms in creating and maintaining employment, many have expressed hopes that this sector will provide a solution to the lingering economic doldrums of the twentieth-first century. Charles Sabel and Jonathan Zeitlin (2004), for instance, argue that production strategies pursued in the past by large firms in the industrially advanced nations are no longer viable. The cornerstone of past strategies was the production of large numbers of standardized products through the use of product-specific equipment. This strategy is sometimes called **Fordism** because of its successful use by Henry Ford to revolutionize the automobile industry. Over time, however, this strategy tends to deskill workers, rendering industries based on Fordism attractive to low-wage competitors such as South Korea, Taiwan, Brazil, and Mexico. International competition over markets for such mass-produced goods has eroded the once dominant position of the North American and West European nations in many industries.

The solution proposed by Sabel and Zeitlin and others is that the industrially advanced nations begin to concentrate on specialized products with high value added that imitators find difficult to copy. Sabel and Zeitlin offer an example of an existing system of this type in central and northeastern Italy. Many small factories employing from five to fifty people have sprung up in this region. These shops specialize in the production of custom textiles, automatic machines, machine tools, and specialty automobiles, buses, and agricultural equipment. They have been able to escape the role of being subcontractors dependent on larger firms through diversifying their marketing operations to include other small producers in the local area and in the broader European market. As a result they have altered their status to become independent innovators, developing new products for a wider market. In the past the subcontractor's customers arrived with a blueprint to execute; now they arrive with a problem for the subcontractor to solve. The magnitude of potential employment growth from such specialization is open to debate, but such strategies are becoming an important part of development plans in many industrialized nations.

Size and Innovation Part of the optimism concerning small firms results from their demonstrated ability to innovate more quickly and less expensively than large bureaucratic organizations. The National Science Foundation finds that small firms

T A B L E 15.2   Some Important Innovations by Independent Inventors and Small Organizations

Invention	Inventor
Xerography	Chester Carlson
Insulin	Frederick Banting
Vacuum tube	Lee De Forest
Penicillin	Alexander Fleming
Cyclotron	Ernest O. Lawrence
Catalytic cracking of petroleum	Eugene Houdry
Automatic transmission	H. F. Hobbs
Jet engine	Frank Whittle and Hans von Ohain
Frequency modulation radio	Edwin Armstrong
Helicopter	Juan De La Cierva, Heinrich Focke, and Igor Sikorsky
Mercury dry cell	Samuel Ruben
Kodachrome	L. Mannes and L. Godowsky, Jr.
Air conditioning	Willis Carrier
Polaroid camera	Edwin Land
Tungsten carbide	Karl Schroeter
Oxygen steel-making process	C. V. Schwarz, J. Miles, and R. Durrer
Video games	Noland Bushnel
Artificial heart	Robert Jarvik
Solar powered car	Hans Tholstrup and Larry Perkins
Computer chip	Jack Kirby
Oral contraceptive	Gregory Pincus
Laser disc	David Paul Gregg
Microprocessor	Marcian Hoff and Frederico Faggin
Human in vitro fertilization	Patrick Christopher Steptoe
Internet	multiple developers

SOURCE: U.S. Department of Commerce, 1967, *Technological Innovations: Its Environment and Management.* Washington, DC: U.S. Government Printing Office. Also, Kevin Desmond, 1986, *Inventions Innovations Discoveries,* London: Constable; and Rodney Carlisle, 2004, *Scientific American Inventions and Discoveries,* Hoboken, NJ: John Wiley, p. 401.

are about six times more effective in creating technological innovations than are large firms (U.S. Senate, 1986). A frequently cited example of this phenomenon comes from the steel industry:

> The oxygen converter is one of the most important advances ever developed in steelmaking. It was invented and used in Europe prior to the Second World War. It was introduced in the United States in the 1950s—not by big firms, but by McLough Steel, a small independent firm based in Detroit. A comparable case is continuous casting, a revolutionary process in steel production, which was introduced by ninth-ranked Allegheny Ludlum. (Reid, 1976:50)

Some other important innovations made by individuals and small firms are listed in Table 15.2.

Public policy encouraging small firms as a source of innovation have gained increasing support in recent years. Great hopes are being placed on this sector and on the belief that "decentralization is the great facilitator of social change" (Naisbitt, 1999). A contemporary example is provided by steel mini-mills that convert scrap metal into specialty steel products. These are being proclaimed as at least a partial answer to the decline of the U.S. steel industry. (See Chapter 8 for a discussion of the role of these mills in the U.S. steel industry.) Retaining high levels of innovation in an economy dominated by giant, globally interconnected and highly bureaucratic corporations is becoming an important challenge facing contemporary societies.

SUMMARY

Large corporations exert great power in relation to workers, their host communities, and smaller companies. In recent years they have further increased their power through mergers, expansion into diverse product lines, interlocking boards of directors, and subcontracting.

The prevalence of large, diversified corporations and the rapid pace of merger activity have many potential negative consequences for workers and communities. Workers may lose jobs without warning and on the basis of decisions in which they had no role and about which they have little or no knowledge. Similarly, the use of subcontracting may lessen the power of workers and communities and increase their vulnerability to shifts in corporate strategies.

Large-scale production in huge corporations will continue to be the dominant way in which most manufactured goods are produced. However, it also seems likely that the small-firm sector will continue to be reproduced in the niches between these corporate giants. Despite its limited productivity, this sector employs a significant number of workers. Whether it can produce technological and organizational innovations that will increase its relative share of production seems less certain. Its significance, however, should not be underestimated; it will continue to play an important role in providing jobs, goods, and services. The fact that a third of the labor force works for firms with fewer than one hundred employees should caution us not to assume that the experiences of workers in the largest corporations are the only or most significant type of work experience in industrially advanced nations.

KEY CONCEPTS

corporate economic power

corporate political power

corporate inflexibility

monopoly

oligopoly

backward linkages

forward linkages

vertical integration

predatory pricing

concentration ratio

limited liability

management-controlled firms

corporate mergers

diversification

cash cow

corporate divisions

lean production

golden parachutes

whistle-blowing

subcontracting

satellite firms

loyal opposition firms

free agent firms

multiplant companies

Fordism

QUESTIONS FOR THOUGHT

1. Identify a large economic enterprise in your community. Research this enterprise and find out as much as you can about who owns it and about the ownership and control linkages between this enterprise and other enterprises. Is this enterprise a subcontractor to some other enterprise? Does it subcontract out any of its own work or support services?

2. What factors encourage the increased size of organizations? What factors, if any, limit organizational size?

3. In what ways do large corporations exercise power over their employees? Over competitors? Over the government? Over communities?

4. Draw a time-line depicting the history of corporate mergers. Note the different types of firms emerging during the different periods.

5. What purposes are served by limited liability for the owners of corporations? Write an essay either defending or criticizing limited liability. How do the sanctions levied against the directors of Trans Union relate to limited liability?

6. How important do you think the small-firm sector will be in the future of advanced industrial society? Would you prefer to work in a small firm, a medium-sized firm, or a large firm?

MULTIMEDIA RESOURCES

Print

Michael Moore. 1997. *Downsize This!* New York: Harper. A witty and disturbing look at the state of the American economy and the values guiding it.

Fligstein, Neil. 2001. *The Architecture of Markets.* Princeton, NJ: Princeton University Press. A penetrating analysis of the influence of financial managers on the economy and on modern society.

Barry Bluestone and Bennett Harrison. 2000. *Growing Prosperity: The Battle for Growth with Equity in the Twenty-First Century.* Boston: Houghton Mifflin. A critique of the economic inefficiencies caused by large corporations. Includes proposals for revitalizing the economy based on distributing control to workers and communities.

Eric Mann. 1987. *Taking on General Motors.* Los Angeles, CA: Center for Labor Research and Education, Institute of Industrial Relations, University of California. A first-hand case study of the union and community effort to keep the GM Van Nuys plant open.

Sanford M. Jacoby. 2004. *Employing Bureaucracy: Managers, Unions, and the Transformation of Work in the 20th Century.* Mahwah, NJ: Lawrence Erlbaum.

A revealing analysis of the rise of the modern economy dominated by large corporations.

Internet

Wal-Mart Watch. *walmartwatch.com* Devoted to supporting the struggles of employees, citizens, consumers, and suppliers to deal with power to Wal-Mart to set wages, reduce benefits, and pressure communities.

Power Structure Research. *darkwing.uoregon.edu/~vburris/whorules* An Internet guide to contemporary social science research on the concentration of economic power and its consequences.

Interlocking Corporate Directorates. *www.theyrule.net* An interactive site that allows users to visualize the interlocks between the largest corporations.

Financial Scandals around the World. *www.ex.ac.uk/~RDavies/arian/scandals* Extensive reports on both business and government fraud and financial scandals.

Forbes Magazine. www.forbes.com A popular magazine devoted to business issues from the perspective of business owners.

Small Business Administration. *www.sbaonline.sba.gov*
     The key federal agency providing support for small
     businesses.

National Business Incubator Association. *www.nbia.org*
     An independent association devoted to helping new
     businesses succeed.

RECOMMENDED FILMS

Enron: The Smartest Guys in the Room (2005). A highly
     acclaimed (and disturbing) documentary on one of
     the largest corporate scandals in American history in
     which top executives from the seventh largest
     company in the country plundered over a billion
     dollars from investors, employees, and retirees.

Wal-Mart: The High Cost of Low Price (2005). Tells the
     story of how Wal-Mart directly pressures its sup-
     pliers to lower their prices below what they can
     afford, resulting in forced movement to lower-wage
     locations in order to survive.

16

Globalization

Three times a day scores of small buses make the six-mile trip north from
Saltillo's working-class neighborhoods to Ramos Arizpe where both
General Motors and Chrysler operate huge assembly plants. At Chrysler,
three shifts per day assemble motors for shipment to the U.S. and other
Chrysler plants in Mexico. At General Motors, less than a mile away,
shift workers turn out complete cars, also for shipment north. Autoworkers
here buy reduced priced meals in the plants, ride free company buses, labor
their way through the seven-step plant hierarchies, and enjoy up to two
weeks of vacation per year. They study family planning, collaborative
working, and a variety of technical mini-courses offered within the plants.
It's almost like working in Detroit. But here the air is clear and the wage
scale about three dollars for a ten-hour day.

(DUBOSE, 1987)

Today we live in a vast, integrated world economy. Workers in one part of the
world make products consumed by workers in others parts of the world. And
because production facilities are increasingly mobile, workers also compete for
the same jobs. This interdependence has been growing. Before the nineteenth
century most goods were produced and consumed locally. By the twentieth
century, increasing numbers of people were producing goods destined for
national, or even international, markets and were consuming goods that had
been produced in other parts of the world. In the twenty-first century these
trends have become an overwhelming reality.

This chapter provides a brief history and overview of the world economy. It
highlights the emergence of world economic dominance by the United States
following the Second World War and the growth of international competition in
the latter decades of the twentieth century. In the second half of the chapter we

examine international differences in how countries organize their economies to confront the rapidly changing global economy. The themes developed in previous chapters concerning the division of labor, organization, technology, trade unions, gender, race, and marginality provide important building blocks for understanding the modern global economy.

In an era of increased international competition, it seems that workers and nations have to work ever faster and harder to maintain their positions as producers of goods and services (Friedman, 2005). In this context, some forces favor already-developed nations, and other forces favor newly developing nations. This constellation of forces produces both winners and losers in the global economy. In recent decades several Pacific Rim nations, such as China and Singapore, have moved forward dramatically on the world economic stage. Understanding the forces leading to success or failure in the global economy is crucial for all workers and nations today.

THEORIES OF INDUSTRIAL DEVELOPMENT

Three major theories have been advanced to explain industrialization—the process through which societies change from a primary reliance on agricultural production to include a greater role for manufacturing and services. We discuss in turn each of these theories: *modernization* theory, *dependency* theory, and *world systems* theory.

Modernization   The earliest theory, **modernization theory**, argues that societies advance from agricultural production to industrial production based largely on *internal dynamics* (Rostow, 2003). These internal dynamics, such as technological advances and rising domestic consumption, help transform economies grounded in agricultural production for local consumption into economies grounded in urban manufacturing for regional and world markets. In this view countries specialize in a certain commodity because of a comparative advantage, such as naturally occurring resources or a conducive climate or location.

This theory of development emphasizes domestic needs and technological advances. Its greatest relevance, however, is for the earliest nations to industrialize—the Netherlands and England. And even for these nations, the theory overlooks the crucial role of international markets in the expansion stage of industrialization. All later developing societies have had to pursue the process of industrialization in competition with existing industrialized societies. Modernization theory tends to overlook the early emergence of a world economy and the incorporation of virtually all nations into that economy, either as manufacturing nations or as suppliers of raw materials.

Table 16.1 lists some of the largest nations of the world ranked by their gross national product (GNP) per capita, which is closely correlated with a country's level of industrial development. Note the extreme range of the inequality between nations and the important role of the timing of industrialization in determining a nation's fate. You might find it instructive to compare this table with Table 4.1, which reports the salaries of the highest paid U.S. executives.

Dependency   The unequal power relationships between industrialized and less industrialized countries are the central focus in a theory of economic development called **dependency theory**. This theory notes that many less-industrially developed nations have long been incorporated in the world economy but in a dependent role. Thus, Cuba was incorporated in the world economy for nearly three hundred years as a sugar-growing colony and

TABLE 16.1 Per Capita GNP Around the World

Countries	U.S. Dollars	Countries	U.S. Dollars
Low-Income Economies		South Africa	2,500
Ethiopia	100	Brazil	2,830
Nigeria	180	Malaysia	3,540
Uganda	240	Venezuela	4,080
Tanzania	290	Argentina	4,220
Kenya	360	Chile	4,250
Sudan	370	Poland	4,570
Bangladesh	380	Hungary	5,290
Pakistan	420	Czech Republic	5,480
Vietnam	430	Mexico	5,920
India	470	**Richest Economies**	
Indonesia	710	Portugal	10,720
Sri Lanka	850	Greece	11,660
China	960	Spain	14,580
Lower Middle-Income Economies		Italy	19,080
Philippines	1,030	Australia	19,539
Syria	1,130	Singapore	20,690
Morocco	1,170	France	22,240
Belarus	1,360	Belgium	22,290
Egypt	1,470	Canada	22,390
Algeria	1,720	Germany	22,740
Iran	1,720	Netherlands	23,390
Bulgaria	1,770	Hong Kong	24,690
Colombia	1,820	United Kingdom	25,510
Romania	1,870	Sweden	25,970
Upper Middle-Income Economies		Japan	34,010
Thailand	2,000	United States	35,400
Peru	2,020	Switzerland	36,170
Russia	2,130		
Turkey	2,490		

SOURCE: U.S. Department of Commerce, Bureau of the Census, 2005, *Statistical Abstract of the U.S.*, p. 852.

has only recently begun to develop a more diversified industrial structure. Similarly, Honduras has long participated in the world economy, but in the limited role of a banana-producing economy.

According to dependency theory, less industrially developed nations do not have the opportunity to advance from agricultural societies to industrial societies as dictated by their internal dynamics.

Rather, their role as less developed **peripheral nations** is maintained by their relationship to the industrially advanced **core nations**, which buy raw materials from them and sell them manufactured goods in return (Cohen and McBride, 2003). This dependency operates in a variety of ways. Along the west coast of Africa, for example, railroad lines were built under the direction of European colonial rulers. These rail lines, however, all run inland rather than along the coast. This rail network thus facilitates exploitation of the raw resources of Africa for export but limits the development of regional trade ties along the coast. The industrial infrastructure of West Africa thus supports the dependent relationship of this region to the world economy, while contributing only minimally to local development.

Additional evidence for dependency theory comes from analysis of data on world capital investments, exports, and economic growth. Export-oriented development has a positive short-term effect on economic growth rates in developing nations. However, on average, this effect is reversed when the country has large amounts of foreign investment, when it is engaged in exporting a raw material, or when the prices of its exports fluctuate drastically (Jaffee, Kloppenburg, and Monroy, 2004). These negative effects occur because foreign investment distorts economic infrastructure, as in the example of railroads in West Africa. In addition, education and professional services often remain underdeveloped. Thus, being involved in the world export trade may help countries develop, but this one-sided "development" maintains their dependent status as suppliers of raw material or as subcontractors in larger manufacturing processes.

World Systems The most recent theoretical advances in the study of economic development have come from **world systems theory**. This theory builds on the notion of dependency between core and peripheral nations by adding the insight that the world economy is and has long been an *integrated economic system*. Proponents of this theory reject any notion of a necessary development of nations from agriculture to manufacturing that is either inevitable (as proposed by modernization theory) or actively blocked by already industrialized countries (as proposed by dependency theory). Instead, they focus on how the roles of nations in the world system are constantly reproduced by existing trading relations (Wallerstein, 2003). World systems theory also highlights the emergence of a "semiperiphery" of developing nations involved in the manufacture of mass production goods and subcomponents. (The countries in the semiperiphery are also referred to as "developing nations.") World systems theory identifies low levels of capital investment as central to understanding continuing underdevelopment, especially in Africa and parts of Central America.

A final insight of world system theory is that the world economy involves not only the movement of goods but also the large-scale migration of people. The European migrations to the East Coast of the Americas in the 1880s and 1920s and the Chinese migration to the West Coast of the Americas in the 1870s are well-known historic examples of such economically inspired population movements. Two major flows of population have been occurring in the world system in more recent decades. One of these flows is between the periphery and the semiperiphery and involves the migration of workers to countries with large oil exports and to cities with rapidly growing industrial bases producing for the world market, such as Singapore and Djakarta, Indonesia. The second major migration is from Asia, South America, and the Caribbean Basin to Western Europe and the United States (Portes, 2001). The example of Germany's "guest worker" program is presented in Box 16.1.

Emergence of an Integrated Global Economy

The first elements of an integrated world economy emerged in the 1600s with the development of the Dutch trading empire. The principal markets for Dutch goods were other European countries and the civilizations of North Africa. The Netherlands became a manufacturing center at this time, with over 50 percent of its population living in urban places (Wallerstein, 2003). The Dutch not only engaged in trade but also, along with Spain, England, and the other imperial nations, seized immense amounts of wealth from their colonial empire in

---

BOX 16.1 Germany's "Guest Workers"

Labor and capital are both mobile—jobs move to cheap labor, and people in need of work move to jobs. One of the most famous, and possibly infamous, examples of an "arranged marriage" between unemployed workers and unfilled jobs is Germany's *Gastarbeiter* program. Starting in the 1960s, because of labor shortages in Germany's booming economy, arrangements were made allowing workers from other countries to enter Germany "temporarily" as guest workers. The assumption was that these workers would return to their home nations within a few years, and other workers would replace them as needed by fluctuating economic demands. Workers moved to Germany in waves, starting with Italy, Greece, Spain, Portugal, Yugoslavia, and later Turkey. The program was sharply curtailed following the oil crisis of 1973 and the resulting economic downturn, although it still continues even today in reduced form. By the 2000s, 7.3 million non-citizens lived in Germany as a result of these programs, with increasing numbers of them born there to parents who had lived in Germany for decades. The two largest groups are Turks (1.9 million) and citizens of the former Yugoslavia (1.1 million).

The *Gastarbeiter* program has produced enduring problems for Germany, and for its guest workers. The underlying source of these problems is that the jobs filled—often low-wage assembly jobs in manufacturing—were not "temporary," even if the workers were intended to be. They needed to be filled on a permanent basis, and the workers, having moved their lives to Germany, were not eager to return home and seek work in a place they had left because of lack of work. This reality created a large caste of people living in Germany on a more-or-less permanent basis, who, nevertheless held less than the full rights of citizenship. Acquiring German citizenship required *fifteen years residence* and renunciation of one's prior citizenship. In 2004 Germany reformed its *Gastarbeiter* legislation, reducing the requirement for citizenship to "only" eight years. If family members attempt to join the worker,

they are prohibited from holding work permits for five years, offering a significant barrier to family unification. After working for thirty-five years in Germany, a guest worker lost his job and applied for unemployment benefits—something forbidden by the guest worker legislation. The worker was subsequently deported to Turkey. His younger brother was shocked by the treatment: "He has never been convicted of any crime, he worked and paid taxes. His brothers and sisters, nieces and nephews live here, and his twenty-five-year-old daughter. He is a stranger in Turkey" (Kent, 2005:2).

The guest worker program has resulted in chronically high rates of poverty, unemployment, crime, victimization, and illiteracy among the workers who moved to Germany. In economic downturns these are the first workers laid off and the last rehired. Their legal status as temporary workers has significantly aggravated these problems. Many workers are unmotivated to assimilate and speak or read little, if any, German. The younger generation born to these workers has a high rate of dropping out from school. Significant numbers of these disenfranchised youths subscribe to hip hop and other youth styles glamorizing rejection of German culture.

Not all guest workers have suffered under the program. A Turkish guest worker reports: "I thought I would be here for ten or fifteen years. But I have saved a little money, and now I have my own business and I'm happy. Things are good." The head of the German Turkish Association notes: "We came here as guest workers and forty years later we are still guest workers. But it will change—the third generation will be German" (Broomby, 2001:1–2).

The underlying message most observers take from Germany's guest worker program is that insisting on temporary status for immigrant workers only delays the inevitable process of integration—a delay with significant costs in terms of human suffering and social problems. Labor and capital may both be mobile, but labor generally moves only once every few generations and ties to the "homeland" are easily lost and not necessarily retrievable.

Africa, the Americas, and the Pacific in the form of precious metals, trade goods, and human slaves. This wealth helped underwrite the expansion of European manufacturing. Marx (1967 [1887]) called this process **primitive accumulation** and considered it a necessary step in the development of capitalism.

The British Empire The British colonial empire held the center stage in the world economy of the 1800s. Britain's colonies were incorporated into this economy as producers of raw products (such as cotton, sugar, rum, tobacco, and rubber) for the homeland and as consumers of industrial products from

British factories (textiles and, later, steel and other manufactured products). The subordinate role of new territories in the British Empire frequently sparked resentment among colonists and indigenous populations because of the unfavorable terms of trade imposed. These terms included high taxes, inflated import prices or high import duties, and low prices paid for commodities exported from the colonies. The "Boston Tea Party" and related episodes leading to the American Revolution provide early examples of angry resistance to unfavorable terms of trade.

Not only did Britain and the other imperialist nations impose unfavorable terms of trade on their colonies, but they also dismantled indigenous industries to open up markets for their own manufactured goods. For instance, loom operators in India who had supplied textiles to Asian and European markets in the 1600s and 1700s were put out of business when the British Parliament in the early 1800s restricted Indian producers to growing raw cotton and stipulated that textiles could be manufactured only in Britain. Later, when this restriction was removed, Indian textiles were subject to high tariffs when exported to Britain, whereas goods made in Britain were admitted into India with little or no import duty (Hobsbawm, 1996).

Specialization Among Less Developed Nations

In their roles as suppliers of raw materials, the colonies of the European nations often became highly specialized. Subsistence agriculture for local consumption was displaced by production of a more restricted range of agricultural products for export. Sometimes whole regions specialized in providing just one product. The American South became primarily a cotton-producing area, and Jamaica and the West Indies primarily produced sugar. As railroads advanced into the interiors of these countries, agricultural production for local consumption was displaced by agricultural production for the export market. Thus, from the very earliest stages of the Industrial Revolution, these nonindustrialized regions played a major role in the world economy, but it was a subordinate role based on the greater economic and military might of the imperial powers.

The transformation from subsistence agriculture to the production of agricultural commodities for the imperial powers was accompanied by a removal of small producers from the land and the concentration of land ownership in large plantations. In Central and South America these plantations were known as *latifundia*. Peasants were transformed by this process from independent producers of agricultural products for personal or local consumption to dependent hirelings (or slaves) of landowners with large holdings. Just as the enclosure movement in England, which we described in Chapter 1, was a long and violent process, so too was the process of centralizing land ownership in the colonies. The labor force for the production of export crops was secured primarily through subjugating native people to various forms of peonage, such as tenant farming, sharecropping, and wage labor, and through importing slave labor. Colonists migrating from the imperial homeland in the search for new land also played a major role in the production of export commodities.

A Captive Market The Industrial Revolution resulted in a dramatic increase in manufacturing productivity in England. The British colonial empire played an essential role in this industrial revolution. Without world markets for the goods mass-produced in England, there would have been much less demand spurring on the development of modern industrial techniques and machines (Hobsbawm, 1975). British industrial productivity rose most dramatically around the mid-1800s. British manufactured goods were the essential medium through which the world economy came to penetrate all the continents of the globe. At the height of Britain's colonial power in the second half of the nineteenth century, its armies and navies controlled the world's seas and a sizable proportion of its inhabited territories. These military forces ensured the ability of Britain to sell its manufactured goods around the world and to purchase raw materials at favorable prices.

Dominance by Britain and the other European nations over their Asian and African colonies continued until the end of the Second World War. The military destruction of the industrial base of Japan and the European nations during the war resulted in a temporary decline in the economic and military power of these nations. No longer able to police worldwide empires, the industrialized nations began to lose control of their previously dependent colonies. Liberation movements demanding national autonomy sprang up in Africa, Asia, the Caribbean, and the Pacific Islands. Almost immediately after the Second World War, France, attempting to hold on to its colonies, became embroiled in disastrous wars in Vietnam and Algeria. Similarly, Britain became involved in conflicts in its former colonies of Malaysia and India, and Belgium became embroiled in conflict in Zaire (formerly the Belgium Congo). Other European nations, realizing their inability to police colonial empires, allowed the transition to political independence to take place more peacefully. The next two decades saw the sometimes peaceful, but often violent, transformation of former colonies into politically independent nations.

The Rise of the United States and Neo-Imperialism In terms of military and civilian casualties and economic devastation, the United States fared better in the Second World War than any other major combatant. At the end of the war it had the only intact industrial apparatus in the world and possessed the largest standing army and navy. The United States was also a creditor to most of the industrial nations, who were heavily in debt for the purchase of American military goods, foodstuffs, and manufactured goods during the war.

The United States was thus in a unique position in the world economy in the 1950s. In some ways its position was similar to that of England at the beginning of the Industrial Revolution. However, the global U.S. empire was based more on economic dominance than on direct political or military control over dependent areas, though these more direct means were used as well. Historians and sociologists give the name **neo-imperialism** to this new version of the world economy in which less-industrialized nations possess at least nominal political independence (Hobsbawm, 1996).

The United States also worked selectively to encourage reindustrialization following the Second

World War in nations it identified as potential allies against the spread of Communism. Economic development policies, such as the Marshall Plan, facilitated the reindustrialization of Germany and Japan, as well as many other nations, with the indirect consequence of sharply limiting the period of uncontested American economic dominance in the world economy.

Neo-imperialism, with the United States at its head, remained unchallenged for only about two decades. During these decades the economies of Europe and Japan were rebuilding, and they began to compete with the United States. Also, some less developed countries had at least limited success in undergoing their own processes of industrialization and provided additional competition in the world market for manufactured goods. Many less developed countries also began to supply some of their own local needs for manufactured goods, thus reducing their dependence on the world market for these goods. Other regions organized into **cartels** to reduce competition among themselves over the pricing of their exports. A *cartel* is a combination of political or economic organizations acting to limit price competition among themselves. The Organization of Petroleum Exporting Countries (OPEC) provides the most successful and widely known example of a cartel pricing arrangement. Similar cartels emerged for regulating the price of such commodities as coffee and cacao. Thus, the world economy at the beginning of the twenty-first century is more complex and diverse than during any prior period in history. It includes large numbers of *industrialized nations, industrializing nations,* and *less-industrialized nations,* many of which are either actively trying to industrialize or are at least trying to protect the prices of their raw export products.

THE WORLD ECONOMY TODAY

Based on the above history and theoretical framework, we can now examine the contemporary world economy and see what is uniquely emergent at this point in history. In the following sections we highlight the important role of multinational corporations in the global economy. We also discuss the causes of slowed rates of economic growth in the industrialized nations and current challenges to development in the less developed nations.

The Role of Multinational Corporations

So far we have discussed the world economy as if international trade were merely a matter of relations between nations. In fact, the world economy is to a large extent orchestrated by the heads of **multinational corporations** (MNCs) rather than by the heads of governments. A MNC is a firm that operates in many countries. MNCs typically have their headquarters in one of the industrially developed countries, but many of their plants are located in less developed nations. For example, for decades IBM's popular personal computer was manufactured in China. In 2005 these production facilities were sold to a Chinese company. IBM now simply sources its "IBM personal computers" from this company. Similarly, the Ford Escort is assembled primarily from parts made outside the United States, including parts made in Austria, Belgium, Britain, Canada, Denmark, France, West Germany, Italy, North Ireland, Japan, Netherlands, Norway, Spain, Sweden, and Switzerland (Bluestone and Harrison, 2000). By the 2000s the sales of U.S.-owned firms abroad were three times the value of United States exports. That is, three-fourths of production by U.S. firms for the world market takes place outside the United States.

The United States is not the only home base for large multinational corporations eager to expand their operations. Japanese automobile makers have also dispersed their production and assembly operations around the world. North America has become an important destination for Japanese automobile assembly and parts plants, many of which were started as joint ventures with North American automakers (Perrucci and Wysong, 2003).

The powerful role of multinational corporations in the world today is based on a high consumption lifestyle for the world's industrialized nations and aspirations for increasing consumption in the less industrialized nations. Without high levels of consumption, fewer goods and services would be mass marketed, and there would be more room for local producers. This reality leads many to implicate MNCs in the world's environmental problems as well as its economic problems (Korten, 1996).

Power without Boundaries   Many social scientists are concerned about the concentration of economic power in the hands of the leaders of a few global corporations. The two hundred largest global corporations control over 80 percent of the assets of the Western world. For the first time in history, there exists an elite group of owners and managers who potentially have the power to manage the world as an integrated unit (Cavanagh and Mander, 2004). The centers from which the world economy is managed are the headquarters of the huge multinational corporations located in such cities as New York, Tokyo, London, Munich, and Paris.

According to many sociologists, the danger that these unimaginably large firms present is based not only on their size but also on their ability to make decisions independently of the nations in which they operate. "The capacity of any government to command a particular firm to undertake a specified task in support of a public policy, such as settling in a backward region or holding down a key price, has been reduced; large firms now have a capacity that they never had before for choice between competing nations" (Vernon, 1998:136–137). Table 16.2 lists the twenty largest MNCs. Note that this list of the very largest global companies contains a mix of manufacturing, energy, and finance. Eight of the top twenty MNCs are headquartered in the United States. Germany, Japan, the United Kingdom, France, the Netherlands, and Belgium also have companies in this elite group.

Multinational corporations sometimes benefit less industrialized nations by transferring technology and expertise to them. The technology transferred, however, is often outdated. In addition, managers and professional workers are more likely to be transferred from the home country than to be recruited locally. Technology transfers thus often fail to significantly

TABLE 16.2 The World's 20 Largest Multinational Corporations

2006 Rank	Corporation	Country of Origin	Sales
1	Exxon Mobil	United States	$339,938,000,000
2	Wal-Mart	United States	315,654,000,000
3	Royal Dutch-Shell	Netherlands	306,731,000,000
4	British Petroleum	United Kingdom	267,600,000,000
5	General Motors	United States	192,604,000,000
6	Chevron	United States	189,481,000,000
7	Daimler-Chrysler	Germany	186,106,000,000
8	Toyota Motor	Japan	185,805,000,000
9	Ford Motor	Japan	177,210,000,000
10	Conoco-Phillips	United States	166,683,000,000
11	General Electric	United States	157,153,000,000
12	Total Energy	Belgium	152,361,000,000
13	ING Insurance	Netherlands	138,235,000,000
14	Citigroup	United States	131,045,000,000
15	AXA Insurance	France	129,839,000,000
16	Allianz Insurance	Germany	121,406,000,000
17	Volkswagon	Germany	118,377,000,000
18	Fortis Finance	Belgium	112,351,000,000
19	Crédit Agricole	France	110,765,000,000
20	Amer. Intl. Insurance	United States	108,905,000,000

SOURCE: *www.fortune.com/fortune/global500* May 2006.

improve the employment prospects or the rate of economic growth in the host country. The potential advantages of the technology and skill transfers that occur are often overshadowed by the loss of economic autonomy suffered by the less industrialized nation, by increased dependence on foreign technology, and by lost opportunities to cultivate the development of local infrastructure, technology, and skills (Dicken, 2003). The movement of plants to new host nations basically exports routinized, dangerous and poorly paid work to developing nations, while modestly expanding clean and well-paid work in core nations (Bornschier, 2005).

In the industrially advanced nations many researchers fear that the actions of multinational corporations create a more segmented job structure.

A small increase in managerial and professional jobs in an industrialized nation may occur as a result of the expansion of MNCs headquartered in that nation. As these companies ship jobs overseas in the search for cheaper labor, however, manufacturing workers lose employment and suffer deteriorating conditions (Dicken, 2003). Even in the industrially advanced nations, manual workers may come to work under conditions not unlike those of workers in less industrially developed countries.

Slowed Growth in the Industrialized Nations

The emergence of a more tightly integrated world economy at the beginning of the twenty-first century has brought with it a number of problems.

In the industrialized world, growth rates have slowed appreciably. The annual growth rate in the United States fell from 3.8 percent in the 1960s to 3 percent or less in subsequent decades (see Figure 16.1). Industrial growth in Japan was once triple that of the United States, but it, too, has dropped dramatically following a collapse of inflated real estate prices and inflated values for productive assets. Growth rates in Germany and France have also steadily declined. Growth rates in the United Kingdom were the lowest of any of the great powers in the early decades following the Second World War but have rebounded toward a level more comparable to the other industrial nations.

These broadly declining growth rates result from increased international competition and from the increased ability of MNCs to move around the world in the search for cheaper labor and more lenient environmental regulations.

The End of U.S. Economic Dominance

The United States remains the largest producer of industrial goods in the world, but its lead has slipped dramatically since the 1960s. The United States no longer leads in automobiles, steel, railroads, or electronic goods. Between 1963 and 2000, the U.S. share of world manufacturing output dropped from 40.3 percent to 22.0 percent. Meanwhile, Japan's share increased from 5.5 percent to 21.4 percent (Dicken, 2003).

Why did the United States decline as the leading industrial nation in the world, especially given all its advantages of market power and technological leadership? One important reason is simply that the United States no longer dominates world markets as a sole supplier. The European nations and Japan have successfully recovered from the devastation of the Second World War and have become major competitors with the United States.

A second reason for the relative decline of the United States as an industrial leader is that its government spends a huge share of its research and development money on military research. Although military research has spin-offs that may increase productivity in some industries, such investments are

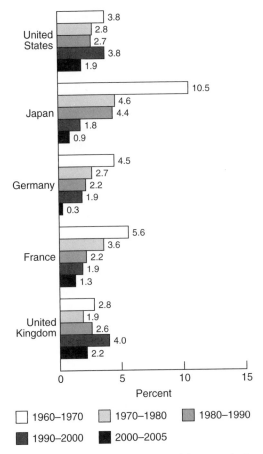

FIGURE 16.1 Declining World Economic Growth Rates (Average annual rates of change for Gross Domestic Products)

SOURCE: U.S. Department of Commerce, Bureau of the Census, 2006, *Statistical Abstract of the U.S.*, Table 1328, p. 874.

generally less economically beneficial than using research and development (R&D) monies directly for industrial research. In 2000 the United States spent 3.2 percent of its GNP on the military. West Germany spent 2.6 percent; Sweden, 2.4 percent; Switzerland, 2.1 percent; and Japan, 1.0 percent (Census, 2005). All of these countries have achieved growth rates in excess of that of the United States. Conversely, the former Soviet Union spent 11.7 percent—a burden that contributed to the Soviet Union's economic and political problems and to its eventual collapse.

A third cause for the declining international dominance of U.S.-made products is that American labor has been undercut by lower-priced labor in less industrialized countries. Even if workers are less productive in these countries, the sharp difference between the wages of workers in less industrially developed counties and those of workers in industrialized nations can make products manufactured in these countries competitive on international markets. One reason wages remain low in these nations is that workers there lack trade unions and have limited ability to strike for higher wages. Unions are typically outlawed or closely controlled by the government. MNCs also typically employ large numbers of female workers, who are paid even less than their male counterparts. Such extreme wage differentials pose a serious threat to the competitiveness of U.S.-made products and to the jobs of workers in the United States.

Even professional workers in industrialized nations run the risk of being undercut by lower-priced labor located elsewhere (Borjas, 2005). For example, U.S.-based Intel Corporation built a new microchip manufacturing plant in Ireland and hired exclusively Irish engineers at salaries only a fraction of those it pays its engineers in the United States (Zachary, 2003).

Protectionism, Free Trade, and Fair Trade

Protectionism—protecting domestic industry through high tariffs on foreign imports—is often proposed as a solution to the challenges of international competition. High tariffs decrease or eliminate the competitive advantage of products made elsewhere in the world. Such a policy would seem to have great potential for protecting troubled industries in the industrialized nations. Japan, China, and what have sometimes been called the "four little tigers"—Hong Kong, Taiwan, South Korea, and Singapore—all used tariff barriers to shelter their domestic industries from the world economy during crucial periods of development. High tariffs on imported manufactured goods were an important component in these countries' successful efforts to promote sustained economic development.

Many people fear, however, that increasing tariffs on foreign goods could provoke other nations to establish similar tariff barriers. High tariffs can thus protect some industries from international competition, but they can also negatively affect the health of other industries that are dependent on foreign markets for sales. Indeed, many historians believe that tariff wars in the 1930s increased the intensity and duration of the Great Depression. Starting in 1947, the General Agreement on Tariffs and Trade (GATT), involving most of the industrialized nations of the world, has slowly reduced tariffs on key goods traded around the world. The seventh (and final) round of the GATT negotiations, called the Uruguay Round, concluded in 1994 and resulted in average tariff reductions of 33 percent on manufactured goods and 36 percent on agricultural products ("World's Biggest Trade Deal," 1993).

In 1995 the World Trade Organization (WTO, *www.wto.org*) was established by representatives of the governments of the leading industrial nations as a permanent body to replace the periodic GATT meetings. Its mandate is to reduce tariffs worldwide, thus increasing trade between nations as a mechanism to promote growth and development. The goal of the WTO has been to encourage **free trade**— trade unrestricted by tariffs and by other governmental restrictions. This policy strategy has resulted in increased global trade and in increased development and employment worldwide. However, these policies have also been disastrous for selected industries in specific nations. These policies have also made it more difficult for nations to reject goods manufactured with child or prison labor or under unsafe conditions. The actions of the WTO have thus undermined the ability of nations to use their purchasing power to pressure other nations to adopt safer and more humane labor conditions and better environmental practices (Gereffi et al., 2002). As a result of this mixed record of achievements, the WTO has often been the object of criticism. Opposition to the actions of the WTO centers on the need for **fair trade** as well as more open markets. *Fair trade* is typically defined as bringing workers' rights, workplace safety, and environment regulations up to the standards enjoyed in industrialized nations rather than reducing them to the lowest common denominator (see also Chapter 8 on the global restructuring of manufacturing).

In addition to the reduction of global barriers to trade, many nations have also entered into regional trading partnerships to promote trade and development. These regional agreements provide additional opportunities to encourage trade and development and also to establish uniform and fair conditions for trade. As we will see, however, increased trade is often easier to negotiate than are protections for workers and the environment.

Trading Blocks: Regional Solutions to Lagging Growth

The governments of Europe, North America, and other world regions have sought to create regional trading blocks to lower tariffs and promote the exchange of goods. The European Economic Community (EEC) and the North American Free Trade Agreement (NAFTA) are examples of these efforts. Responses to these efforts at integration have been mixed and have included both great optimism and great trepidation. Proponents of these measures argue that the industrially dominant nations will find expanded markets for their goods with the lowering of trade barriers and that less industrialized nations will likewise find expanded markets for their products, leading to growth for all nations in the block. Opponents argue that, with the lowering of trade barriers, corporations will quickly move to the less industrialized nations in the block to lower their wage bills and escape environmental regulations at home. This movement of jobs will reduce employment in the core nations without necessarily raising the standard of living in the developing nations.

The primary motivation for the development of the EEC and NAFTA has been the desire to stimulate economic growth, but motivation has also come from a desire to reduce the flow of immigration from less industrialized nations to more industrialized nations within trading blocks. Proponents of trading blocks argue that by stimu-

lating growth in the peripheral nations, immigration to the core nations will be reduced. The EEC, operating under this theory, has developed explicit policies to promote economic convergence among EEC members (Metcalfe, 2000). Under NAFTA such outcomes are left up to the invisible hand of the market, in spite of a much wider divergence in GNP per capita between the United States and Mexico than between, for example, Germany and Portugal. Because NAFTA addresses only trade issues and fails to include safeguards for working conditions or the environment, unions in both the United States and Mexico have generally opposed it (Dreiling, 2000). The economic, political, and environmental consequences of the EEC and NAFTA will take decades to be realized, and positive outcomes may depend on safeguards still waiting to be introduced.

Combined and Uneven Development in Less Developed Nations

With increased involvement in the world economy, many less industrialized nations have experienced **combined and uneven development** in which the export sector is highly developed, but the domestic economy is underdeveloped. One important cause of combined and uneven development is the tendency of multinational corporations to borrow most of their investment capital locally. U.S.-based global corporations operating in Latin America are estimated to secure 87 percent of their financing locally (Cavanagh and Mander, 2004). Borrowing by multinational corporations absorbs resources that could otherwise be used for economic development of the domestic economy. Thus, the presence of MNCs contributes directly to the underdevelopment of the domestic economy. This process creates a few jobs in the capital-intensive export sector but at the expense of many more jobs that might have been created in the labor-intensive domestic consumption sector. For example, a few job openings may occur in automobile parts manufacturing but at the expense of many opportunities lost in bicycle manufacturing, food processing, and residential construction.

Combined and uneven development contributes to inequalities in the economies of the nations involved (Boswell and Chase-Dunn, 2000). A small wealthy class of owners in the export sector consumes imported luxury goods in the face of mass poverty and the underdevelopment of production for domestic consumption. This class often becomes significantly estranged from the broader population and its concerns. The encapsulation of the export sector from the broader society can also erode the power of the state to encourage domestic development because it has little say in this key part of the economy and thus little ability to use it as a source of funding for development, health, or education (Evans, 1995). Partly as a result of such patterns of combined and uneven development, world unemployment levels are at an all time high, exceeding 30 percent in many developing nations, and poverty is increasing in many less developed nations (Cavanagh and Mander, 2004).

The **commodification of agriculture**—that is, the change from agricultural production for local consumption to production for international commodity markets—also contributes to combined and uneven development. This transformation creates food scarcities in local markets that drive up the price of the necessities. Commodification of agricultural production has also brought with it increased mechanization and more centralized ownership. These have in turn displaced large numbers of peasants, who swell the ranks of homeless people living on the outskirts of major cities in developing nations, such as Mexico City and Bombay, India.

A final cause of combined and uneven development is the use of **transfer prices** that do not accurately reflect the value of the goods produced in less developed nations. Transfer prices are the prices that large corporations charge when a product is transferred from one division to another. For example, a U.S. automobile manufacturer may build its engine blocks in Mexico and then export them to Detroit for final assembly. The transfer price is the amount that the Ford Motor Company pays its Mexican plant for the engine blocks shipped to Detroit. These prices may bear little or no relationship to the actual value of the component. Instead,

transfer prices can be an accounting mechanism for shifting profits wherever desired to avoid taxes, to leverage nations or communities into economic concessions, or simply to conceal the export of profits to be reinvested elsewhere. Before the creation of OPEC, the giant oil companies successfully transferred profits from oil-exporting nations to their domestic operations by undervaluing crude oil exports while simultaneously overvaluing refined oil products, such as gasoline and heating oil (Claes, 2000). Raw material prices, in general, are particularly vulnerable to manipulation—a large multinational may control the production all the way from the farm or mine to the retail outlet and can establish internal transfer prices that support its own advantage rather than reflecting any realistic or market assessment of value. Unrealistic transfer prices reduce the earnings of less developed nations desperately in need of capital to finance economic development.

Developing nations involved in global markets thus often fill very specific and vulnerable niches. Often their role involves extractive activities based on mining or agriculture. Other involvements include the manufacture of subcomponents that are later assembled into finished products elsewhere in the world. In this way, developing nations become involved in **commodity chains** that may stretch around the world (Frenkel, 2001). The increasing prevalence of such lengthy commodity chains is allowed by a dramatic upsurge in the *fragmentation* of the production process for many products into detailed stages or subcomponents that can be distributed around the world. The ease of coordinating and integrating production using virtually instantaneous transmission of information and cheap transport for subcomponents underwrite increasing fragmentation. These processes have brought about a "collapse of space" so that places on opposite sides of the world can be involved in a coordinated production process as if they were in adjacent buildings (Arndt and Kierzkowski, 2001).

Border Plants The plants on Mexico's northern border, known as *maquiladoras*, exemplify the contradictory effects of international investment on the economies of developing nations. More than

1,150 factories employing 500,000 workers have opened along the border since 1990 (Bandy, 2004). Overwhelmingly, these plants are subsidiaries of United States-based MNCS, manufacturing everything from television sets to automobile parts to Levi pants to brassieres. These plants generate about seven billion dollars a year in exports for Mexico, second only to oil. Over 90 percent of the products are exported directly back to the United States. Wages average about $1.75 per hour, and over 80 percent of the labor force is female. The plants have brought large numbers of Mexican workers into the international economy, and the *maquiladora* sector is responsible for an increasing share of Mexico's export trade.

There is also evidence, however, that the *maquiladoras* have many negative consequences for the Mexican economy. For example, unemployment has increased in the host cities, partly because more workers are drawn to these areas in the search for jobs than actually find employment. Few manufacturing or service facilities have been spun off, because most of the components and services are imported directly from the United States. Even more damaging is the vulnerability of the plants to international fluctuations in the demand for their products and in the cost of manufacturing them elsewhere. Border plants have a notoriously short life expectancy with many plants already closed and operations moved to even lower wage nations in South Asia (Bacon, 2000).

The Debt Crisis Many less developed nations have borrowed heavily to fuel industrialization and domestic consumption. Because of the dynamics of combined and uneven development, however, these debts have been very difficult to repay. These developments have produced a **debt crisis** in the world economy. The less developed nations owe the industrially developed nations over one trillion dollars. As the costs of manufactured goods have risen in recent years and the costs of raw commodities have fallen, many of these nations have fallen into seemingly unrepayable debt (Edgar, 2000).

Rising Stars

In spite of the many challenges to succeeding in the global economy, some industrialized nations are successfully holding their place as efficient producers of quality goods. And some developing nations, such as China, are rapidly assuming key roles in the global economy. To understand the success and failure of nations in the global economy, we need to take a closer look at the social organization of production across nations.

DIFFERING WORK PRACTICES AROUND THE GLOBE

The level of industrialization in a country is one of the most important determinants of work practices and working conditions. Accordingly, we discuss working conditions first in the **least developed nations** and then in somewhat more advanced **developing nations**. Industrial relations in developed economies also vary widely. We discuss examples from state-regulated capitalism (France and Canada), codetermination (Germany), autonomous work groups (Sweden and Norway), macroplanning (Japan), and post-communist societies (China, Eastern Europe, and the former Soviet Union). For each type of economy we begin with the question, "Who makes the key economic decisions?" and examine the consequences of the answer for economic growth and working conditions.

Least Developed Nations

Pakistan, Mozambique, Uganda, Kenya, Honduras, and El Salvador are among the least industrially developed nations in the world. In such countries economic power is held to a large degree by agricultural interests—often by owners of latifundia and other large plantations producing for the international market. The patron of such an estate holds enormous power over the peasants who till the land. These relations are sometimes considered feudal in nature, because they may entail obligatory labor on the patron's land in exchange for the peasant's right to till a small plot for private consumption (Scott and Bhatt, 2001). Besides such export-oriented agricultural enterprises, the economies of less developed nations also include substantial amounts of subsistence agriculture. They are also frequently typified by large migrations of workers from rural to urban areas as subsistence agriculture is displaced by export-oriented agriculture. In such societies many goods consumed locally are still produced in the **informal sector** by local artisans and sold by small shop keepers and street vendors (de Soto, 2000).

One of the most striking features of less industrialized nations is the stark contrast between the poverty of the rural poor and the consumption of Western consumer goods by urban elites. Western clothes, food, soft drinks, and even cigarettes are promoted on billboards, radio, and television. The transformation of local tastes to include these expensive commodities provides a market for Western consumer goods but drains significant amounts of money from the market for local goods. The introduction of Western products and culture into less developed nations is met with both enthusiasm and distrust and has sometimes served as a catalyst for anti-Western sentiment as both local goods and local values are displaced.

One particularly difficult development challenge for many least developed nations is the high level of inequality in the society. Many least developed nations are characterized by extreme inequality between landed and industrial elites and the mass of poor peasants and slum dwellers surrounding the large cities. In such situations, elites are often more interested in maintaining their position than in promoting development through education, health services, and related social programs. With few ways to earn an honest living, corruption is often rife. Worse, elites often use violence and torture to maintain their prerogatives and to silence those seeking change (You and Khagram, 2005). In addition to the direct suffering that results, corruption and violence undermine both internal development and foreign investment.

Box 16.2 reports on the violence and abuse associated with debt servitude and child labor in India.

Developing Nations

In contrast to the often stagnant situation in the least developed nations, Venezuela, Algeria, Turkey, Brazil, and Mexico, for example, have made significant, although often regionally localized, advances along the road to industrialization. Such countries are often referred to as developing nations. In developing nations the power shifts from landowners, who produce raw materials for export, to wealthy industrialists, who have ties to multinational capital (Toledo, 2002).

The Philippines provides a good example of this type of dependent industrial system. It was a U.S. colony between 1898 and 1946. During that time the economy slowly moved from agricultural exports of sugar and rubber to a manufacturing base. In the 1970s and 1980s the Philippines became a major exporter of textiles, apparel, and electronic components, mainly to the United States. This pattern of **dependent development** has been a mixed blessing for the people of the Philippines. The Philippines has one of the lowest average wage rates in the Pacific basin. Under the late President Ferdinand Marcos it also gained a reputation as one of the most graft-ridden economies in the world and has a history of political repression and unrest. Partly as a consequence of these factors, the Philippines had significant emigration, especially to the United States and to nearby Hong Kong.

Working conditions in many factories in developing nations are reminiscent of conditions in the now industrialized nations at the beginning of the twentieth century. The labor of children is commonly used in industrial pursuits, and unsafe working conditions are commonplace (Schlemmer, 2000). For example, in 1993 a fire in a toy factory fifteen miles west of Bangkok, Thailand, resulted in the deaths of 200 workers. The factory had only two narrow fire escapes for *over 4,000 employees*. The fire was uncannily reminiscent of the infamous New York City Triangle Waist Company fire

---

BOX 16.2 Child Labor in India

Millions of children in India toil as virtual slaves, unable to escape the work that will leave them impoverished, illiterate, and often crippled by the time they reach adulthood. These are India's bonded child laborers.... Denied an education and a normal childhood, some children are confined and beaten, reduced to slavery. The International Labour Organization (ILO) has estimated that 250 million children between the ages of five and fourteen work in developing nations, at least 120 million on a full-time basis....

Bonded labor takes place when a family receives an advance payment (sometimes as little as U.S. $15) to hand a child—boy or girl—over to an employer. In most cases the child cannot work off the debt, nor can the family raise enough money to buy the child back. The workplace is often structured so that "expenses" and/or "interest" are deducted from a child's earnings in such amounts that it is almost impossible for a child to repay the debt....

My sister is ten years old. Every morning at seven she goes to the bonded labor man, and every night at nine she comes home. He treats her badly; he hits her if he thinks she is working slowly or if she talks to the other children, he yells a her, he comes looking for her if she is sick and cannot go to work. I feel this is very difficult for her.

I don't care about school or playing. I don't care about any of that. All I want is to bring my sister home from the bonded labor man. For 600 rupees I can bring her home—that is our only chance to get her back.

We don't have 600 rupees ... we will never have 600 rupees.

Lakshmi, nine-year-old beedi (cigarette) roller. Six hundred rupees is the equivalent of approximately $17.20.

SOURCES: Human Rights Watch, 1996, *The Small Hands of Slavery.* www.hrw.org/reports/1996/India3.htm and Human Rights Watch, 2004, *Child Labor.* www.hrw.org/children/labor.htm

---

eighty-two years earlier in 1911, which we discussed in Chapter 6. Table 16.3 reports average hourly wages for manufacturing workers in a variety of least industrialized, industrializing, and industrialized nations. Note that there is over a one hundred-fold difference between the wages at the top and the bottom of this scale.

Unions often play an important role in the process of development. In nations first undergoing industrialization and in developing nations trapped in dependent development, labor movements are important organizers of resistance against low wages and oppressive working conditions. In such settings labor unions are often key actors in social movement coalitions struggling to achieve goals of livability, justice, and fundamental rights of citizenship (Frundt, 2005).

Consumers in the industrially advanced countries can also be a source of pressure toward improving working conditions in developing nations. Consumer boycotts of goods manufac-

tured with child labor or under harsh and exploitive conditions bring attention to these issues and can be an important force for change. This is especially the case where highly visible brands or retail chains in the industrially advanced nations wish to avoid the negative publicity associated with having goods produced under exploitive conditions.

State-Regulated Capitalism

In many industrial capitalist societies the owners and managers of large corporations hold the economic power. The state, however, regulates the activities of corporations so that they do not undermine the interests of the nation as a whole. Government regulations cover such issues as minimum wages, fair hours, child labor, how retirement accounts can be invested, monopolistic practices, and environmental protection (Kaufman, 2004). Although these societies frequently proclaim an

TABLE 16.3 Hourly Wages for Manufacturing Workers in Various Countries 🌐 💲

Country	U.S. Dollars 2003
India	$ .26
Pakistan	.41
Sri Lanka	.42
Kenya	.43
China	.66
Philippines	.71
Guatemala	1.45
Brazil	1.81
Mexico	2.21
Czech Republic	3.42
Poland	3.56
Portugal	5.01
Taiwan	5.01
Hong Kong	5.07
Singapore	6.45
South Korea	7.60
Israel	9.86
Italy	12.45
France	14.57
Canada	15.88
Australia	16.17
United Kingdom	16.59
Japan	16.68
United States	17.12
Sweden	18.06
Belgium	19.54
Germany	22.78
Switzerland	23.14
Norway	26.10
Denmark	29.13

SOURCE: United Nations, *Statistical Yearbook*, 44th edition, 2003, New York: United Nations; U.S. Bureau of Labor Statistics, *www.bls.gov/fls*; International Labour Organization, *laborsta.ilo.org*.

ideology of laissez-faire (unregulated) capitalism, all of the industrially advanced capitalist nations can be more correctly regarded as examples of **state-regulated capitalism**.

Canada Labor relations in Canada are broadly similar to those in the United States. Both countries rely on collective bargaining between unions and companies in preference to direct government regulation of working conditions. However, the Canadian economy holds a different place in the world economic system than the United States. Canada, a sparsely populated nation, has never been a major world economic power and has long struggled against economic domination by the United States. It has pursued what some economists call an **import substitution** development strategy, in which a nation attempts to develop industries that produce goods it has previously imported. This policy has made Canada a self-sufficient producer of many of the manufactured goods consumed in the country (Krahn and Lowe, 1998). More recently, Canada has also been successful in becoming a thriving member of the world *export* economy. Prior to the 1980s its exports were primarily those of field and forest. However, in the 2000s Canada has also become a major exporter of automobiles, aircraft, and advanced telecommunications equipment (Giles and Belanger, 2002).

France The French economy is dominated, on the one hand, by a few industrial giants and, on the other, by a great many small producers. This curious organization can be traced both to the early bourgeois revolutions in France, which created political structures designed to support the bourgeoisie (small producers), and to later state-sponsored efforts to industrialize through creating giant modern corporations. Thus, the French economy is simultaneously both more centralized and more decentralized than the U.S. economy.

The overriding importance of a few large industrial giants in France has led to a cartel relationship among these enterprises as they have organized to divide available markets among themselves. Large French firms include Renault, Peugeot-Citroen, and Thompson Electronics (which now owns the U.S. firm RCA). The Economic Ministry institutes regular five-year plans. Although these plans are not blueprints for production, they outline the general directions in which the government will encourage the economy to develop. The government exerts its control of the economy through state ownership of the banks, with resulting control of investment decisions. Many wage rates also fall under legislative control.

Collective bargaining in France differs from that in the United States and Canada. Multiple unions exist within every major company, and a comprehensive written agreement is not the final product of collective bargaining. Instead, a process somewhere between consultation and collective bargaining takes place. If the outcome is not agreeable to both the unions and the company, there will be a strike or lockout as they attempt to pressure each other toward their respective goals. This bargaining system is responsible for a paradox: France has one of the numerically weakest labor movements in the industrially advanced societies, yet it has one of the most volatile and strike-prone systems of industrial relations (Desmarez, 2002). Current economic policy in France includes an attempt to increase the formality of collective bargaining in hopes of making labor relations more consistent across enterprises and industrial sectors. Efforts are also being undertaken to supplement contractual arrangements with legislative standards (DiPrete, 2005).

German Codetermination

In unified Germany the principal holders of economic power are the owners and managers of large corporations. However, these corporations share power with the government, unions, and **works councils**. Works councils are comprised of worker representatives (typically selected through the union) who sit on management boards and represent the workers in the day-to-day affairs of the enterprise. Working together, owners and workers collectively **codetermine** production practices and employment policies. German workers have a much greater say in the management of their enterprises than do North American workers. In large enterprises workers have 50 percent representation on the management board. The works councils, which date back to the violent social revolutions of 1848, are a unique form of worker participation and have attracted a great deal of interest outside of Germany (Muller-Jentsch, 2002).

Works councils have a strong influence on personnel matters, such as scheduling, vacations, and pensions. They also influence, though they do not have the final say on, recruitment, training, job transfers, and discharges. Works councils have the least say in economic matters, such as manufacturing techniques, the introduction of new technology, plant layout, and plant closings and openings. In the area of economic policy, their role is bolstered by union-operated banks, which influence investment decisions and, therefore, directions for economic growth. Unions also exert influence on the German economy through training programs collectively organized by unions, management, and the state. These programs include rigorous apprenticeship systems, as well as ten-month paid sabbaticals for workers selected for retraining in needed areas of technological growth (DiPrete, 2005).

Based on its extensive apprenticeship program and a highly skilled and committed work force, Germany has been able to control world markets for selected high quality goods and to produce these goods in volume. The archetype for this achievement is the Mercedes-Benz line of cars and trucks, which are recognized for quality throughout the world. Quality standards are ensured by using highly trained and skilled workers. This strategy also allows German manufactures to introduce new design features quickly and effectively—little new training is required because the workers are already highly skilled and actively involved in ongoing training (Streeck and Thelen, 2005).

The system of codetermination is responsible for a low strike rate in Germany compared with that in the United States or France. This low strike rate is based on German workers having additional effective channels to voice their concerns. One of the most important concerns of German workers over the last decade has been the demand for a thirty-five-hour workweek with limited reduction in pay. Significant movement in this direction has been made, especially in the important metal trades where a shorter workweek has been defended as a strategy for preserving employment in the face of technological and market changes (Wolf, 2002).

In the economic slowdown experienced by the advanced industrial nations in recent decades, the German industrial system has fared relatively well. The existence of a stable, mutually agreed-on system in which workers have a regular voice in the operations of enterprises is an important ingredient in this success. Early retirements have also been used to maintain employment levels for younger workers—a strategy made possible by historically high wages and a well-funded and secure retirement system. These institutional structures have also helped ease the disruptions associated with the reunification of East and West Germany (Fitzgerald and Stirling, 2004).

Scandinavian Autonomous Work Groups

The most distinctive aspect of labor relations in Sweden and Norway is the large degree of power held at the shop-floor level by **autonomous work groups**. The owners and managers of large corporations hold the greatest share of economic decision-making power; however, as in Germany, they do so in combination with the government, trade unions, and workers. In the Scandinavian countries workers in many companies are organized in small shop-floor groups (rather than in councils operating at the level of the enterprise as in Germany). Legislation in the 1970s substantially strengthened workers' roles in economic decisions in Sweden. The 1976 Codetermination Law was particularly important in this regard. This law extended the areas that could be legally covered by collective bargaining to virtually any decision affecting employees (Levinson, 2000). Partly as a

result of this and related laws, Sweden has been a leader in innovative work practices that have employee well-being and job security at their center.

The Scandinavian system is in some ways similar to the German system, but its focus is on autonomous work groups at the point of production. A classic example of the Swedish system is provided by the Volvo automotive plant at Kalmar, where groups of fifteen to twenty workers build entire cars (or major subsystems) rather than each worker being assigned a highly specific task on a rapidly moving assembly line. Some of the roles of work groups in the Kalmar Volvo assembly plant are described in Box 16.3.

The Scandinavian countries were late to industrialize but have been a major success story in the world economy. Only a century ago these countries were primarily agricultural, and the majority of their industrial development has occurred since the Second World War. The economies of Sweden and Norway depend heavily on export trade; over 30 percent of the gross national product of these countries comes from exports (Whyman, 2003). Most of these exports are high-quality machine parts, machine tools, lumber, and finished wood products. Scandinavian exports have a reputation for quality and workmanship.

In Sweden 91 percent of eligible workers, including most white-collar employees, are members of trade unions. This high level of participation has brought labor stability and peace to the Scandinavian countries. Workers have successfully gained access to economic power through the political process, trade unions, and autonomous work groups. In this context most workplace issues become matters of information, negotiation, and cooperation rather than issues to be resolved by dispute and confrontation. The proportion of work hours lost because of strikes in Sweden is only about 1.4 percent of that lost in the United States.

The Social Democratic Party in Sweden has been a powerful representative of workers' interests at the national level and has contributed

BOX 16.3   Work Groups at the Kalmar Volvo Plant

Throughout the company, we have a hierarchy of works councils, with representatives from both management and the employees. Some of these councils—one per plant—have been required by collective agreement since 1946. Others, like the Corporate Works Council, have been created on a voluntary basis to meet our own needs for consultation. . . .

Job rotation . . . started in the upholstery department, for very practical reasons. Employees complained of sore muscles from doing the same operation over and over. They discovered that if they paired off and traded jobs every day or so, they were able to use different sets of muscles. From that ergonomic, down-to-earth beginning have grown most of the other changes that focus on the quality of work life more generally. Soon the upholsterers (mainly women) were trading among three or four, and the relief from wrist, arm, shoulder, and back pains was far greater than anything they had been able to get from the small army of supervisors, doctors, safety specialists, and industrial engineers who had developed methods or mechanical aids to relieve their complaints. . . .

Eventually the workers in the upholstery department were given responsibility for planning all their own work, a fairly complex task because the blue seat with the special fabric has to leave the upholstery group in time to reach the assembly line at the right moment to meet the blue car with the special headliner. The planning had been a white-collar job in the past, but the group members, who alternated the paperwork task, had no trouble absorbing the planning. They get information on the assembly requirements from a teleprinter, and have to translate this into individual assignments. They have also taken over checking the quality of incoming upholstery materials, a job that used to require four inspectors. All the operators attended a short training program to learn how to find and identify all the kinds of defects and what to do about them.

As these changes took place, rather gradually, there was a concomitant increase in team spirit and individual commitment. Employee turnover and absenteeism changed dramatically, and the upholstery quality improved because team members understood the entire process and felt much more responsible for the product of their work.

SOURCES: Pehr G. Gyllenhammar, 1977, *People at Work*. Reading, Massachusetts: Addison-Wesley Publishing Company, pp. 81–82 and 85. Reprinted by permission of the author.

---

several unique features to Swedish industrial relations (von Otter, 2002). These include national labor courts that review workers' grievances. In the United States only unionized workers, government workers, and some professional workers, such as academics, have access to this sort of third-party review of workplace complaints. The Swedish government has also developed the most comprehensive full-employment, job-retraining, and relief-work programs of any nation in the world. There seems little doubt that the labor peace and job security provided by these programs have been important contributors to the success of Swedish industry in producing high quality products for the world market. In Sweden the average hours worked per week have dropped in the twentieth century from sixty-four to thirty-six, and wages have risen almost fourfold in this same period.

Overwork across the Western Industrialized Nations

Workers in the Western industrialized nations very greatly in annual hours worked—Americans work more hours on average than workers in any other Western nation. The American average of 1,976 hours per year is 111 percent of what Canadians work and 115 percent of the average for the United Kingdom. Comparisons with the nations of continental Europe are even more striking: France (123%), Germany (127%), and The Netherlands (144%). Surveys of Americans reveal a number of problems from overwork, including "feeling used up at the end of the work day" and "not enough time for self and family because of the job" (Jacobs and Gerson, 2004). For dual-earner couples, Americans are the only nation that exceeds an average of eighty hours of combined paid work per week, with 68 percent of dual-earner couples working more than eighty hours per week. These figures contrast particularly sharply with those from the Netherlands and Sweden where only 15.8 percent and 6.6 percent of couples, respectively, work more than eighty combined hours per week (see Table 16.4). Such numbers suggest an uncomfortably hectic life, especially for dual-earner couples with children.

Macroplanning in Japan

In Japan economic power is held by large corporations, especially those involved in the export sector, and by government planning ministries. In the 1980s Japan overtook the United States as the largest producer of cars in the world (Sey, 2000). This and other successes are attributable to the unique economic and industrial system that Japan has developed since the Second World War. The Japanese system of **macroplanning** includes economic planning at the economy level and job security and teamwork at the enterprise level.

Long-Term Planning The Ministry of International Trade and Industry organizes the Japanese economy to compete effectively in the international market for a variety of key goods, including steel, automobiles, and electronics. The ministry has identified each of these areas as a target for development. It coordinates the efforts of different industries, as well as the educational system and the financial system, to facilitate these success stories. To protect selected domestic sectors, Japan has also vigorously upheld restrictive import practices. These trade restrictions have been an important component in the policy of sheltering domestic industry to a point at which it can effectively compete in the international market (Kono and Clegg, 2001).

Work Practice Innovations Japan is also an innovative leader in workplace practices. Central to these innovations is a strong reliance on work groups. Work groups, organized into **Quality**

TABLE 16.4 Hours Worked in Western Nations

Country	Dual-Earner Couples	
	Mean hours per week	% 80+ hours
United States	81.2	68.2
Belgium	79.0	35.0
Italy	78.2	47.6
Finland	77.4	25.1
Canada	77.0	45.6
France	76.3	32.9
Germany	75.1	42.7
United Kingdom	74.3	34.4
Sweden	69.3	6.6
Netherlands	64.0	15.8

SOURCE: Jerry A. Jacobs and Kathleen Gerson, 2004, *The Time Divide: Work, Family, and Gender Inequality*, Cambridge, MA: Harvard University Press, Appendix Table A7, page 216.

Control Circles, take initiative for developing incremental changes in production techniques that help increase productivity through **continuous improvement** in production techniques (Deming, 1986). Such systems of teamwork depend on first establishing trust between the worker and the enterprise. This is achieved through the assurance of lifetime employment for many workers, something virtually unheard of in the United States. Once loyalty and trust have been established, workers are more willing to cooperate with their teammates and supervisors to promote productivity. The system is in some ways similar to the German and Scandinavian systems of worker attachment and commitment to the firm and has many of the same positive consequences for productivity and labor relations. The Japanese system differs from the German and Scandinavian systems, however, in that workers have less of an *autonomous* voice in enterprise affairs. Box 16.4 presents a corporate song and morning recitation that highlight differences between Japanese and American work orientations.

BOX 16.4   Japanese Corporate Spirit

The following song and recitations start every working day at the large Uedagin Bank in Japan:

> A falcon pierces the clouds
> A bright dawn is now breaking.
> The precious flower of our unity blossoms here.
> Uedagin Uedagin
> Our pride in her name ever grows.
>
> Smiling in our hearts with glory,
> For we carry the responsibility for tomorrow's
> Independent Japan.
> Our towns and villages prosper
> Under our banner of idealism raised on high.
> Uedagin Uedagin
> Our hopes inspired by her name.
>
> Marching forward to the new day
> With strength unbounded,
> We continue forward step by step.
> Oh, the happiness of productive people.
> Uedagin Uedagin
> Brilliantly radiates her name.

They remain standing to recite in unison the "Uedagin Principles" *(Uedagin koryo)*:

> Constantly abiding by the ideas of cooperative banking, we will, together with the general populace, advance in our mission to serve as an instrument of small and medium business enterprises.
>   Intent in the spirit of service, we will contribute to public welfare and social prosperity. Emphasizing trust and possessing an enterprising spirit we will advance scientific administration.
>   With mutual respect and affection, we will work with diligence, employed in maintaining systematic order.

Possessing a spirit of love for Uedagin, we pledge to plan for the prosperity of the bank and for the public welfare and to make the bank the greatest in Japan.

And finally, before sitting down, a second catechism is recited, this one known as the "President's Teachings" *(shachokun)*:

> Harmony *(wa)*. The bank is our lifelong place of work, let us make it a pleasant place, starting with our greetings to each other each morning.
>
> Sincerity *(seijitsu)*. Sincerity is the foundation of trust, let us deal with our customers with a serious and earnest attitude.
>
> Kindness *(shinsetsu)*. Have a warm heart. Be scrupulously kind.
>
> Spirit *(tamashii)*. Putting our heart and soul into it, let us work with all our strength.
>
> Unity *(danketsu)*. Strong unity is the source of energy for our business.
>
> Responsibility *(sekinin)*. Responsibility makes rights possible; first let us develop responsibility.
>
> Originality *(soi)*. In addition, let us think creatively and advance making each day a new day.
>
> Purity *(seiketsu)*. Have a noble character and proper behavior.
>
> Health *(kenko)*. With ever growing pride let us fulfill the Uedagin dream.

SOURCES: Thomas P. Rohlen, *For Harmony and Strength*, p. 36. Copyright 1974, The Regents of the University of California. Used with permission of The University of California Press.

The teamwork system in large corporations has had many positive consequences both for productivity and for the quality of work life in Japan. Large enterprises are more concerned with employee welfare than is typical in most Western state-regulated capitalist societies. Japanese firms spend about 8.5 percent of total labor costs on housing, medical facilities, food service, transportation services, sports, social activities, and special welfare grants. In contrast, British firms, for example, spend only about 2.5 percent of their payrolls on these employee services (Dore, 2001).

The Japanese system of employment security and team organization has made it possible for enterprises to integrate quality control operations into production processes. Because workers are committed to the success of the firm, they are more eager to improve quality than workers in other advanced capitalist economics, who are rewarded only for the work they directly perform. American and British firms do most of their quality control checks after production is complete. They then have to go through the costly process of fixing mistakes that the Japanese system encourages workers to avoid making in the first place. In addition, guarantees of employment security mean that workers no longer have to fear technological innovation and can instead work to implement innovations as effectively as possible.

The Underside of the Japanese Miracle The much-acclaimed system of lifetime employment security in Japan applies only to part of the Japanese industrial system. Workers in such enterprises make up only about a third of the Japanese labor force. The other two-thirds is made up of workers in peripheral enterprises, which are often subcontractors to the core firms, and women, who work as secondary workers in all sectors of the economy (Ogasawara, 1998). These workers do not have lifetime employment or any of the other benefits of workers in the core manufacturing sector. The exploitation of these workers has been an essential, though hidden, aspect of the Japanese "economic miracle" (Brinton, 2001). In many ways the Japanese economy is a contradiction, containing elements of both highly advanced industrial economies and developing economies. High levels of efficiency, the latest innovations in computer technology, and lifetime employment security exist alongside a large sector of underpaid and highly vulnerable workers.

The organizational innovation of **"just-in-time" delivery** systems, first developed in the Japanese automobile industry, builds on the unequal relations between core and peripheral enterprises. In just-in-time delivery systems, suppliers are required to deliver specified lots of parts to the principal manufacturer on a very exact but rapidly changing time schedule. This allows the manufacturer to drastically reduce inventory and related costs. The subcontracting supplier absorbs these uncertainties by paying lower wages and utilizing flexible employment levels.

Japanese workers also work longer hours and have fewer holidays than workers in many other industrially advanced nations. Pressures for unrecorded and unpaid overtime are also widespread (Mouer and Kawanishi, 2005). Death by overwork *(karoshi)* has been recognized as a national health problem in Japan. Japanese and American workers average around 1,800 hours per year, in contrast to an average of 1,500 hours for European workers (Mishel et al., 2005:415). Vacation times are similar with American and Japanese workers averaging two and a half weeks and European workers averaging five weeks. An organization of Japanese workers airs some of these complaints in a newspaper advertisement presented in Box 16.5.

Japanese unions provide only limited voice for workers in their enterprises. Union membership rose steadily after the Second World War, reaching a peak of 35 percent of the labor force in 1970 (Kaufman, 2004). Membership has since declined to about 24 percent. Japanese unions are specific to each enterprise, are typically dominated by management, and are thus typically referred to as **company unions**. On the positive side, enterprises are often willing to share planning information with their unions because of the cooperative spirit that pervades Japanese labor relations. Because many workers have a guarantee against being laid off, the union can work with management to devise the best plan for adjusting the company's labor force to meet changing economic circumstances. Strikes in Japan are infrequent. They generally address highly specific issues and are of short duration. The Japanese system of industrial relations, somewhat ironically, demonstrates that worker participation can encourage productivity and product quality without necessarily entailing increased worker power, autonomy, or satisfaction (Lincoln and Gerlach, 2004).

The Four Little Tigers

The Pacific Rim is also home to what are sometimes called the **four little tigers** of Asia: Hong Kong, Singapore, South Korea, and Taiwan. These nations have all experienced dramatic economic

BOX 16.5 Japanese Workers Protest Work Load

WHO SAID,
"YOU'VE COME A LONG WAY, BABY!"
?

**We haven't really.
And someone is making our way ahead
longer and tougher.**

Deceptive new bill now before Diet

A bill to renew the "Labor Standards Law" is now being proposed by the Japanese Government for legislation during the present Diet session. It calls for a 40-hour work week, instead of the 48-hour work week under the current law. This sounds like an advancement, but in fact it is deceptive.

To start with, no timetable is given for the enforcement of the 40-hour work week. On top of that, an unbelievably generous exemption is being granted to all businesses with less than 300 employees, which account for 85% of the total labor force in Japan, for an unspecified period of time.

This means that the new bill does not guarantee equality under the law, one of the principal rights guaranteed by our Constitution.

Annual Working Hours in 1985

Holidays and Paid Leaves in 1985

How many of you would look at the long working hours in Japan and say, "A great way to win in international competition. Let them have another decade of this." ?

We believe that these long hours, which have been a feature of the Japanese working environment for decades, constitute one of the major causes behind the current trade imbalance between Japan and the U.S., and we fear that this new legislation may well worsen the situation. Management will be entitled to override the principle of up to 8 hours work per day, without overtime pay.

Legalizing longer normal working hours (even 10 to 12 hours a day) at peak periods is sure to make it difficult for female employees with household chores, particularly career women, to retain their jobs.

With all these privileges, Japanese business and industry are likely to be in an even more competitive position with respect to international trade.

How many of you would look at the short vacations taken in Japan and say, "A reasonable way to reduce labor costs. Let their employers enjoy further immunity so that they can strengthen their competitiveness." ?

A minimum of 10 days of paid annual leave sounds like an improvement from the 6 days under the current law. Deceptive again, because the paid annual leave that an employee can take at his or her own exclusive discretion without losing pay will be reduced to only 5 days. The remaining "leave" may be allocated at the convenience of the employer.

Such a system may simply eliminate the need for replacement personnel, eliminate additional employment even though the nominal "vacation" may increase, all at the sacrifice of the rights of the employees.

Dwindling labor's relative share* of income in Japan

After all, can you stand yet another deceptive move by the Japanese Government that will affect you through industrial competition ?

We are by no means born-to-be workaholics. We are totally dissatisfied with the fact that we are deprived of pay and time for ourselves and our families which we believe we deserve. The compensation we receive as a percentage of corporate income, for example, is the smallest among the five major advanced countries.

Voice Your Concern, Now

Not against the people of Japan, but in support of the people of Japan, and to protect your own established rights for fair competition.

Express your view on this issue in your own way to the Japanese Embassy and through your own channels to the government and the Diet (congress) of Japan.

CONCERNED LABOR OF JAPAN
c/o Japan Federation of Newspaper Workers' Unions
5-6, Misakicho 3-chome, Chiyoda-ku, Tokyo, 101 Japan

This message is brought to you by the committee formed and supported by individuals and organizations covering a broad spectrum of professions and industries, including the print media, broadcasting, publications, advertising, printing, medical and clinical services, transportation, distribution, education, national and municipal public services, agricultural cooperatives, legal services and jurisprudence, together with contributions from private citizens.

We call for international solidarity to rectify the labor policy of the Japanese Government so that we will have

NO MORE UNFAIR COMPETITION
and a solution to the Japan-U.S. trade imbalance.

SOURCE: Used with permission of Concerned Labor of Japan.

---

growth in recent decades. The rapidly rising productivity of these nations, coupled with that of Japan and China, is the basis for projections that the twenty-first century will be economically defined by events in Asia rather than in Europe or North America.

All four nations have succeeded in growing despite limited natural resource bases. Instead of relying on natural resources, these nations have invested heavily in their educational systems and succeeded through cultivating the development of their people's human capital. Given their relatively small size and limited domestic markets, they have all had to succeed through the manufacture of export goods (Holzer, 2000).

Development strategies among the four tigers have been diverse, suggesting that there is more than one path to economic success in the 2000s.

Hong Kong has relied on family enterprises, inherited from its Chinese cultural tradition, as the basis for economic growth. Similarly, Taiwan has successfully developed on the basis of small and medium-sized manufacturing establishments combined in networks through extensive subcontracting systems (Lee, 2004). South Korea, in sharp contrast, has succeeded in its development project by investing in a few giant corporations that compete on the world market in basic commodities such as steel and automobiles. Singapore has relied heavily on direct investment by foreign multinational corporations and serves as a major subassembly center for world manufacturing. Singapore is also characterized by a strong role of the central state in directing development. Although all four of these nations are nominally democratic, none enjoy the full range of freedoms taken for granted in the industrialized West. It is noteworthy that in the twenty-first century, four of the five ports with the greatest shipping tonnage are in Asia: Singapore, Hong Kong, Shanghai, and Nagoya (Japan). The fifth member of this elite group is Rotterdam (The Netherlands).

China

China is the largest country in the world with a population of over 1.2 billion. China is, of course, the "big tiger" of Asia—a term suggesting its tremendous potential for economic and political power. In recent decades it has experienced an annual economic growth rate in excess of 10 percent in contrast with the 2 percent to 4 percent growth rates typical of Western industrialized nations. This rapid growth rate results from internal changes in the Chinese political and economic system as well as from external forces.

Internally, China has moved at a measured pace toward privatization of significant parts of its economy. In China many hybrid enterprises exist that are partly public and partly private (Hassard, Morris, and Sheehan, 2004). Market relations have emerged between these enterprises, but even these relations often operate within networks defined by past administrative linkages (Keister,

2000). In addition, many small private firms have emerged to fill the niches between these larger enterprises.

External forces adding to these changes include heavy direct foreign investment in China by multinational corporations. This investment has been motivated by a desire to take advantage of China's cheap labor and also by a desire to secure access to China's immense potential as a market for goods. These markets are both for imports and for goods made in China by foreign firms producing directly for the Chinese market (Xiao, 2004).

The new jobs in China often pay somewhat better than the meager earnings secured in rural areas, resulting in significant internal migration within China from rural areas to boom areas in the south and, more recently, in the north around Shanghai. Conditions at many of these factories are far from ideal. The story of one worker who migrated from rural China to work in one of these newly opened factories is presented in Box 16.6.

Chinese factory workers did not earn much under a centralized economy. Nor did they have the opportunity to move in the search for better employment. This labor system has been called "organized dependence" by China observers. The new system replacing it entails greater freedom to move about and to change employers. However, previous protections and securities have also been removed and replaced by oftentimes harsh exploitation—a system sometimes referred to as "disorganized despotism" (Chan, 2001).

With the advance of privatization, China's commitment to equality appears to be eroding. Free college education for the most qualified students, regardless of ability to pay, is now a thing of the past. Rapidly rising tuitions have increasingly made college an option only for the children of the newly emerging entrepreneurial class. Similar trends are taking place in housing and health care. Such trends may result in many Chinese rethinking the particular mix of planned and private enterprise that will eventually be the China of the twenty-first century. Currently, however, these trends toward increasing inequality are occurring without significant gains in democratization that would be needed

BOX 16.6   Lucky Gem Factory in Shenzhen, China

$

Mr. Wu was the middle of five children living in a small town in a mountainous area of Southwest China. The peasants there grew grain and vegetables but gave up much of the grain in taxes. They had vegetables to eat, but no one to sell them to since everyone grew the same vegetables. Money was extremely scarce.

Mr. Wu's family eventually scraped enough money together to buy tickets for him and his younger cousin to go to the newly booming city of Shenzhen just up the Pearl River from Hong Kong. There he found employment in the "Lucky Gem" factory recently relocated from Hong Kong and employing about 200 people. His job was to grind semiprecious stones (primarily jade) for twelve hours a day, seven days a week, with either one or two days off each month, depending on how busy the season was. He was paid one hundred Yuan a month (about $12), spent sixty on living expenses, and sent the rest home. His wife eventually moved to Shenzhen to be with him, although they had to live in separate dormitories for

two years before they could afford their own room. His two children never made the trip, and his wife eventually returned home.

Mr. Wu worked at Lucky Gem for ten years until he developed chronic coughing and breathlessness. He was initially diagnosed with tuberculosis, but after seeking a second opinion at the Guangzhou Occupational Disease Center, he was told he had silicosis—a breakdown of the small air sacs in the lungs and an incurable condition. Forty-five workers from Lucky Gem would eventually come down with the disease. Their attempt to sue the company for damages in Shenzhen courts failed on the grounds that the company had subsequently moved to Huizhou. In Huizhou, the court heard the case and eventually ruled in favor of the workers but did not award any compensation. When asked about the treatment of factory workers in China, Mr. Wu observed, "In China, it is a death sentence to be a worker."

SOURCES: Isabel Hilton, 2005, "Made in China," *Granta*, 89:13–54.

---

in order for a serious public debate about China's future to take place (Taylor, Chang and Qi, 2003).

Eastern Europe and Russia

In the past the power to make economic decisions in the Soviet Union (as well as Communist China) was held jointly by the Communist Party and by government administrative bodies. The government developed annual plans and five-year plans for the economy as a whole. At the national level these plans specified only broad goals for each sector of the economy. It was left to local and regional government agencies and enterprises to develop detailed blueprints for achieving these goals. The Soviet model of industrial organization emphasized clear lines of top-down authority and incorporated no formal role for workers' councils at the enterprise level or for autonomous work groups at the shop-floor level. Industrial relations at the plant level were thus similar to those in many state-regulated capitalist economies; however,

central planning rather than market forces drove the decisions about what to produce (Lengyel and Neumann, 2002).

Trade unions existed in every Soviet industry and were an important mechanism through which workers' voices were heard on enterprise management boards and in government ministries. Trade unions were important in the areas of labor recruitment, training, safety and health, and cultural activities. They were also actively involved in welfare functions for workers such as managing health care and retirement systems. Although trade unions in the Soviet Union represented workers' interests to the government and to enterprise managers, they were not autonomous in the same sense as unions in the state-regulated capitalist economies. The leaders of Soviet unions were members of the Communist party (as were the majority of people in leadership · positions in all aspects of Soviet society). Through this mechanism the party exerted a strong influence over union activities, and union demands rarely fell outside party agendas. In spite of party control, strikes were not unknown in the Soviet Union, including walkouts by thousands of workers (Silverman and Yanowitch, 2000). Strikes, however, tended to address highly specific local issues and to be of short duration for fear of government reprisal.

Under the leadership of President Mikhail Gorbachev the Soviet Union instituted reforms calling for *glasnost* (openness) and *perestroika* (restructuring) in the late 1980s. This increased openness set the stage for a political devolution of the Soviet Union into independent states. Today many republics of the former Soviet Union are fully independent, and others have significant autonomy. The remaining core of the old Soviet Union, composed of Russia, Siberia, and nineteen smaller republics, is today organized politically as the Russian Federation.

Economic reforms have created a climate for change and have set the groundwork for many subsequent changes in Russian life and in the Russian economic system. Privatization of the economy has become a central focus of these reforms. Little knowledge existed in the Soviet Union, however, about how to proceed with privatization. As a result, many early reforms were exploited as opportunities for illicit gain. The result might more aptly be described as *"pirate capitalism"* than as "industrial capitalism." Unemployment, poverty, and shortages have increased dramatically.

At present, the economies of the states of the former Soviet Union and those of its former satellites in Eastern Europe combine capitalism entrepreneurialism (especially in the small firm and service sectors), limited elements of state ownership retained from the communist period, and some worker collective ownership of factories sold off by the state (Standing, 1996). Backlashes have occurred in response to the excesses and failures of the transition period, and the future economic structure and health of the economies of the states of the former Soviet Union, including the Russian Federation, remain uncertain (Hedlund, 2005). In many regions, despair has become widespread as the promises of privatization have been replaced by the realities of rising unemployment and falling incomes (Kolodko, 2000). Alcoholism has increased and suicides have risen to record levels.

In regions of the former Soviet sphere where peaceful political transitions can be achieved without civil war between rival nationalities, there appear to be reasonable opportunities for the ferment of competing strategies of economic development to yield improvements in the standard of living. Poland, Hungary, and the Czech Republic are potential examples of such successes. Chronically high unemployment rates and increased poverty for many, however, appear likely for the foreseeable future.

Competing Organizational Forms

Japanese automobile plants are being transplanted to the United States and Canada. The United States builds Ford Falcons in Australia. German automobile plants are being transplanted to the United States and vice versa. British petroleum corporations are opening retail gasoline stations around the world. French pharmaceutical companies are building factories worldwide with a special focus on former French colonies. As large economic organizations become increasingly global in their operations, they take their organizational cultures with them to their new host nations. Thus, in the world of the twenty-first century, there is a regular commerce in organizational forms as well as in the products produced by these organizations (Kaufman, 2004).

Do global corporations take on the organizational features, human relations policies, and other characteristics of their new host nations, or do they impose their own organizational structures and human relations policies on their host nations and on their new labor forces? These questions are crucial for the future of work in many nations.

Probably the most widespread organizational transplants in the world today are those of Japanese automobile and electronics companies, which have opened operations in South Asia, North America, and Europe. These plants generally seek to reestablish the quality control circles and other innovations that are core components of the Japanese system of production based on continuous improvement.

However, these plants also have to make adjustments to local conditions and cultures (Kenney and Florida, 2004). Resistance to Japanese corporate cultures has been significant in North America and Europe where workers expect to maintain greater independence and autonomy from their employers than do Japanese workers (Delbridge, 1998).

The organizational forms producing the highest quality products most cheaply and demonstrating the most flexibility in adjusting to rapid market changes will win in the world of heightened global competition. Organizations that respect workers' rights and motivate them to give their best efforts to the success of the enterprise are likely to have a significant advantage in the increasingly competitive environment of the twenty-first century (O'Reilly and Pfeffer, 2000).

International Labor Solidarity

Workers have many resources for power too, even in the modern interconnected world economic system. In the face of increasing organizational complexity and interconnections, workers have attempted to develop new strategies for securing a share of the benefits of economic development. Labor unions in the United States have successfully supported federal legislation requiring the Overseas Private Investment Corporation to refuse insurance to U.S. corporations operating in countries that fail to protect "internationally recognized workers' rights." These rights include "the right of association; the right to organize and bargain collectively; a prohibition on any form of forced or compulsory labor; a minimum age for the employment of children; and acceptable conditions of work with respect to minimum wages, hours of work, and occupational safety and health" (International Labour Organization, *www.ilo.org*). Such legislation is intended not only to promote more humane working conditions in less developed countries but also to keep workers in the United States from having to compete against workers who labor under exploitive conditions. Such legislative efforts are an important way in which workers in the industrially advanced nations can find common ground with

workers in developing nations. Box 16.7 presents an appeal by the United Automobile Workers to support additional legislation of this sort.

The International Labour Organisation (ILO) The United Nations conducts research and develops policy guidelines on labor issues through **the International Labour Organisation (ILO)**. The ILO does not involve itself in contract negotiations or other direct interventions in labor relations. Instead, it serves as a research institute on international labor issues and as a clearinghouse for information of value to international unions and other workers' groups. The ILO was founded in 1919 by the Treaty of Versailles and is the only remaining institution founded by that Treaty, and thus it predates the UN. Perhaps the most lasting contribution of the ILO is its establishment and promotion of international standards for the rights of workers.

World Labor Confederations   Three major international labor organizations *are* directly involved in supporting and coordinating union activities around the world. They are the World Federation of Trade Unions (WFTU), the International Confederation of Free Trade Unions (ICFTU), and the World Confederation of Labour (WCL). The WFTU (*www.wftucentral.org/*), the largest of these groups, has a membership of over 200 million. About two thirds of these members are in Eastern Europe and the former Soviet Union. The ICFTU (*geneva.ch/icftu.htm*) claims an affiliated membership of about 105 million, mostly in Western Europe and in former European colonies. Both of these organizations have grown in recent decades and have memberships that dwarf that of the American labor movement. The WCL (*www.cmt-wcl.org*) was founded in 1920 as a Catholic labor organization. It aggressively opposes the current role of MNCs in the world economy. Although the WCL has lost membership in Western Europe, it has grown in many developing nations and has maintained a large membership in Canada. Its largest membership today is in Latin America where it enrolls about 19 million workers—similar to the number of North American workers enrolled by the AFL-CIO.

BOX 16.7   UAW Support for Fair Trade   **$**

Many "Cheap" Foreign Imports Come At Too High A Price.

It's a simple fact that many Americans don't want to face. Workers' rights are being trampled upon abroad to keep labor and other production costs down. The price tags we see on goods from those countries may be lower, but that's because someone else has already paid dearly.

Jail terms, beatings, kidnappings, torture are all-too-common features of life in nations like Chile, Paraguay, South Korea, and Taiwan. Workers are paying for their efforts to form unions, bargain collectively, or seek some small measure of safety and health protection.

These violations of basic human rights should not enable foreign nations or multi-national corporations to prosper. That's why we need strong worker rights provisions in our trade laws.

The Pease Worker Rights provision in the House Trade Bill—H.R. 3—would define such practices as unfair under U.S. law. It would allow action to stop employers and governments from profiteering at the expense of basic worker rights. Contact your U.S. Representative and urge support for the Worker Rights provision in the House Trade Bill.

Don't let them trade on repression.
Support the Worker Rights Provision in the House Trade Bill—HR 3.

International Union, UAW   Owen Bieber, President

These international labor confederations have developed a number of strategies for improving the position of labor in the increasingly complex world economy. They have promoted United Nations guidelines for labor standards. They have promoted codetermination arrangements in which union representatives have voting rights on the boards of directors of enterprises. In important developing nations, such as Brazil and South Africa, labor unions have played a leadership role in social movements promoting broad agendas of social justice, livability, and democratic citizenship rights (Jakobsen, 2001).

The Challenges of Global Coordination for Unions   Attempts to coordinate workers' demands on a world scale face significant problems that have limited the effectiveness of coordinated international efforts by workers to improve their conditions. In brief, workers are most immediately concerned with their own wages than with those of workers in other nations. Problems also arise from trying to coordinate the activities of workers in societies with different levels of unionization. Where only a small proportion of the labor force is unionized, union recognition and survival are central concerns. Where a majority of the labor force is unionized, concerns for job security, safety and health, and worker training take precedence. Workers in different countries may also face different legal frameworks, some of which are very repressive (Brecher, 2000).

In spite of the difficulties involved, workers in different nations can support one another and coordinate their efforts to improve their working conditions (Gordon and Turner, 2000). Direct involvement by international unions can have a significant positive influence on the outcome of local or regional labor negotiations. Workers can also exert influence through pressuring their governments to support improved working conditions in other countries and through mobilizing public opinion in support of fair working conditions.

SUMMARY

The world is increasingly integrated as one interconnected economy. Theories of modernization, dependency, and world systems each contribute to our understanding of the global economy. This economy is characterized by a dominant role for multinational corporations, reduced growth rates in the industrialized nations, increasing privatization in former Communist nations, uneven development in less developed countries, and a high degree of diversity among nations in the organization of work.

Britain established a colonial empire stretching around the world. The colonies supplied raw materials for British factories and markets for British manufactured goods. As the Industrial Revolution spread to other European nations, they too established colonial empires to support their industrialization. This system lasted until the Second World War, after which it was replaced by a new system of neo-imperialism with the United States as its head. Neo-imperialism was based on economic control of the underdeveloped areas rather than on direct political control. The same patterns of dependent development continued under this new system of imperialism, with nonindustrialized nations supplying raw materials to the industrialized nations and providing a market for manufactured goods.

Multinational corporations occupy a central role in the global economy at the beginning of the twenty-first century. MNCs are the central institutional arrangement through which economic transactions occur in the global economy. They have encouraged the development of export-oriented economies in less-industrialized nations. This type of industrialization encourages underdevelopment in the domestic manufacturing sector by absorbing resources that could have been used for the production of goods to be consumed locally. Export-oriented industrialization has created a development process called combined and uneven development. Such development strategies have as yet been unable to lift the majority of people in these developing societies out of poverty. Some nations, such as China, Japan, and South Korea, have achieved greater successes. They have encouraged the development of industries producing for domestic consumption by protecting domestic markets from foreign goods. Simultaneously, they have developed their own export sectors.

The world economy at the beginning of the twenty-first century is more diverse than at any previous time. Not only have less industrialized nations begun to take a variety of paths to development, but great diversity in industrial organization also exists within the industrially developed nations. Individuals, governments, or workers may have ownership over economic enterprises. Economic planning decisions may be made by the managers of large enterprises, by the state, by a planning body of enterprises and the state (with or without union participation), or by some form of codetermination in which workers and owners are both represented. One particularly bright spot in the world economy is the economic successes associated with increased levels of worker participation. These successes suggest that the most effective systems of economic production in advanced industrial societies may also be those that are the most democratic and humane. Such settings allow human beings to apply the full range of their creative talents to their productive activities.

The diversity of successful forms of economic organization suggests that many ways of organizing an industrial society are possible. The continuation of such diversity is probably good for the future of the world. No single way of organizing work is best for every situation; each has its own strengths and limitations. Only through accepting and encouraging experimentation and innovation can we hope to improve our systems of industrial production and the quality of life for all.

KEY CONCEPTS

modernization theory	free trade	developing nations	macroplanning
dependency theory	fair trade	informal sector	Quality Control Circles
peripheral nation	combined and uneven development	dependent development	continuous improvement
core nation	commodification of agriculture	state-regulated capitalism	"just-in-time" delivery
world systems theory			
primitive accumulation	transfer prices	import substitution	company unions
neo-imperialism	commodity chains	works councils	four little tigers
cartel	*maquiladoras*	codetermination	central planning
multinational corporation	debt crisis	autonomous work groups	International Labour Organisation
protectionism	least developed nations		

QUESTIONS FOR THOUGHT

1. Consider your daily round of activities. Make a list of where the products you use are made. Be aware that the components of a product assembled in one country may be made in other countries.

2. Describe the manner in which some less industrialized nations become dependent on the export of one or more raw agricultural or extractive products. How might a nation move out of this role? What difficulties would the nation be likely to encounter in trying to move out of this role?

3. What factors have contributed to a reduced rate of economic growth in North America and Europe in recent decades? What growth rate do you think we can expect in coming decades? What might be done to increase this growth rate? Is an increased growth rate even desirable?

4. Describe some of the positive and negative roles played by MNCs in the world economy. What policies might be effective in limiting the negative behaviors of MNCs while retaining their positive effects?

5. Compare and contrast Scandinavian autonomous work groups, German codetermination, American collective bargaining, and Japanese Quality Control Circles as systems for incorporating workers into workplace decisions. Which systems do you think will be most likely to emerge as dominant in the next twenty-five years?

6. What is the role of international labor organizations in the modern world economy? Do you think that such organizations will have a significant effect on working conditions in the twenty-first century? Why or why not?

MULTIMEDIA RESOURCES

Print

Joseph E. Stiglitz. 2003. *Globalization and Its Discontents.* New York: W. W. Norton. A comprehensive tour of issues concerning globalization, free trade, and international development policies.

Terry Boswell and Christopher Chase-Dunn. 2000. *The Spiral of Capitalism and Socialism: Toward Global Democracy.* Boulder, CO: Lynne Rienner. An examination of the tension between market dynamics and social justice. Advances the thesis that this tension is useful for the pursuit of democracy and desirable social outcomes.

Friedman, Thomas L. 2005. *The World is Flat: A Brief History of the Twenty-First Century.* New York: Farrar, Straus and Giroux. Presents a provocative thesis that modern digital telecommunications have leveled the playing field for international competition, setting vast forces of change into motion.

Ronald P. Dore and Mari Sako. 1998. *How the Japanese Learn to Work.* London: Routledge. The best available examination of Japanese training strategies in the workplace and their role in producing quality products.

Ching Kwan Lee. 1998. *Gender and the South China Miracle: Two Worlds of Factory Women.* Berkeley: University of California Press. A study of two General Electric-Liton factories—one located in Hong Kong and the other across the Chinese border in Shenzhen.

Internet

United Nations Human Development Report. *www.undp.org* Crucial data and insights on the economic condition and quality of life of people around the world.

Multinational Monitor. *multinationalmonitor.org* Private watchdog group reporting on violations of human, civil, and workers' rights by multinational corporations.

EarthRights International. *www.earthrights.org* Dedicated to defending human rights and the environment. Includes current news items from around the world.

Institute of International Studies, University of California at Berkeley. *globetrotter.berkeley.edu*

Links to research and policy statements on global social issues of all kinds by leading international figures.

International Labour Organization. *www.ilo.org* The oldest and most important public body promoting the interests of workers world wide.

RECOMMENDED FILM

Mi Familia (1995). Starring Edward James Olmos. A multigenerational epic of the Sanchez family starting in the 1920s with grandfather Sanchez's long walk from Mexico to Los Angeles.

17

The Future of Work

The media provide mounting evidence of "time poverty," overwork, and
a squeeze on time. Nationwide, people report their leisure time has
declined by as much as one third since the early 1970s. Predictably, they
are spending less time on the basics, like sleeping and eating. Parents are
devoting less attention to their children. Stress is on the rise, partly owing
to the "balancing act" of reconciling the demands of work and family life.
The experts were unable to predict or even see these trends. I
suspect they were blinded by the power of technology—seduced by
futuristic visions of automated factories effortlessly churning out products. . . . To understand why forty years of increasing productivity have
failed to liberate us from work, [it is necessary] to abandon a naive faith
in technological potential and analyze the social, economic, and political
context in which technology is put to use.

(SCHOR, 1991, *THE OVERWORKED AMERICAN*, PAGES 5–6)

Changes of great significance are occurring in the nature and organization of
work. For the first time in history we can meaningfully speak of a global
workplace. Telecommunication networks link distant work sites. Night shifts in
North America communicate instantaneously with morning shifts in Europe and
with afternoon shifts in Asia. Multinational corporations link production sites
around the world with little regard for geographic or political boundaries.

It is tempting to view these transformations in terms of technological change
alone. But each aspect of change—new technologies, new work organizations,
and a new workforce—has its own unique dynamics. Innovations in the organization of work, such as the assembly line, have historically contributed at least as
much to productivity as have mechanical inventions.

It is also tempting to view these changes as beyond the control of the average
worker. Even chief executives may feel helpless in the face of "world competition"

or "the system." The welfare of workers and of all of us depends on who controls the nature and direction of change. Workers around the world are deeply involved in this process and are far from powerless. **Workers' power** is based on their knowledge of the technology and the product and on their collective organizations, such as unions and professional associations.

The difficulty, for workers and analysts alike, is anticipating the consequences of these changes in the nature of work. All changes have multiple effects. Some are likely to be benign, others not so benign. In this chapter we examine three master trends that will be of crucial importance in determining the nature of work in the twenty-first century. We project, based on these trends, the possibility of a future society in which there is an increasing divide between a highly innovative and productive sector and a marginal sector. The emergence of such a divided society is a possibility but not a certainty. Therefore, we also discuss mechanisms for increasing innovation and reducing marginality.

PIVOTAL WORK TRENDS

The beginning decades of the twenty-first century are witnessing many important changes in the nature of work. Three trends are particularly important: (1) the spread of microprocessor-based technology throughout the workplace, (2) increased competition in the world economy, and (3) the increased movement of women into paid labor. In this section we review each of these trends and discuss how they are shaping the nature of work in the twenty-first century.

Computer Technology

During the twentieth century advances in technology and organization greatly increased productivity in manufacturing. Initially, these advances involved a finer division of labor into more and more minute tasks and the organization of jobs along moving assembly lines. New technologies in the twenty-first century involve the use of automated machines, computer-controlled robotics, and automated information processing. These electronic technologies are having an even greater impact on a broader range of industries and occupations than did the mass-production technologies of the twentieth century.

On the positive side, computer technologies have increased productivity and created skilled jobs. They have also shortened the time needed to develop new products. Thus, the pace of change has increased, giving rise to potentially faster economic growth. Microprocessor technologies have also encouraged greater employee participation, which is required to ensure the successful use of sophisticated technologies. On the negative side, new technologies have created new stresses because of the deskilling of some workers, the displacement of others, and the electronic monitoring of many others.

In sum, the accelerated pace of technological change associated with the extensive use of microprocessors has created rapid change in the workplace and in workplace relations. These changes have destabilized older patterns of work and relationships among workers and between workers and managers. Technology by itself, however, does not determine the direction of these changes. Other factors are equally important for determining changes in the nature of work.

An Integrated World Economy

An integrated world economy emerged following the devastation of the Second World War (see Chapter 16). Today, this integrated world economy includes tremendous diversity among nations. Countries vary in their level of industrialization and in how they are integrated into the global economy. Some are dependent suppliers of raw materials or partially finished components to more industrialized nations; others are autonomous producers of finished goods and services.

Recent developments in the world economy have greatly increased competition. After the Second World War, Japan and the European nations rebuilt with newer, more modern factories and reentered the world market. Many developing nations have also industrialized and have become important producers of manufactured goods. For example, South Korea now sells cars in the American market, something undreamed of twenty years ago. In addition, much of the work in low-wage, labor-intensive industries is now performed in less-developed nations, further intensifying the competitive pressure on workers in the industrialized countries.

Thus, changes in the world economy have also contributed to a period of change and flux in the workplace. Increased international competition has heightened threats to the jobs of workers in the industrialized nations. Greater competitive pressures require organizational change and innovation for economic survival.

Similar technological, organizational, and market forces have caused a convergence between capitalist and socialist nations. Formerly socialist nations today incorporate market forces to a significant degree in their economies. And the most successful capitalist nations have moved toward greater economic planning (Kenworthy, 2004). Such planning includes targeting specific industries for expansion, identifying the training needs of the labor force for these industries, and developing programs to meet these needs. The next stage of industrial society will transcend the distinction between capitalism and socialism. Other distinctions, however, are becoming more salient—such as those between more innovative and less-innovative industrial systems and those between the developed global North and the less developed global South.

Labor Force Diversity

Women have always worked. However, the transformation of their work into paid labor outside the home occurred somewhat later for women than it did for men, and it is still continuing today. The decline of high fertility and the reduction of child-rearing duties have been crucial in this process. In addition, employment opportunities have expanded in areas that have traditionally employed female workers, especially in clerical and service occupations. In combination, these factors have resulted in a rapid increase in the proportion of female workers in the labor force. This process has had tremendous consequences for the family. It can no longer be assumed that women will remain at home for child-care and home-tending duties. It would be incorrect, however, to assume that women's and men's paid jobs are similar. Women are still segregated by occupation and industry, and they are much more likely than men to be part-time and part-year workers.

Since the passage of the Civil Rights Act of 1964 and related legislation, female and minority workers in the United States have made important strides in the workplace, as well as in other spheres of society. Opportunities for female and minority workers have increased significantly. Full equality, however, is still far away. For minority workers, centuries of oppression have become embedded in class inequalities that make it difficult for many to take advantage of increased opportunities. Similarly, continuing assumptions on the part of both men and women that home and child-care duties should fall more on women's shoulders make it difficult for many women to take advantage of increased opportunities. In addition, for both female and minority workers, blatant and subtle forms of prejudice and discrimination continue to create barriers. Thus, in spite of improved conditions, women continue to suffer segregation into lower-status and lower-paying jobs, and blacks and other minorities still differentially occupy the lowest positions in society. Fully incorporating and taking advantage of a diverse labor force remains a challenge for organizations and society.

WORK IN THE TWENTY-FIRST CENTURY

Technological change, increased world market competition, and the increased proportion of women and minority workers in the labor force are major changes that will determine the nature of work in the future. But what will work be like in the twenty-first century? It appears possible that the economy will be characterized by two very different employment sectors. In one, which we call the **innovative sector**, the response to heightened international competition and technological change will be the development of technological and organizational innovations leading to increased productivity. In the innovative sector technological and organizational innovation will be continuous, jobs will be reasonably secure, pay will be adequate, job conditions will be more or less pleasant, and, perhaps most importantly, workers will have an increased say in determining the conditions of their work.

In the other sector, which we call the **marginal sector**, employers will respond to heightened international competition and technological change by lowering labor costs through reducing wages and benefits. In the marginal sector, innovation will be slow, jobs will be insecure, pay will be low, and conditions will be unpleasant and even hazardous. The organizing principle of the marginal sector will be to achieve economic viability by driving down wages rather than by promoting technical and organizational innovation. Workers will have little say in determining the conditions of their employment or the policies of their organizations. Female and minority workers will likely be disproportionately employed in the marginal sector. Thus, we do not believe as some have proposed that work will disappear in the high-technology future (Rifkin, 2004). Rather, work is here to stay, but the nature of work appears to be diverging between two increasingly distinct sectors of employment.

In this section we discuss the reasons for a divergence between these two distinct ways of organizing work and the characteristics of work in each sector. Box 17.1 depicts the three trends in work that we have described and possible future scenarios to which these trends may lead.

The Innovative Sector

What factors encourage the growth of an innovative sector? On what basis do we project its continued and increasing importance? The growth of innovation results from the development of microprocessor technologies that facilitate such innovation and from increased international competition and worker pressure that demand it in order to protect profits and jobs (Bluestone and Harrison, 2000). Many recent workplace innovations can be seen as attempts to use new technologies to adapt to increased competition in the world economy through utilizing a high-skill labor force. Another impetus toward workplace innovations has been the desire to reduce the inefficiencies of bureaucratic and hierarchical arrangements of work (Vallas, 2003). In this sense, workplace relations in the innovative sector can be better described as **postbureaucratic** than as postindustrial. Because human beings are central to the process of production, innovations in the social organization of work will continue to be at least as important for the success of this sector as technological innovations.

The defining characteristics of work in the innovative sector are *increased worker education* and *participation*. Increased worker education and participation create the conditions for *continuous learning* and continual *job redesign* (Appelbaum et al., 2000). Continual **job redesign** will be necessary because of the pace of technological change and the highly competitive and rapidly changing global economy. The specific ways in which jobs will be redesigned are impossible to predict in great detail because they will be unique to each setting and each technology. However, some general principles are relatively clear (Kenworthy, 2004).

The Centrality of Participation Workers hold the power to make organizational and technological advances succeed or fail through formal and

BOX 17.1 A Crystal Cube of the Workplace

Social scientists do not have crystal balls and cannot predict the future. They can, however, examine trends and describe what might happen if certain trends continue. Throughout this book we have tried to identify trends that may affect the workplace of the future. Three master trends that we have discussed are the role of technology and organization in the workplace, the competitive climate, and the changing composition of the labor force. These three trends may combine in many ways to produce many possible outcomes.

The cube in Figure A shows how the three master trends might interact. Imagine North American jobs right now as centered somewhere in the middle. A choice to use organization and technology in a way that increases the utilization of workers' skills would represent a shift to the left; a choice to use organization and technology to simplify or eliminate jobs would represent a shift to the right. Similarly, if increased productivity alleviates the competitive threat to North American industry, jobs would shift downward in the cube. As we have noted, women and members of minority groups will come to be a larger proportion of the labor force, but whether their labor power is adequately utilized depends on the types of jobs available to them. A "nearer" point in the three dimensional space represents more access for female and minority workers to full-time, year-round jobs in occupations and industries comparable to those of white men. It is possible that women and minorities could be less adequately utilized than they are today; this is the "farther" end of the third dimension.

These options could, of course, be combined in many, many ways, but there are two points in our hypothetical space that we wish to discuss further. One is Point I on the figure; at this point technology and organization are used to complement workers' skills, competitive threats are lessened by productivity increases, and women and minorities move closer to equality. Even if most jobs in North America fail to move toward this point, some jobs will probably approach Point I. In this chapter we refer to such jobs as comprising the innovative sector.

On the other hand, organization and technology may be used to deskill or to eliminate jobs, especially in response to competitive pressure. Under such conditions, pressures will persist to use women and members of minority groups as reserve workers, calling them up for part-time or seasonal work as needed or paying them low wages to undercut the wage demands of higher-paid workers. These conditions are represented schematically as Point II in the diagram. Even if most jobs do not move toward Point II, some jobs will. In this chapter we refer to such jobs as representing the marginal sector.

It is difficult to predict with any accuracy the relative size of the marginal sector and the innovative sector. Most jobs will probably still lie somewhere between these extremes. But there are reasons to think that these two sectors will represent significant numbers of jobs in the twenty-first century. It is useful to discuss these possibilities as "best-case" and "worst-case" scenarios representing endpoints in the master trends we have identified.

---

informal **employee participation**. Some observations by industrial sociologist Robert Guest provide a good example of the importance of employee input. Guest was a visitor at a steel mill in the process of making a major technological change. A new steel process was delayed for six months because of differences between the company and the union on the incentive plan that would distribute part of the benefits of the new technology to the workers. After the incentive

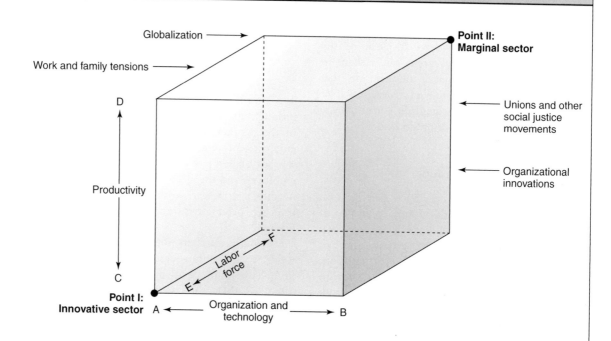

Organization and technology:

   A. They are used to increase productivity by complementing the abilities of skilled workers.
   B. They are used to replace workers and deskill jobs.

Productivity:

   C. Innovation and productivity increase.
   D. Productivity stagnates.

Labor force:

   E. Women and minority groups approach equity with majority men.
   F. Women and minority groups disproportionately occupy low-wage and part-time jobs.

FIGURE A Three Master Trends in the Workplace

question was finally settled, Guest got a call from a worker and was told that he would see something interesting if he came down to the mill at the start of the midnight shift. Guest reports the following events:

At precisely midnight a loud klaxon sounded. The lead man raised his arm and in a loud voice called out, "Let 'er roll!" The red hot billets spit out of the helical rolls at a speed I have never seen before. There were no delays or breakdowns on the shift, and within a month capacity had gone up over twenty percent. (Guest, 1987:5)

The moral of the story is that workers hold the key to the success of programs of technological and organizational redesign. To engage workers' fullest abilities requires that they have a piece of the action—not just a share of profits but a share in decision making. The sectors of the economy that succeed in introducing effective job redesign will be the sectors that include a leading role for employee participation at every stage of production.

Work Groups

One important form of employee participation occurs through small work groups, typically of eight to twelve workers, who are given collective responsibility for a task. Work groups offer an important venue for employee participation, though they often limit the topics of discussion to product quality or to minor aspects of the work environment. Many college students work in settings which have some aspects of team organization, or which at least claim team organization by calling employees "associates" or "partners." Work groups can be an important source of innovation (O'Reilly and Pfeffer, 2000). They can also be important for improving the **quality of work life**. However, as we will see, they can also be used to intensify work and heighten pressures on the job.

Team systems of production based on significant degrees of self-management by work groups have become increasingly important in contemporary organizations. Team-based production systems, however, actually have a long history in the workplace. Miners, seafarers, and other skilled trades have long relied on teams to coordinate work in situations involving complex, difficult, or dangerous tasks.

The increased importance of teams in the modern workplace reflects many forces, including increased skill demands associated with sophisticated technologies, new management theories about how best to organize production, and worker demands for increased voice at the workplace (Bills, 2004).

Japanese companies and their affiliates around the world have led the way toward increased utilization of team-based production systems. Under Japanese team production systems, employees organized in *Quality Control Circles* are expected to be ever vigilant for opportunities to work more effectively by identifying and eliminating underutilization of time and resources.

An important underpinning of Japanese Quality Control Circles and other initiatives to improve productivity and increase quality has been the tying of the worker to the company through lifetime employment and through finely graded systems of seniority-based pay (Dore, 2001). The tying of the employee and the firm together in a lifelong partnership encourages workers to use their skills to improve productivity and thus ensure the firm's future.

Other researchers note, however, that Japanese workers are not necessarily enthusiastic about involvement in team-based production (Endo, 1994). Rather, they see participation in Quality Control Circles and related team activities as a requirement for the economic success of their enterprise. Japanese workers thus participate in problem-solving activities with honesty and candor, but not generally with great enthusiasm or a sense of personal gratification.

In many workplaces Japanese-style teams have been associated with work intensification, increased pressures for production, employee monitoring of peers, and antiunion campaigns (Kono and Clegg, 2001). It is also a mistake to assume that managers and supervisors disappear in team production settings. Under Japanese team-based systems, frontline supervisors continue to play an active role in controlling and evaluating workers. In many ways, workers are more tightly controlled in team settings than in traditional supervisory settings. The power of the supervisor is not removed; rather, it is extended through allocating additional supervisory functions to the team as a whole.

Team organizations of work are today well established as a source of innovation, creativity, and heightened productivity (Smith, 2001). When combined with a plan that allows workers

BOX 17.2 Success through Participation

Powered by brilliant engineers, mathematicians, and technological visionaries, Google Inc. ferociously pushes the limits of everything it undertakes.... Their goal: to organize all the world's information and make it universally accessible.... Despite its growing head count of more than 4,000 employees worldwide, Google maintains the pace of innovation ... by continuing to work in teams of three to five, no matter how big the undertaking....

Google has also created a new kind of work environment. It serves three free meals a day to employees (known as Googlers) so that they can remain on-site and spend more time working. It gives them free on-site medical and dental care and haircuts, as well as washers and dryers. It charters buses with wireless Web access between San Francisco and Silicon Valley so employees can toil en route to the office. To encourage innovation, it gives employees one day a week to work on anything that interests them.

To eliminate the distinction between work and play—and to keep the Googlers happy at the Googleplex—they have volleyball, foosball, puzzles, free drinks and snacks, vibrating massage chairs, and a culture encouraging Googlers to bring their dogs to work. (No cats allowed.)

Meanwhile, the Googlers spend countless hours tweaking Google's hardware and software to reliably deliver search results in a fraction of a second.

SOURCE: David A. Vise, 2005, "What lurks in its soul?" *The Washington Post*, November 13, p. B1. Copyright © 2005 Washington Post Writers Group, Reprinted with permission.

a share in the benefits of their heightened efforts, they can be a powerful tool for organizational success. Box 17.2 describes the team organization and working environment of Google—one of the most important Internet companies in the world.

Co-Determination and Joint Union-Management Programs Worker participation can also occur through formal consultation with workers at every level of the organization, from the shop floor to the boardroom. In Western Europe various forms of employee participation are widespread. These forms include the workers' councils in Germany, which act as an autonomous board to review management policy, and technology stewards in Norway, who review and advise on technological change.

Additional forms of participation occur through joint union-management initiated programs. **Joint union-management programs** are based on explicit collectively negotiated agreements between unions and management to jointly sponsor programs that include employee involvement. In the United Kingdom such programs are relatively commonplace across a wide range of industries (Biagi, 2002). In North America such programs are concentrated in the automobile and telecommunications industries (Cutcher-Gershenfeld and Ford, 2005). Other well-established programs exist in steel, construction, and the public sector. The key focus of many of these programs is on improved worker training to meet the challenges of automation and global competition.

In joint union-management programs, the issues to be discussed are not necessarily restricted to management-defined agendas. Workers in the automobile industry have successfully bargained for various forms of accelerated training under joint union-management programs and voice a great deal of satisfaction with these programs (Cutcher-Gershenfeld and Ford, 2005). In these programs, workers receive additional training as part of an exchange for their greater involvement in the workplace and their increased contributions to productivity. The programs can involve supplemental training both on and off company time. Joint union-management programs thus typically include a consideration of employees and their rights and interests rather than focusing solely on production related issues as is typical, for example, of Japanese Quality Control Circles.

Workers in joint union-management programs are also increasingly allowed to go on purchasing and sales trips previously reserved for management and sales personnel. Workers provide valuable hands-on information in negotiations to secure the best components and new technologies. They also work directly with customers to learn how to improve quality and meet customer needs. The new knowledge and flexibility generated by such programs provide workers with opportunities to develop better relationships with their coworkers and with workers up and down the production chain. The opportunities provided by joint programs thus encourage employees to construct their organizational roles more actively. This active orientation generates new roles and new ideas that are often missing when work roles are unilaterally prescribed by management. In general, workers have been very enthusiastic about joint union-management programs and about participating in decision-making processes historically reserved for management (Ferraro et al, 2005).

The bilateral nature of joint initiatives provides a legitimacy to these programs that is sometimes missing when programs are initiated unilaterally by management. This legitimacy has been identified as a significant foundation for the success of joint union-management programs in stimulating productivity and improving working conditions. The initiatives emerging from joint union-management programs are also often more complementary with the public purpose than unilateral management initiatives because they include a focus on the preservation of employment and on the quality of employment as well as on increased profits (Cutcher-Gershenfeld and Ford, 2005).

Employee Ownership An additional type of employee participation is based on **employee ownership**. Worker ownership can be either total or partial through employee stock ownership plans **(ESOPs)**. In 2000, 13 million U.S. workers, or about 10 percent of the labor force, participated in ESOPs (Census, 2006).

Employee ownership generally results in improved productivity and improved employee satisfaction (Appelbaum et al., 2000). The underlying reason for these improvements is that worker-owned enterprises are simply more concerned with the well-being of their employees than organizations owned by outside shareholders. They are able to solicit high levels of worker involvement and participation because of the genuine overlap between the goals of the enterprise and those of the employees (Rothschild, 2000). Improved communication, teamwork, and participation are important underpinnings for the success of worker-owned enterprises. The greatest participation occurs in worker-owned cooperatives in which workers not only own the firm but also actively manage its day-to-day affairs (Logue and Yates, 2001).

Worker buyouts of existing companies have often been initiated in an effort to preserve jobs. Worker-owner enterprises thus often face precarious circumstances because of external factors. Worker ownership often results from an employee buyout of a plant in a last-ditch effort to save the plant and the jobs it represents. In such situations market forces may already be working against the enterprise. The market niche it serves may be shrinking, or its technology and equipment may be outdated. Employee buyouts thus often face a precarious future because of the circumstances of their birth (Logue and Yates, 2001). Worker ownership offers no necessary panacea to troubled firms, but even tested in this harsh environment, it has had a good record of success. The story of an employee owned and a highly successful Ceramics factory in Argentina is presented in Box 17.3.

Job Security An essential foundation for all forms of heightened employee participation is **job security**. Without job guarantees, both on paper and in a history of commitment, workers are reluctant to give their best efforts to increasing productivity. This reluctance is especially strong in areas of active technological change, where the possibility of displacement for large numbers of workers is very real. Only when there is a strong commitment by the organization to maintain employment levels will workers give their full support to overcoming the inevitable problems associated with technological and organizational innovations. *Loyalty by the company to its employees*

BOX 17.3 **Another World is Possible—The Ceramics of Zanon**

The workers of a ceramics factory who took control of the company and have been functioning as a cooperative for four years now, demonstrate that even working for a large, high-tech business, it is possible to create another life.... Zanon is Argentina's most important ceramics factory, covering almost twenty acres and utilizing the latest technology.

[Following a company proposal to drastically downsize], in March of 2002, 220 of the 330 workers decided to occupy the factory and began production under worker control.... From the point of view of production, the development of Zanon under the ownership of its workers has been a success: at the moment they are producing 300,000 square meters of tile, exceeding previous production levels, and they expect to reach 400,000 square meters in the near future.... One notable difference between Zanon and other businesses recovered by their employees is its relationship with the community. The factory has ... started making a countless number of donations: to hospitals, schools, nursing homes, soup kitchens for the general public and for children, to indigenous groups and groups with disabilities, firefighters, and the Red Cross....

The differences between the current and previous managements are striking. A former employee recalls,

"We were not allowed to leave or go to the bathroom. The pathways were marked out with different colors.... Back then, the kiln operators had to wear red clothes, electricians green, and so forth. That way, they could tell if someone from another sector was somewhere they weren't supposed to be. It was like a jail." The managers were on the upper floor, in offices with glass windows so they could keep watch....

Once a month the factory calls a day-long meeting in which every member participates. It covers all topics—from the type of footgear necessary for each section, to the purchases they will make, and external actions of solidarity in which they will participate. "The social, political, and production aspects are all discussed. For each point, we have a specific order that we go in, and we will not adjourn the meeting until every last issue is agreed upon," recount the workers. This way of functioning—democratic, participatory, and horizontal—requires a lot of energy: "It is exhausting, but it is productive because you find solutions to all of the problems, debating them with everyone. It is worse to let time go by because everyone ends up with doubts."

SOURCE: Raúl Zibechi, 2006, "Another World is Possible: The Ceramics of Zanon," (Silver City, NM: International Relations Center). *http://americas.irc-online.org/am/3078.* Reprinted by permission.

---

is thus an essential precondition for the realization of the full benefits of employee participation.

Training

Training programs for employees are essential if they are to have the knowledge necessary to take a leading role in a more innovative workplace (Bills, 2004). **Worker education** programs have grown dramatically in the 2000s in community colleges and training institutions around the country. Community colleges provide flexible course sequences tailored to the needs of local industry. In addition, there has been increased interest in expanding traditional apprenticeship programs for the skilled trades. This interest has been sparked by unfavorable comparisons between the U.S. system

of apprenticeship and the more developed German system, which many observers credit with making German products world renowned for their quality (Streeck and Thelen, 2005). In-house training programs have also been expanding as employers and unions seek to expand the skills of workers as a means of increasing productivity, saving jobs, and increasing profits. Such programs create an environment for **continuous learning** for an organization's workforce.

Distributing Profits A final key to successful job redesign is the development of mechanisms for distributing some of the profits of innovation back to workers (McHugh, Cutcher-Gershenfeld and Bridge, 2005). The redistribution of profits is important for maintaining worker enthusiasm and commitment.

The most innovative organizations are aware of this and have developed various means to redistribute increased earnings to workers. These include higher pay, production bonuses, and profit sharing.

The increased attention to workers' needs in innovative organizations may result in an improvement in women's relative position in this sector. Because of reduced discrimination and less traditional career choices on the part of women, female workers may move increasingly into traditionally male-typed jobs, which are more likely to be in the innovative sector. The greater flexibility of women in their career choices as a result of lessened child-rearing duties may further facilitate these trends (England, Thompson, and Aman, 2001).

Barriers to Innovative Job Redesign   Job redesign also faces certain barriers that may limit its effectiveness. Chief among these is the problem of limited commitment by large corporations to their workers. Employee participation programs in the United States have often been superficial and have been accused of being more window dressing than substance (Vallas, 2003). U.S. corporations have often acted as if *proclaiming* their allegiance to employee participation somehow constitutes an adequate solution to lagging productivity and to lack of management loyalty and commitment to their workers. Developing programs that actually incorporate workers in active roles at all levels of decision making and becoming committed to the long-term interests of employees requires more than just a public relations announcement. In addition, large organizations may experiment with job redesign in one plant with great success only to cancel the experiment because of changes in organizational strategy initiated from the top (Rothschild, 2000).

A second potential point of conflict can arise because increased employee participation often reduces the need for managers by incorporating managerial and supervisory activities within shop-floor groups. Employees may even be asked to evaluate their manager's performance. These changes may threaten the jobs of supervisors and middle-level managers and may stimulate resistance

on their part (Smith, 2001). When combined with agendas of corporate restructuring and theories of "lean production" that stress downsizing, job redesign programs put middle managers at significant risk of being laid off. Middle-level managers are well positioned to either facilitate job redesign programs or to sabotage them through subtle noncompliance and other tactics that workers themselves occasionally use with great effectiveness. (Recall our discussion in Chapter 3 of the resistance strategies used by alienated workers.)

The ways in which employee participation can be incorporated into job redesign and technological innovation are extremely diverse. In this chapter we have described some of the forms of employee participation. The major forms of employee participation are presented in Figure 17.1 along with the corresponding issues that are open to negotiation at the various levels of employee participation. At one extreme, workers are involved only in decisions about how to improve product quality and efficiency. At the other extreme they are involved in investment decisions about when and where to build new factories and what new endeavors to pursue. Workers have shown themselves able to participate effectively in decisions about their own working conditions, in decisions about the production process, and in decisions about investment (Logue and Yates, 2001). All of these forms of employee participation are important, all have been proven to be effective in some circumstances, and all have problems. No one form or level of participation is right for all circumstances. In industries with rapidly changing technologies, employee participation in job design may be most important. In industries with rapidly changing market situations, employee participation in investment decisions may be essential for continued economic viability.

The persistence of enthusiasm for employee participation and job redesign demonstrates that these programs are here to stay. Greater employee participation can make important contributions to productivity and competitiveness. Such programs are precursors to new systems of industrial relations that will characterize a significant share of employment positions in the twenty-first century.

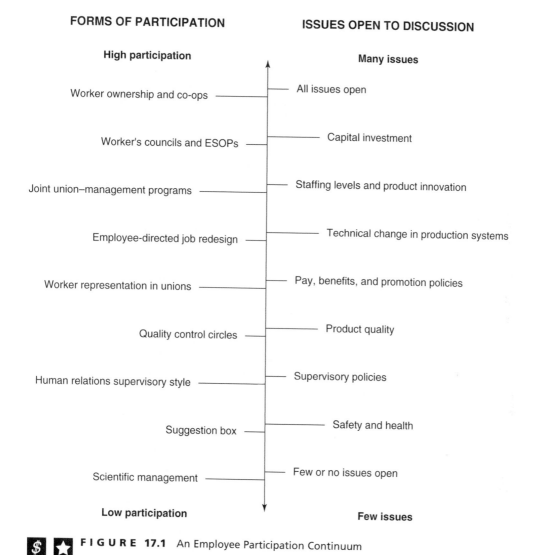

FORMS OF PARTICIPATION

High participation

Worker ownership and co-ops

Worker's councils and ESOPs

Joint union–management programs

Employee-directed job redesign

Worker representation in unions

Quality control circles

Human relations supervisory style

Suggestion box

Scientific management

Low participation

ISSUES OPEN TO DISCUSSION

Many issues

All issues open

Capital investment

Staffing levels and product innovation

Technical change in production systems

Pay, benefits, and promotion policies

Product quality

Supervisory policies

Safety and health

Few or no issues open

Few issues

FIGURE 17.1   An Employee Participation Continuum

The Marginal Sector

In the economy of the twenty-first century it is also likely that a large sector of marginal employment will exist. The existence of such a sector alongside a highly innovative sector with increased employee participation suggests a more divergent economic structure in the next century. Why might a marginal sector grow in the future? Driving down wages is one possible response of organizations to competition. By cutting wages, enterprises can remain competitive, at least for a time. In sectors where technological change is slow, such as services, such a strategy may appear more attractive than strategies for increasing productivity.

Low Pay and Few Benefits   An increase in the marginal sector would cause an intensification of current social problems. Workers in this sector would have reduced buying power, thus limiting the demand for the goods produced by other workers and slowing the growth of the economy as a whole. Without a national health care plan, workers and families would lose health insurance and other benefits, thus putting additional pressure on already pinched social services and welfare budgets. The growth of marginalized jobs is deeply implicated in increasing inequality in the United States over recent decades (Gottschalk and Danziger, 2005). Continued expansion of the marginal sector would further increase poverty and homelessness.

What will work be like in the marginal sector of the future? Based on the characteristics of the current marginal sector, which we explored in Chapter 14, we can anticipate that many aspects of long-term, stable employment will be missing. Pay will be low, and part-time and temporary work will be common. This sector will be typified by extensive subcontracting and frequent use of temporary workers (Tilly, 2004). Employers will have little interest in or commitment to employees. Probationary periods of employment with reduced rights will also be common. Two-tier pay scales with lower pay for newer employees might be used to bring in lower wage levels, with senior employees on the higher scale being rapidly phased out. Workers in the informal economy of unreported (and untaxed) work will also be part of this expanded marginal sector. Basic benefits and employment security are largely missing from this employment sector.

Modern Sweatshops Some aspects of work in the marginal sector of the twenty-first century will be similar to work in the factories of the early nineteenth-century Industrial Revolution described in Chapter 1—but with some new twists. There will be fewer safeguards against hazardous working conditions than in the innovative sector. Enthusiastic commitment to work will be rare. Discontent, subtle forms of noncompliance and even sabotage will be relatively common. Many workers in this sector will be underemployed, and many others will experience periodic unemployment as temporary jobs come and go. Young people will have a particularly

hard time locating permanent jobs and may spend an increasing portion of their work lives in marginal employment. Similarly, disregard of minimum wage restrictions and fair time and hours standards will be all too common. Mandatory drug testing, electronic surveillance, and other forms of monitoring may be widespread. Workers in the marginal sector will represent a transient **underclass**. It is an unfortunate but realistic projection that such an underclass may be a sizable component of the North American workforce in the twenty-first century (Wilson, 2003).

An Absence of Employer Commitment In what parts of the economy will marginal employment grow? Marginal jobs, with unilateral managerial control and without the benefits of employee participation, are most likely to occur in settings that are not unionized and that are technologically stagnant (Clawson, 2003). In these industries, employees lack sufficient bargaining strength to demand participation in managerial decisions and technological and market forces allow low-wage strategies to replace high-productivity strategies. The marginal sector is thus likely to grow in the absence of organized worker power and of competition based on technological change or foreign producers. Service industries, such as restaurants, and retail trade provide examples of such situations. It is also possible that the marginal sector will grow where advanced technology is used to reduce the need for skilled workers.

Without technologically or market-based competition there is little reason to initiate job redesign and employee participation. Without organized worker power there is often inadequate reason for managers to respond to heightened competition with workplace innovation rather than with lower wages and reduced **employer commitment** to workers. In situations where either condition is absent, it is likely that marginal jobs will increase. Such situations may represent a sizable or even majority share of new employment positions in the economy of the twenty-first century. Indeed, if the trend continues for new areas of job growth to be outside traditional union strongholds and outside the traditional professions, elements of marginality may typify many areas of new employment. In areas of growing employment, such as services and microprocessor manufacturing, marginality can be expected to increase. Box 17.4 describes a hypothetical marginal job of the future.

Continuing Disadvantages for Female and Minority Workers

How will female and minority workers fare in the economy of the twenty-first century? This question is difficult to answer. The recent past has seen improvements in employment conditions for female workers, but the situation for minority workers has been more mixed. If women's earnings continue to advance against men's earnings, due to reduced discrimination in hiring and promotion and less traditional occupational choices by women, the prognosis for the future of women's employment is relatively good (England et al., 2001). If minority workers continue to make occupational advances into fields previously dominated by whites, the prognosis for minority employment is reasonably good. However, any remaining prejudices and discrimination against women and minority workers will result in their disproportionate representation in the marginal work sector of the twenty-first century. If the burden for child care continues to fall disproportionately on women and if the divorce rate continues at a high level or even increases, then poverty associated with family dissolution will continue to plague many women and will seriously constrain their employment options. Similarly, the reemergence and growth of a black underclass reminds us that, even if some minority workers have made progress in recent decades, many others have been left behind (Wilson, 2003).

Increased immigrant populations in many industrialized nations have added to the possibilities for sharply divided societies. New immigrants generally assimilate into a society, but "guest worker" programs in Europe have contributed to the creation of a seemingly permanent underclass of workers without the benefits of full citizenship rights (see Chapter 16). Similar programs are also periodically discussed in the United States, seemingly without full appreciation of their implications for creating a permanent underclass.

We have predicted the emergence of an increasingly divided economy and society with large innovative and marginal sectors. Such a bifurcation could result in today's industrialized nations having class structures that more closely resemble those of periphery or semi-periphery nations with extreme inequality between the rich and the poor. Social and political forces, many of which are outside the scope of this book, will influence the relative size of these two employment sectors. Similarly, the relative representation of minority groups and women in each sector will be contingent on the ability of these groups to further reduce discrimination through mass movements and legislative agendas. Nevertheless, there are many relevant factors that will influence the relative development of these two sectors that we *can* consider.

WORKING TOWARD A BRIGHTER FUTURE

The emergence of an economy divided between innovative and marginal sectors is not inevitable. The problem before society is how to increase the size of the innovative sector and how to decrease the size and diminish the negative consequences of the marginal sector. How might these goals be achieved?

Increasing Innovation

What conditions foster innovative job redesign and employee participation? Workers in industrially advanced nations today are more highly educated than ever before in history. They are also more interested in safety and health issues and in improving the conditions of their employment. They have come to have high expectations about the satisfactions and rewards that work can provide. These expectations are perhaps the most important precondition for the growth of a highly innovative

BOX 17.4   A Marginal-Employment Scenario

It had snowed all night, but the snow-jets had not yet cleared the lot. No matter, thought Debra as she braked her bike. Like most other parking lots built in the 1990s, this one was too big. *Way too big,* she thought, *for my bike.* If today was typical, hers would be the only vehicle parked there through the ten-hour shift. Marge Henry's little van was there now, but it was nearly shift-change time. Debra eased her right leg over the bike and into the ankle-deep snow. This was no time to slip and break something: the company had already cut back the health benefits to the government-required minimum, and Debra had nothing extra for the 50 percent coinsurance.

The snow muffled all sounds; it was almost as quiet here in the parking lot as it would be inside. The winter sun was nearly up over the horizon, and Debra looked over her shoulder at the lightening gray clouds in the east. In the winter she saw the sun only on Sunday. Her suntan, like her marriage, was a casualty of the company's shift policy. Ten hours a day, six days a week, take it or leave it. Many, many would take it if she left it. Certainly poor Jim. It was hard having an unemployed husband. It *had been* hard, she corrected herself. She pulled off a mitten and pushed her thumb against the glass for print recognition.

As the glass doors rolled back, she stepped across the laser beam on the threshold. It was easier than punching a clock but unrelenting in its accounting of her time. All personal areas of the shop had been fenced in with laser beams, so any "nonproductive" time was deducted from your paycheck. No tolerance, she thought wryly, if you had morning sickness, a touch of the "bug," or just too much spicy food last night. That was one reason for showing up a little early—it gave you a cushion of time if you needed an extra stop in the personal room. She crossed the corridor to the control room and pressed her thumb against another print-reader. Marge, she knew, would be locked inside. The company kept tight security, mainly for the sake of the robots (which were expensive).

"Any problems, Marge?" Debra asked.

"Number thirty-two is acting up again. Quiet night otherwise."

Marge crossed the laser beam to the dressing room and sat down to pull on her snow boots. She resented Debra. Both women worked ten hours a day, but Marge worked the night shift, and so she received only minimum wage. Debra made ten cents an hour

more. Education had nothing to do with it; both jobs required a degree in robotics. The difference, the company claimed, was in the volume of work. This warehouse filled orders for toys—or rather, its forty-five robots did. The orders came in on the electronic LINCOM, and the robots located the proper storage bin, selected the proper number of boxes, issued address tags, and loaded the boxes into huge shipping containers. Twice a day, trucks came to pick up these containers. During the day shift, orders came in from across the United States and Canada, a few from Mexico. During the night shift, most of the orders came from Europe, Japan, and Africa. Although the toy market was growing there (in inverse proportion, it seemed, to declining birthrates), the volume of night business was much lower. Ample reason, according to the company, for Marge's paycheck to be a dollar less every day than Debra's even though the same forty-five robots needed routine and preventive maintenance, and there were still occasional foul-ups.

Then there was the matter of the swing shift. Between 6 P.M. and 10 P.M. a young college graduate came in to work the short evening shift. It was so hard for new graduates to find jobs these days that some companies viewed these little part-time jobs as favors they did the worker, not as jobs that required a paycheck. This company paid an honorarium—half of minimum wage—for the decided advantage of gaining work experience. The evening shift wasn't all that easy, either, what with the boom in the Pacific Basin.

Marge was ready to leave. Debra saw her out, then checked the security system. The LINCOM was already humming, and she saw two more robots move off to fetch toys. Thirty-two was idle and would bear watching. As she punched up the maintenance record on thirty-two, she saw the reflection of a blue light blinking overhead. The blue light was the "spy in the sky," the electronic monitoring device that ensured that employees did not alleviate their boredom with transistor videos, personal message units, or other distractions. She was glad that someone, somewhere, knew she was here, even if only to check up on her. The workday had begun, but except for the humming of the LINCOM and the soft whir of the robots' wheels, you might never have known.

SOURCE: Adapted from Teresa A. Sullivan, 1989, "Women and Minority Workers in the New Economy: Optimistic, Pessimistic, and Mixed Scenarios." *Work and Occupations* 16, 4 (November):393–395. Used by permission of Sage Publications, Inc.

sector based on heightened participation and continuous learning.

Organizational and Market Imperatives   Not all industries, however, are equally likely to pursue a strategy of workplace innovation. Innovation has tended to emerge when "both labor and management perceive a necessity for technological innovation; . . . and workers have the bargaining strength to demand participation in managerial decision making" (Cornfield, 1987:333). When these conditions are missing, innovation is unlikely to occur, and work practices will more closely resemble those of the mass-production industries of the past or those of the marginal sector.

Intensified competition also provides an impetus to workplace innovation. Labor and management do not typically cooperate unless competitive pressures force them to do so. Without these pressures they will be more likely to pursue their separate goals by engaging in adversarial bargaining over their respective shares of the economic pie.

Increased Worker Power   An important precondition for innovation is that workers have sufficient power to demand participation and innovation. When workers are unable to demand participation, managers may favor a strategy of reducing wages, benefits, and job security in order to remain competitive rather than following the more uncertain, complex, and demanding route of workplace innovation. In the long run, however, the option of remaining competitive by reducing wages may be unrealistic in many industries (even ignoring the ethical hazards involved in such a strategy). As a result there is reason to hope that a significant portion of organizations in industrially advanced societies will opt for workplace innovation. Professional, semi-professional, and craft workers are in particularly good positions to leverage their scare skills into higher levels of participation, thus setting the stage for productivity enhancing innovations in these sectors.

The Role of Unions   Employee participation is most likely to take place in unionized firms. The concepts of direct employee participation and cooperation with management, however, are to some extent antithetical to traditional union approaches to industrial relations. On topics ranging from work rules to grievance procedures, unions have increased workers' rights by negotiating explicit rules and ensuring their enforcement. This approach has often translated into a formalistic and legalistic approach to workplace issues, exactly the sort of approach that can be stifling to innovation (Diamond and Freeman, 2002).

Unions have also been cautious in advocating innovative workplace changes because they fear that altering the status quo may undermine their ability to deliver the package of benefits and protections that have been their traditional offering to workers. To survive in innovative settings, however, unions will have to stop *reacting* to changing circumstances and begin to take an *active role* in developing proposals for increasing productivity. In the past, union leaders have often dismissed employee participation as "a ruse to increase productivity and cut jobs" (Vallas, 2003). Continuing to dismiss employee participation in this manner will undermine the role of unions in the workplace of the future. The emergence of joint labor-management groups at various levels of the organization is necessary to increase productivity in rapidly changing and highly competitive industries. If the innovative sector is to grow and prosper, it will have to include the unionized industries, but unions will have to change to accommodate and promote innovation (Fantasia and Voss, 2004). These changes are threatening to established union practices in many ways. Unions in the industrially advanced European nations, however, have taken a leading role in promoting increased employee participation. Unions and employee participation are not inherently antithetical; indeed, they may be dependent on each other.

Education and Training   Educating the workforce for innovation will require an increased commitment not only to college education but also to **vocational education** and to ongoing **retraining** for workers already in the labor force. Many European nations have instituted programs for mid-career retraining. Observers credit a substantial part of these countries' economic vitality to such programs. North Americans must move beyond retraining programs designed to address the needs of the currently unemployed. Such programs include the Comprehensive Employment Training Act and Job Training Partnership Act. Although these programs are an important safeguard for workers otherwise destined for work in the marginal sector, they are largely inadequate for training the highly skilled labor force necessary for the innovative sector (Bills, 2004). Besides providing college for the middle classes and job training for the unemployed, North America must provide its existing skilled workers with the education and continuing training necessary to compete effectively in the global economy.

Workplace Experimentation Increased **workplace experimentation** is also needed to stimulate the growth of an innovative sector. *Successful job redesign* comes only from experimentation. It cannot be fully specified in abstract formulas—it requires a constant process of review and change. Unfortunately, sustained programs of experimentation in job redesign and employee participation are not widespread in the United States and Canada.

To increase the innovative sector, the preconditions for its existence must be broadened. These preconditions include workplace experimentation, increased worker power, and technological innovation. Greater worker power can come about either on an individual basis, through increased education, training, and professional development, or on a collective basis, through increased representation in unions and professional associations. Encouraging education and unions fosters the growth of an innovative sector and decreases the prevalence of marginality. The impetus for these changes must come from workers, unions, and professional associations. Reforms of this sort will not occur unless workers and their organizations demand them (Diamond and Freeman, 2002).

Reducing Marginal Employment

Besides supporting workers' abilities to demand technological innovation and competitiveness and pro-

viding a labor force educated to meet the challenge of innovation, society can also set up roadblocks against encroaching marginality. These roadblocks could include an increased minimum wage, restrictions on mergers and plant closings, and standardization of tax, labor, and environmental laws across the country so that states are not forced to compete with one another by offering companies the most minimal employment standards (Malecki, 2004). In many cases higher minimum wages have been shown to have significant positive effects on employment stability and the acquisition of additional training without substantial negative effects on new hiring (Rodgers and Rubery, 2003). Roadblocks to capital flight would keep capital at home and in place where it can be used to foster innovation and increased productivity as preferred responses to competitive pressures. Programs encouraging worker buyouts, ESOPS, and cooperatives can also reduce marginality. Banks could be encouraged to lessen their resistance to extending needed credit to worker-owned enterprises (Logue and Yates, 2001).

Expanded worker training programs are important not only for facilitating the spread of innovation but also for reducing marginal employment. The need for such programs is increased by the heightened pace of technological change. Fewer and fewer workers can expect to work at just one job throughout their careers. With more rapid job changes, improved job placement and matching services also become increasingly important if jobs are to be filled by qualified applicants, if workers' abilities are to be used effectively, and if workers are to avoid periods of marginality. Employment of the labor force at nearer its full potential would increase productivity and reduce welfare expenses immensely.

Reducing the marginal aspects of available jobs would help lessen underemployment, unstable employment, and blocked mobility. In conjunction with education and training programs and the expansion of the innovative sector, a reduction of marginal employment positions would facilitate the incorporation of all Americans in the mainstream of society. The success of such agendas will also depend on the vigorous enforcement of laws protecting the rights of workers.

Expanding Leisure

Expanded leisure hours could also make an important contribution to improving the quality of work life and distributing available employment. The average hours worked per week dropped from near seventy in 1850 to about forty by the 1930s. Starting in the 1970s, however, this trend began to be reversed. Fully employed U.S. workers found that their workweek increased by about 5 percent between 1970 and 2000. Simultaneously paid holidays and vacation days fell by 15 percent (Schor, 2003). This change occurred simultaneously with rising unemployment. In the 2000s full-time work has become harder to get, and it is often highly demanding when it *is* secured. Reducing the hours of work would be an effective way to distribute available employment in a period of rapidly increasing productivity and technological change. Reduced hours of work would also help prevent polarization of society between those with too much work and those with inadequate work or no work at all (Hunnicutt, 2003).

A related possibility is that people will combine work, leisure, and education throughout their lives rather than completing education before starting work and saving their leisure years for retirement. As we discussed in Chapter 5, such a pattern would deviate from the so-called normative career pattern. A career pattern with greater integration of work, education, and leisure would include periodic breaks for retraining and renewal, breaks that would be extremely important for sustaining the innovative spirit in workers. Lifestyles that allow for a greater integration of work, education, and leisure across the life cycle would also help resolve the dilemma of fewer workers being needed in core industries as technologically based advances in productivity continue to accumulate.

Public Goods and Reduced Consumption Lifestyles

Over the history of industrial society the provision of "public goods" has increased. **Public goods** are products or services to which the citizens or residents of a society are entitled without direct payment or at a nominal fee. Such public goods include education, clean air, public parks, libraries, social security retirement income, and, to some extent, health care. If the provision of public goods expands, workers who remain in the marginal sector of employment will at least be spared some of the most debilitating consequences of poverty and marginality. The provision of public goods can thus be an important preventive against the reproduction of a marginal sector. The provision of public goods also creates jobs, providing additional escape routes from marginality. Alternatively, if such services are provided only privately as parts of benefit packages restricted to the innovative sector or on a fee-for-service basis, many in society will have to do without them.

Extremely high private health care costs, in particular, restrict the expansion of the innovative sector and limit economic growth in general. Health care benefits become part of the costs of businesses significantly reducing the competitiveness of many American businesses. Providing health care as a public good, or otherwise constraining rising costs in health care, could be a powerful incentive to economic growth and innovation.

Greater provision of public goods also provides an alternative to the "work-and-spend" treadmill that traps many Americans in a high-work–high-consumption lifestyle. Average consumption expenditures for Americans have doubled in the last thirty years from $11,171 to $22,152 (Schor, 2003). The effort of earning this much money, shopping to spend it, and then tending these goods leaves many middle-income Americans too exhausted to enjoy their lives. Worse, for many, high private consumption patterns result in excessive debt leading to bankruptcy and severe disruptions of personal and family lifestyles (Sullivan et al., 2000).

The provision of public goods and the training of workers for employment in the innovative sector are among the most important strategies for avoiding a deeply class-divided society in the twenty-first century and for providing a better quality of life for all. Whether we pursue such agendas as a society depends on the importance we attach to the creation of a just and inclusive society (Ackerman, 2000).

SUMMARY

The rapidly changing economic, technological, and organizational realities of today's global economy are setting the stage for the world of work in the twenty-first century. Alternative scenarios are available, but the choice among them depends on political and social actions we take today. The dilemmas described in this book can be resolved in a positive manner. Workers and other members of society can pressure organizations to respond to competition by increasing innovation rather than by cutting wages and fostering marginality. They can also pressure labor unions and other workers' associations to adapt to the new conditions of a competitive environment and to provide leadership in the areas of technology, organizational innovation, and employee training. Finally, we can all pressure the government to enact laws and programs that will foster innovation, job creation, employee involvement, expanded leisure,

and a better quality of life for all. Significant advances on all of these dimensions are needed if we are to avoid an increasingly divided society in the twenty-first century.

Industrial societies have the capacity to constantly increase productivity through technological and organizational innovation. This capacity has never been greater than it is today. This very capacity though creates the dilemma of how to distribute available work when more and more goods and services can be produced by fewer and fewer workers. These, then, are the central challenges of contemporary industrial society: How do we continue to increase productivity? How do we distribute available work? And how do we distribute the goods and services produced? The manner in which we resolve these dilemmas will determine the political, social, and economic landscape of the twenty-first century.

KEY CONCEPTS

worker power

innovative sector

marginal sector

postbureaucratic

job redesign

employee participation

quality of work life

joint union-management programs

employee ownership

ESOPs

job security

worker education

continuous learning

underclass

employer commitment

vocational education

worker retraining

workplace experimentation

expanded leisure

public goods

QUESTIONS FOR THOUGHT

1. Consider your chosen (or likely) field of work. Identify one way in which you think each of the factors of technological change, increased world competition, and increased employee participation might influence the nature of your work and career.

2. What developments, other than those discussed in this chapter, do you think will significantly influence the nature of work and society in the future? (HINT: Environment? Population? Immigration? Something else?)

3. What factors might continue to channel female and minority workers into the marginal sector in the twenty-first century? Do you think women will gain full equality in the workplace and in society? Will minorities gain equality? Why or why not?

4. To what extent do you think the United States will be successful in expanding the innovative sector? What factors are likely to play a leading role in the success or failure of this expansion?

5. How optimistic are you about the possibilities for reducing marginality in the twenty-first century? What might be some of the consequences of failing to address the issue of marginality?

MULTIMEDIA RESOURCES

Print

John de Graaf (editor). 2003. *Take Back Your Time*. San Francisco: Berrett-Koehler. Provocative essays on overwork, leisure, and the quality of life.

Eileen Appelbaum, Thomas Bailey, Peter Berg and Arne L. Kalleberg. 2000. *Manufacturing Advantage: Why High-Performance Work Systems Pay Off*. Ithaca, N.Y.: Industrial and Labor Relations Press. A systematic analysis of work systems involving greater employee involvement and their benefits for productivity and enhanced worker well-being.

James R. Barker. 1999. *The Discipline of Teamwork: Participation and Concertive Control*. Thousand Oaks, CA: Sage. An ethnographic account of the use of a team-based approach to increase productivity. Reveals both the benefits and the pressures of teamwork.

Internet

Employee Involvement Association. *www.eianet.org* Dedicated to promoting workplace change through employee involvement.

Team Builders Plus. *www.teambuildersplus.com* A commercial site for those involved in developing, managing, or working in self-directed teams. Many resources and examples provided.

Ohio Employee Ownership Center. *dept.kent.edu/oeoc* Devoted to helping employees plan and manage the transition to employee ownership. Extensive links.

Canadian Government Human Resources Department. *www.hrsdc.gc.ca/en/home.shtml* Workplace innovations support site with great links.

RECOMMENDED FILM

Gattaca (1997). In a future that may not be far away, genetic modification creates humans without disease or imperfections. Those who are unmodified are discriminated against and relegated to menial labor.

APPENDIX TABLE 1 Employed Civilians, by Occupation, Sex, Race, and Hispanic Origin: 2005
[**141,730 represents 141,730,000.** For civilian noninstitutional population 16 years old and over. Annual average of monthly figures. Based on Current Population Survey; see text, Section 1, Population, and Appendix III. Occupational classifications are those used in the 2000 Census and are not comparable to those used in the 1990 census.]

OCCUPATION	Total employed (1,000)	Percent of total			
		Female	Black or African American[1]	Asian	Hispanic or Latino[2]
Total, 16 years and over	141,730	46.4	10.8	4.4	13.1
Management, professional, and related occupations	49,245	50.6	8.1	5.9	6.4
Management, business, and financial operations occupations	20,450	42.5	7.1	4.8	6.5
Management occupations[3]	14,685	37.2	6.0	4.2	6.3
Chief executives	1,644	23.8	3.0	3.9	3.8
General and operations managers	872	29.9	6.8	3.6	6.2
Advertising and promotions managers	67	56.1	3.0	1.6	7.5
Marketing and sales managers	798	41.0	3.8	4.2	6.0
Administrative services managers	80	30.2	6.2	2.3	6.4
Computer and information systems managers	351	32.3	5.9	6.3	3.0
Financial managers	1,045	51.7	7.4	6.3	6.0
Human resources managers	272	72.7	8.5	2.9	9.4
Industrial production managers	310	17.2	6.7	2.4	6.5
Purchasing managers	183	39.7	5.7	2.1	5.5
Transportation, storage, and distribution managers	244	11.7	9.1	2.6	8.7
Farm, ranch, and other agricultural managers	195	19.3	1.4	0.2	8.9
Farmers and ranchers	827	25.6	0.6	0.9	1.1
Construction managers	881	6.3	2.7	2.3	5.8
Education administrators	805	63.4	13.4	2.4	5.4
Engineering managers	97	9.5	1.7	10.0	3.8
Food service managers	929	44.2	6.0	10.6	12.0
Lodging managers	159	49.8	5.5	12.2	4.6
Medical and health services managers	470	71.2	9.0	2.7	5.5
Property, real estate, and community association managers	626	48.7	8.0	2.6	13.0
Social and community service managers	307	69.3	11.7	2.0	6.2
Business and financial operations occupations[3]	5,765	55.9	9.8	6.5	7.0
Wholesale and retail buyers, except farm products	213	50.7	7.5	1.9	7.9
Purchasing agents, except wholesale, retail, and farm products	281	55.0	8.2	3.3	7.8
Claims adjusters, appraisers, examiners, and investigators	303	61.6	13.8	3.7	6.8
Compliance officers, except agriculture, construction, healthand safety, and transportation	139	47.2	15.8	3.6	12.5
Cost estimators	112	17.4	3.0	1.1	4.0
Human resources, training, and labor relations specialists	660	70.9	13.1	5.2	7.8
Management analysts	549	40.9	8.4	7.6	3.4
Accountants and auditors	1,683	61.9	8.4	9.8	6.4
Appraisers and assessors of real estate	125	34.7	2.2	1.9	4.0
Financial analysts	96	39.0	7.3	13.8	3.7
Personal financial advisors	380	30.8	5.1	6.2	4.4
Insurance underwriters	110	71.9	12.0	2.6	5.8
Loan counselors and officers	439	54.1	11.9	3.8	12.7
Tax examiners, collectors, and revenue agents	76	58.4	20.7	4.9	10.4
Tax preparers	93	68.2	13.6	6.3	14.9

OCCUPATION	Total employed (1,000)	Percent of total			
		Female	Black or African American[1]	Asian	Hispanic or Latino[2]
Professional and related occupations	28,795	56.3	8.8	6.6	6.4
Computer and mathematical occupations[3]	3,246	27.0	6.9	14.7	5.3
Computer scientists and systems analysts	745	30.3	8.3	11.4	4.8
Computer programmers	581	26.0	4.6	18.0	5.7
Computer software engineers	832	21.9	5.0	24.6	3.9
Computer support specialists	334	33.2	9.1	5.5	9.7
Database administrators	89	32.3	13.1	6.3	5.1
Network and computer systems administrators	200	18.4	4.9	10.4	4.8
Network systems and data communications analysts	322	24.6	9.3	9.4	6.6
Operations research analysts	86	50.5	10.1	6.2	2.3
Architecture and engineering occupations[3]	2,793	13.8	5.1	9.9	6.1
Architects, except naval	235	24.4	3.2	11.0	8.9
Aerospace engineers	90	13.3	4.3	6.7	8.2
Civil engineers	319	13.2	6.1	10.3	3.5
Computer hardware engineers	81	10.8	8.4	23.1	5.5
Electrical and electronics engineers	352	7.1	5.6	15.6	3.4
Industrial engineers, including health and safety	189	14.9	5.5	7.0	3.8
Mechanical engineers	318	5.8	2.5	8.7	4.0
Drafters	212	21.6	4.2	6.5	10.1
Engineering technicians, except drafters	410	20.2	8.9	7.0	10.1
Surveying and mapping technicians	95	2.7	3.5	2.1	11.8
Life, physical, and social science occupations[3]	1,406	42.5	5.0	10.1	4.5
Biological scientists	119	48.7	2.9	10.2	4.2
Medical scientists	125	45.9	3.0	29.6	2.6
Chemists and materials scientists	116	35.3	5.6	14.0	2.7
Environmental scientists and geoscientists	99	22.5	3.2	1.9	3.0
Market and survey researchers	131	55.3	5.1	7.3	4.3
Psychologists	188	67.3	2.8	0.3	4.7
Chemical technicians	86	28.1	18.8	6.7	7.9
Community and social services occupations[3]	2,138	61.3	17.1	2.8	9.8
Counselors	626	69.8	18.6	1.5	10.1
Social workers	670	80.1	19.0	3.3	9.6
Miscellaneous community and social service specialists	276	67.0	24.3	2.4	16.2
Clergy	435	15.5	9.7	4.4	5.8
Legal occupations	1,614	49.4	6.0	2.6	6.9
Lawyers	961	30.2	4.7	2.0	3.5
Judges, magistrates, and other judicial workers	70	41.2	7.0	4.6	5.9
Paralegals and legal assistants	353	86.4	8.2	2.5	11.5
Miscellaneous legal support workers	229	75.3	7.6	4.3	14.2
Education, training, and library occupations[3]	8,114	73.8	9.9	3.7	6.9
Postsecondary teachers	1,185	44.4	6.7	11.2	4.1
Preschool and kindergarten teachers	719	97.7	14.7	3.0	10.4
Elementary and middle school teachers	2,616	82.2	9.6	2.4	5.9
Secondary school teachers	1,136	56.8	7.3	1.1	5.2
Special education teachers	421	85.3	10.8	2.5	5.7
Other teachers and instructors	709	65.9	11.0	5.3	6.5
Librarians	214	84.9	5.8	2.5	4.6
Teacher assistants	947	90.9	14.1	1.8	14.0

OCCUPATION	Total employed (1,000)	Percent of total			
		Female	Black or African American[1]	Asian	Hispanic or Latino[2]
Arts, design, entertainment, sports, and media occupations[3]	2,736	47.8	5.9	4.3	7.1
Artists and related workers	234	52.6	2.4	5.3	6.7
Designers	774	55.0	2.1	6.8	6.6
Producers and directors	124	35.1	7.5	0.7	7.4
Athletes, coaches, umpires, and related workers	273	30.8	10.2	2.3	6.3
Musicians, singers, and related workers	213	37.6	8.2	3.8	7.8
News analysts, reporters and correspondents	89	56.5	6.3	3.3	6.0
Public relations specialists	154	61.4	6.5	2.9	4.7
Editors	150	58.1	5.8	2.3	5.8
Writers and authors	178	62.8	5.6	0.5	2.2
Broadcast and sound engineering technicians and radio operators	92	13.6	8.5	1.8	11.2
Photographers	150	39.0	8.0	3.9	9.2
Healthcare practitioner and technical occupations[3]	6,748	73.2	9.9	7.4	5.4
Chiropractors	82	21.8	1.4	4.0	2.5
Dentists	164	22.5	3.6	14.1	4.4
Dietitians and nutritionists	68	95.3	18.4	6.0	5.9
Pharmacists	248	48.3	6.3	10.4	5.5
Physicians and surgeons	830	32.3	5.3	17.3	5.2
Physician assistants	74	62.1	9.0	7.5	11.1
Registered nurses	2,416	92.3	10.0	6.4	4.3
Occupational therapists	85	92.9	3.6	12.0	1.4
Physical therapists	177	68.9	4.6	6.0	5.1
Respiratory therapists	94	58.6	11.4	2.6	6.0
Speech-language pathologists	98	92.0	5.0	0.8	2.8
Clinical laboratory technologists and technicians	334	74.2	11.4	12.1	7.5
Dental hygienists	132	97.1	1.0	3.3	8.5
Diagnostic-related technologists and technicians	276	72.0	12.5	1.3	5.5
Emergency medical technicians and paramedics	155	31.3	8.1	1.5	10.1
Health diagnosing and treating practitioner support technicians	390	81.8	11.8	5.7	8.5
Licensed practical and licensed vocational nurses	510	93.4	21.6	2.6	5.5
Medical records and health information technicians	121	86.6	19.0	2.3	9.0
Service occupations	23,133	57.3	15.8	4.2	19.2
Healthcare support occupations[3]	3,092	89.0	24.8	4.1	13.8
Nursing, psychiatric, and home health aides	1,900	88.7	32.5	4.6	15.0
Massage therapists	111	82.0	5.5	4.3	3.1
Dental assistants	259	96.1	6.3	3.3	13.5
Protective service occupations[3]	2,894	22.4	19.3	1.6	10.4
First-line supervisors/managers of police and detectives	122	12.5	8.2	0.2	6.3
Fire-fighters	243	3.3	10.6	1.4	7.9
Bailiffs, correctional officers, and jailers	424	29.1	23.7	1.3	10.8
Detectives and criminal investigators	123	24.0	13.8	1.4	8.9
Police and sheriff's patrol officers	677	14.3	13.7	1.1	11.3
Private detectives and investigators	89	37.7	12.8	2.8	6.4
Security guards and gaming surveillance officers	814	24.7	31.3	2.5	12.1
Food preparation and serving related occupations	7,374	56.6	11.6	5.4	20.6
Chefs and head cooks	317	20.4	11.8	19.3	18.1
First-line supervisors/managers of food preparation and serving workers	574	58.5	14.9	3.6	15.5
Cooks	1,838	42.3	16.5	5.0	29.3

OCCUPATION	Total employed (1,000)	Percent of total			
		Female	Black or African American[1]	Asian	Hispanic or Latino[2]
Food-preparation workers	664	61.2	12.2	6.8	24.4
Bartenders	384	56.8	3.2	1.7	8.6
Combined food preparation and serving workers, including fast food	305	71.5	12.5	3.5	13.0
Counter attendants, cafeteria, food concession, and coffee shop	321	65.7	10.3	4.4	11.4
Waiters and waitresses	1,927	71.8	7.1	5.2	15.3
Food servers, nonrestaurant	147	69.1	23.4	6.4	19.4
Dining room and cafeteria attendants and bartender helpers	373	46.6	9.6	6.7	30.4
Dishwashers	264	23.8	15.4	1.5	35.4
Hosts and hostesses, restaurant, lounge, and coffee shop	256	84.6	6.5	3.3	12.8
Building and grounds cleaning and maintenance occupations	5,241	40.6	15.8	2.7	30.6
First-line supervisors/managers of housekeeping and janitorial workers	291	36.3	15.4	2.8	20.0
First-line supervisors/managers of landscaping, lawn service, and grounds-keeping workers	238	6.2	6.3	0.3	13.2
Janitors and building cleaners	2,074	32.8	18.3	2.8	27.3
Maids and housekeeping cleaners	1,382	89.5	20.1	4.2	35.2
Pest control workers	69	2.6	7.8	2.4	27.8
Grounds maintenance workers	1,187	7.5	9.0	1.3	37.4
Personal care and service occupations[3]	4,531	78.3	14.2	6.0	12.9
First-line supervisors/managers of gaming workers	128	48.3	3.7	3.1	5.3
First-line supervisors/managers of personal service workers	152	64.4	7.8	15.0	4.3
Nonfarm animal caretakers	146	69.8	1.4	1.0	10.3
Gaming services workers	98	49.8	6.6	20.0	9.8
Barbers	106	19.2	31.5	4.9	12.3
Hairdressers, hairstylists, and cosmetologists	738	92.0	11.3	4.0	10.3
Miscellaneous personal appearance workers	216	85.1	7.4	43.2	8.9
Baggage porters, bellhops, and concierges	69	15.7	22.8	2.6	21.9
Transportation attendants	103	74.5	14.0	3.3	9.4
Child care workers	1,329	94.8	15.5	2.5	18.1
Personal and home care aides	668	85.9	24.1	5.3	15.6
Recreation and fitness workers	337	67.0	11.3	1.9	7.5
Sales and office occupations	**35,962**	**63.3**	**11.2**	**4.0**	**11.1**
Sales and related occupations[3]	16,433	49.1	9.2	4.4	10.6
First-line supervisors/managers of retail sales workers	3,523	41.5	7.5	5.2	9.0
First-line supervisors/managers of non retail sales workers	1,414	25.5	5.4	5.1	9.8
Cashiers	3,075	75.9	15.5	5.4	16.2
Counter and rental clerks	159	51.9	12.5	1.9	9.8
Parts salespersons	139	12.0	5.9	1.7	10.3
Retail salespersons	3,248	51.9	10.5	4.2	11.8
Advertising sales agents	235	50.2	7.6	2.9	5.9
Insurance sales agents	531	45.2	7.9	2.6	6.2
Securities, commodities, and financial services sales agents	392	28.7	6.2	5.0	7.6
Travel agents	91	76.3	7.8	11.3	7.5
Sales representatives, services, all other	474	32.9	9.3	1.6	7.3
Sales representatives, wholesale and manufacturing	1,379	27.2	3.4	2.7	7.0
Real estate brokers and sales agents	995	57.1	5.9	4.7	7.4
Telemarketers	179	64.4	21.0	1.4	15.5
Door-to-door sales workers, news and street vendors, and related workers	279	62.7	6.6	3.3	13.3

OCCUPATION	Total employed (1,000)	Percent of total			
		Female	Black or African American[1]	Asian	Hispanic or Latino[2]
Office and administrative support occupations[3]	19,529	75.3	12.9	3.8	11.6
First-line supervisors/managers of office and administrative support workers	1,598	69.8	9.8	3.4	9.0
Bill and account collectors	207	72.2	22.7	3.0	14.1
Billing and posting clerks and machine operators	427	89.0	15.9	3.8	9.0
Bookkeeping, accounting, and auditing clerks	1,456	91.3	6.9	3.4	8.4
Payroll and timekeeping clerks	164	91.4	10.7	3.7	11.2
Tellers	418	87.3	12.8	5.9	11.0
Court, municipal, and license clerks	106	81.4	11.3	4.6	7.8
Customer service representatives	1,833	68.7	16.2	3.6	13.1
File clerks	376	81.4	16.2	4.6	13.7
Hotel, motel, and resort desk clerks	138	65.9	15.4	6.9	14.0
Interviewers, except eligibility and loan	167	79.4	17.0	2.6	15.3
Library assistants, clerical	130	85.4	11.0	7.1	13.4
Loan interviewers and clerks	198	77.0	11.4	4.8	12.9
Order clerks	143	67.2	11.4	4.2	13.2
Human resources assistants, except payroll and timekeeping	66	89.3	16.6	2.6	10.5
Receptionists and information clerks	1,376	92.4	11.3	3.8	12.2
Reservation and transportation ticket agents and travel clerks	169	64.0	18.7	4.4	13.9
Couriers and messengers	309	15.8	13.3	3.5	14.7
Dispatchers	291	59.1	13.6	1.2	9.4
Postal service clerks	146	41.7	20.1	12.1	11.9
Postal service mail carriers	348	39.1	15.2	5.4	9.9
Postal service mail sorters, processors, and processing machine operators	111	45.1	30.8	14.8	11.1
Production, planning, and expediting clerks	294	57.1	7.1	2.2	10.3
Shipping, receiving, and traffic clerks	543	31.9	15.2	3.1	20.7
Stock clerks and order fillers	1,461	37.1	15.3	3.4	17.0
Weighers, measurers, checkers, and samplers, recordkeeping	71	52.6	15.6	1.9	16.3
Secretaries and administrative assistants	3,499	97.3	9.2	2.0	8.6
Computer operators	184	55.3	14.5	6.2	11.0
Data entry keyers	495	82.2	16.8	5.8	14.2
Word processors and typists	295	95.0	20.0	2.7	12.5
Insurance claims and policy processing clerks	263	86.9	19.4	4.4	8.6
Mail clerks and mail machine operators, except postal service	132	55.5	26.3	6.5	9.5
Office clerks, general	965	84.5	13.8	6.4	13.1
Natural resources, construction, and maintenance occupations	**15,348**	**4.6**	**7.1**	**1.8**	**23.1**
Farming, fishing, and forestry occupations[3]	976	22.5	5.1	1.6	40.3
Graders and sorters, agricultural products	69	69.4	17.8	3.5	40.5
Logging workers	85	1.7	9.2	-	10.7
Construction and extraction occupations[3]	9,145	3.0	7.0	1.2	26.8
First-line supervisors/managers of construction trades and extraction workers	1,012	2.9	5.2	0.7	12.6
Brickmasons, blockmasons, and stonemasons	245	0.9	8.7	0.6	33.7
Carpenters	1,797	1.9	4.8	1.1	24.4
Carpet, floor, and tile installers and finishers	297	2.3	4.5	1.4	40.0
Cement masons, concrete finishers, and terrazzo workers	119	1.7	7.5	0.4	54.4
Construction laborers	1,491	3.6	10.5	1.2	40.8
Operating engineers and other construction equipment operators	414	2.7	6.2	1.1	10.1
Drywall installers, ceiling tile installers, and tapers	252	0.8	3.2	1.1	46.8

OCCUPATION	Total employed (1,000)	Percent of total			
		Female	Black or African American[1]	Asian	Hispanic or Latino[2]
Electricians	852	2.6	8.1	1.7	15.1
Painters, construction and maintenance	689	7.5	7.4	1.8	35.0
Pipelayers, plumbers, pipefitters, and steamfitters	661	1.2	8.9	0.4	18.1
Roofers	274	2.4	7.2	0.7	42.0
Sheet metal workers	147	3.6	3.5	2.4	17.2
Helpers, construction trades	113	3.2	8.9	0.9	38.6
Construction and building inspectors	107	8.5	12.6	3.8	8.1
Highway maintenance workers	94	4.6	10.9	0.1	13.6
Installation, maintenance, and repair occupations[3]	5,226	4.2	7.5	3.0	13.6
First-line supervisors/managers of mechanics, installers, and repairers	314	5.7	7.8	1.1	7.2
Computer, automated teller, and office machine repairers	384	13.7	9.2	5.3	10.4
Radio and telecommunications equipment installers and repairers	214	13.5	15.1	3.8	9.3
Aircraft mechanics and service technicians	140	4.5	7.0	5.4	9.0
Automotive body and related repairers	168	1.9	3.2	3.2	18.5
Automotive service technicians and mechanics	954	1.8	7.3	4.2	17.7
Bus and truck mechanics and diesel engine specialists	365	0.5	4.9	2.3	13.3
Heavy vehicle and mobile equipment service technicians and mechanics	224	1.0	3.6	1.3	14.3
Small engine mechanics	63	1.8	2.7	0.3	3.1
Heating, air conditioning, and refrigeration mechanics and installers	393	1.4	5.0	2.0	12.9
Industrial and refractory machinery mechanics	444	2.6	7.1	3.1	13.8
Maintenance and repair workers, general	344	4.1	10.7	1.7	14.9
Millwrights	65	2.0	4.6	-	6.2
Electrical power-line installers and repairers	110	3.5	9.8	0.7	11.1
Telecommunications line installers and repairers	152	6.0	12.1	1.0	13.5
Precision instrument and equipment repairers	69	7.7	2.6	11.5	7.0
Production, transportation, and material occupations	**18,041**	**22.9**	**14.1**	**3.5**	**19.2**
Production occupations[3]	9,378	30.3	12.4	4.9	20.0
First-line supervisors/managers of production and operating workers	868	20.9	10.4	2.9	12.5
Electrical, electronics, and electromechanical assemblers	209	58.7	10.6	16.1	21.1
Bakers	183	54.0	9.8	8.1	24.6
Butchers and other meat, poultry, and fish processing workers	292	22.3	9.2	3.3	42.0
Food batchmakers	85	46.4	10.2	4.6	21.5
Computer control programmers and operators	51	11.2	6.0	6.7	5.7
Cutting, punching, and press machine setters, operators, and tenders, metal and plastic	123	22.8	15.7	1.1	11.0
Grinding, lapping, polishing, and buffing machine tool setters, operators, and tenders, metal and plastic	60	10.3	13.6	1.9	18.9
Machinists	420	6.8	7.6	3.6	11.7
Molders and molding machine setters, operators, and tenders, metal and plastic	60	20.3	14.3	0.9	20.4
Tool and die makers	90	1.1	2.1	-	2.6
Welding, soldering, and brazing workers	576	5.8	8.1	2.7	20.0
Printing machine operators	218	15.6	10.8	3.3	16.7
Laundry and dry-cleaning workers	178	62.8	20.3	7.0	29.4
Pressers, textile, garment, and related materials	70	76.0	29.3	8.5	35.7
Sewing machine operators	269	77.4	13.8	14.7	33.6
Tailors, dressmakers, and sewers	91	79.3	5.0	9.3	24.3
Cabinetmakers and bench carpenters	82	4.9	2.5	0.5	7.7

OCCUPATION	Total employed (1,000)	Percent of total			
		Female	Black or African American[1]	Asian	Hispanic or Latino[2]
Stationary engineers and boiler operators	116	3.0	16.2	3.1	11.0
Water and liquid waste treatment plant and system operators	79	4.6	4.2	-	9.3
Crushing, grinding, polishing, mixing, and blending workers	93	10.8	12.8	4.4	24.1
Cutting workers	100	21.2	10.7	1.8	28.6
Inspectors, testers, sorters, samplers, and weighers	684	38.5	12.0	6.2	13.7
Medical, dental, and ophthalmic laboratory technicians	103	46.7	9.3	13.8	9.3
Packaging and filling machine operators and tenders	303	55.6	21.6	4.3	37.6
Painting workers	198	15.0	9.5	2.0	25.7
Photographic process workers and processing machine operators	76	62.9	8.1	1.1	10.5
Transportation and material-moving occupations[3]	8,664	14.8	16.1	2.0	18.4
Supervisors, transportation and material-moving workers	202	18.1	14.7	1.9	8.6
Aircraft pilots and flight engineers	121	5.2	2.5	0.3	5.7
Bus drivers	591	48.7	23.1	1.4	12.7
Driver/sales workers and truck drivers	3,409	4.5	14.5	1.1	16.4
Taxi drivers and chauffeurs	291	15.5	28.7	9.7	16.6
Parking lot attendants	65	10.0	16.3	6.8	25.8
Service station attendants	100	14.2	6.7	7.1	11.3
Crane and tower operators	72	2.8	9.0	1.4	7.8
Dredge, excavating, and loading machine operators	86	0.2	3.7	-	11.4
Industrial truck and tractor operators	541	6.3	22.8	0.8	23.8
Cleaners of vehicles and equipment	347	14.1	14.0	1.2	33.5
Laborers and freight, stock, and material movers, hand	1,806	17.3	15.7	2.4	19.6
Packers and packagers, hand	452	61.1	16.8	4.8	41.6
Refuse and recyclable material collectors	73	8.7	27.4	1.4	19.1

- Represents or rounds to zero.

[1] The 2003 Current Population Survey (CPS) allowed respondents to choose more than one race. Data represent persons who selected this race group only and exclude persons reporting more than one race. The CPS in prior years only allowed respondents to report one race group. See also comments on race in the text for Section 1, Population.

[2] Persons of Hispanic or Latino ethnicity may be of any race.

[3] Includes other occupations, not shown separately.

SOURCE: U.S. Bureau of Labor Statistics, Employment and Earnings, monthy, January 2006 issue. See Internet site http://www.bls.gov/cps/home.htm

Glossary

achieved characteristics characteristics of workers that are developed through a worker's choices and experiences, such as level of education

administrator a top manager in a government agency or in nonprofit firms

advanced industrial society a society with highly productive extractive and manufacturing sectors and a large labor-intensive service sector; also called "postindustrial society"

affirmative action government mandated hiring procedures that attempt to compensate for past discrimination through hiring goals, preferential consideration among otherwise equal candidates, or active recruitment of women or minority workers

alienating work work that provides inadequate sustenance, meaning, and identity

American Federation of Labor (AFL) oldest existing union organization in North America; serves as an umbrella organization enrolling member unions and coordinating their activities

apprenticeship a combination of classroom instruction and supervised on-the-job training that typically lasts from two to six years

aquaculture the raising of fish in ponds or in large floating cages in the sea

artisan a specialist in a skilled trade; used especially to refer to skilled workers in medieval cities

ascribed characteristics characteristics with which workers were born and therefore cannot easily alter, such as gender or race

assembly line a power-driven system that moves objects past workers as each does a small repetitive task toward the assembly of a finished product

autonomous work groups Swedish industrial system in which significant power over procedures is held by groups of 12–15 employees in each area of the factory

backward linkages control by a firm of important sources of supplies

biotechnology use of biological knowledge about genetic, molecular, and cellular processes to develop more commercially useful plant, animal, and organic products

boundaryless career a career that may be pursued with multiple employers and builds on the experiences and learning in a variety of work contexts

bullying to intentionally inflict harm on others in the workplace by humiliating, intimidating, tormenting, pressuring, or mocking

bureaucracy a hierarchical power arrangement based on fixed areas of expertise and written rules

burnout chronic stress on a job that leads to deteriorating performance or even quitting the job

business services services that are marketed to assist individuals, firms, and organizations in carrying out their economic functions, such as temporary help agencies or advertising

capital mobility the easy movement of money to fund investments around nations and around the globe

career the sequence of events within a person's work history

case study a study of a workplace that uses multiple sources of information and seeks to understand the perspectives of multiple actors in the workplace

cartel a group of political or economic organizations that act to limit price competition among themselves

cash cow a profitable company acquired as a subsidiary to generate cash for expansion in other areas

central planning detailed planning for economic output goals by the central government

centralization of control concentration of power by the top executives in an organization and by the largest companies in the economy

chemical sensitization heightened allergic reactions due to extended exposure to toxic chemicals at work

class structure the economic classes that constitute a society and their relative wealth and power (see *social classes*)

client firm a firm that is dependent for most of its business on another, usually larger firm

codetermination the sharing of economic paper in Germany by large corporations, the government, unions, and works councils of employees in each factory

collective bargaining bargaining between an employer and workers over wages, benefits, and conditions of employment through use of a union as a representative of the workers

combined and uneven development a highly developed export sector in combination with underdevelopment of the domestic consumption sector

commodification of agriculture the movement from agricultural production for local consumption to crops sold on commodity markets

commodity chains the geographic, financial, and social linkages that trace raw products and subcomponents as they are combined into finished goods

commitment an attachment between workers and their employer that develops when workers perceive that the goals and values of their employer are compatible with their own goals and values

company unions unions dominated by management that do not provide an autonomous voice for workers

comparable worth discrimination occupational differences in pay that include unequal rewards for equal qualifications and job demands between jobs that have typically been filled by men and those that have typically been filled by women

comparison group a specific group used as a standard of comparison; important for evaluating one's own level of job satisfaction

concentration ratio the percentage of sales in an industry by the largest firms

conglomerate a firm operating in several non-overlapping industries

Congress of Industrial Organizations (CIO) splintered from the AFL in 1935 to organize semi-skilled factory workers in the newly emerging mass production industries; reintegrated in 1955 to form the AFL-CIO

contingency theory argues that organizational structure results from the environment that the organization faces

contingent employment non-permanent employment whose continuity depends upon the availability of work

continuous improvement the key goal of Japanese Quality Control Circles

continuous learning ongoing education and training throughout people's work lives

continuous-process technology a production process involving a continuous flow of product through the manufacturing stages, such as in oil refineries or chemical plants

core and peripheral jobs and industries in dual labor market theory, the principal division that separates stable employment (core) from unstable employment (peripheral)

core nation industrialized nations that sell manufactured goods and buy raw products from peripheral nations

corporate accountability the effort to hold corporations accountable for illegal activities such as price fixing, environmental pollution, false advertising, bribery, tax evasion, political payoffs, and the sale of unsafe products

corporate campaign a union pressure tactic that targets financial backers of the company, consumers, and the public through informational campaigns about the corporation's undesirable behaviors

corporate divisions semi-independent product lines within a large corporation

corporate economic power the power of large companies to exclude smaller competitors from the market and influence overall rates of economic growth

corporate inflexibility an inability to change rapidly, adapt to new situations, or innovate resulting from excessive management hierarchy and bureaucracy

corporate mergers corporate growth through the acquisition of other firms

corporate political power corporate power exercised through campaign contributions to promote legislation favorable to the industry; also frequently exercised through industry control and staffing of the governmental bodies intended to regulate the industry

craft technology skilled handicraft production based on the manufacture and use of specialized tools for each trade

craft union association of skilled workers in a given craft or trade

cultural division of labor historical and cultural patterns that channel minorities and majorities and men and women toward different occupational roles

cyclical unemployment the gap between available work and people needing work during a temporary economic downturn

data stewards union show stewards in Sweden whose job is to keep abreast of the latest technology, consult with the company concerning its deployment, and protect workers' legislative and contractual rights

dead-end jobs entry-level jobs that require relatively little skill, often have a high turnover, and rarely lead to promotions

debt crisis the heavy burden of debt acquired by less-developed nations in frequently unsuccessful attempts at industrialization; high interest payments further undercut development

declining middle class a reduction in the proportion of middle-income earners due to decreased job security and stagnant or falling wages

demographic characteristics the characteristics of workers, such as age, gender, race, or level of education

dependency theory argues that many less industrially developed nations are thoroughly incorporated in the world economy but in a dependent role

dependent development a pattern of industrialization in which a nation's manufacturing activities fill niches as determined by the needs of stronger nations and multinational companies

deprofessionalization the process of weakening or eliminating the professional characteristics of an occupational group

developing nations nations with some industrialization; often serve as subcomponent manufacturers for large multinational corporations

deviant occupations occupations whose job content is defined as illegal or immoral

deskilling a reduction in needed job skills due to automation and job simplification

direct personal control the company owner shows the workers how to do the work, dictates the appropriate pace, and evaluates the workers' performances

discouraged workers people who give up actively seeking employment because of discouragement and a perceived absence of opportunities

dispersion of information the thesis that computer technology tends to make information more widely available

distributive services services that bring about the geographic dispersion of goods, such as transportation and warehousing

diversification movement of a company into new but technologically similar product lines

division of labor the division of work into its component tasks and the specialization of workers in these different tasks

downsizing reduction in a company's workforce; sometimes achieved through increased outsourcing or the movement of work to cheaper locations

dual labor markets job markets that operate to preserve jobs with better pay and greater security for dominant groups

economies of scale cost savings for large producers resulting from lower per unit costs for large batches

electronic surveillance the monitoring of the activities of individual workers through their interface with a computer

emotion work the process of managing one's emotional responses in interactions with customers or supervisors

employee ownership majority ownership of a company by its employees

employee participation the inclusion of workers or their representatives in all levels of organizational decision making

employer commitment a sincere employer pledge to respect the rights of workers, their job security, and their benefits

enclosure movement the fencing of grazing land previously held in common by peasants and landlords for the exclusive use of the landlord in raising sheep for wool or other cash crops

entry ports entry-level jobs that offer the worker the possibility for training and promotion

environmental degradation erosion of the quality of the air, water, or landscape by dangerous industrial practices and insufficient attention to reducing or containing waste products

ESOPs employee stock ownership plans which entail partial worker ownership, sometimes with only limited rights of participation and control

establishment the physical building and location of an organization; one company may have many different establishments

establishment survey a survey in which the resonding units are workplaces

ethnography (as used in this text) a careful analysis of a work situation by a knowledgeable observer who may be a participant or nonparticipant in the workplace

executive a worker at the very top of a workplace bureaucracy

expanded leisure reducing the hours of work as a way of improving the quality of work life and distributing available employment

experimental bias a change in the results of an experiment or an observation that results from workers changing their behaviors in response to the presence of the experimenter or observer

externalization of costs the ability of companies to lower their expenses by unlawfully dumping dangerous chemical wastes, continuing dangerous work practices, producing unsafe products, and bargaining with local governments for tax abatements

extractive industries industries such as agriculture, forestry, fishing, and mining in which a raw product is taken from the environment

extrinsic rewards rewards that are realized off the job, including such things as pay, fringe benefits, and job security

face-work the management of facial expression required in a service interaction; see also emotion work

fair trade a doctrine that argues that "free trade" must also include common standards for the rights of workers and for environmental protection

family life cycle the stages of formation, growth, and dissolution of the nuclear family

feminization of clerical work process by which clerical work changed from predominantly male to predominantly female workers

feudal society agricultural societies in which peasants were obligated to stay on the land and to give a share of their crops to the wealthy landowners

firm an employing organization, generally incorporated as a legal entity

first line supervisor the lowest level manager in a workplace

flexibility corporate strategy based on downsizing to facilitate more rapid matching of resources to market opportunities

flextime allows workers to set their own hours, within some limits

Fordism the production of large numbers of standardized products using power-driven equipment and assembly lines; named after Henry Ford, a pioneer in this production strategy

foreman's control a hired foremen performs the duties of recruiting, training, and supervising workers

forward linkages control by a firm of the next manufacturing stage for its product, potentially including wholesale or retail marketing of its products or services

four tigers the nations of Hong Kong, Taiwan, South Korea, and Singapore in recognition of their rapid industrial development

free agent firms small firms that occupy less profitable market niches unexploited by larger corporations

free trade international trade unrestricted by tariffs or quotas

frictional unemployment unemployment that results from unavoidable delays between jobs, such as when a person moves to a new location

fringe benefits nonwage compensation and perquisites, such as pensions, health and other forms of insurance, sick leave, and paid vacation

gender-typing the categorization of certain occupations as appropriate for women (or men) because women (or men) have historically been responsible for these types of activities

general strike coordinated simultaneous strikes involving several trades

general union a union that enrolls all workers, regardless of occupation or industry

golden parachutes employment contracts that guarantee generous severance packages to top executives if their firm is acquired

grievance procedure a set of procedures for handling workers' complaints about violations of their rights under an employment contract

guilds medieval associations of skilled workers for regulating the hours and quality standards of an occupation; precursors to modern trade unions and professional associations

Hawthorne effect the observation that job redesign tends to increase short-term productivity, regardless of its content, because it entails added social attention. See experimental bias.

high-technology industry industries with high proportions of engineers and scientists, such as electronics, ordnance, chemicals, pharmaceuticals, and genetic engineering

hired hand relatively permanent position in agriculture or tending livestock, sometimes compensated with both money and payment in kind (such as room and board)

home production the non-market production of goods and services, usually for the family but occasionally on a volunteer basis for schools, churches, or other groups

horizontal dimension of bureaucracy within an organization, functional units that perform discreet tasks

human capital the valuable productive skills possessed by workers

human relations school a tradition that argues that a supportive social context maximizes workers' productivity

ideology a system of ideas that specifies a given economic and political arrangement as appropriate and desirable

import substitution a development strategy in which a nation attempts to develop industries that produce goods it has previously imported

income squeeze the phenomenon that families often have greater income needs at the career stage that the worker's salary increases flatten out

indentured labor workers under contract to work for a certain amount of time—typically eight to ten years—for a set price or as part of their penalty for being found guilty for a crime such as petty theft or vagabondage

indoor air pollution dust, chemicals, gases, molds, and bacteria trapped inside air-tight or poorly ventilated buildings

indulgency pattern a pattern of supervisors overlooking rule violations by employees in exchange for greater effort on other fronts

Industrial Revolution the often violent transition between subsistence agriculture and the widespread production of manufactured goods for sale

industrial union a union that organizes all the various trades in a given industry

Industrial Workers of the World a radical general union active in North America in the early 1900s; supported replacement of capitalists by committees of workers

industry a branch of economic activity devoted to the production of a particular good or service

informal sector small independent artisans, shop keepers, and street vendors who produce and sell goods for local markets

informal work culture social relationships and expectations that emerge among people who work together

injunction a court order to cease and desist a specific action, such as a strike; frequently used against unions well into the 20th century and even occasionally today

innovative sector industries in which heightened international competition, technological change, and increased employee participation lead to organizational innovations and increased productivity

inside game union pressure tactic that operates within the establishment during normal operations to pressure management; includes rigidly following rules and conspicuous displays of symbols of worker solidarity

intergenerational mobility a child's attainment of an occupation of greater or lesser status than that of his or her parents

interlocking directorates a set of companies with overlapping boards of directors; facilitates oligopolistic practices

internal labor markets job ladders within organizations or occupations

International Labour Organization a United Nations agency that promotes international standards for the rights of workers

intragenerational mobility a change in occupations or jobs within the career of a single worker

intrinsic rewards rewards that are realized while at work, including such things as the freedom to plan one's own work and the chance to use one's abilities

inverted-U curve of technology and alienation the observation that alienation increases following the displacement of craft organizations of work by machine pacing and assembly-line technologies, but that it decreases again under more advanced continuous-process technologies

job autonomy extent to which workers control their own job tasks and organization of work

job complexity the level of manual, interpersonal, or mental skills required by a job

job diversity the number of different tasks and responsibilities required by a job

job enlargement process of varying job content through additional tasks, often requiring additional skills

job enrichment process of improving jobs through developing higher level skills in workers

job instability fluctuating wages or hours and periods of unemployment associated with specific jobs

job ladder the additional position to which one might be promoted within the firm

job redesign change in job content and organization; key current components include employee participation and team production systems

job satisfaction the summary evaluation that people make of their jobs

job security a reasonable expectation that workers can keep their jobs over time

job stress chronic tension resulting from job conditions such as noise, time pressure, abusive or incompetent bosses, excessive responsibilities, and long hours

job-worker fit the degree of fit between a worker's values and needs and the content of his or her job

joint union-management programs jointly sponsored union-management programs that include employee involvement, such as new training programs

journeyman a skilled craft worker who works as an employee rather than opening his or her own business

just-in-time delivery an invention of the Japanese automobile industry in which parts suppliers are required to deliver exactly the right number of parts at the exactly right time, minimizing the need for warehousing at the core assembly plant

Knights of Labor a progressive general union in North America active in the decades before 1900; supported immigrants and women

labor force all of the employed and unemployed workers in a geographic area at a specific time

labor force participation rate the percentage of people eligible to be counted as part of the labor force who are either employed or unemployed

layoffs large scale temporary or permanent employment terminations because of falling sales, downsizings, or plant closings

least developed nations nations characterized by subsistence agriculture and raw material production for export

lean production a corporate strategy for flexibility based on reducing its core labor force and using more short-term and temporary employment contracts

life cycle the ordering of roles in a human life beginning with infancy and ending with death

lifetime employment guaranteed job security for life for workers as a strategy for securing commitment and extra effort; popular in Japan

limited liability corporate liabilities such as debts, bankruptcy, and lawsuits are limited to the corporation itself, rather than extending to its owners

line manager a supervisor of workers directly involved in the production process

long-term and indirect effects indirect consequences of technology that may generate job growth through increased productivity and economic expansion

loyal opposition firms smaller firms that provide competition to larger corporations in their own industries

macroplanning coordination between corporations, banks, governments, and educational institutions to target certain industries for development

major occupational group broad categories of occupations

making out creative efforts by workers to organize their tasks to get the work done without exhausting themselves in the process

management-controlled firms corporations without large controlling stockholders; differences in firm behavior between management-controlled firms and owner-controlled firms are an object of debate

manager someone whose job requires leadership and supervision of a workplace

manufacturing industries industries that process raw products into finished goods

maquiladoras U.S. factories on Mexico's northern border that manufacture goods for export to the United States

marginal jobs jobs generally viewed as inadequate because they do not meet the community norms for wages, hours, or use of worker skills; see underemployment

marginal sector industries in which a lack of worker power allows employers to respond to heightened international competition and technological change by reducing labor costs

mass-production technology production of large batches of products through the integration of specialized tools and equipment in a mechanized production operation

mass strike a coordinated general strike of workers with support from the general population through demonstrations, marches, and insurrections

master social status a master social status is one that affects many aspects of one's life and interactions with others

matrix organization a bureaucratic system in which each worker reports to a division supervisor (e.g., electrical engineering) and to a project supervisor (e.g., a certain aircraft project)

mentor a more experienced worker who serves as a role model, protector, adviser, and cheerleader for a newer worker, typically one at a lower level within the workplace

merchant capitalism the earliest form of capitalism based on buying and selling finished goods and agricultural products in intercity market networks

meritocracy a system in which rewards are based on achievement

microchip technology computer technology based on circuits chemically etched on miniaturized silicon chips

middle managers layers of management above first-line and below executives

migrant farm workers farm workers who move with the agricultural cycle and work for many employers in the course of the season

mini-mills small steel mills that use electric furnaces to turn scrap into basic steel products

MIS a computerized "management information system" that records production information and potentially monitors individual output and workers' activities

modernization theory argues that societies advance from agricultural production to industrial production based largely on internal dynamics

monopoly exclusive or dominant control of a market

multiplant companies companies with establishments in many different locations

multinational corporation a firm that operates in many countries

nanotechnology use of chemical and atomic properties to position molecules and even atoms to manufacture or improve the performance of commercial products

National Labor Relations Act (1935) lays groundwork for union-management relations in the United States; provides for secret ballot vote by workers for or against union representation

NC and CNC systems that operate metalworking lathes, drills, and cutting tools automatically from numeric codes (NC) or computer software (CNC)

neo-imperialism the use of economic dominance, rather than direct military subjugation, to control and exploit poorer nations

new middle class middle-income workers whose earnings come from knowledge-based services, such as professions, rather than from ownership of physical capital

NILF abbreviation for "not in the labor force"

nonparticipant observation a type of ethnography in which a trained researcher observes but does not actually become a part of the work group

norm a rule that a group develops for thinking, feeling, or behaving

occupation a cluster of job-related activities constituting a single economic role that is usually directed toward making a living

occupational communities a distinct occupational subculture that provides its members with a shared identity

occupational disease diseases resulting from improper working conditions or exposure to toxins at work

occupational segregation the channeling of minority groups or women into restricted sets of occupations

on-the-job training formal and informal training that occurs at the workplace

oligopoly a market controlled by few enough firms that they can engage in collaborative price fixing

opportunity structure the occupations and jobs available within a society at a particular point in time

organizational structure the architecture or pattern of formal and informal relations in an organization

outsourcing purchasing goods and services that could be made or provided internally by the company from outside sources; often involves hiring cheaper labor in less developed nations

paper entrepreneurialism buying and selling companies and manipulating balance sheets to realize profits from the sale of productive assets rather than from the sale of products

paraprofession an occupation that assists an established profession

parent company a firm that owns and controls another firm

payment in kind payment in goods instead of money

peripheral jobs and industries see core jobs and industries

peripheral nation nations that sell raw materials and buy manufactured goods from industrialized nations

personal services services that are produced principally for a family or for individuals, such as restaurants, hotels and motels, or entertainment

plant closings plant shutdowns due to lost competitiveness or to the movement to cheaper locations

population the group to which a researcher wishes to generalize the results of a randomly selected sample

postbureaucratic possible organizational structure of the future in which hierarchy and bureaucracy are replaced by employee participation and innovation

postindustrial society a society in which high manufacturing productivity allows more and more workers to be engaged in service provision

predatory pricing intentional short-term underbidding of competitors by large corporations to drive them out of the market

primitive accumulation the seizure of wealth from the European colonial empires in Africa, the Americas, and the Pacific in the form of precious metals, trade goods, and slaves; important for financing European industrialization

producer services services that are necessary to help other industries create their products or services, such as electricity, gas, water, sanitation, and other utilities

profession a high-status occupation marked by abstract knowledge, authority, autonomy, and altruism

professionalism competence and effectiveness of workers in their job performance

professionalization a process by which an occupation seeks to more closely resemble a profession

professional culture the characteristic jargon, behaviors, and lifestyles associated with the practitioners of a profession

professional services knowledge-based services for clients who may be individuals or firms; examples are medical services and legal services

protectionism the protection of domestic industry through high tariffs or restrictive limits on foreign imports

Protestant work ethic a religious view that identifies success in the pursuit of one's occupational calling with spiritual grace

public goods products or services to which the citizens of a society are entitled without direct payment or at a nominal fee, such as public schools and parks

putting-out system a system of production in which a merchant capitalist "puts out" raw materials to be worked up by people in their homes and later collects the finished products to be sold

Quality Control Circles Japanese system in which work groups identify production techniques that can be changed to improve quality and productivity

quality of work life a set of factors that make for satisfying and fulfilling work, including adequate pay, safety, and autonomy

rationalization the process by which ends are more closely fitted to means

reliability a characteristic of research methods that produce the same results repeatedly or that produce the same results when used by a different researcher

response error errors in sample surveys that result from vague or erroneous answers

right-to-know legislation laws allowing workers to be informed about the chemical hazards in their workplaces

role a set of behaviors or expectations associated with a particular position in society

role conflict stress from contradictory demands arising from different roles a person occupies

role overload stress caused by having too many different roles, each with a set of demands

role strain stress from contradictory demands arising from a single role

routinization the development of a standardized procedure for doing a work task

sabotage intentional acts of disruption and destruction against employers for vengeance or leverage

sample a mathematically selected subgroup representative of a population

sample survey a research technique that uses questionnaires to seek information from a representative group of people

satellite firms dependent firms engaged in subcontracting relationships with larger firms

scientific management the idea that there is one best way to do every task and that this way can be discovered through scientific techniques of observation and experimentation

secondary boycott unions supporting each other by refusing to handle struck goods

sectoral transformation of the labor force the shift over time in the principal industry of employment from agricultural through manufacturing to services

self-actualization the realization of one's broadest human needs through work

semiprofession an occupation that has achieved some of the characteristics of a profession

semiskilled work jobs requiring less than two weeks of formal training

seniority allocating protection from layoffs, first choice of shifts, access to training programs, etc., on the basis of years of service

sequential life plan when applied to work, the normative ordering of roles

services acts provided in return for payment

service industries industries that provide useful activities instead of goods, including personal service (waitressing, hairdressing, entertainment), educational services (teaching), medical services (nurses, doctors), and business services (accounting)

severance pay a lump-sum payment to workers at the time of a permanent layoff; paid only if mandated by an union agreement or other employment contract

sexual harassment repeated, unwelcome behavior with a sexual content when submission to such behavior is explicitly or implicitly a condition for the person's hiring or for other employment decisions, or when such behavior creates a hostile, intimidating, or offensive working environment

shadow economy production of goods and services that is not in cash, is not taxed, or is unregulated by the government

sharecropper an agricultural worker whose pay consists of a portion of the harvest

shop steward a worker who is allowed under a collective bargaining agreement limited time away from work to inform fellow workers of their contractual rights and to handle their grievances about violations of these rights

shrinking middle class hypothesis that the class structure tends to increase the wealthy and the poor at the expense of the middle of the class structure

sit-down strike a strike in which workers stay in their places but stop working; prevents the use of replacement workers

skilled craft work trades requiring extensive training and experience, such as machinists, mechanics, and electricians

social classes groups in society defined by the possession of different amounts of wealth and power

social relations of production the rights and obligations of people toward each other at every level of production

social services services that benefit the community, and often are financed by the government, such as protective services or government services

social stratification the unequal distribution of rewards and power in society

social structure the architecture of social relations in formal and informal groups and in the larger society

socialization the process of learning norms, roles, and skills either informally or formally by instructors, which are sometimes termed socialization agents

soldiering intentionally working well below one's capacity

solidarity mutual defense and support in times of crisis or challenge

span of control the number of persons that one manager supervises

staff and line positions bureaucratic positions entailing ancillary expertise (staff positions) versus chain of command (line positions)

staff manager a supervisor of workers who perform a service that is necessary to the organization but is not the organization's major product or service, such as purchasing or maintenance

state-regulated capitalism an economic system in which large corporations hold economic power but the state regulates their activities to protect the interests of the nation as a whole; includes the U.S., Canada, and most industrialized nations

statistical discrimination discrimination that arises when an employer bases hiring or promotion decisions on the *average* qualifications of a group, rather than on an individual's qualifications

steepness of hierarchy the number of levels of supervision

structural unemployment the chronic gap between the number of jobs the economy provides and the number of people seeking work

subcontracting (see outsourcing) purchasing goods and services that could be made by the company from outside sources

subminimum wage a wage below the minimum wage that may be legally paid to workers in specified categories (such as teenagers)

subsidiary a firm that is owned by another firm

Taft-Hartley Amendments (1947) amendments to the National Labor Relations Act that outlaw secondary boycotts, allow states to outlaw union shops, and allow the President to declare a state of emergency and force workers and employers to resume production

technical control control of work tasks by machine pacing and job design as on an assembly line

technological determinism a theory that organizational structure and the nature of work are rigidly determined by technology

technological displacement elimination of jobs resulting from the introduction of new technologies; a current concern because of the widespread application of microprocessor technologies

technology knowledge, skills, and tools used for the achievement of practical purposes

telecommuting doing information processing work at home on a personal computer or remote terminal

tertiarization the development of a service economy without a manufacturing base, usually in a developing country

Theory X a theory that employees work hard only if coerced (by threat of firing) or bribed (by promises of pay raises) into doing so

Theory Y a theory that workers will be more productive if they receive more humane consideration and attention

Theory Z a theory that views productivity as embedded in workers, in their skills, and in their cooperative attitudes

time-boundedness a characteristic of services that are provided at culturally determined times

tokenism the hiring of a few minorities or women in an occupation as token adherence to diversity goals

top management the executive leadership of an organization

trade association an organization of firms within the same industry

transfer prices the prices that large corporations charge when a product is transferred from one division to another

underclass an economic class with access only to marginal employment at best

underemployment employment that is inadequate for the worker in terms of income, hours of work, or use of the worker's skills

unemployed persons labor force participants who are not at work for even one hour during the reference week, who are actively seeking work or awaiting results of a recent search, and who are available to work

unemployment compensation limited term payments (typically 26 weeks) to workers by states following layoffs

unemployment rate the proportion of the labor force unemployed at a specific time

unskilled work jobs requiring little or no formal training

validity a characteristic of research methods that produce information that fits closely with reality

VDT a video display terminal for a personal computer or operated as a remote station from a mainframe computer

vertical dimension of bureaucracy the bureaucratic hierarchy of the workplace

vertical integration control by a firm of both its supplies (backward linkages) and marketing (forward linkages)

vocational education education specifically targeted toward usable workplace skills; typically one or two years in duration or as part of a longer apprenticeship program

whistle-blowing employees reporting illegal activities inside the organization to law enforcement or regulatory agencies

worker education ongoing adult and on-the-job education necessary for the innovative workplace of the future

worker power worker involvement in economic decisions within the firm; essential for generating change and innovation

worker retraining additional education for workers already in the labor force to train them in new areas or occupations

worker resistance intentional withdrawal of cooperation by workers ranging from soldiering to sabotage

workplace experimentation a sustained process of developing and evaluating new forms of work organization

work reorganization changes in job content, supervision, and/or relations among workers that are intended to substantially affect how a worker performs a job

works councils worker representatives in Germany who sit on management boards and represent workers in the day-to-day affairs of the enterprise

world systems theory a theory that the world is an integrated economic system with difficult-to-change roles for peripheral nations

References

Abbott, Andrew D. 1986. "Jurisdictional conflicts—A new approach to the development of the legal professions." *American Bar Foundation Research Journal* 2 (Spring): 187–224.

Abbott, Andrew. 1988. *The System of Professions: An Essay on the Division of Expert Labor.* Chicago: University of Chicago Press.

Abbott, Andrew D. 2005. "Linked ecologies: States and universities as environments for professions." *Sociological Theory* 23: 245–274.

Adams, Walter. 2004. *The Biggest Complex: Industry, Labor and Government in the American Economy.* Stanford, CA: Stanford Economics and Finance.

Ackerman, Frank. 2000. *The Politics of Inequality.* Washington, DC: Island Press.

Adler, Paul S. 2005. "The evolving object of software development." *Organization* 12: 401–435.

Adler, Paul S., and Bryan Borys. 1996. "Two Types of Bureaucracy: Enabling and Coercive." *Administrative Science Quarterly* 41, 1 (March) 61–89.

Adler, William M. 2000. *Mollie's Job: A Story of Life and Work on the Global Assembly Line.* New York: Scribner.

Allen, Steven G. 2001. "Technology and the wage structure." *Journal of Labor Economics* 19: 440–483.

Aminzade, Ronald. 2001. *Silence and Voice in the Study of Contentious Politics.* Cambridge: Cambridge University Press.

Anderson, Sarah D., and John Cavanagh. 2005. *Field Guide to the Global Economy.* New York: W.W. Norton.

Anderson-Connolly, Richard, Leon Grunberg, Edward S. Greenberg, and Sarah Moore. 2002. "Is lean mean? Workplace transformation and employee well-being." *Work, Employment and Society* 16, 2 389–413.

Andersson, Fredrik, Harry J. Holzer, and Julia I. Lane. 2005. *Moving Up or Moving On: Who Advances in the Low Wage Labor Market?* New York: Russell Sage Foundation.

Apostle, Richard, et al. 1998. *Community, State and Market on the North Atlantic Rim.* Toronto: University of Toronto Press.

Appelbaum, Eileen. 1984. "The impact of technology on skill requirements and occupational structure in the insurance industry, 1960–90." Cited in Philip Kraft, "A review of empirical studies of the consequences of technological change on work and workers in the U.S." Washington, DC: National Research Council.

Appelbaum, Eileen, Thomas Bailey, Peter Berg, and Arne L. Kalleberg. 2000. *Manufacturing Advantage: Why High-Performance Work Systems Pay Off.* Ithaca, NY: Industrial and Labor Relations Press.

Appelbaum, Eileen, Annette Bernhardt, and Richard J. Murnane (editors). 2003. *Low Wage America: How Employers are Reshaping Opportunity in the Workplace.* New York: Russell Sage Foundation.

Appier, Janis. 1998. *Policing Women: The Sexual Politics of Law Enforcement and the LAPD.* Philadelphia: Temple University Press.

Applebaum, Herbert. 1999. *Construction Workers, U.S.A.* Westport, CT: Greenwood.

Applegath, John. 1982. "What's good about the home office?" *The Futurist* 16, 3 (June): 46.

Argote, Linda. 1999. *Organizational Learning*. Boston: Kluwer.

Argyris, Chris. 2003. "A life full of learning." *Organization Studies* 24: 178–192.

Arndt, Sven W., and Henryk Kierzkowski. 2001. *Fragmentation: New Production Patterns in the World Economy*. Oxford: Oxford University Press.

Ash, Mary Kay. 1981. *Mary Kay*. New York: Harper and Row.

Ash, Mary Kay. 1984. *Mary Kay on People Management*. New York: Warner.

Ashkanasy, Neal M., Celeste Wilderom, and Mark F. Peterson (editors). 2000. *Handbook of Organizational Culture and Climate*. Thousand Oaks, CA: Sage.

Aughinbaugh, A. 2000. "Reapplication and extension: intergenerational mobility in the United States." *Labour Economics* 7, 6 (November): 785–796.

Auletta, Ken. 1982. *The Underclass*. New York: McGraw-Hill.

Averitt, Robert T. 1968. *The Dual Economy: The Dynamics of American Industrial Structure*. New York: Norton.

Bacon, David. 2000. "Tijuana troubles: NAFTA is failing Mexican workers." *In These Times*, August 21, p. 23.

Baker, Ted, and Howard E. Aldrich. 1996. "Prometheus stretches: Building identity and cumulative knowledge in multiemployer careers." Pp. 132–149 in Michael B. Arthur and Denise M. Rousseau (editors), *The Boundaryless Career*. New York: Oxford University Press.

Baldi, Stephanie, and Debra Branch McBrier. 1997. "Do the determinants of promotion differ for Blacks and Whites?" *Work and Occupations* 24: 478–497.

Baldoz, Rick, Charles Koeber, and Philip Kraft. 2001. *The Critical Study of Work: Labor, Technology, and Global Production*. Philadelphia: Temple University Press.

Bandy, J. 2004. "Paradoxes of transnational civil societies under neoliberalism." *Social Problems* 51: 410–431.

Banwell, Suzanne Stout. 2001. *Vendors' Voices*. Phnom Penh, Cambodia: The Asia Foundation.

Baran, Barbara. 1986. "The technological transformation of white-collar work: A case study of the insurance industry." Pp. 25–62 in Heidi I. Hartmann (editor), *Computer Chips and Paper Clips: Technology and Women's Employment*. Volume 2. Washington, DC: National Academy Press.

Baratz, Morton S. 1971. "Corporate giants and the power structure." Pp. 146–155 in Richard Gillam (editor), *Power in Post-War America*. Boston: Little Brown.

Barber, Bernard. 1965. "Some problems in the sociology of the professions." Pp. 15–34 in Kenneth S. Lynn and the editors of Daedalus (editors), *The Professions in America*. Boston: Houghton Mifflin.

Barker, James R. 1999. *The Discipline of Teamwork: Participation and Concertive Control*. Thousand Oaks, CA: Sage.

Barnet, Richard J., and John Cavanagh. 1994. *Global Dreams: Imperial Corporations and the New World Order*. New York: Simon and Schuster.

Barnett, William, James N. Baron, and Toby E. Stuart. 2000. "Avenues of attainment." *American Journal of Sociology* 106: 88–144.

Baron, James N., Brian S. Mittman, and Andrew E. Newman. 1991. "Targets of opportunity." *American Journal of Sociology* 96: 1362–1401.

Bartos, Otomar J. and Paul Wehr. 2002. *Using conflict theory*. Cambridge: Cambridge University Press.

Baxandall, Rosalyn, Linda Gordon, and Susan Reverby. 1976. *America's Working Women*. New York: Random House.

Baxter, Vern K. 1994. *Labor and Politics in the U.S. Postal Service*. New York: Plenum.

Bayard de Volo, Lorraine. 2003. "Service and surveillance: Infrapolitics at work among casino cocktail waitresses." *Social Politics* 10, 3: 346–376.

Becker, Howard S., Blanche Geer, Everett C. Hughes, and Anselm L. Strauss. 1961. *Boys in White: Student Culture in Medical School*. Chicago: University of Chicago Press.

Bebchuk, Lucian. 2006. "How much does the boss make?" *The Wall Street Journal*, January 18, p. A10.

Bednarzik, Robert W. 2000. "The role of entrepreneurship in U.S. and Canadian job growth." *Monthly Labor Review* 123, 3: 3–17.

Belanger, Jacques, Paul K. Edwards, and Larry Haiven. 1994. *Workplace Industrial Relations and the Global Challenge*. Ithaca, NY: Industrial and Labor Relations Press.

Bell, Daniel. 1976. *The Coming of Post-Industrial Society*. New York: Basic Books.

Belman, D., and Paula B. Voos. 2004. "Changes in union wage effects by industry." *Industrial Relations* 43: 491–519.

Ben-David, Joseph. 1984. *The Scientist's Role in Society: A Comparative Study*. Chicago: University of Chicago Press.

Benokraitis, Nijole V., and Joe R. Feagin. 1986. *Modern Sexism: Blatant, Subtle and Covert Discrimination*. Englewood Cliffs, NJ: Prentice Hall.

Benson, Susan Porter. 1986. *Counter Cultures: Saleswomen, Managers, and Customers in American Department Stores, 1890–1940*. Urbana, IL: University of Illinois Press.

Berger, Alan. 2002. *Reclaiming the American West*. New York: Princeton University Press.

Berle, Adolf A., and Gardiner C. Means. 1932. *The Modern Corporation and Private Property*. New York: Harcourt Brace Jovanovich.

Bernstein, Paul. 1997. *American Work Values: Their Origin and Development*. Albany, NY: State University of New York Press.

Bersoff, David, and Faye Crosby. 1984. "Job satisfaction and family status." *Personality and Social Psychology Bulletin* 10, 1: 79–83.

Besser, Terry L. 1996. *Team Toyota: Transplanting the Toyota Culture to the Camry Plant in Kentucky*. Albany, NY: State University of New York Press.

Best, Fred, and Barry Stern. 1977. "Education, work, and leisure: Must they come in that order?" *Monthly Labor Review* 100, 7: 3–10.

Biagi, Marco. 2002. *Quality of Work and Employee Involvement in Europe*. New York: Kluwer.

Bianchi, Suzanne M., John P. Robinson, and Melissa A. Milkie. 2006. *Changing Rhythms of American Family Life*. New York: Russell Sage Foundation.

Biggart, Nicole Woolsey. 1989. *Charismatic Capitalism: Direct Selling Organizations in America*. Chicago: University of Chicago Press.

Bills, David B. 2004. *The Sociology of Education and Work*. Malden, MA: Blackwell.

Birch, David. 1987. *Job Creation in America: How Our Smallest Companies Put the Most People to Work*. New York: Free Press.

Bittman, Michael, and Judy Wajcman. 2000. "The rush hour: The character of leisure." *Social Forces* 79: 165–189.

Blair-Loy, Mary. 1999. "Career patterns of executive women in finance." *American Journal of Sociology* 104, 5: 1346–1397.

Blau, Francine D., Marianne A. Ferber, and Anne E. Winkler. 1998. *The Economics of Women, Men, and Work*. Upper Saddle River, NJ: Prentice Hall.

Blau, Judith R., and Alberto Moncada. 2005. *Human Rights: Beyond the Liberal Vision*. Lanham, MD: Rowman & Littlefield.

Blau, Peter M., and Otis Dudley Duncan. 1967. *The American Occupational Structure*. New York: Wiley.

Blau, Peter M., and Marshall W. Meyer. 1971. *Bureaucracy in Modern Society*, 2nd edition. New York: Random House.

Blauner, Robert. 1964. *Alienation and Freedom*. Chicago: University of Chicago Press.

Bloom, David E., and Adi Brender. 1993. "Labor and the emerging world economy." *Population Bulletin* 48, 2.

Bluestone, Barry, and Bennett Harrison. 1982. *The Deindustrialization of America*. New York: Basic Books.

Bluestone, Barry, and Bennett Harrison. 2000. *Growing Prosperity: The Battle for Growth with Equity in the Twenty-First Century*. Boston: Houghton Mifflin.

Boddy, David, and David Buchanan. 1981. "Information technology and the experience of work." In *Information Technology: Impact on the Way of Life*. Dublin, Ireland: Tycooly International Publishers.

Bohen, Halcyone H., and Anamaria Viveros-Long. 1981. *Balancing Jobs and Family Life: Do Flexible Work Schedules Help*? Philadelphia: Temple University Press.

Bok, Derek C. 1993. *The Cost of Talent: How Executives and Professionals are Paid and How It Affects America*. New York: Free Press.

Bonacich, Edna. 2001. "The challenge of organizing in a globalized/flexible industry: The case of the apparel industry in Los Angeles." Pp. 155–178 in Richard Baldoz, Charles Koeber, and Philip Kraft (editors), *The Critical Study of Labor*. Philadelphia: Temple University Press.

Booze, Regena Michelle. 2000. "Protecting our essence: Coping strategies of African American female faculty at predominantly white colleges and universities." *Dissertation Abstracts International, The Humanities and Social Sciences* 60, 12: 4618A–4619A.

Borjas, George Jesus. 2005. *Labor Economics*, 3rd edition. New York: McGraw-Hill.

Bornschier, Volker. 2005. *Culture and Politics in Economic Development*. London: Routledge.

Bose, Christine, Philip L. Bereano, and Mary Malloy. 1984. "Household technology and the social construction of housework." *Technology and Culture* 15: 53–82.

Bosk, Charles L. 2003. *Forgive and Remember: Managing Medical Failure*, 2nd edition. Chicago: University of Chicago Press.

Boswell, Terry, and Christopher Chase-Dunn. 2000. *The Spiral of Capitalism and Socialism: Toward Global Democracy*. Boulder, CO: Lynne Rienner.

Bowe, John, Marisa Bowe, and Sabin Streeter. 2000. *Gig: Americans Talk About Their Jobs*. New York: Crown Publishers.

Boyer, Richard O., and Herbert M. Morais. 1970. *Labor's Untold Story*. New York: United Electrical, Radio and Machine Workers of America.

Boyett, Joseph H., with Himmie T. Boyett. 1995. *Beyond Workplace 2000: Essential Strategies for the New American Corporation*. New York: Dutton.

Bradshaw, York W., and Michael Wallace. 1996. *Global Inequalities*. Thousand Oaks, CA: Pine Forge.

Brady, David and Michael Wallace. 2004. "Specialization, direct investment, and labor outcomes in the American States." *Work and Occupations* 31: 3–37.

Braun, Dennis D. 1997. *The Rich Get Richer*, 2nd edition. Chicago: Nelson-Hall.

Braverman, Harry. 1974. *Labor and Monopoly Capital*. New York: Monthly Review Press.

Brecher, Jeremy. 1997. *Strike!* Boston: South End Press.

Brecher, Jeremy. 2000. *Globalization from Below*. Cambridge, MA: South End Press.

Bridges, William P. 1982. "The sexual segregation of occupations: Theories of labor stratification in industry." *American Journal of Sociology* 88: 270–295.

Brint, Stephen G. 1984. "New class and cumulative trend explanations of the liberal attitudes of professionals." *American Journal of Sociology* 90: 30–71.

Brinton, Mary C. (editor). 2001. *Women's Working Lives in East Asia*. Sanford, CA: Stanford University Press.

Britton, Dana. 1997. "Gendered organizational logic: Policy and practice in men's and women's prisons." *Gender and Society* 11: 796–818.

Brody, David. 2005. *Labor Embattled: History, Power, Rights*. Urbana: University of Illinois Press.

Bronfenbrenner, Kate (editor). 1998. *Organizing to Win*. Ithaca, NY: Industrial Relations Press.

Broomby, Rob. 2001. "Germany's guest workers mark 40 years." BBC News, October 30.

Brown, Charles. 1982. "Dead-end jobs and youth unemployment." Pp. 427–444 in Richard B. Freeman and David A. Wise (editors), *The Youth Labor Market Problem*. Chicago: University of Chicago Press.

Browne, Irene (editor). 1999. *Latinas and African American Women at Work*. New York: Russell Sage Foundation.

Browne, Irene, and Joya Misra. 2003. "The intersection of gender and race in the labor market." *Annual Review of Sociology* 29: 487–513.

Browning, Harley, and Joachim Singelmann. 1975. *The Emergence of a Service Society*. Report to the U.S. Department of Labor, Manpower Administration.

Brownstone, Ellen Gail. 2000. "A qualitative study of the occupational status and culture of dental hygiene in Canada." *Dissertation Abstracts International, The Humanities and Social Sciences* 60, 12: 4627–4627A.

Bruce-Briggs, B. (editor). 1979. *The New Class?* New Brunswick, NY: Transaction Books.

Brueggemann, John, and Cliff Brown. 2003. "The decline of industrial unionism in the meatpacking industry." *Work and Occupations* 30: 327–360.

Bryant, Clifton D., and Kenneth B. Perkins. 1986. "Containing work disaffection: The poultry processing worker." Pp. 155–165 in Lauri Perman (editor), *Work in Modern Society*. Dubuque, IA: Kendall Hunt.

Buchanan, David A. 1997. "The limitations and opportunities of business processing re-engineering in a politicized organizational climate." *Human Relations* 50: 51–72.

Buchmann, Claudia, and Thomas A. DiPrete. 2006. "The growing female advantage in college completion: The role of family background and academic achievement." *American Sociological Review* 71: 515–541.

Burawoy, Michael. 1979. *Manufacturing Consent*. Chicago: University of Chicago Press.

Burawoy, Michael. 2000. *Global Ethnography*. Berkeley: University of California Press.

Burawoy, Michael. 2001. "Neoclassical sociology: From the end of Communism to the end of class." *American Journal of Sociology* 106: 1099–1120.

Burris, Beverly H. 1983. *No Room at the Top.* New York: Praeger.

Burris, Beverly H. 1993. *Technocracy at Work.* Albany, NY: State University of New York Press.

Butler, John S. 1991. *Entrepreneurship and Self-Help Among Black Americans: A Reconsideration of Race and Economics.* Albany, NY: State University of New York Press.

Butler, Suellen, and William Snizek 1976. "The waitress-diner relationship: A multimethod approach to the study of subordinate influence." *Sociology of Work and Occupations* 3, 2: 209–222.

Carlisle, Rodney. 2004. *Scientific American: Inventions and Discoveries.* Hoboken, NJ: John Wiley.

Carruthers, Bruce C., and Sarah L. Babb. 2000. *Economy/Society: Markets, Meanings and Social Structure.* Thousand Oaks, CA: Pine Forge.

Carter, Michael J., and Susan B. Carter. 1985. "Internal labor markets in retailing: The early years." *Industrial and Labor Relations Review* 38: 586–598.

Cartledge, Paul, Edward E. Cohen, and Lin Foxhall. 2002. *Money, Labour and Land: Approaches to the Economics of Ancient Greece.* London: Routledge.

Casey, Bernard, and Michael Gold. 2000. *Social Partnership and Economic Performance.* Cheltenham, England: Edward Elgar.

Casey, C. 1996. "Corporate transformations: Designer culture, designer employees and post-occupational solidarity." *Organization* 3: 317–339.

Cassell, Joan. 1991. *Expected Miracles: Surgeons at Work.* Philadelphia: Temple University Press.

Castells, Manuel. 2000. *The Rise of the Network Society.* Oxford: Blackwell.

Catholic Campaign for Human Development. 2007. "Budgeting for poverty." Retrieved from www.nccbuscc.org/cchd/povertyusa

Cattan, Peter. 1991. "Child-care problems: An obstacle to work." *Monthly Labor Review* 114, 10: 3–9.

Cavanagh, John, and Jerry Mander (editors). 2004. *Alternatives to Economic Globalization.* San Francisco: Berrett-Koehler.

Cavendish, Ruth. 1982. *Women on the Line.* London: Routledge and Kegan Paul.

Chalcraft, David J., and Austin Harrington. 2001. *The Protestant Work Ethic Debate.* Translated by Austin Harrington and Mary Shields. Liverpool: Liverpool University Press.

Chan, Anita. 2001. *China's Workers Under Assault: Exploitation and Abuse in a Globalising Economy.* New York: M.E. Sharpe.

Chandler, Alfred D., Jr. 1980. "The United States: Seedbed of managerial capitalism." In Alfred du Pont Chandler, Jr., and Herman Daems (editors), *Managerial Hierarchies: Comparative Perspectives on the Rise of Modern Industrial Enterprise.* Cambridge, MA: Harvard University Press.

Chandler, Alfred D., Jr., Franco Amatori, and Takashi Kikino. 1997. *Big Business and the Wealth of Nations.* New York: Cambridge University Press.

Chaney, Elsa M., and Mary Garcia Castro (editors). 1989. *Muchachas No More: Household Workers in Latin America and the Caribbean.* Philadelphia: Temple University Press.

Chang, Clara, and Constance Sorrentino. 1991. "Union membership statistics in 12 countries." *Monthly Labor Review* 114, 12: 46–53.

Charles, Maria, and David B. Grusky. 2004. *Occupational Ghettos: The Worldwide Segregation of Women and Men.* Stanford, CA: Stanford University Press.

Chase-Dunn, Christopher, Yukio Kawano, and Benjamin D. Brewer. 2000. "Trade globalization since 1795: Waves of integration in the world-system." *American Sociological Review* 65: 77–95.

Chen, May Ying. 1993. "Reaching for their rights: Asian workers in New York City." Pp. 133–150 in Glenn Adler and Doris Suarez (editors), *Union Voices.* Albany, NY: State University of New York Press.

Cherlin, Andrew J., Frank F. Furstenberg, Jr., P. Lindsay Chase-Lansdale, Kathleen E. Kiernan, Philip K. Robins, Donna Ruane Morrison, and Julien O. Teitler. 1991. "Longitudinal studies on effects of divorce on children in Great Britain and the United States." *Science* 252, 5011 (June): 1386–1390.

Cherry, Mike. 1974. *On High Steel: The Education of an Ironworker.* New York: Quadrangle.

Claes, Dag Harald. 2000. *The Politics of Oil-Producer Cooperation.* Boulder, CO: Westview.

Chomsky, Noam. 2001. *9–11.* New York: Seven Stories.

Cipolla, Carlo. 1980. *Before the Industrial Revolution,* 2nd edition. New York: W.W. Norton.

Clawson, Dan. 2003. *The Next Upsurge: Labor and the New Social Movements.* Ithaca, NY: Cornell University Press.

Cleveland, Harlan, and W. T. Anderson. 1999. "Transnational: A word whose time has come." Futures 31, 9–10: 879–885.

Clogg, Clifford C. 1979. *Measuring Underemployment*. New York: Academic Press.

Clogg, Clifford C., and Teresa A. Sullivan. 1983. "Labor force composition and underemployment trends, 1969–1980." *Social Indicators Research* 12: 117–152.

Cohen, Marjorie Griffin, and Stephen McBride. 2003. *Global Turbulence*. Burlington, VT: Ashgate.

Cohn, Samuel Ross. 1985a. "Clerical labor intensity and the feminization of clerical labor in Great Britain, 1857–1937." *Social Forces* 63: 1060–1068.

Cohn, Samuel. 1985b. *The Process of Occupational Sex-Typing: The Feminization of Clerical Labor in Great Britain*. Philadelphia: Temple University Press.

Colclough, Glenna, and Charles M. Tolbert, II. 1992. *Work in the Fast Lane*. Albany, NY: State University of New York Press.

Coleman, James W. 1998. *The Criminal Elite: The Sociology of White Collar Crime*, 4th edition. New York: St. Martins.

Colestock, Harry. 2005. *Industrial Robots*. New York: McGraw-Hill.

Colgan, Fiona, and Sue Ledwith. 2002. *Gender, Diversity and Trade Unions*. London: Routledge.

Collins, Sharon M. 1997. *Black Corporate Executives: The Making and Breaking of the Black Middle Class*. Philadelphia: Temple University Press.

Collinson, David L. 2003. "Identities and insecurities: Selves at work." *Organization* 10: 527–547.

Company, Sarah O., and Martin E. Personick. 1992. "Profiles in safety and health: Retail grocery stores." *Monthly Labor Review* 115, 9: 9–25.

Connor, John M. 2001. *Global Price Fixing*. Boston: Kluwer Academic.

Constable, Nicole. 2003. "Filipina workers in Hong Kong homes: Household rules and relations." Pp. 115–141 in Barbara Ehrenreich and Arlie Russell Hochschild (editors), *Global Woman: Nannies, Maids, and Sex Workers in the New Economy*. New York: Metropolitan Books.

Cook, Karen S. (editor). 2001. *Trust in Society*. New York: Russell Sage Foundation.

Cooley, Mike. 1999. "Human-centered design." Pp. 59–82 in Robert Jacobson (editor), *Information Design*. Cambridge, MA: MIT Press.

Carleton S. Coon (editor). 1977. *A Reader in Cultural Anthropology*. Huntington, NY: Krieger.

Cornfield, Daniel B. (editor). 1987. *Workers, Managers, and Technological Change: Emerging Patterns of Labor Relations*. New York: Plenum.

Cornfield, Daniel B., and Holly J. McCammon (editors). 2003. *Labor Revitalization: Global Perspectives and New Initiatives*. Amsterdam: Elsevier-JAI.

Cornfield, Daniel B., and Teresa A. Sullivan. 1983. "Fieldwork in the oligopoly: Protecting the corporate subject." *Human Organization* 42, 3 (Fall): 258–263.

Cortright, Joseph, and Heike Mayer. 2002. "Signs of life: The growth of biotechnology centers in the U.S." Washington, DC: Brookings Institute. Retrieved from http://www.brookings.edu/dybdocroot/es/urban/publications/biotech.pdf

Coscia, Stephen. 1998. *Tele-Stress: Relief for Call Center Stress Syndrome*, 2nd edition. New York: CMP Books.

Costello, Cynthia. 1984. "Women's work in the office." *Social Science Journal* 21, 4 (October): 116–121.

Costello, Cynthia. 1991. *We're Worth It! Women and Collective Action in the Insurance Workplace*. Urbana, IL: University of Illinois Press.

Coughlin, Ellen K. 1989. "Making a business of belief: Sociologist examines the direct-selling industry in America." *The Chronicle of Higher Education* 35, 45 (July 19): 4–6.

Creighton, Sean, and Randy Hodson. 1995. "Whose side are they on? Technical workers and management ideology." In Steven Barley (editor), *Technical Workers and Technical Work*. Ithaca, NY: Cornell University Press.

Crompton, Rosemary, and Gareth Stedman Jones. 1984. *White-Collar Proletariat: Deskilling and Gender in Clerical Work*. Philadelphia: Temple University Press.

Crowley, Martha, and Vincent J. Roscigno. 2004. "Farm concentration, political-economic process, and stratification." *Journal of Political and Military Sociology* 32: 133–155.

Crowston, Clare Haru. 2001. *Fabricating Women: The Seamstresses of Old Regime France, 1675–1791*. Durham, NC: Duke University Press.

Cullen, John B. 1985. "Professional differentiation and occupational earnings." *Work and Occupations* 12, 3 (August): 351–372.

Cutcher-Gershenfeld, Joel, and J. Kevin Ford. 2005. *Valuable Disconnects in Organizational Learning Systems.* New York: Oxford University Press.

Cutler, Jonathan. 2005. "Labor's civil war." *In Critical Solidarity* 5, 2: 1, 5–6.

Daday, Gerhard, and Beverly Burris. 2001. "Technocratic teamwork." Pp. 241–262 in Steven Vallas (editor), *The Transformation of Work, Volume 10 of Research in the Sociology of Work.* Greenwich, CT: Elsevier.

Dahl, C. 2004. "What went wrong at Enron?" *Energy Journal* 25: 115–134.

Dalaker, Joe. 2005. "Alternative poverty estimates in the US: 2003.*" Current Population Reports.* P60-227 (June).

Dalton, Frederick John. 2003. *The Moral Vision of César Chávez.* Maryknoll, NY: Orbis.

Danziger, James N. 2004. "Innovation in innovation." *Social Science Computer Review* 22: 100–110.

Davey, Caroline L., and Marilyn J. Davidson. 2000. "The right of passage? The experiences of female pilots in commercial aviation." *Feminism and Psychology* 10, 2 (May): 195–225.

Davidson, Kenneth M. 1985. *Megamergers: Corporate America's Billion-Dollar Takeovers.* Cambridge, MA: Ballinger.

Davidson, Marilyn J., and Ronald J. Burke (editors). 2004. *Women in Management Worldwide: Facts, Figures, and Analysis.* Aldershot, Hants, England and Burlington VT: Ashgate.

Davies, Margery W. 1982. *Woman's Place Is at the Typewriter: Office Work and Office Workers, 1870–1930.* Philadelphia: Temple University Press.

Davies, Matt. 2005. "The public spheres of unprotected workers." *Global Society* 19: 131–154.

Davis-Floyd, Robbie. 1998. "The ups, downs, and interlinkages of nurse- and direct-entry midwifery: Status, practice, and education." Pp. 67–118 *in Pathways to Becoming a Midwife: Getting an Education.* Eugene, OR: Midwifery Today.

DeAnda, Roberto M. 1996. "Falling back: Mexican-origin men and women in the U.S. economy." Pp. 41–50 in Roberto M. DeAnda (editor), *Chicanas and Chicanos in Contemporary Society.* Boston: Allyn and Bacon.

Deckard, Barbara Sinclair. 1983. *The Women's Movement,* 2nd edition. New York: Harper and Row.

DeFreitas, Gregory. 1993. "Unionization among racial and ethnic minorities." *Industrial and Labor Relations Review* 46, 2 (January) 284–301.

De Graff, John (editor). 2003. *Take Back Your Time: Fighting Overwork and Time Poverty in America.* San Francisco: Berrett-Koehler.

DeGrazia, Raffaele. 1980. "Clandestine employment: A problem of our times." *International Labour Review* 119, 5 (September–October): 549–563.

Delbridge, Rick. 1998. *Life on the Line in Contemporary Manufacturing.* Oxford: Oxford University Press.

Deming, W. Edwards. 1986. *Out of Crisis.* Cambridge, MA: MIT Press.

Denby, Charles. 1978. *Indignant Heart.* Boston: South End Press.

Derber, Charles (editor). 1982. *Professionals as Workers: Mental Labor in Advanced Capitalism.* Boston: G. K. Hall.

Desmarez, Pierre. 2002. "The sociology of work: From work to the firm and beyond." In Daniel B. Cornfield and Randy Hodson (editors), *Worlds of Work.* New York: Plenum.

Desmond, Kevin. 1986. *Inventions Innovations Discoveries.* London: Constable.

de Soto, Hernando. 2000. *The Mystery of Capital: Why Capitalism Triumphs in the West and Fails Everywhere Else.* New York: Bantam.

Devinatz, Victor G. 1999. *High Tech Betrayal: Working and Organizing on the Shop Floor.* East Lansing: Michigan University Press.

Diamond, W. J., and Richard B. Freeman. 2002. "Will unionism prosper in cyberspace?" *British Journal of Industrial Relations* 40: 569–596.

Dicken, Peter. 2003. *Global Shift,* 4th edition. New York: Guilford.

Dinnerstein, Myra. 1992. *Women Between Two Worlds: Midlife Reflections on Work and Family.* Philadelphia: Temple University Press.

DiPrete, Thomas A. 2002. "Life course risks, mobility regimes, and mobility consequences: A Comparison of Sweden, Germany, and the United States." *American Journal of Sociology* 108: 267–309.

DiPrete, Thomas A. 2005. "Labor markets, inequality and change: A European perspective." *Work and Occupations* 32: 119–139.

DiTomaso, Nancy, and R. Hooijberg. 1996. "Diversity and the demands of leadership." *Leadership Quarterly* 7, 2 (Summer): 163–187.

Dixon, Marc. 2005. *The Politics of Union Decline.* Ph.D. dissertation, Ohio State University, Department of Sociology.

Dixon, Roland B. 1977. "The Northern Maidu." Pp. 262–291 in Carleton S. Coon (editor), *A Reader in Cultural Anthropology*. Huntington, NY: Krieger.

Dobbin, Frank, John R. Sutton, John W. Meyer, and W. Richard Scott. 1993. "Equal opportunity law and the construction of internal labor markets." *American Journal of Sociology* 99: 396–427.

Doeringer, Peter B., Christine Evans-Klock, and David G. Terkla. 2002. *Start-up Factories*. New York: Oxford University Press.

Dohm, Arlene. 2000. "Gauging the labor force effects of retiring baby boomers." *Monthly Labor Review* 123, 1: 17–25.

Domhoff, G. William. 2002. *Who Rules America?*, 4th edition. Boston, McGraw-Hill.

Donaldson, Lex. 2001. The Contingency Theory of Organizations. Thousand Oaks, CA: Sage.

Dong, Ok Lee. 1992. "Commodification of ethnicity: The sociospacial reproduction of immigrant entrepreneurs." *Urban Affairs Quarterly* 28: 258–276.

Dore, Ronald P. 2001. *What It Means to Take Japan Seriously*. Northampton, MA: Edward Elgar.

Dore, Ronald P., and Mari Sako. 1998. *How the Japanese Learn to Work*. London: Routledge.

Dreeben, Robert. 1968. *On What Is Learned in School*. Reading, MA: Addison-Wesley.

Dreiling, Michael C. 2000. "Corporate political action leadership in the defense of NAFTA." *Social Problems* 47: 21–48.

Drucker, Peter F. 2003. *A Functioning Society*. New Brunswick, NJ: Transaction.

Dubose, Louis. 1987. "Letter from northern Mexico." *Texas Observer*, March 20, p. 23.

Duchin, Faye. 1998. *Structural Economics: Measuring Change in Technology, Lifestyles, and the Environment*. Washington, DC: Island.

Dulles, Foster Rhea, and Melvyn Dubofsky. 2004. *Labor in America*, 7th edition. Arlington Heights, IL: Harlan Davidson.

Duncan, O. D. 1961. "A socioeconomic index for all occupations." Pp. 109–138 in A. J. Reiss et al. (editors), *Occupations and Social Status*. New York: Free Press.

Duneier, Mitchell. 1999. *Sidewalk*. New York: Farrar, Straus and Giroux.

Durkheim, Emile. 1966 [1897]. *Suicide*. (Translated by John A. Spaulding and George Simpson.) New York: Free Press.

Durkheim, Emile. 1984 [1933]. *The Division of Labor in Society* (Translated by W. D. Halls). New York: Free Press.

Dyer-Witheford, Nick. 1999. *Cyber-Marx: Cycles and Circuits of Struggle in High-Technology Capitalism*. Urbana: University of Illinois Press.

Ebaugh, Helen Rose Fuchs. 1988. *Becoming an Ex: The Process of Role Exit*. Chicago: University of Chicago Press.

Edelman, Lauren B. 1990. "Legal environments and organizational governance: The expansion of due process in the American workplace." *American Journal of Sociology* 95, 6 (May) 1401–1440.

Edgar, R.W. 2000. "Jubilee 2000: Paying our debts." *Nation*, April 24, pp. 20–21.

Edin, Kathryn, and Laura Lein. 1997. *Making Ends Meet: How Single Mothers Survive Welfare and Low-Wage Work*. New York: Russell Sage Foundation.

Edwards, Richard C. 1979. *Contested Terrain*. New York: Basic Books.

Edwards, Richard C. 1993. *Rights at Work: Employment Relation in the Post-Union Era*. Washington, DC: Brookings Institute.

Ehlers, Tracy Bachrach. 2000. *Silent Looms: Women and Production in a Guatemalan Town*. Austin: University of Texas Press.

Ehrenreich, Barbara. 1989. *Fear of Falling: The Inner Life of the Middle Class*. New York: Pantheon.

Ehrenreich, Barbara. 2003. "Maid to order." Pp. 85–103 in Barbara Ehrenreich and Arlie Russell Hochschild (editors), *Global Woman: Nannies, Maids, and Sex Workers in the New Economy*. New York: Metropolitan Books.

Ehrenreich, Rosa. 1999. "Dignity and discrimination: Toward a pluralistic understanding of workplace harassment." *Georgetown Law Review* (November): 1–47.

Elliott, Carole, and Sharon Turnbull. 2005. *Critical Thinking in Human Resource Development*. London: Routledge.

Ellison, Nicole B. 2004. *Telework and Social Change*. Westport, CT: Praeger.

Ellul, Jacques. 1964. *The Technological Society*. New York: Vintage.

Employee Ownership Options. 2005. Case study of Williams Commercial Body worker buy-out. Retrieved from www.employee-ownership.org.uk/eoowills.htm

Endo, Koshi. 1994. "*Satei* (personal assessment) and interworker competition in Japanese firms." *Industrial Relations* 33, 1 (January): 70–82.

England, Paula. 1992. *Comparable Worth*. New York: Aldine de Gruyter.

England, Paula, Jennifer Thompson, and Carolyn Aman. 2001. "The sex gap in pay and comparable worth." Pp. 551–567 in Ivar Berg and Arne L. Kalleberg (editors), *Sourcebook of Labor Markets*. New York: Kluwer/Academic.

Epstein, Cynthia Fuchs. 1970. *Woman's Place: Options and Limits in Professional Careers*. Berkeley: University of California Press.

Epstein, Cynthia Fuchs. 1982. "Ambiguity as social control: Women in professional elites." Pp. 61–72 in Phyllis L. Stewart and Muriel G. Cantor (editors), *Varieties of Work*. Beverly Hills, CA: Sage.

Epstein, Cynthia Fuchs, and Arne L. Kalleberg. 2004. *Fighting for Time: Shifting Boundaries of Work and Social Life*. New York: Russell Sage.

Erickson, Bonnie H., Patricia Albanese, and Slobodan Drakulic. 2000. "Gender on a jagged edge: The security industry, its clients, and the reproduction and revision of gender." *Work and Occupations* 27, 3 (August): 294–318.

Erikson, Erik H. 1963. *Childhood and Society*, 2nd edition. New York: Norton.

Ermann, M. David, and Richard J. Lundman. 2001. *Corporate and Governmental Deviance*, 7th edition. New York: Oxford University Press.

Essed, Philomena. 1991. *Understanding Everyday Racism*. Newbury Park, CA: Sage.

Evans, Peter B. 1995. *Embedded Autonomy: States and Industrial Transformation*. Princeton, NJ: Princeton University Press.

Everett, Melissa. 1999. *Making a Living While Making a Difference: A Guide to Creating Careers with a Conscience*. Gabriola Island, British Columbia: New Society.

Falk, William W., Michael D. Schulman, and Ann R. Tickamyer. 2003. *Communities of Work: Rural Restructuring in Local and Global Contexts*. Athens: Ohio University Press.

Fantasia, Rick. 1988. *Cultures of Solidarity*. Berkeley: University of California Press.

Fantasia, Rick, and Kim Voss. 2004. *Hard Work: Remaking the American Labor Movement*. Berkeley: University of California Press.

Faunce, William A. 1981. *Problems of an Industrial Society*. New York: McGraw-Hill.

Feagin, Joe R. 1991. "The continuing significance of race: Antiblack discrimination in public places." *American Sociological Review* 56, 1 (February): 101–116.

Feagin, Joe R., Anthony M. Orum, and Gideon Sjoberg. 1991. *A Case for the Case Study*. Chapel Hill, NC: University of North Carolina Press.

Feldman, D.C., and M. C. Bolin. 1998. "Moving on out: When are employees willing to follow their organization during corporate relocation?" *Journal of Organizational Behavior* 19, 3: 275–288.

Ferber, Marianne A., and Brigid O'Farrell (editors). 1991. *Work and Family: Policies for a Changing Work Force*. Washington, DC: National Academy Press.

Fernandez, John P., and Julie Davis. 1999. *Race, Gender, and Rhetoric: The True State of Race and Gender Relations in Corporate America*. New York: McGraw-Hill.

Fernandez, R. M., E. J. Castilla, and P. Moore. 2000. "Social capital at work: Networks and employment in a phone center." *American Journal of Sociology*. 105: 1288–1356.

Ferraro, F., Jeffery Pfeffer, and R. I. Sutton. 2005. "Economics language and assumptions: How theories can become self-fulfilling." *Academy of Management Review* 30: 8–24.

Ferree, Myra Marx. 1984. "Class, housework, and happiness: Women's work and life satisfaction." *Sex Roles* 11: 1057–1074.

Fine, Gary Alan. 1999. "Field labor and ethnographic reality." *Journal of Contemporary Ethnography* 28: 532–539.

Fine, Gary Alan. 1996. *Kitchens: The Culture of Restaurant Work*. Berkeley: University of California Press.

Fink, Deborah. 1998. *Cutting into the Meatpacking Line*. Chapel Hill, NC: University of North Carolina Press.

Fishman, Paula M. 1978. "Interaction: The work women do." *Social Problems* 75: 397–406.

Fitzgerald, Ian, and John Stirling. 2004. *European Works Councils*. London: Routledge.

Fleischer, Arthur, Jr., Geoffrey C. Hazard, Jr., and Miriam Z. Klipper. 1988. *Board Games: The Changing Shape of Corporate Power*. Boston: Little Brown.

Fletcher, Joyce K. 1999. *Disappearing Acts: Gender, Power and Relational Practice at Work*. Cambridge, MA: MIT Press.

Fligstein, Neil. 2001. *The Architecture of Markets*. Princeton, NJ: Princeton University Press.

Florida, Richard. 2004. *Cities and the Creative Class*. New York: Routledge.

Foner, Philip S. 1982. *Organized Labor and the Black Worker, 1619–1981*. New York: International Publishers.

Forbes. 2006. "Richest Americans," May 15.

Form, William, and David Byron McMillen. 1983. "Women, men, and machines." *Work and Occupations* 10, 2: 147–178.

Forsyth, Patrick B., and Thomas J. Danisiewicz. 1985. "Toward a theory of professionalization." *Work and Occupations* 12, 1 (February): 59–76.

Fox, Bonnie, and Pamela Sugiman. 1999. "Flexible work, flexible workers: The restructuring of clerical work in a large telecommunications company." *Studies in Political Economy* 60 (Autumn): 59–84.

Francis, Arthur. 1994. "The design-manufacturing interface." Pp. 63–79 in John Storey (editor), *New Wave Manufacturing Strategies*. London: P. Chapman.

Freel, M. S. 2003. "Sectoral patterns of small firm innovation, networking and proximity." *Research Policy* 32: 751–770.

Freeman, Richard B. (editor). 1994. *Working Under Different Rules*. New York: Russell Sage.

Freeman, Richard B., and James L. Medoff. 1984. *What Do Unions Do?* New York: Basic Books.

Freeman, Richard B., and Joel Rogers. 1999. *What Workers Want*. Ithaca, NY: Industrial and Labor Relations Press.

Freidson, Eliot. 1970. *The Profession of Medicine: A Study in the Sociology of Applied Knowledge*. New York: Harper and Row.

Freidson, Eliot. 1986. *Professional Powers*. Chicago: University of Chicago Press.

French, Wendell L., Fremont E. Kast, and James E. Rosenzweig. 1985. *Understanding Human Behavior in Organizations*. New York: Harper and Row.

Frenkel, Stephen J. 2001. "Globalization, athletic footwear commodity chains and employment relations in China" *Organization Studies* 22: 531–562.

Frenkel, Stephen J., Marek Korczynski, Karen A. Shire, and May Tam. 1999. *On the Front Line: Organization of Work in the Information Society*. Ithaca, NY: Cornell University Press.

Freund, Caroline L., and Diana Weinhold. 2004. "The effect of the Internet on international trade." *Journal of International Economics* 62: 171–189.

Friedman, Thomas L. 2005. *The World Is Flat: A Brief History of the Twenty-First Century*. New York: Farrar, Straus, and Giroux.

Frundt, H. J. 2005. "Toward a hegemonic resolution in the banana trade." *International Political Science Review* 26: 215–237.

Fuchs, Victor. 1968. *The Service Economy*. New York: National Bureau of Economic Research.

Fuller, Linda, and Vicki Smith. 1996. "Consumers' reports: Management by customers in a changing economy." Pp. 74–90 in Macdonald, Cameron Lynn, and Carmen Sirianni (editors), *Working in the Service Society*. Philadelphia: Temple University Press.

Galenson, David W. 1984. "The rise and fall of indentured servitude in the Americas." *Journal of Economic History* 71, 1 (March): 1–26.

Galle, Omer R., Candace Hinson Wiswell, and Jeffrey A. Burr. 1985. "Racial mix and industrial productivity." *American Sociological Review* 50: 20–33.

Gamst, Frederick C. (editor). 1995. *Meanings of Work*. Albany, NY: State University of New York Press.

Gannon, Martin J. 1984. "Preferences of temporary workers: Time, variety, and flexibility." *Monthly Labor Review* 107, 8 (August): 26–28.

Gardner, John. 1993. "Community Organizing: Seeds of Justice." *In These Times*, May 17, pp. 18–19.

Gatta, Mary Lizabeth. 2001. *Juggling Food and Feelings: Emotional Balance in the Workplace*. Lanham, MD: Lexington.

Gaventa, John. 1980. *Power and Powerlessness: Quiescence and Rebellion in an Appalachian Valley*. Urbana, IL: University of Illinois Press.

Gecas, Victor. 1979. "The influence of social class on socialization." In W. R. Burr et al. (editors), *Contemporary Theories about the Family*. Volume 1. New York: Free Press.

Geison, Gerald L. (editor). 1983. *Professions and Professional Ideologies in America*. Chapel Hill, NC: University of North Carolina Press.

Geisst, Charles R. 2004. *Deals of the Century: Wall Street, Mergers and the Making of Modern America*. Hoboken, NY: Wiley.

General Social Survey. 2002. James A. Davis, Tom W. Smith, and Peter V. Marsden, Principal Investigators. Retrieved from www.icpsr.umich.edu

Gereffi, Gary, David Spener, and Jennifer Bair (editors). 2002. *Free Trade and Uneven Development*. Philadelphia: Temple University Press.

Gerson, Kathleen. 1985. *Hard Choices: How Women Decide About Work, Career, and Motherhood*. Berkeley: University of California Press.

Gevitz, Norman. 1982. *The D.O.'s: Osteopathic Medicine in America*. Baltimore: Johns Hopkins University Press.

Gifford, Court (editor). 2002. *Directory of U.S. Labor Organizations*. Washington, DC: Bureau of National Affairs.

Giles, Anthony, and Jacques Belanger. 2002. "The sociology of work in Canada." In Daniel B. Cornfield and Randy Hodson (editors), *Worlds of Work*. New York: Plenum.

Gillespie, Richard. 1991. *Manufacturing Knowledge: A History of the Hawthorne Experiments*. New York: Cambridge University Press.

Giuffre, Patti A., and Christine L. Williams. 1994. "The boundary lines: Labeling sexual harassment in restaurants." *Gender & Society* 8: 378–401.

Glass, Jennifer. 2000. "Envisioning the integration of family and work: Toward a kinder, gentler workplace." *Contemporary Sociology* 29, 1 (January): 129–143.

Glass, Jennifer, and Valerie Camarigg. 1992. "Gender, parenthood, and job-family compatibility." *American Journal of Sociology* 98, 1 (July): 131–151.

Glenn, Evelyn Nakano, and Roslyn L. Feldberg. 1977. "Degraded and deskilled: The proletarianization of clerical work." *Social Problems* 25, 1 (October): 52–64.

Gobert, James, and Maurice Punch. 2003. *Rethinking Corporate Crime*. London: Butterfield.

Goffman, Erving. 1955. "On Face-Work: An Analysis of Ritual Elements in Social Interaction." *Psychiatry* 18: 213–231.

Gordon, David M. 1996. *Fat and Mean: The Corporate Squeeze of Working Americans and the Myth of Managerial Downsizing*. New York: Free Press.

Gordon, Michael E., and Lowell Turner. 2000. *Transnational Cooperation Among Labor Unions*. Ithaca, NY: Industrial and Labor Relations Press.

Gottschalk, P., and Sheldon Danziger. 2005. "Inequality of wage rates, earnings and family income in the United States, 1975–2002." *Review of Income and Wealth* 2: 231–254.

Gouldner, Alvin W. 1964. *Patterns of Industrial Bureaucracy*. New York: Free Press.

Gouldner, Alvin W. 1979. *The Future of the Intellectuals and the Rise of the New Class*. London: Macmillan Publishing.

Goyder, John. 2005. "The dynamics of occupational prestige: 1975–2000." *The Canadian Review of Sociology and Anthropology/La Revue Canadienne de Sociologie et d'Anthropologie* 42: 1–23.

Graham, Laurie. 1995. *On the Line at Subaru-Isuzu*. Ithaca, NY: Industrial and Labor Relations Press.

Granovetter, Mark. 1995. *Getting A Job: A Study of Contacts and Careers*, 2nd edition. Chicago: University of Chicago Press.

Granovetter, Mark. 2000. "The economic sociology of firms and entrepreneurs." Pp. 244–276 in Richard Swedberg (editor), *Entrepreneurship*. New York: Oxford University Press.

Green, James R. 2000. *Taking History to Heart*. Amherst: University of Massachusetts.

Greenbaum, Joan M. 2004. *Windows on the Workplace: Technology, Jobs and the Organization of Office Work*. New York: Monthly Review Press.

Greenberg, Edward S. 1980. "Participation in industrial decision making and work satisfaction: The case of producer cooperatives." *Social Science Quarterly* 60, 4 (March): 551–569.

Greenberg, Edward S. 1986. *Workplace Democracy*. Ithaca, NY: Cornell University Press.

Greenhouse, Steven. 2005a. "At a small shop in Colorado, Wal-Mart beats a union once more." *The New York Times*, February 26, 2005.

Greenhouse, Steven. 2005b. "Wal-Mart workers are finding a voice without a union." *The New York Times*. September 3.

Greenwood, R., and R. Lachman. 1996. "Change as an underlying theme in professional service organizations." *Organization Studies* 17, 4: 563–572.

Grenier, Guillermo J. 1988. *Inhuman Relations: Quality Circles and Anti-Unionism in American Industry*. Philadelphia: Temple University Press.

Griffin, Larry J., Philip J. O'Connell, and Holly J. McCammon. 1989. "National variation in the context of struggle." *Canadian Review of Sociology and Anthropology* 26, 1 (February): 37–68.

Griffith, David. 1993. *Jones's Minimal: Low-Wage Labor in the United States*. Albany, NY: State University of New York Press.

Grint, Keith. 1991. *The Sociology of Work*. Cambridge: Basil Blackwell.

Gruber, James E., and Lars Bjorn. 1986. "Women's responses to sexual harassment." *Social Science Quarterly* 67, 4 (December): 814–826.

Grunberg, Leon, R. Anderson-Connolly, and Edward S. Greenberg. 2000. "Surviving layoffs." *Work and Occupations* 27: 7–31.

Guerrier, Yvonne, and Amel S. Adib. 2000. "'No, we don't provide that service': The harassment of hotel employees by customers." *Work, Employment, and Society* 14: 689–705.

Guest, Robert H. 1987. "Industrial sociology: The competitive edge." *American Sociological Association Footnotes* 15, 1 (January): 5.

Gullickson, Gay L. 2002. *Spinners and Weavers of Auffay*. Cambridge: Cambridge University Press.

Gunderson, Morley, and Allen Ponak. 2000. *Union-Management Relations in Canada*, 4th edition. Reading, MA: Addison-Wesley.

Gutek, Barbara A. 1985. *Sex and the Workplace*. San Francisco: Jossey-Bass.

Guthrie, Douglas, and L. M. Roth. 1999. "The states, courts, and equal opportunities for female CEOs." *Social Forces* 78: 511–542.

Guy, Mary E. 1985. *Professionals in Organizations: Debunking a Myth*. New York: Praeger Press.

Gyllenhammar, Pehr G. 1977. *People at Work*. Reading, MA: Addison-Wesley.

Haber, Samuel. 1991. *The Quest for Authority and Honor in the American Professions, 1750–1900*. Chicago: University of Chicago Press.

Hage, Jerald, and Michael Aiken. 1970. *Social Change in Complex Organizations*. New York: Random House.

Hakim, Danny. 2005. "Automaker plans closings." *The New York Times News Service*, June 8.

Halaby, Charles N. 2003. "Where job values come from." *American Sociological Review* 68: 251–278.

Hall, Richard H., and Pamela S. Tolbert. 2004. *Organizations: Structures, Processes, and Outcomes*, 9th edition. New York: Prentice Hall.

Halle, David. 1984. *America's Working Man*. Chicago: University of Chicago Press.

Haller, Max, Wolfgang Konig, Peter Krause, and Karin Kurz. 1985. "Patterns of career mobility and structural positions in advanced capitalist societies: A comparison of men in Austria, France, and the United States." *American Sociological Review* 50, 5 (October) 579–603.

Halliday, Terence C. 1987. *Beyond Monopoly: Lawyers, State Crises, and Professional Empowerment*. Chicago: University of Chicago Press.

Halliday, Terence C., and Lucien Karpik. 1997. "Postscript: Lawyers, political liberalism, and globalization." Pp. 350–370 in Terence C. Halliday and Lucien Karpik (editors), *Lawyers and the Rise of Western Political Liberalism*. New York: Oxford University Press.

Halliday, Terence C., Michael J. Powell, and Mark Granfors. 1993. "After minimalism: Transformations of state bar associations, 1918–1950." *American Sociological Review* 58, 4 (August): 515–535.

Halpern, Sydney A. 1992. "Dynamics of professional control: Internal coalitions and crossprofessional boundaries." *American Journal of Sociology* 97, 4 (January): 994–1022.

Hamilton, Richard F., and James D. Wright. 1986. *The State of the Masses*. New York: Aldine.

Hamper, Ben. 1991. *Rivethead: Tales from the Assembly Line*. New York: Warner.

Handel, Michael J. 2003. "Skills mismatch in the labor market." *Annual Review of Sociology* 29: 135–165.

Handlin, Oscar. 1973. *The Uprooted*, 2nd edition. Boston: Little, Brown.

Hansmann, Henry. 1996. *The Ownership of Enterprise*. Cambridge, MA: Harvard University Press.

Harper, Douglas. 1987. *Working Knowledge*. Chicago: University of Chicago Press.

Harriman, Ann. 1985. *Women/Men/Management*. New York: Praeger.

Harris, M., and A. Raviv. 2002. "Organization design." *Management Science* 48: 852–865.

Harrison, Bennett. 1994. *Lean and Mean: The Changing Landscape of Corporate Power in the Age of Flexibility*. New York: Basic Books.

Hartmann, Heidi. 2000. "Closing the gap: Women's economic progress and future prospects." Pp. 119–130 in Ray Marshall (editor), *Back to Shared Prosperity*. Armonk, NY: M.E. Sharpe.

Hartmann, Heidi, Robert E. Kraut, and Louise A. Tilly (editors). 1986. *Computer Chips and Paper Clips: Technology and Women's Employment*. Washington, DC: National Academy Press.

Hartmann, Heidi, Patricia A. Roos, and Donald J. Treiman. 1985. "An agenda of basic research on comparable worth." Pp. 3–33 in Heidi I. Hartmann (editor), *Comparable Worth*. Washington, DC: National Academy Press.

Hassard, John, Jonathan Morris, and Jackie Sheehan. 2004. "The 'third way': The future of work and organization in a 'corporatized' Chinese economy." *International Journal of Human Resource Management* 15: 314–330.

Hauser, Robert M., and David L. Featherman. 1977. *The Process of Stratification: Trends and Analyses*. New York: Academic Press.

Hayano, David M. 1982. *Poker Faces: The Life and Work of Professional Card Players*. Berkeley: University of California Press.

Hayes, Robert H., and William J. Abernathy. 1980. "Managing our way to economic decline." *Harvard Business Review* 58, 4 (July–August): 67–77.

Healey, James R. 1992. "U.S. Steel learns from experience." *USA Today*, April 10, pp. B1–B2.

Hecker, Daniel E. 2005. "Occupational employment projections to 2014." *Monthly Labor Review* 128, 11 (November): 70–101.

Hedlund, Stefan. 2005. *Russian Path Dependence*. London: Routledge.

Heinz, John P., and Edward O. Laumann. 1982. *Chicago Lawyers: The Social Structure of the Bar*. New York and Chicago: Russell Sage Foundation and American Bar Association.

Helmick, Sandra A., and Judith D. Zimmerman. 1984. "Trends in the distribution of children among households and families." *Child Welfare* 62: 401–409.

Henson, Kevin D. 1996. *Just a Temp*. Philadelphia: Temple University Press.

Herzberg, Frederick. 2003. "One more time: How do we motivate employees?" *Harvard Business Review* 81, 1: 87–97.

Hill, Reuben. 1986. "Life cycle stages for types of single parent families: Of family development theory." *Family Relations* 35: 19–29.

Hilton, Isabel. 2005. "Made in China." *Granta* 89: 13–54.

Himmelweit, Susan (editor). 2000. *Inside the Household: From Labour to Cave*. New York: St. Martin's Press.

Hirsch, Paul M. 1986. "From ambushes to golden parachutes: Corporate takeovers as an instance of cultural framing and institutional integration." *American Journal of Sociology* 91: 800–837.

Hirschhorn, Larry. 2002. "Campaigning for change." *Harvard Business Review* 80, 7: 98–104.

Hobsbawm, Eric J. 1975. *The Age of Capital*. New York: Scirbner.

Hobsbawm, Eric J. 1996. *The Age of Extremes*. New York: Vintage.

Hochschild, Arlie. 1983. *The Managed Heart: Commercialization of Human Feeling*. Berkeley: University of California Press.

Hochschild, Arlie. 1989. *The Second Shift: Working Parents and the Revolution at Home*. New York: Viking.

Hochschild, Arlie. 1997a. *The Time Bind: When Work Becomes Home and Home Becomes Work*. New York: Metropolitan Books.

Hochschild, Arlie. 1997b. "Work: The great escape." *The New York Times Magazine*, April 20, pp. 51–55, 81, 84.

Hochschild, Arlie, and Anne Machung. 1997. *The Second Shift: Working Parents and the Revolution at Home*. New York: Avon.

Hodge, Robert W., Paul M. Siegel, and Peter M. Rossi. 1964. "Occupational prestige in the United States, 1925–1963." *American Journal of Sociology* 70, 3 (November): 286–302.

Hodson, Randy. 1983. *Workers' Earnings and Corporate Economic Structure*. New York: Academic Press.

Hodson, Randy. 1989. "Gender differences in job satisfaction: Why aren't women more dissatisfied?" *Sociological Quarterly* 30, 3 (September): 385–399.

Hodson, Randy. 1991. "Good soldiers, smooth operators, and saboteurs: A model of workplace behaviors." *Work and Occupations* 18, 3 (August): 271–290.

Hodson, Randy. 1999. "Management citizenship behavior: A new concept and an empirical test." *Social Problems* 46, 3: 460–478.

Hodson, Randy. 2001. *Dignity at Work*. New York: Cambridge University Press.

Hodson, Randy. 2002. "Worker participation and teams: New evidence from analyzing organizational ethnographies." *Economic and Industrial Democracy* 23: 491–528.

Hodson, Randy, and Robert L. Kaufman. 1982. "Economic dualism: A critical review." *American Sociological Review* 47, 6 (December): 727–739.

Hodson, Randy, and Teresa A. Sullivan. 1985. "Totem or tyrant? Monopoly, regional, and local sector effects on worker commitment." *Social Forces* 63: 716–731.

Hoffman, Lily M. 1989. *The Politics of Knowledge: Activist Movements in Medicine and Planning*. Albany, NY: State University of New York Press.

Hogan, Dennis P. 1981. *Transitions and Social Change: The Early Lives of American Men*. New York: Academic Press.

Holzer, B. 2000. "Miracles with a system: The economic rise of East Asia and the role of sociocultural patterns." *International Sociology* 15: 455–478.

Hood, Jane C. 1988. "From night to day: Timing and the management of custodial work." *Journal of Contemporary Ethnography* 17, 1 (April): 96–116.

Hooks, Gregory, and C. L. Smith. 2004. "The treadmill of destruction: National sacrifice areas and Native Americans." *American Sociological Review* 69: 558–575.

House, James S. 1981. *Work Stress and Social Support*. Reading, MA: Addison-Wesley.

Howe, Louise Kapp. 1977. *Pink Collar Workers: Inside the World of Women's Work*. New York: G.P. Putnam's Sons Publishing.

Hsiung, Ping-Chun. 1996. *Living Rooms as Factories: Class, Gender and the Satellite Factory System in Taiwan*. Philadelphia: Temple University Press.

Huang, Shirlena, Brenda S.A. Yeoh, and Noor Abdul Rahman (editors). 2005. *Asian Women as Transnational Domestic Workers*. London: Marshall Cavendish.

Hughes, Everett Cherrington. 1958. *Men and Their Work*. Glencoe, IL: Free Press.

Hughes, Everett Cherrington. 1965. "Professions." Pp. 1–14 in Kenneth S. Lynn and the editors of Daedalus (editors), *The Professions in America*. Boston: Houghton Mifflin.

Hughes, Thomas P. 2004. *Human-Built World: How We Think about Technology and Culture*. Chicago: University of Chicago Press.

Human Development Report. 1998. New York: Oxford University Press.

Hunnicutt, Benjamin K. 2003. "When we had the time." Pp. 114–122 in John de Graaf (editor), *Take Back Your Time*. San Francisco: Berrett-Koehler.

Hunt, G., and D. Rayside. 2000. "Labor union response to diversity in Canada and the United States." *Industrial Relations* 39: 401–444.

Iams, Howard. 1993. "Working wives and social security: Many still rely on husband's benefit." *Population Today* 21, 11 (November): 6–7.

Illich, Ivan. 1977. *Disabling Professions*. London: Marion Boyars.

Industrial Workers of the World. 1984. *Songs of the Workers*. Chicago: Industrial Workers of the World.

International Iron and Steel Institute. 2005. worldsteel. org. Brussels.

International Labour Office, United Nations. 2006. "Global employment trends brief, January." Retrieved from http://www.ilo.org/public/english/employment/strat/download/getb06en.pdf

Jackall, Robert. 1978. *Workers in a Labyrinth: Jobs and Survival in a Bank Bureaucracy*. Montclair, NJ: Allanheld and Osmun.

Jackall, Robert. 1988. *Moral Mazes: The World of Corporate Managers*. New York: Oxford University Press.

Jacobs, Jerry A., and Kathleen Gerson. 2004. *The Time Divide: Work, Family, and Gender Inequality*. Cambridge, MA: Harvard University Press.

Jacoby, Sanford M. 2004. *Employing Bureaucracy: Managers, Unions, and the Transformation of Work in the 20th Century*. Mahwah, NJ: Lawrence Erlbaum.

Jaffee, David, J. R. Kloppenburg, and M. B. Monroy. 2004. "Bringing the 'moral change' home: Fair trade within the north and within the south." *Rural Sociology* 69: 169–196.

Jakobsen, K. A. 2001. "Rethinking the International Confederation of Free Trade Unions and its Inter-American Regional Organization." *Antipode* 33: 363–383.

Jasso, Guellermina, Douglas S. Massey, Mark R. Rosenzweig, and James P. Smith. 2000. "The new immigrant survey pilot." *Demography* 37, 1 (February): 127–138.

Jaynes, Gerald David. 1986. *Branches without Roots: Genesis of the Black Working Class in the American South, 1862–1882*. New York: Oxford University Press.

Jensen, Richard J., and John C. Hammerback. 2002. *The Words of César Chávez* [electronic resource]. College Station: Texas A&M University Press.

Johnson, Julia Overturf. 2005. "Who's minding the kids? Child care arrangements: Winter 2002." *Current Population Reports* P70–101 (October).

Johnson, Roberta Ann. 2003. *Whistleblowing: When It Works and Why*. Boulder, CO: Lynne Rienner.

Jones, Stephen R. G. 1992. "Was there a Hawthorne effect?" *American Journal of Sociology* 98, 3 (November): 451–468.

Jordan, Brigitte. 1992. *Birth in Four Cultures: A Cross-cultural Investigation of Childbirth in Yucatan, Holland, Sweden, and the United States*, 4th edition. Long Grove, IL: Waveland.

Juravich, Tom. 1985. *Chaos on the Shop Floor: A Worker's View of Quality, Productivity, and Management.* Philadelphia: Temple University Press.

Kahn, Robert L. 1972. "The meaning of work: Interpretation and proposals for measurement." Pp. 159–203 in Angus S. Campbell and Philip E. Converse (editors), *The Human Meaning of Social Change.* New York: Russell Sage Foundation.

Kalev, Alexandra, Frank Dobbin, and Erin Kelly. 2006. "Best practices or best guesses? Assessing the efficacy of corporate affirmative action and diversity policies." *American Sociological Review* 71: 589–617.

Kalleberg, Arne L. 2006. *The Mismatched Worker.* New York: W.W. Norton.

Kalleberg, Arne L., David Knoke, Peter V. Marsden, and Joe L. Spaeth. 1996. *Organizations in America.* Thousand Oaks, CA: Sage.

Kalleberg, Arne L., Barbara F. Reskin, and Ken Hudson. 2000. "Bad jobs in America: Standard and non-standard employment relations and job quality in the United States." *American Sociological Review* 65, 2 (April): 256–278.

Kamata, Satoshi. 1982. *Japan in the Passing Lane.* New York: Pantheon Books. Translated by Tatsuru Akimoto.

Kamerman, Sheila B. 1986. "Maternity, paternity, and patenting policies: How does the United States compare?" Pp. 53–66 in Sylvia Ann Hewlett, Alice S. Ilchman, and John J. Sweeney (editors), *Family and Work: Bridging the Gap.* Cambridge, MA: Ballinger.

Kanter, Rosabeth Moss. 1977. *Men and Women of the Corporation.* New York: Basic Books.

Kanter, Rosabeth Moss. 1983. *The Change Masters: Innovation for Productivity in the American Corporation.* New York: Simon and Schuster.

Kanter, Rosabeth Moss. 2001. *Evolve! Succeeding in the Digital Culture of Tomorrow.* Boston: Harvard Business School Press.

Kaplan, Robert E., Michael M. Lombardo, and Mignon S. Mazique. 1985. "A mirror for managers: Using simulation to develop management teams." *Journal of Applied Behavioral Science* 21, 3: 241–253.

Katz, Harry C., Rosemary Batt, and Jeffrey H. Keefe. 2003. The revitalization of the CWA: Integrating collective bargaining, political action, and organizing." *Industrial and Labor Relations Review* 56: 573–589.

Kauffman, Scott. 1979. "Plywood strikers back fired woman." *In These Times*, December 5–11, p. 5.

Kaufman, Bruce E. 2004. *The Global Evolution of Industrial Relations.* Geneva: International Labour Organization.

Kay, Tamara. 2005. "Labor transnationalism and global governance." *American Journal of Sociology* 111, 3: 715–756.

Keister, Lisa A. 2000. *Chinese Business Groups.* New York: Oxford University Press.

Keister, Lisa A. 2005. *Entrepreneurship. Volume 15 of Research in the Sociology of Work.* Amsterdam: Elsevier/JAI.

Keister, Lisa A. 2006. *Getting Rich: A Study of Wealth Mobility in America.* New York: Cambridge University Press.

Kelley, Robert E. 1985. *The Gold Collar Worker: Harnessing the Brain Power of the New Work Force.* Reading, MA: Addison-Wesley.

Kellner, Hansfried, and Frank W. Heuberger (editors). 1992. *Hidden Technocrats: The New Class and New Capitalism.* New Brunswick, NJ: Transaction.

Kendall, Diana. 2002. *The Power of Good Deeds: Privileged Women and the Social Reproduction of the Upper Class.* Lanham, MD: Rowman and Littlefield.

Kenney, Martin, and Richard Florida. 2004. *Locating Global Advantage.* Stanford: Stanford University Press.

Kent, Bülent. 2005. "Germany: Turkish worker deported for drawing welfare benefits." World Socialist Web Site, August 3.

Kenworthy, Lane. 2004. *Egalitarian Capitalism: Jobs, Incomes and Growth in Affluent Countries.* New York: Russell Sage.

Kenyon, R. 2000. "Changes in unemployment insurance legislation." *Monthly Labor Review* 123: 27–35.

Kessler-Harris, Alice. 2001. *In Pursuit of Equality: Women, Men, and the Quest for Economic Citizenship in 20th Century America.* New York: Oxford University. A detailed history of the experiences and struggles of working women in North America.

Kidder, Tracy. 1981. *The Soul of a New Machine.* New York: Avon Books.

Kim, Kwang Chung, Won Moo Hurh, and Marilyn Fernandez. 1989. "Intra-group differences in business participation: Three Asian immigrant groups." *International Migration Review* 23, 1 (Spring): 73–96.

Kimeldorf, Howard. 1999. *Battling for American Labor*. Berkeley: University of California Press.

Kingsolver, Anne E. (editor). 1998. *More Than Class: Studying Power in U.S. Workplaces*. Albany, NY: State University of New York Press.

Kleinman, Sherryl. 1996. *Opposing Ambitions: Gender and Identity in an Alternative Organization*. Chicago: University of Chicago Press.

Kluegel, James R., and Lawrence Bobo. 1993. "Opposition to race-targeting." *American Sociological Review* 58, 4 (August): 443–464.

Kobayashi-Hillary, Mark. 2005. *Outsourcing to India: The Offshore Advantage*, 2nd edition. London: Springer-Verlag.

Kohn, Melvin L. 1976. "Social class and parental values: Another confirmation of the relationship." *American Sociological Review* 41, 3 (June): 538–545.

Kohn, Melvin L. 1990. *Social Structure and Self-Direction*. Cambridge, MA: Blackwell.

Kohn, Melvin L., Atsushi Naoi, Carrie Schoenbach, Carmi Schooler, and Kazimierz M. Slomczynski. 1990. "Position in the class structure and psychological functioning in the United States, Japan, and Poland." *American Journal of Sociology* 95, 4 (January): 964–1008.

Kohn, Melvin L., and Carmi Schooler. 1983. *Work and Personality*. Norwood, NJ: Ablex.

Kolodko, Grzegorz W. 2000. *From Shock to Therapy: The Political Economy of Postsocialist Transformation*. Oxford: Oxford University Press.

Kondo, Dorinne K. 1990. *Crafting Selves: Power, Gender and Discourses of Identify in a Japanese Workplace*. Chicago: University of Chicago Press.

Kono, Toyohiro, and Stewart Clegg. 2001. *Trends in Japanese Management*. New York: Palgrave.

Korten, David. 1996. "The mythic victory of market capitalism." Pp. 183–191 in Jerry Mander and Edward Goldsmith (editors), *The Case against the Global Economy*. San Francisco: Sierra Club Books.

Kotter, John P. 1982. *The General Managers*. New York: Free Press.

Krahn, Harvey J., and Graham S. Lowe. 1998. *Work, Industry, and Canadian Society*. Toronto: Nelson.

Krantz, Matt. 2006. "Let's make a deal: Merger and acquisitions activity on a track to hit record." *USA Today*, December 5.

Kranzberg, Melvin, and Joseph Gies. 1986. "Medieval work: Guilds and the putting-out system." Pp. 2–6 in Lauri Perman (editor), *Work in Modern Society*. Dubuque, IA: Kendall-Hunt.

Krause, Elliott A. 2001. "Professional group power in developing societies." *Current Sociology* 49, 4: 149–175.

Krier, Dan. 2005. *Speculative Management: Stock Market Power and Corporate Change*. Albany, NY: State University of New York Press.

Kritzer, Herbert M. 1999. "The professions are dead, long live the professions: Legal practice in a post-professional world." *Law and Society Review* 33, 3: (October) 713–759.

Kunda, Gideon. 1992. *Engineering Culture*. Philadelphia: Temple University Press.

Kusterer, Ken C. 1978. *Know-How on the Job: The Important Working Knowledge of "Unskilled" Workers*. Boulder, CO: Westview Press.

Kuttner, Robert. 1996. *Everything for Sale: The Virtues and Limits of Markets*. New York: Alfred A. Knopf.

Lamphere, Louise. (editor) 1992. *Structuring Diversity: Ethnographic Perspectives on the New Immigration*. Chicago: University of Chicago Press.

Lander, Guy P. 2004. *What Is Sarbanes-Oxley?* New York: McGraw-Hill Professional.

Langfred, C.W. 2000. "The paradox of self-management: Individual and group autonomy in work groups." *Journal of Organizational Behavior* 21: 563–585.

Lapan, R.T., Albert Adams, S. Turner, and J.M. Hinkelman. 2000. "Seventh graders' vocational expectations." *Journal of Career Development*. 26, 3: 215–229.

Larson, Magali Sarfatti. 1977. *The Rise of Professionalism: A Sociological Analysis*. Berkeley: University of California Press.

Lawrence, Thomas B. 2004. "Rituals and resistance: Membership dynamics in professional fields." *Human Relations* 57, 2: 115–143.

Leach, B. 2005. "Agency and the gendered imagination: Women's actions and local culture in steelworker families." *Identities* 12, 1: 1–22.

Lee, Anru. 2004. *In the Name of Harmony and Prosperity: Labor and Gender Politics in Taiwan's Economic Restructuring*. Albany, NY: State University of New York Press.

Lee, Ching Kwan. 1998. *Gender and the South China Miracle: Two Worlds of Factory Women.* Berkeley: University of California Press.

Lee, Ching Kwan. 1999. "From organized dependence to disorganized despotism." *China Quarterly* 157: 44–71.

Lee, Richard B. 1981. "Politics, sexual and nonsexual, in an equalitarian society: The !Kung San." Pp. 83–102 in General D. Berreman (editor), *Social Inequality: Comparative and Developmental Approaches.* New York: Academic Press.

Lehmann, Wolfgang. 2005. "'I'm still scrubbing the floors': Experiencing youth apprenticeships in Canada and Germany." *Work, Employment and Society* 19: 107–129.

Leicht, Kevin T., and Mary L. Fennell. 2001. *Professional Work: A Sociological Approach.* Malden, MA: Blackwell.

Leicht, Kevin T., and Scott T. Fitzgerald. 2006. *Post-Industrial Peasants: The Illusion of Middle-Class Prosperity.* New York: Worth.

Leidner, Robin. 1993. *Fast Food, Fast Talk: Service Work and the Routinization of Everyday Life.* Berkeley: University of California Press.

Leigh, Duane E., and Kirk D. Gifford. 1999. "Workplace transformation and worker upskilling." *Industrial Relations* 38, 2: 174–191.

Lengyel, Gyorgy, and Laszlo Neumann. 2002. "The sociology of work in Hungary." In Daniel B. Cornfield and Randy Hodson (editors), *Worlds of Work.* New York: Plenum.

Lenski, Gerhard. 1984. *Power and Privilege: A Theory of Social Stratification.* Chapel Hill: University of North Carolina Press.

Leonhardt, David. 2006. "Scant progress on closing gap in women's pay." *The New York Times* December 24, 1, 1.

Leventman, Paula Goldman. 1981. *Professionals Out of Work.* New York: Free Press.

Levine, David I. 1995. *Reinventing the Workplace: How Business and Employees Can Both Win.* Washington, DC: Brookings Institute.

Levinson, K. 2000. "Codetermination in Sweden." *Economic and Industrial Democracy* 21: 457–473.

Levitan, Sar A., and Isaac Shapiro. 1987. *Working but Poor: America's Contradiction.* Baltimore: Johns Hopkins University Press.

Levy, Frank. 1987. *Dollars and Dreams: The Changing American Income Distribution.* New York: Russell Sage Foundation.

Levy, Frank, and Richard Murnane. 2005. *The New Division of Labor.* Princeton, NJ: Princeton University Press.

Lewin, Tamar. 1984. "Atari layoff case highlights question of workers' rights." *Austin American Statesman,* July 25, p. G6.

Lewis, Patricia. 2005. "Suppression or expression: An exploration of emotion management in a special care baby unit." *Work, Employment, and Society* 19: 565–581.

Li, J. H., and Joachim Singelmann. 1998. "Gender differences in class mobility: A comparative study of the United States, Sweden, and West Germany." *Acta Sociologica* 41, 4: 315–333.

Liebow, Elliot. 1967. *Tally's Corner.* Boston: Little, Brown.

Light, Ivan H. 1972. *Ethnic Enterprise in America: Business and Welfare Among Chinese, Japanese, and Blacks.* Berkeley: University of California Press.

Liker, Jeffrey K., W. Mark Fruin, and Paul S. Adler. 1999. *Remade in America: Transplanting and Transforming Japanese Management Systems.* New York: Oxford University Press.

Lilley, Keith D. 2002. *Urban Life in The Middle Ages, 1000–1450.* New York: Palgrave.

Lincoln, James R., and Michael L. Gerlach. 2004. *Japan's Network Economy.* Cambridge: Cambridge University Press.

Lincoln, James R., and Arne L. Kalleberg. 1996. "Commitment, Quits, and Work Organization in Japanese and U.S. Plants." *Industrial and Labor Relations Review* 50: 39–59.

Lipset, Seymour Martin, and Gary Marks. 2000. *It Didn't Happen Here: Why Socialism Failed in the U.S.* New York: W.W. Norton.

Lipset, Seymour Martin, Martin A. Trow, and James S. Coleman. 1956. *Union Democracy.* Glencoe, IL: Free Press.

Litan, Robert E., and Alice M. Rivlin. 2001. "Projecting the economic impact of the Internet." *The American Economic Review* 91, 2: 313–317.

Little, Malcolm [Malcolm X] as told to Alex Haley. 1965. *The Autobiography of Malcolm X.* New York: Grove Press.

Lloyd, Timothy C., and Patrick B. Mullen. 1990. *Lake Erie Fishermen: Work Identity and Tradition*. Urbana, IL: University of Illinois Press.

Lo, Jeannie. 1990. *Office Ladies Factory Women: Life and Work at a Japanese Company*. Armonk, NY: M.E. Sharpe.

Lobao, Linda, and Katherine Meyer. 2001. "The great agricultural transition." *Annual Review of Sociology* 27: 103–124.

Logue, John, and Jacquelyn S. Yates. 2001. *The Real World of Employee Ownership*. Ithaca, NY: Industrial and Labor Relations Press.

Lopez, Steven H. 2004. *Reorganizing the Rust Belt*. Berkeley: University of California Press.

Lorber, Judith. 1984. *Women Physicians: Careers, Status, and Power*. New York: Tavistock.

Louie, Miriam Ching Yoon. 2001. *Sweatshop Warriors*. Cambridge, MA: South End Press.

"The Lowdown on Literacy." 1994. *American Demographics* 16, 6: 6.

Lowe, Graham S. 1987. *Women in the Administrative Revolution: The Feminization of Clerical Work*. Toronto: University of Toronto Press.

Luger, Michael Ian. 1996. "Quality-of-life differences and urban and regional outcomes." *Housing Policy Debate* 7: 749–771.

Luthans, Fred, and D. L. Lockwood. 1984. "Toward an observation system for measuring leadership in natural settings." Pp. 117–141 in James G. Hunt, D. Hosking, C. Schriesheim, and R. Stewart (editors), *Leaders and Managers: International Perspectives on Managerial Behavior and Leadership*. New York: Pergamon.

Luthans, Fred, Stuart A. Rosenkrantz, and Harry W. Hennessey. 1985. "What do successful managers really do? An observation study of management activities." *Journal of Applied Behavioral Science* 21, 3: 255–270.

MacCorquodale, Patricia M., and Gary Jensen. 1993. "Women in the law: Partners or tokens?" *Gender and Society* 7, 4: 582–593.

Madu, Christian N. 1996. *Managing Green Technologies for Global Competitiveness*. Westport, CT: Quorum Books.

Majka, Linda C., and Theo J. Majka. 1982. *Farm Workers, Agribusiness, and the State*. Philadelphia: Temple University Press.

Maklan, D. M. 1977a. *The Four-Day Workweek: Blue Collar Adjustment to a Nonconventional Arrangement of Work and Leisure Time*. New York: Praeger.

Maklan, D. M. 1977b. "How blue collar workers on 4-day work weeks use their time." *Monthly Labor Review* 100, 8: 18–26.

Malambre, Alfred L., Jr., and Lindley H. Clark, Jr. 1992. "Productivity Statistics for the Service Sector May Understate Gains." *The Wall Street Journal* (August 12): 1.

Malecki, Edward J. 2004. "Jockeying for position: What it means and why it matters to regional development policy when places compete." *Regional Studies* 38: 1101–1120.

Mangum, Garth, Donald Mayall, and Kristen Nelson. 1985. "The temporary help industry: A response to the dual internal labor market." *Industrial and Labor Relations Review* 38, 4 (July): 599–611.

Mann, Eric. 1987. *Taking on General Motors*. Los Angeles, CA: Center for Labor Research and Education, Institute of Industrial Relations, University of California.

Marchak, M. Patricia. 1996. *Logging the Globe*. Montreal: McGill-Queen's University Press.

Margolis, Diane Rothbard. 1979. *The Managers: Corporate Life in America*. New York: Morrow.

Maril, Robert Lee. 1995. *The Bay Shrimpers of Texas*. Lawrence: University of Kansas Press.

Mars, Gerald. 1994. *Cheats at Work*. London: Aldershot.

Martin, Andrew W. 2005. "Addressing the selection bias in media coverage of strikes." *Research in Social Movements, Conflicts, and Change* 26: 143–178.

Martin, Andrew W. 2004. "From institutionalization to confrontation: Describing the current range of union organizing repertoires." Paper presented at the annual meeting of the American Sociological Association, San Francisco, California.

Martin, Patricia Yancey. 2005. *Rape Work*. New York: Routledge.

Martin, Susan Ehrlich. 1980. *Breaking and Entering: Policewomen on Patrol*. Berkeley: University of California Press.

Marx, Gary T. 2003. "A tack in the shoe: Neutralizing and resisting the new surveillance." *Journal of Social Issues* 59: 369–390.

Marx, Karl. 1959 [1844]. *The Economic and Philosophic Manuscripts of 1844*. Translated by Martin Milligan. Moscow: International Publishers.

Marx, Karl. 1967 [1887]. *Capital.* Volume 1. New York: International Publishers.

Maslow, Abraham H. 1999. "Classic Motivation Theories." Pp. 169–171 in John A. Shtogren (editor), *Skyhooks for Leadership.* New York: Amacom.

Massey, Douglas S. 1996. "Inequality in the age of extremes." *Demography* 33: 1–20.

Massey, Douglas S. and Nancy A. Denton. 1993. *American Apartheid.* Cambridge, MA: Harvard University Press.

Maume, David J., Jr. 1999. "Glass ceilings and glass escalators." *Work and Occupations* 26: 483–509.

McCarthy, Michael J. 1988. "Managers face dilemma with 'temps.'" *The Wall Street Journal*, April 5, p. 27.

McElroy, Michael P., and Michael B. Hazzard. 1994. "Occupational staffing patterns at the four-digit SIC level." *Monthly Labor Review* 117, 2 (February): 30–33.

McGee, Leo, and Robert Boone, 1979. *The Black Rural Landowner.* Westport, CT: Greenwood Press.

McGregor, Douglas Murray. 1985. *The Professional Manager.* New York: McGraw-Hill.

McHugh, P. P., Joel Cutcher-Gershenfeld, and D. L. Bridge. 2005. "Examining structure and process in ESOP firms." *Personnel Review* 34: 277–93.

McMahon, Patrick J., and John H. Tschetter. 1986. "The declining middle class: A further analysis." *Monthly Labor Review* 109, 9 (September): 22–27.

Meiksins, Peter F., and Peter Whalley. 2002. *Putting Work in Its Place.* Ithaca, NY: Industrial and Labor Relations Press.

Mendoza, Martha. 1997. "Federal work rules see many changes." *Columbus Dispatch*, December 14.

Menon, Kalyani, and Laurette Dube. 2004. "Service provider responses to anxious and angry customers: Different challenges, different payoffs." *Journal of Retailing* 80: 229–237.

Menon, Tanya, and Jeffrey Pfeffer. 2003. "Valuing internal vs. external knowledge: Explaining the preference for outsiders." *Management Science* 49: 497–513.

Mers, Gilbert. 1988. *Working the Waterfront: The Ups and Downs of a Rebel Longshoreman.* Austin: University of Texas Press.

Merton, Robert K. 1968. *Social Theory and Social Structure.* New York: Free Press.

Metcalfe, L. 2000. "European integration and globalization." *International Review of Administrative Sciences* 66,1 (March): 119–124.

Milkman, Ruth (editor). 1985. *Women, Work and Protest: A Century of U.S. Women's Labor History.* Boston: Routledge and Kegan Paul.

Milkman, Ruth. 1997. *Farewell to the Factory: Auto Workers in the Late Twentieth Century.* Berkeley: University of California Press.

Milkman, Ruth, and Cydney Pullman. 1991. "Technological change in an auto assembly plant." *Work and Occupations* 18, 2 (May): 123–147.

Miller, Delbert C., and William H. Form. 1980. *Industrial Sociology*, 3d edition. New York: Harper and Row.

Miller, Gale. 1978. *Odd Jobs: The World of Deviant Work.* Englewood Cliffs, NJ: Prentice-Hall.

Mills, C. Wright. 1951. *White Collar: The American Middle Classes.* New York: Oxford University Press.

Mills, C. Wright. 1959. *The Sociological Imagination.* London: Oxford University Press.

Minchin, Timothy J. 1999. *Hiring the Black Worker.* Chapel Hill: University of North Carolina Press.

Mintzberg, Henry. 1973. *The Nature of Managerial Work.* New York: Harper and Row.

Mirowsky, John, and Catherine E. Ross. 2003. *Social Causes of Psychological Distress.* New York: Aldine de Gruyter.

Mishel, Lawrence, Jared Bernstein, and John Schmitt. 2005. *The State of Working America.* Ithaca, NY: Cornell University Press.

Mishel, Lawrence and Paula B. Voos (editors). 1992. *Unions and Economic Competitiveness.* Armonk, NY: Sharpe.

Mizruchi, Mark S. 2004. "Berle and Means revisited: The governance and power of large U.S. corporations." *Theory and Society* 33: 579–617.

Moberg, David. 1994. "Buy the friendly skies." *In These Times*, February 21, p. 26.

Molstad, Clark. 1986. "Choosing and coping with boring work," *Urban Life* 15, 2 (July): 215–236.

Moody's Industrial Manual. 2000. New York: Mergent.

Moore, Michael. 1997. *Downsize This!* New York: Harper.

Moore, Richard, and Elizabeth Marsis. 1984. "Telecommuting: Sweatshop at home sweet home?" *In These Times*, April 18, pp. 12–13.

Morrill, Calvin. 1995. *The Executive Way: Conflict Management in Organizations.* Chicago: University of Chicago Press.

Morris, Charles J. 2005. *The Blue Eagle at Work: Reclaiming Democratic Rights in the American Workplace*. Ithaca, NY: Cornell University Press.

Mort, Jo-Ann. 2000. *Not Your Father's Union Movement*. New York: Verso.

Mortimer, Jeylan T., and Jayne London. 1984. "The varying linkages of work and family." Pp. 20–35 in Patricia Voydanoff (editor), *Work and Family: Changing Roles of Men and Women*. Palo Alto, CA: Mayfield.

Mortimer, Jeylan T., Jon Lorence, and Donald S. Kumka. 1986. *Work, Family, and Personality: Transition to Adulthood*. Norwood, NJ: Ablex.

Mouer, Ross, and Hirosuke Kawanishi. 2005. *A Sociology of Work in Japan*. Cambridge: Cambridge University Press.

Mowshowitz, Abbe. 2002. *Virtual Organization: Toward a Theory of Societal Transformation Stimulated by Information Technology*. Westport, CT: Quorum.

Mueller, Charles W., Ashley Finley, Roderick D. Iverson, and James D. Price. 1999. "The effects of group racial composition on job satisfaction, organizational commitment, and career commitment." *Work and Occupations* 26: 187–219.

Muller-Jentsch, Walther. 2002. "The sociology of work in Germany." In Daniel B. Cornfield and Randy Hodson (editors), *Worlds of Work*. New York: Plenum.

Multinational Monitor. 2004. "The ten worst corporations of 2004." Volume 25 (December): Retrieved from http://multinationalmonitor.org

Mundra, Kusum, Andrew Moellmer, and Waldo Lopez-Aqueres. 2003. *Hispanic Journal of Behavioral Science* 25: 513–529.

Murphree, Mary C. 1981. *Rationalization and Satisfaction in Clerical Work: A Case Study of Wall Street Legal Secretaries*. Ph.D. dissertation, Department of Sociology, Columbia University.

Murphree, Mary C. 1987. "New technology and office tradition: The not-so-changing world of the secretary." Pp. 98–135 in Heidi I. Hartmann (editor), *Computer Chips and Paper Clips: Technology and Women's Employment*, Volume 2. Washington, DC: National Academy Press.

Mutchler, Jan E., Jeffrey A. Burr, and Francis G. Caro. 2003. "From paid worker to volunteer: Leaving the paid workforce and volunteering." *Social Forces* 81: 1267–1294.

Naisbitt, John. 1999. *High Tech/High Touch*. New York: Broadway.

Nakao, Keiko and Judith Treas. 1994. "Updating occupational prestige and socioeconoomic scores: How the new measures measure up." *Sociological Methodology* 24: 1–72.

Nash, June C. 1989. *From Tank Town to High Tech*. Albany, NY: State University of New York Press.

National Commission for Employment Policy. 1979. *Final Report of the American Assembly on Youth Employment*. Harriman, NY: Arden House.

National Commission on Employment and Unemployment Statistics. 1979. *Counting the Labor Force*. Washington, DC: U.S. Government Printing Office.

Naughton, Keith, and Mark Hossenball. 2000. "Ford *vs.* Firestone." *Newsweek*, September 11. Retrieved from www.msnbc.com/news/457861

Nelson, Robert L., and William P. Bridges. 1999. *Legalizing Gender Inequality: Courts, Markets and Unequal Pay for Women in the United States*. Cambridge: Cambridge University Press.

Ness, Immanuel, and Stuart Eimer. 2001. *Central Labor Councils and the Revival of American Unionism*. Amonk, NY: M.E. Sharpe.

Neumark, David and Deborah Reed. 2004. "Employment relationships in the new economy." *Labour Economics* 11: 1–31.

Newman, Katherine S. 1988. *Falling from Grace: The Experience of Downward Mobility in the American Middle Class*. New York: Vintage.

Newman, Katherine S. 1993. *Declining Fortunes: The Withering of the American Dream*. New York: Basic.

Newman, Katherine S. 1999. *Falling from Grace: Downward Mobility in the Age of Affluence*. New York: Vintage.

Newman, Katherine S. 2000. *No Shame in My Game*. New York: Vintage.

Newman, Katherine S. 2006. *Chutes and Ladders: Navigating the Low-wage Labor Market*. Cambridge, MA: Harvard University Press.

Nickson, David. 2004. *Remote Working*. Oxford, England: Elsevier Butterworth-Heinemann.

Nippert-Eng, Christena E. 1996. *Home and Work: Negotiating Boundaries Through Everyday Life*. Chicago: University of Chicago Press.

Noble, David F. 1997. *The Religion of Technology: The Divinity of Man and the Spirit of Invention*. New York: Knopf.

Noble, Kenneth B. 1986. "Unions push retraining plans." *The New York Times*, March 23 (Employment Outlook Supplement), pp. 10–11.

Nordheimer, Jon. 1988. "Older migrants' years of toil in sun end in cold twilight." *The New York Times*, May 29, pp. 1, 14.

Norr, James L., and Kathleen L. Norr. 1997. "Women's status in peasant-level fishing." Pp. 65–84 in Carolyn E. Sachs (editor), *Women Working in the Environment*. Washington, DC: Taylor & Francis.

North, C. C., and P. K. Hatt. 1947. "Jobs and occupations: A popular evaluation." *Public Opinion News* 9, 3 (September 1).

North, Douglass Cecil. 2005. *Understanding the Process of Economic Change*. Princeton: Princeton University Press.

Norwood, Stephen H. 1990. *Labor's Flaming Youth: Telephone Operators and Worker Militancy*. Urbana, IL: University of Illinois Press.

Oakes, Guy. 1990. *The Soul of the Salesman: The Moral Ethos of Personal Sales*. London: Humanities Press International.

Oakley, Ann. 1975. *The Sociology of Housework*. New York: Pantheon.

O'Bannon, Yolanda. 2006. "Living what you do every day." Retrieved from http://www.npr.org/tem plates/story/story.php?storyId=6431548

O'Connell, Martin, and David E. Bloom. 1987. "Juggling jobs and babies: America's child care challenge." *Population Trends and Public Policy* 12 (February).

O'Connor, James. 1998. *Natural Causes*. New York: Gilford.

Ogasawara, Yuko. 1998. *Office Ladies and Salaried Men: Power, Gender, and Work in Japanese Companies*. Berkeley: University of California Press.

Ong, Aihwa. 1983. "Global industries and Malay peasants in Peninsular Malaysia." Pp. 426–441 in June Nash and Maria Patricia Fernandez-Kelly (editors), *Women, Men, and the International Division of Labor*. Albany, NY: State University of New York Press.

Oppenheimer, Andres. 1986. "Haitians regret cashing in pigs for 'rich' U.S. hogs." *Austin American Statesman*, June 8, p. K7.

Oppenheimer, Valerie Kincade. 1970. *The Female Labor Force in the United States*. Population Monograph Series, No. 5. Berkeley: Institute of International Studies, University of California.

Oppenheimer, Valerie Kincade. 1974. "The life cycle squeeze." *Demography* 11, 2 (May): 227–246.

O'Reilly, Charles A, and Jeffrey Pfeffer. 2000. *Hidden Value: How Great Companies Achieve Extraordinary Results with Ordinary People*. Boston: Harvard Business School.

Orbach, Michael K. 1977. *Hunters, Seamen, and Entrepreneurs: The Tuna Seinermen of San Diego*. Berkeley: University of California Press.

Orwell, George. 1933. *Down and Out in London and Paris*. New York: Harcourt Brace Jovanovich.

Ose, S. O. 2005. "Working conditions, compensation, and absenteeism." *Journal of Health Economics* 24: 161–188.

Ospina, Sonia. 1996. *Illusions of Opportunity*. Ithaca, NY: Cornell University Press.

Osterman, Paul. 1986. "The impact of computers on the employment of clerks and managers." *Industrial and Labor Relations Review* 39, 2 (January): 175–186.

Osterman, Paul. 1999. *Securing Prosperity: The American Labor Market: How it Has Changed and What to Do About It*. Princeton: Princeton University Press.

Ouchi, William G. 1981. *Theory Z: How American Business Can Meet the Japanese Challenge*. Reading, MA: Addison-Wesley.

Owings, Alison. 2002. *Hey Waitress! The USA from the Other Side of the Tray*. Berkeley: University of California Press.

Ozaki, Muneto (editor). 1999. *Negotiating Flexibility*. Geneva, Switzerland: International Labour Office.

Pacheco, Ferdie. 1997. *Pacheco's Art of Ybor City*. Gainesville, FL: University of Florida Press.

Padgett, B. L. 2005. "After Dot-Com, after WorldCom, after Enron, after capitalism." *Business Ethics Quarterly* 15: 329–340.

Palmer, Bryan D. 1992. *Working-Class Experience: The Rise of Canadian Labor, 1800–1991*. Toronto: McClelland & Stewart.

Parcel, Toby (editor). 1999. *Work and Family. Research in the Sociology of Work*, Volume 7. Greenwich, CT: JAI Press.

Parker, Mike. 1985. *Inside the Circle: A Union Guide to QWL*. Boston: South End Press.

Parker, Patricia S. 2002. "Negotiating identity in raced and gendered workplace interactions: The use of strategic communication by African American senior executives within dominant culture organizations." *Communication Quarterly* 50, 3&4 (Summer–Fall): 251–268.

Parker, Robert E. 1994. *Flesh Peddlers and Warm Bodies: The Temporary Help Industry and Its Workers*. New Brunswick, NJ: Rutgers University Press.

Parsons, Talcott. 1968. "Professions." In David L. Sills (editor), *International Encyclopedia of the Social Sciences*. New York: Macmillan.

Patterson, Stephanie L. 2004. "Ontario midwives: Reflections on a decade of regulated midwifery." *Canadian Woman Studies* 24, 1: 153–58.

Paules, Greta Foff. 1991. *Dishing It Out: Power and Resistance among Waitresses in a New Jersey Restaurant*. Philadelphia: Temple University Press.

Pavalko, Eliza K., and Glen H. Elder, Jr. 1993. "Women behind the men: Variations in wives' support of husbands' careers." *Gender and Society* 7, 4 (December): 548–567.

Pelled, Lisa Hope, Kathleen M. Eisenhardt, and Katherine R. Xin. 1999. "Exploring the black box: An analysis of work group diversity, conflict, and performance." *Administrative Science Quarterly*, 44, 1 (March): 1–28.

Pencavel, John. 2002. *Worker Participation: Lessons from the Worker Co-ops of the Pacific Northwest*. New York: Russell Sage.

Penn, Roger, Michael Rose, and Jill Rubery. 1994. *Skill and Occupational Change*. New York: Oxford University Press.

Pereira, Alexius A. 2000. "McAunties and McUncles: Older crew members in Singapore's fast food industry." Pp. 129–145 in Randy Hodson, *Marginal Employment*, Volume 9 in *Research in the Sociology of Work*. Stamford, CT: JAI Press.

Perkins, John. 2004. *Confessions of an Economic Hit Man*. New York: Penguin.

Perlow, Leslie. 1997. *Finding Time: How Corporations, Individuals, and Families Can Benefit from New Work Practices*. Ithaca, NY: Cornell University Press.

Perrow, Charles. 1986. *Complex Organizations: A Critical Essay*, 3rd edition. Glenview, IL: Scott, Foresman.

Perrow, Charles. 1984. *Normal Accidents: Living with High-Risk Technologies*. New York: Basic.

Perrucci, Robert, and Earl Wysong. 2003. *The New Class Society*. Lanham, MD: Rowman and Littlefield.

Perry, Stewart E. 1978. *San Francisco Scavengers: Dirty Work and the Pride of Ownership*. Berkeley: University of California Press.

Pfeffer, Jeffrey. 1998. *The Human Equation: Building Profits by Putting People First*. Boston: Harvard Business School Press.

Pfeffer, Jeffrey, and Robert I. Sutton. 2006. "Evidence-based management." *Harvard Business Review* 84,1: 62–74.

Phillips, Brian. 1998. *Global Production and Domestic Decay*. New York: Garland.

Picou, J. Steven, B. K. Marshall, and Duane A. Gill. 2004. "Disaster, litigation, and the corrosive community." *Social Forces* 82: 1493–1522.

Pisati, M. 1997. "Mobility regimes and generative mechanisms: A comparative analysis of Italy and the United States." *European Journal of Sociology* 13, 2 (September): 179–198.

Polanyi, Karl. 1957. *The Great Transformation*. Boston: Beacon.

Pollard, Sidney. 1963. "Factory discipline in the industrial revolution." *Economic History Review* 16: 254–271.

Polsky, Andrew J. 1991. *The Rise of the Therapeutic State*. Princeton, NJ: Princeton University Press.

Portes, Alejandro. 2001. *Legacies: The Story of the Immigrant Second Generation*. Berkeley: University of California Press.

Portes, Alejandro, and Leif Jensen. 1989. "The enclave and the entrants: Patterns of ethnic enterprise in Miami before and after Mariel." *American Sociological Review* 54, 6 (December): 929–950.

Portes, Alejandro, and Saskia Sassen-Koob. 1987. "Making it underground: Comparative material on the informal sector in western market economies." *American Journal of Sociology* 93, 1 (July): 30–61.

Powell, Gary N. (editor). 1999. *Handbook of Gender and Work*. Thousand Oaks, CA: Sage.

Prahalad, C. K. 2006. *The Fortune at the Bottom of the Pyramid: Eradicating Poverty through Profits*. Upper Saddle River, NJ: Wharton.

Presser, Harriet B. 1988. "Shift work and child care among young dual-earner American parents." *Journal of Marriage and the Family* 50,1 (February): 133–149.

Presser, Harriet B. 2000. "Nonstandard work schedules and marital instability." *Journal of Marriage and the Family*. 62: 93–110.

Pringle, Rosemary. 1988. *Secretaries Talk: Sexuality, Power, and Work*. Sydney: Allen and Unwin.

Rabinow, Paul. 1996. *Making PCR: A Story of Biotechnology*. Chicago: University of Chicago Press.

Rabinow, Paul. 2003. *Anthropos Today: Reflections on Modern Equipment*. Princeton, NJ: Princeton University Press.

Radcliffe-Brown, A. 1922. *The Andaman Islanders*. Cambridge: Cambridge University Press.

Radforth, Ian. 1986. "Logging pulpwood in Northern Ontario." Pp. 245–280 in Craig Heron and Robert Storey (editors), *On the Job: Confronting the Labour Process in Canada*. Kingston: McGill-Queen's University Press.

Raeburn, Nicole C. 2002. *Inside Out: The Struggle for Lesbian, Gay, and Bisexual Rights in the Workplace*. Minneapolis: University of Minnesota Press.

Rajan, Amin. 1985. "Office technology and clerical skills." *Futures: The Journal of Forecasting and Planning* 17, 4 (August): 410–413.

Randall, Adrian, and Andrew Charlesworth (editors). 2000. *Moral Economy and Popular Protest*. New York: St. Martin's.

Rayner, C., H. Hoel, and C. L. Cooper. 2002. *Workplace Bullying*. London: Taylor and Francis.

Reeves, Joy B., and Ray Darville. 1986. "Female clerical workers in academic settings: An empirical test of the gender model." *Sociological Inquiry* 56, 1 (Winter): 105–124.

Reich, Robert B. 2002. *I'll Be Short: Essentials for a Decent Working Society*. Boston: Beacon.

Reid, Samuel Richardson. 1976. *The New Industrial Order: Concentration, Regulation, and Public Policy*. New York: McGraw-Hill.

Reiter, Ester. 1991. *Making Fast Food: From the Frying Pan into the Fryer*. Montreal: McGill-Queen's University Press.

Reskin, Barbara F. 1998. *The Realities of Affirmative Action in Employment*. Washington, DC: American Sociological Association.

Reskin, Barbara F. 2000. "Getting it right: Sex and race inequality in work organizations." *Annual Review of Sociology* 26: 707–709.

Reskin, Barbara F., and Debra Branch McBrier. 2000. "Why not ascription? Organizations' employment of male and female managers." *American Sociological Review* 65, 2 (April): 210–233.

Reskin, Barbara F., and Irene Padavic. 1994. *Women and Men at Work*. Thousand Oaks, CA: Pine Forge.

Reynolds, Jeremy. 2003. "You can't always get the hours you want: Mismatches between actual and preferred work hours in the U.S." *Social Forces* 81: 1171–1199.

Riemer, Jeffrey. 1977. "Becoming a journeyman electrician." *Work and Occupations* 4: 87–98.

Rifkin, Jeremy. 2004. *The End of Work*. New York: Penguin.

Rinehart, James, Christopher Huxley, and David Robertson. 1997. *Just Another Car Factory? Lean Production and Its Discontents*. Ithaca, NY: Cornell University Press.

Risen, James. 1984. "Lacking skills, displaced auto workers feel crunch." *Los Angeles Times*, December 30, p. 119.

Ritzer, George. 2004. *The McDonaldization of Society*. Thousand Oaks, CA: Pine Forge.

Ritzer, George A., and David Walczak. 1988. "Rationalization and the deprofessionalization of physicians." *Social Forces* 67, 1 (September): 1–22.

Roberts, Glenda S. 1994. *Staying on the Line: Blue-Collar Women in Contemporary Japan*. Honolulu: University of Hawaii Press.

Rodgers, J., and Jill Rubery. 2003. "The minimum wage as a tool to combat discrimination." *International Labour Review* 142: 543–556.

Roethlisberger, F. J., and William J. Dickson. 1939. *Management and the Worker*. Cambridge, MA: Harvard University Press.

Rohlen, Thomas P. 1974. *For Harmony and Strength*. Berkeley: University of California Press.

Rollins, Judith. 1985. *Between Women: Domestics and Their Employers*. Philadelphia: Temple University Press.

Romero, Mary. 1992. *Maid in the USA*. New York: Routledge.

Roscigno, Vincent J., and William F. Danaher. 2001. "Media and mobilization: The case of radio and Southern textile worker insurgency, 1929–1934." *American Sociological Review* 66, 1 (February): 21–48.

Roscigno, Vincent J., and Randy Hodson. 2004. "The Organizational and Social Foundations of Worker Resistance." *American Sociological Review* 69, 1: 14–39.

Rosenthal, Neal H. 1985. "The shrinking middle class: Myth or reality?" *Monthly Labor Review* 108, 3 (March): 3–10.

Ross, Bob. 1993. "Bottom line for topless dancers." *USA Today*, June 4, p. 2.

Ross, Catherine E., and Maryann P. Wright. 1998. "Women's work, men's work and the sense of control." *Work and Occupations* 25: 333–335.

Rossi, Peter M. 1989. *Down and Out in America: The Origins of Homelessness in the United States*. Chicago: University of Chicago Press.

Rostow, Walt W. 2003. *Concept and Controversy*. Austin: University of Texas Press.

Rothman, Robert A. 1998. *Working: Sociological Perspectives*, 2nd edition. Englewood Cliffs, NJ: Prentice-Hall.

Rothschild, Joyce. 2000. "Creating a just and democratic workplace." *Contemporary Sociology* 29; 1 (January): 195–213.

Rothschild, Joyce, and Marjukka Ollilainen. 1999. "Obscuring but not reducing managerial control: Does TQM measure up to democracy standards?" *Economic and Industrial Democracy* 20: 583–624.

Roy, Donald. 1952. "Quota restriction and goldbricking in a machine shop." *American Sociological Review* 57, 5 (March): 427–442.

Rueschemeyer, Dietrich. 1983. "Professional autonomy and the social control of expertise." Pp. 38–58 in Robert Dingwall and Philip Lewis (editors), *The Sociology of the Professions*. New York: St. Martin's.

Rumberger, Russell W. 1981. "The changing skill requirements of jobs in the U.S. economy." *Industrial and Labor Relations Review* 34, 4: 578–589.

Sabel, Charles F. 1982. *Work and Politics*. Cambridge: Cambridge University Press.

Sabel, Charles F., and Michael J. Piore. 1984. *The Second Industrial Divide*. New York: Basic Books.

Sabel, Charles F., and Jonathan Zeitlin. 2004. "Neither modularity nor relational contracting: Inter-firm collaboration in the new economy." *Enterprise & Society* 5: 388–403.

Sanday, Peggy Reeves. 2002. *Women at the Center: Life in a Modern Matriarchy*. Ithaca, NY: Cornell University Press.

Sandoval, Ralph Armbruster. 2005. *Globalization and Cross-Border Labor Solidarity in the Americas*. London: Routledge.

Santiago, Jack. 1989. *Miles of Smiles, Years of Struggle: Stories of Black Pullman Porters*. Urbana, IL: University of Illinois Press.

Saunders, Conrad. 1981. *Social Stigma of Occupations*. Westmead, England: Gower.

Sayer, Liana C. 2005. "Gender, time, and inequality." *Social Forces* 84, 1: 285–303.

Scherer, Frederick M. 2000. *Competition Policy, Domestic and International*. Northampton, MA: Edward Elgar.

Schervish, Paul G., Planton E. Coutsoukis, and Ethan Lewis. 1994. *Gospels of Wealth*, Westport, CT: Praeger.

Schimmelfenning, Frank, and Ulrich Sedelmeier. 2005. *The Europeanization of Central and Eastern Europe*. Ithaca, NY: Cornell University Press.

Schlemmer, Bernard (editor). 2000. *The Exploited Child*. New York: Zed Books.

Schor, Juliet B. 1991. *The Overworked American*. New York: Basic Books.

Schor, Juliet B. 2003. "The (even more) overworked American." Pp. 6–11 in John de Graaf (editor), *Take Back Your Time*. San Francisco: Berrett-Koehler.

Schor, Juliet B. 2004. *Born to Buy: The Commercialized Child and the New Consumer Culture*. New York: Scribner.

Schroedel, Jean Reith. 1985. *Alone in a Crowd: Women in the Trades Tell Their Stories*. Philadelphia: Temple University Press.

Schwartzman, Helen B. 1993. *Ethnography in Organizations. Qualitative Research Methods Series, #27*. Newbury Park, CA: Sage.

Scott, James C., and Nina Bhatt. 2001. *Agrarian Studies*. New Haven, CT: Yale University Press.

Scott, W. Richard. 1998. *Organizations: Rational, Natural, and Open Systems*, 4th edition. Englewood Cliffs, NJ: Prentice-Hall.

Sears, David O., Jim Sidanius, and Lawrence Bobo (editors). 2000. *Racialized Politics*. Chicago: University of Chicago Press.

Sears, Robert R., Eleanor E. Maccoby, and Harry Levin. 1957. *Patterns of Child Rearing*. New York: Harper and Row.

Seeman, Melvin. 1993. "A Historical Perspective on Attitude Research." Pp. 3–21 in Dagmar Krebs and Peter Schmidt (editors), *New Direction in Attitude Research*, . New York: W. de Gruyter.

Seidman, Gay. 2001. "South Africa: When the system has fallen." Pp. 177–218 in Howard Winant (editor), *The World Is a Ghetto*. New York: Basic.

Sennett, Richard. 1998. *The Corrosion of Character: The Personal Consequences of Work in the New Capitalism.* New York: Norton.

Sennett, Richard, and Jonathan Cobb. 1972. *The Hidden Injuries of Class.* New York: Random House.

Sey, A. 2000. "Team work in Japan: Revolution, evolution or no change at all?" *Economic and Industrial Democracy* 21: 475–503.

Shaiken, Harley, Steven H. Lopez, and I. Mankita. 1997. "Two routes to team production: Saturn and Chrysler compared." *Industrial Relations* 31, 1 (January): 17–45.

Shenhav, Yehouda A. 1988. "Abandoning the research bench: Individual, organizational, and environmental accounts." *Work and Occupations* 15, 1 (February): 5–23.

Shipler, David K. 2005. *The Working Poor: Invisible in America.* New York: Vintage.

Shover, Neal, and John Paul Wright. 2001. *Crimes of Privilege.* New York: Oxford University Press.

Siegel, Lenny, and John Markoff. 1985. *The High Cost Of High Tech: The Dark Side of the Chip.* New York: Harper and Row.

Sikora, Martin. 1993. "Megadeal response to market pressures." *Mergers and Acquisitions* 28, 2 (September–October): 6–8.

Silverman, Bertram, and Murray Yanowitch. 2000. *New Rich, New Poor, New Russia.* Armonk, NY: M.E. Sharpe.

Silvestri, George T. 1993. "Occupational employment: Wide variations in growth." *Monthly Labor Review* 116, 11 (November): 61–73.

Simpson, Richard L. 1985. "Social control of occupations and work." *Annual Review of Sociology* 11: 415–436.

Skocpol, Theda. 2000. *The Missing Middle: Working Families and the Future of American Social Policy.* New York: W.W. Norton.

Smith, Adam. 1937 [1776]. *The Wealth of Nations.* New York: Random House.

Smith, John P. 1987. "The social and economic correlates of bankruptcy during the farm fiscal crisis 1970–1987." *Mid-American Review of Sociology* 12 (Spring): 35–53.

Smith, Roy C., and Ingo Walter. 2003. *Global Banking.* New York: Oxford University Press.

Smith, Ryan A. 2002. "Race, gender, and authority in the workplace: Theory and research." *Annual Review of Sociology* 28: 509–542.

Smith, Timothy K. 1993. "Chiropractors seeking to expand practices take aim at children." *The Wall Street Journal*, March 18, A1.

Smith, Vicki. 1990. *Managing the Corporate Interest: Control and Resistance in an American Bank.* Berkeley: University of California Press.

Smith, Vicki. 2001. *Crossing the Great Divide: Worker Risk and Opportunity in the New Economy.* Ithaca, NY: Cornell University Press.

Snow, David A., and Leon Anderson. 1993. *Down on their Luck: A Study of Homeless People.* Berkeley: University of California Press.

Spell, C. S. 2001. "Organizational technologies and human resource management." *Human Relations* 54: 193–213.

Soupata, L. 2005. "Engaging employees in company success: The UPS approach to a winning team." *Human Resource Management* 44: 95–98.

Spain, Daphne, and Suzanne M. Bianchi. 1996. *Balancing Act: Motherhood, Marriage and Employment Among American Women.* New York: Russell Sage.

Spanier, Graham G., and Paul C. Glick. 1980. "The life cycle of American families: An expanded analysis." *Journal of Family History*, 5, 1 (Spring): 97–111.

Spencer, Charles. 1977. *Blue Collar.* Chicago: Lakeside Charter Books.

Spenner, Kenneth I. 1990. "Skill: Meanings, methods, and measures." *Work and Occupations* 17, 4 (November): 339–421.

Standing, Guy. 1996. *Russian Unemployment and Enterprise Restructuring.* New York: St. Martin's.

Starr, Paul. 1982. *The Social Transformation of American Medicine.* New York: Basic Books.

Staudohar, Paul D., and Holly E. Brown. 1987. *Deindustrialization and Plant Closure.* Lexington, MA: D. C. Heath.

Stearns, Linda Brewster, and Kenneth D. Allan. 1996. "Economic behavior in institutional environments: The corporate merger wage of the 1980s." *American Sociological Review* 61, 4 (August): 699–718.

Steele, Paul D., and Robert L. Hubbard. 1985. "Management styles, perceptions of substance abuse, and employee assistance programs in organizations." *Journal of Applied Behavioral Science* 21, 3: 271–286.

Steinbeck, John. 1939. *The Grapes of Wrath.* New York: Viking.

Stevens, Fred, Frans van der Horst, Frans Nijhuis, and Silvia Bours. 2000. "The division of labour in vision care: Professional competence in a system of professions." *Sociology of Health and Illness* 22: 431–452.

Stevens, Gillian, and Joo Hyun Cho. 1985. "Socioeconomic indexes and the new 1980 Census occupational classification scheme." *Social Science Research* 14, 2 (June): 142–168.

Steward, Barbara. 2000. "Changing times: The meaning, measurement, and use of time in teleworking." *Time and Society* 9, 1 (March): 57–74.

Stewart, Paul (editor). 2005. *Employment, Trade Union Renewal and the Future of Work*. New York: Palgrave.

Stiglitz, Joseph E. 2003. *Globalization and Its Discontents*. New York: W. W. Norton.

Stovel, Katherine, and Mike Savage. 2006. "Mergers and mobility: organizational growth and the origins of career migration at Lloyds Bank." *American Journal of Sociology* 111: 1080–1122.

Strauss, A. 1978. *Negotiations*. San Francisco: Jossey-Bass.

Streeck, Wolfgang, and Kathleen Thelen. 2005. *Beyond Continuity: Institutional Change in Advanced Political Economies*. Oxford: Oxford University Press.

Street, Annette Fay. 1992. *Inside Nursing*. Albany, NY: State University of New York Press.

Strober, Myra H,. and Agnes Miling Kaneko Chan. 1999. *The Road Winds Uphill all the Way: Gender, Work, and Family in the United States and Japan*. Cambridge, MA: MIT Press.

Stroh, Linda K. 1999. "Does relocation still benefit *corporations and employees? An overview of the litera-ture.*" *Human Resource Management Review* 9, 3: 279–308.

Strom, Sharon Hartman. 1986. "'Machines instead of clerks': Technology and the feminization of book-keeping, 1910–1950." Pp. 63–97 in Heidi I. Hartmann (editor), *Computer Chips and Paper Clips: Technology and Women's Employment*. Volume 2. Washington, DC: National Academy Press.

Strom, Sharon Hartman. 1992. *Beyond the Typewriter: Gender, Class, and the Origins of Modern American Office Work, 1900–1930*. Urbana, IL: University of Illinois Press.

Strople, Michael H. 2006. "From supermarkets to supercenters: Employment shifts to the one-stop shop." *Monthly Labor Review* 129, 2 (February): 39–46.

Suggs, David N., and Andrew W. Miracle (editors). 1999. *Culture, Biology and Sexuality*. Athens: University of Georgia Press.

Sullivan, Teresa A. 1978. *Marginal Workers, Marginal Jobs: The Underutilization of American Workers*. Austin: University of Texas Press.

Sullivan, Teresa A. 1989. "Women and minority workers in the new economy: Optimistic, pessimistic, and mixed scenarios." *Work and Occupations* 16, 4 (November): 395–415.

Sullivan, Teresa A. 1990. "The decline of occupation." Pp. 1–31 in M. Hallinan et al. (editors), *Change in Societal Institutions*. New York: Plenum.

Sullivan, Teresa A. 1996. "The cashier complex and the transformation of occupations." Pp. 127–141 in Dennis L. Peck and J. Selwyn Hollingsworth (editors), *Demographic and Structural Change: The Effects of the 1980s on American Society*. Westport, CT: Greenwood.

Sullivan, Teresa A. 2004. "Work-related social problems." Pp. 193–208 in George Ritzer (editor), *Handbook of Social Problems*. Beverly Hills, CA: Sage.

Sullivan, Teresa A. 2005. "Labor Force." Pp. 209–225 in Dudley L. Poston and Michael Micklin (editors), *Handbook of Population*. New York: Springer.

Sullivan, Teresa A., and Jan E. Mutchler. 1985. "Equal pay or equal work? The implications of underem-ployment for comparable worth studies." Pp. 175–192 in R. L. Simpson and I. H. Simpson (editors), *Research in the Sociology of Work*. Greenwich, CT: JAI Press.

Sullivan, Teresa A., Elizabeth Warren, and Jay L. West-brook. 2000. *The Fragile Middle Class: Americans in Debt*. New Haven, CT: Yale University Press.

Sullivan, William M. 2005. *Work and Integrity: The Crisis and Promise of Professionalism in America*. San Francisco: Jossey-Bass.

Sussman, Marvin B., Suzanne K. Steinmetz, and Gary W. Peterson. 1999. *Handbook of Marriage and the Family*. New York: Plenum.

Symonds, William G., and Robert Berner. 2005. "In a lather over the Gillette deal." *Business Week online*, June 16.

Taggart, James H., Maureen Berry, and Michael C. McDermott. 2001. *Multinationals in a New Era*. New York: Palgrave.

Tausky, Carl. 1984. *Work and Society*. Itasca, IL: F. E. Peacock.

Taylor, Bill, Kai Chang, and Li Qi. 2003. *Industrial Relations in China*. Cheltenham, UK: Edward Elgar.

Taylor, Frederick W. 1911. *The Principles of Scientific Management*. New York: Harper and Row.

Terkel, Studs. 1974. *Working*. New York: Avon.

Thompson, E. P. 1967. "Time, work-discipline, and industrial capitalism." *Past and Present* 38 (December):56–97.

Tilly, Chris. 1996. *Half a Job: Bad and Good Part-Time Jobs in a Changing Labor Market*. Philadelphia: Temple University Press.

Tilly, Chris. 1998. *Durable Inequality*. Berkeley: University of California Press.

Tilly, Chris. 2004. "Raw deal for workers." *International Journal of Health Services* 34: 305–311.

Tilly, Chris, and Charles Tilly. 1998. *Work Under Capitalism*. Boulder, CO: Westview.

Tilly, Louise A., and Joan W. Scott. 1978. *Women, Work and Family*. New York: Holt, Rinehart and Winston.

Timmermans, Stefan. 2005. "From autonomy to accountability: The role of clinical practice guidelines in professional power." *Perspectives in Biology and Medicine* 48: 490–501.

Tivendell, J., and C. Bourbonnais. 2000. "Job insecurity in a sample of Canadian civil servants as a function of personality and perceived job characteristics." *Psychological Reports* 87: 55–60.

Tolbert, Charles M., Thomas A. Lyson, and Michael D. Irwin. 1998. "Local capitalism, civic engagement, and socioeconomic well-being." *Social Forces* 77, 2 (December): 401–427.

Tolbert, Pamela S., and Phyllis Moen. 1998. "Men's and women's definitions of 'good' jobs." *Work and Occupations* 25: 168–194.

Toledo, Enrique de la Garza. 2002. "The sociology of work in Mexico." In Daniel B. Cornfield and Randy Hodson (editors), *Worlds of Work*. New York: Plenum.

Tomaskovic-Devey, Donald. 1993. *Gender and Racial Inequality at Work*. Ithaca, NY: Industrial Relations Review Press.

Torstendahl, Rolf, and Michael Burrage (editors). 1990. *The Formation of Professions: Knowledge, State, and Strategy*. Newbury Park, CA: Sage.

Treiman, Donald J. 1977. *Occupational Prestige in Comparative Perspective*. New York: Academic Press.

Trice, Harrison M. 1993. *Occupational Subcultures in the Workplace*. Ithaca, NY: Industrial and Labor Relations.

Trice, Harrison H., and Janice M. Beyer. 1992. *The Cultures of Work Organizations*. Englewood Cliffs, NJ: Prentice Hall.

Troyer, Lisa, Charles W. Mueller, and Pavel I. Osinsky. 2000. "Who's the boss? A role-theoretic analysis of customer work." *Work and Occupations* 27, 3 (August): 406–427.

Tucker, James. 1999. *The Therapeutic Corporation*. New York: Oxford University Press.

Tulin, Roger C. 1984. *A Machinist's Semi-Automated Life*. San Pedro, CA: Singlejack.

Turnley, W. H., Mark C. Bolino, Scott W. Lester, and James M. Bloodgood. 2003. "The impact of psychological contract fulfillment on the performance of in-role and organizational citizenship behaviors." *Journal of Management*, 29 (2): 187–206.

Turner, Lowell. 2005. "From transformation to revitalization." *Work and Occupations* 32, 4: 383–399.

Uchitelle, Louis. 1994. "Job extinction evolving into a fact of life in U.S." *The New York Times*, March 22, pp.A1, D5.

United Nations. 2001. *Human Development Report*. New York: United Nations.

United Nations. 2003. *Statistical Yearbook*, 44th edition. New York: United Nations.

United Nations, Industrial Development Organization. 2000. *International Yearbook of Industrial Statistics*. Vienna: U.N. Industrial Development Organization.

U.S. Bureau of the Census. 1943. *U.S. Census of Population, 1940: Comparative Occupational Statistics, 1870–1940*. Authored by Alba M. Edwards. Washington, DC: U.S. Government Printing Office.

U.S. Bureau of the Census. 1975. *Historical Statistics of the United States, Colonial Times to 1970*. Bicentennial edition, part 1. Washington, DC: U.S. Government Printing Office.

U.S. Bureau of the Census. 1993. *Statistical Abstract of the United States*, 113th edition. Washington, DC: U.S. Government Printing Office.

U.S. Bureau of the Census. 1993b. "Preparing for retirement: Who had pension coverage in 1991?" *Statistical Brief* SB 93–6. April.

U.S. Bureau of the Census. 1999. *Statistical Abstract of the United States*, 119th edition. Washington: U.S. Government Printing Office.

U.S. Bureau of the Census. 2000. *Quarterly Financial Report for Manufacturing*. Washington: U.S. Government Printing Office.

U.S. Bureau of the Census. 2004. *Statistical Abstract of the United States*, 124th edition Washington: U.S. Government Printing Office.

U.S. Bureau of the Census. 2005. *Statistical Abstract of the United States*. Washington, DC: Government Printing Office.

U.S. Bureau of the Census. 2006a. *Statistical Abstract of the United States*. Washington, DC: Government Printing Office. <http://www.census.gov/statab/www/>

U.S. Bureau of the Census. 2006b. *Concentration Ratios: 2002*. EC02–31SR–1. Washington, DC: Bureau of Census.

U.S. Bureau of the Census. 2007. *Statistical Abstract of the United States*. Washington, DC: Government Printing Office. <http://www.census.gov/statab/www/>

U.S. Congress. 1901. *Report of the Industrial Commission on the Relations of Capital and Labor*, 56th Congress, House Document 495. Washington, DC: U. S. Government Printing Office.

U.S. Congress, Office of Technology Assessment. 1985. *Automation of America's Offices, 1985–2000*. OTA-CIT–287, December. Washington, DC: U.S. Government Printing Office.

U.S. Department of Commerce. 1967. *Technological Innovations: Its Environment and Management*. Washington, DC: U.S. Government Printing Office.

U.S. Department of Health, Education, and Welfare. 1973. *Work in America*. Cambridge, MA: MIT.

U.S. Department of Labor, Bureau of Labor Statistics. 1998. "Displaced workers summary." Press Release 98–347, August 19.

U.S. Department of Labor, Bureau of Labor Statistics. 2001a. "How is the unemployment rate related to unemployment insurance claims?" Retrieved from http://www.bls.gov/cps/uiclaims.htm

U.S. Bureau of Labor Statistics. 2001b. "How the Government measures unemployment." Report 864. Retrieved from http://www.bls.gov/cps/cps_htgm.htm

U.S. Bureau of Labor Statistics. 2001c. "Report on the American workforce 2001." Retrieved from http://www.bls.gov/opub/rtaw/rtawhome.htm

U.S. Bureau of Labor Statistics. 2004a. "North American Industry Classification System (NAICS) at BLS." Retrieved from http://www.bls.gov/bls/naics.htm

U.S. Bureau of Labor Statistics. 2004b. "Standard Occupational Classification (SOC) User Guide." Retrieved from http://www.bls.gov/soc/socguide.htm

U.S. Bureau of Labor Statistics. 2005a. Retrieved from http://www.bls.gov

U.S. Bureau of Labor Statistics. 2005b. "Workers on flexible and shift schedules in 2004: Summary." USDL 05–1198, July 1.

U.S. Bureau of Labor Statistics. 2005c. "Contingent and alternative employment arrangements, February." USDL release 05–1433. Retrieved from http://www.bls.gov/news.release/conemp.nr0.htm

U.S. Bureau of Labor Statistics. 2006. *Employment and Earnings*. 53, 1 (January).

U.S. Bureau of Labor Statistics. 2006a. "Earnings of production or nonsupervisory workers on private nonfarm payrolls in current and constant dollars by industry." Retrieved from http://www.bls.gov/news.release/realer.t02.htm

U.S. Bureau of Labor Statistics. 2006b. www.bls.gov

U.S. Bureau of Labor Statistics. 2006c. "Median weekly earnings of full-time wage and salary workers by detailed occupation and sex." Retrieved from http://www.bls.gov/cps/cpsaat39.pdf

U.S. Bureau of Labor Statistics. 2006d. "Worker displacement: 2003–2005." USDL 06–1454, August 17. Retrieved from http://www.bls.gov/news.release/disp.nr0.htm

U.S. Senate. 1986. Hearings before the Subcommittee on Innovation and Technology. Oversight on the Small Business and Innovation and Research Program. 99th Congress. Washington, DC: U.S. Government Printing Office.

Useem, Michael. 1996. *Investor Capitalism: How Money Managers are Changing the Face of Corporate America*. New York: Basic Books.

Useem, Michael. 2004. "The problem with proxies." *Harvard Business Review* 82, 11: 20, 24.

Useem, Michael, and Jerome Karabel. 1986. "Pathways to top corporate management." *American Sociological Review* 51, 2 (April): 184–200.

Vagg, P. R., C. D. Spielberger, and C. F. Wassala. 2002. "Effects of organizational level and gender on stress in the workplace." *International Journal of Stress Management* 9: 243–261.

Valian, Virginia. 1999. *Why So Slow? The Advancement of Women*. Cambridge, MA: MIT Press.

Vallas, Steven Peter. 1999. "Rethinking post-Fordism: The meaning of workplace flexibility." *Sociological Theory* 17, 1 (March): 68–101.

Vallas, Steven P. 2003. "Why teamwork fails: Obstacles to workplace change in four manufacturing plants." *American Sociological Review* 68: 223–250.

Vallas, Steven Peter. 2006. "Empowerment redux: Structure, agency, and the remaking of managerial authority." *American Journal of Sociology* 111: 1677–1717.

Vanek, Joann. 1988. "Housewives as workers." Pp. 392–414 in Ann H. Stromberg and Shirley Harkess (editors), *Working Women*, 2nd edition. Mountain View, CA: Mayfield.

Vaughan, Diane. 1996. *The Challenger Launch Decision: Risky Technology, Culture, and Deviance at NASA*. Chicago: University of Chicago Press.

Veblen, Thorstein. 1921. *The Engineers and the Price System*. New York: Viking.

Vecsey, George. 1974. *One Sunset a Week: The Story of a Coal Miner*. New York: E. P. Dutton.

Venkatesh, Sudhir Alladi. 2006. *Off the Books: The Underground Economy of the Urban Poor*. Cambridge, MA: Harvard University Press.

Verma, Rohit, and Kenneth K. Boyer. 2000. "Service classification and management challenges." *Journal of Business Strategies* 17, 1 (Spring): 5–24.

Vernon, Raymond. 1998. *In the Hurricane's Eye: The Troubled Prospects of Multinational Enterprises*. Cambridge, MA: Harvard University Press.

Vincent, Peggy. 2002. *Baby Catcher: Chronicles of a Modern Midwife*. New York: Scribner.

Vise, David A. 2005. "What lurks in its soul?" *The Washington Post*, November 13. Retrieved from www.washingtonpost.com/wp-dyn/content/article/2005/11/11/AR2005111101644.html

Vollmer, H. 2004. "Consequences and functions of organized calculation." *Zeitschrift fur Soziologie* 33: 450–470.

Vonnegut, Kurt, Jr. 1952. *Player Piano*. New York: Dell.

von Otter, Casten. 2002. "The sociology of work in Sweden." In Daniel B. Cornfield and Randy Hodson (editors), *Worlds of Work*. New York: Plenum.

Wajcman, Judy. 1998. *Managing Like a Man: Women and Men in Corporate Management*. Cambridge: Polity.

Waldfogel, Jane. 1997. "Working mothers then and now." Pp. 82–126 in Francine Blau and Ronald Ehrenberg (editors), *Gender and Family Issues in the Workplace*. New York: Russell Sage.

Waldinger, Roger, and Michael Lichter. 2003. *How the Other Half Works: Immigration and the Social Organization of Labor*. Berkeley: University of California Press.

Walker, Charles R., and Robert H. Guest. 1952. *The Man on the Assembly Line*. Cambridge, MA: Harvard University Press.

Wallace, Michael. 1998. "Downsizing the American Dream: Corporate restructuring, work and family at century's end." In Dana Vannoy and Paula Dubeck (editors), *Challenges for Work and Family in the 21st Century*. Hawthorne, NY: Aldine de Gruyter.

Wallace, Michael, Kevin T. Leicht, and Lawrence E. Raffalovich. 1999. "Unions, strikes and labor's share of income." *Social Science Research* 28: 265–88.

Wallace, Phyllis A. (editor). 1982. *Women in the Workplace*. Boston: Auburn.

Wallerstein, Immanuel M. 2003. *The Decline of American Power*. New York: Free Press.

Walsh, Joan. 1985. "Comparable worth advocates optimistic." *In These Times*, October 9, p. 5.

Walsh, Joan. 1986. "Layoff settlement zaps Atari." *In These Times*, June 25, p. 7.

Walsh, John P. 1993. *Supermarkets Transformed: Understanding Organizational and Technological Innovations*. New Brunswick, NJ: Rutgers University Press.

Walsh, John P., and Shu-fen Tseng. 1998. "The Effects of Job Characteristics on Active Effort at Work." *Work and Occupations* 25, 1 (February): 74–96.

Walsh, Mary Williams. 2006. "More companies ending promises for retirement." *The New York Times*, January 9, p. A1.

Wanner, R.A., and B. C. Hayes. 1996. "Intergenerational occupational mobility among men in Canada and Australia." *Canadian Journal of Sociology/Cahiers Canadiens de Sociologie* 21, 1: 43–76.

Ward, Kathryn (editor). 1990. *Women Workers and Global Restructuring*. Ithaca, NY: Cornell University School of Industrial and Labor Relations Press.

Warren, Christian. 2000. *Brush with Death*. Baltimore: John Hopkins University Press.

Warren, Kenneth. 2001. *Big Steel: The First Century of The United States Steel Corporation, 1901–2000.* Pittsburgh: University of Pittsburgh Press.

Watson, Tony J. 2003. *Sociology, Work and Industry*, 4th edition. London: Routledge and Kegan Paul.

Weaver, Charles N. 2002. "The effects of generational status on the work attitudes of Mexican Americans." *Hispanic Journal of Behavioral Sciences* 24: 63–73.

Weber, Max. 1946. "Bureaucracy." In H. H. Gerth and C. Wright Mills (editors), *From Max Weber.* New York: Oxford University Press.

Weber, Max. 1958 [1904]. *The Protestant Ethic and the Spirit of Capitalism.* New York: Scribner's.

Webster, Juliet. 2000. "Today's second sex and tomorrow's first? Women and work in the European information society." Pp. 119–140 in Ken Ducatel, Juliet Webster, and Werner Herrmann (editors), *The Information Society in Europe.* Lanham, MD: Rowman and Litlefield.

Weinfeld, Morton. 1999. "The challenge of ethnic match: Minority-origin professionals in health and social services." Pp. 117–141 in Harold Troper and Morton Weinfeld (editors), *Ethnicity, Politics, and Public Policy: Case Studies in Canadian Diversity.* Toronto: University of Toronto Press.

Wells, Miriam J. 1996. *Strawberry Fields: Politics, Class, and Work in California Agriculture.* Ithaca, NY: Cornell University Press.

Werhane, Patricia H., Tara J. Radin, and Norman E. Bowie. 2004. *Employment and Employee Rights.* Malden, MA: Blackwell Publishing.

Wharton, Amy S. 1993. "The affective consequences of service work." *Work and Occupations* 20, 2 (May): 205–232.

Whyman, Philip. 2003. *Sweden and the "Third Way."* Burlington, VT: Ashgate.

Wickham, James, and Grainne Collins. 2004. "The call centre: A nursery for new forms of work organization?" *Services Industries Journal* 24: 1–18.

Wilensky, Harold, and Charles N. Lebeaux. 1986. "The early impact of industrialization on society." Pp. 28–39 in Lauri Perman (editor), *Work in Modern Society.* Dubuque, IA: Kendall-Hunt.

Williams, Bruce B. 1987. *Black Workers in an Industrial Suburb.* New Brunswick, NJ: Rutgers University Press.

Williams, Christine L. 1995. *Still a Man's World.* Berkeley: University of California Press.

Williams, Christine L. 1992. "The glass escalator: Hidden advantages for men in the 'female' professions." *Social Problems* 39, 3: 253–268.

Williams, Christine L., Patti A. Giuffre, and Kirsten Dellinger. 1999. "Sexuality in the workplace: Organizational control, sexual harassment, and the pursuit of pleasure." *Annual Review of Sociology* 25 (August): 73–93.

Wilson, F. 2000. "The social construction of sexual harassment and assault of university students." *Journal of Gender Studies* 9, 2: 171–187.

Wilson, William Julius. 1987. *The Truly Disadvantaged: The Inner City, the Underclass and Public Policy.* Chicago: University of Chicago Press.

Wilson, William Julius. 1997. *When Work Disappears: The World of the New Urban Poor.* (Reprinted edition.) New York: Vintage.

Wilson, William Julius. 2003. "Race, class and urban poverty." *Ethnic and Racial Studies* 26: 1096–1114.

Winslow, Ron. 1993. "Shortage of primary-care doctors spurs schools, HMOs to try to redress balance." *The Wall Street Journal*, February 24, p. B1.

Winter, Michael P. 1988. *The Culture and Control of Expertise: Toward a Sociological Understanding of Librarianship.* New York: Greenwood.

Wolf, E. 2002. "Lower wage rates for fewer hours? A simultaneous wage-hours model for Germany." *Labour Economics* 9: 643–663.

Wolff, Edward N. 2000. "How persistent is industry specialization over time in industrialized countries?" *International Journal of Technology Management* 19: 194–205.

Woodard, Michael D. 1997. *Black Entrepreneurs in America: Stories of Struggle and Success.* New Brunswick, NJ: Rutgers University Press.

Woodward, Joan. 1980. *Industrial Organization: Theory and Practice.* Oxford: Oxford University Press.

Woolnough, Roisin. 2000. "Racism reinforces the glass ceiling." *The Guardian* (London): 301.

"World's biggest trade deal approved." 1993. *Herald-Times*, December 16, p. A3.

Wright, John W. 2000. *The American Almanac of Jobs and Salaries.* New York: Avon.

Wysocki, Bernard, Jr. 1998. "Oil megadeal signals a world of low inflation." *The Wall Street Journal*, December 2, p. 112.

Xiao, Qin. 2004. *The Theory of the Firm and Chinese Enterprise Reform.* London: Routledge.

Yoon, In-Jin. 1991. "The changing significance of ethnic and class resources in immigrant businesses: The case of Korean immigrant businesses in Chicago." *International Migration Review* 25, 2 (Summer): 303–333.

You, Jong-sung, and Sanjeev Khagram. 2005. "A comparative study of inequality and corruption." *American Sociological Review* 70: 136–157.

Yount, Kristen R. 1991. "Ladies, flirts, and tomboys: Strategies for managing sexual harassment in an underground coal mine." *Journal of Contemporary Ethnography* 19, 4 (January): 396–422.

Zachary, G. Pascal. 2003. *The Diversity Advantage: Multicultural Advantage in the New World Order.* Boulder, CO: Westview.

Zachary, G. Pascal, and Bob Ortega. 1993. "Workplace revolution boosts productivity at cost of job security." *The Wall Street Journal*, March 10, pp. A1, A8.

Zaucher, Sabine, Christian Korunka, A. Weiss, and A. Kafka-Lutzow. 2000. "Gender-related effects of information technology implementation." *Gender, Work and Organization* 7, 2 (April): 119–132.

Zeitlin, Maurice. 1974. "Corporate ownership and control." *American Journal of Sociology* 79: 1073–1119.

Zelizer, Viviana A. Rotman. 1979. *Morals and Markets: The Development of Life Insurance in the United States.* New York: Columbia University Press.

Zeller, Tom, Jr. 2006. "House member criticizes Internet companies for practices in China," *The New York Times*, February 15. Retrieved from www.nytimes.com/2006/02/15/technology/15cnd-internet.html

Zey, Mary. 1998. *Rational Choice theory and Organizational Theory.* Thousand Oaks, CA: Sage.

Zhou, Xueguang. 1993. "Occupational power, state capacities, and the diffusion of licensing in the American states, 1890 to 1950." *American Sociological Review* 58: 536–552.

Zibechi, Raúl. 2006. "Another world is possible: The ceramics of Zanon." Silver City, NM: International Relations Center. Retrieved from http://americas.irc-online.org/am/3078

Zuboff, Shoshana. 1988. *In the Age of the Smart Machine.* New York: Basic.

Zuboff, Shoshana, and James Maxmin. 2004. *The Support Economy: Why Corporations are Failing Individuals and the Next Episode of Capitalism.* New York: Viking.

Zucker, Lynne G., and Michael R. Darby. 2005. "Socio-economic impact of Nanoscale Science." Cambridge, MA: National Bureau of Economic Research. Retrieved from http://www.nber.org/papers/w11181

Zunz, Olivier. 1998. *Why the American Century?* Chicago: University of Chicago Press.

Zwerdling, Daniel. 1980. *Workplace Democracy.* New York: Harper and Row.

Index